ENVIRON
POLITICS
FOR A CHANGING
WORLD

POWER, PERSPECTIVES, AND PRACTICE

SECOND EDITION

RONNIE D. LIPSCHUTZ
UNIVERSITY OF CALIFORNIA, SANTA CRUZ

DOREEN STABINSKY
COLLEGE OF THE ATLANTIC

ROWMAN & LITTLEFIELD
Lanham • Boulder • New York • London

Executive Editor: Traci Crowell
Assistant Editor: Mary Malley
Senior Marketing Manager: Amy Whitaker
Interior Designer: Pro Production Graphic Services
Cover Designer: Dustin Watson

Credits and acknowledgments for material borrowed from other sources, and reproduced with permission, appear on the appropriate page within the text.

Published by Rowman & Littlefield
An imprint of The Rowman & Littlefield Publishing Group, Inc.
4501 Forbes Boulevard, Suite 200, Lanham, Maryland 20706
www.rowman.com

Unit A, Whitacre Mews, 26-34 Stannary Street, London SE11 4AB, United Kingdom
Copyright © 2019 by The Rowman & Littlefield Publishing Group, Inc.
First edition 2004.

British Library Cataloguing in Publication Information Available

Library of Congress Cataloging-in-Publication Data
Names: Lipschutz, Ronnie D., author. | Stabinsky, Doreen, author.
Title: Environmental politics for a changing world : power, perspectives, and practice / Ronnie D. Lipschutz, University of California, Santa Cruz, Doreen Stabinsky, College of the Atlantic.
Other titles: Global environmental politics
Description: Second Edition. | Lanham, Maryland : Rowman & Littlefield, [2018] | "First edition © 2004 by CQ Press"—T.p. verso. | Includes bibliographical references and index.
Identifiers: LCCN 2018014332 (print) | LCCN 2018026508 (ebook) | ISBN 9781538105115 (electronic) | ISBN 9781538105092 (cloth : alk. paper) | ISBN 9781538105108 (paper : alk. paper)
Subjects: LCSH: Environmental policy—International cooperation.
Classification: LCC GE170 (ebook) | LCC GE170 .L55 2018 (print) | DDC 363.7/0561—dc23
LC record available at https://lccn.loc.gov/2018014332

Printed in the United States of America

Brief Contents

Contents

Illustrations

★ ★ ★

PHOTOS

FIGURES

TABLES

TEXTBOXES

Preface

Our all-time favorite comic appeared on September 17, 1985. Drawn by Richard Guindon, it showed two startled scientists staring at their computers as a colleague ran into the lab amid a flurry of sheets of paper. "The ecosystem collapsed again! Frogs! We need more frogs!" he was shouting. A few years later, in a case of life imitating art, a *New York Times* article reported that "Scientists Confront an Alarming Mystery: The Vanishing Frog." Biologists and ecologists were finding that "amphibians are rapidly disappearing from many ponds, rivers, mountains and rain forests around the world." A professor from the University of California at Berkeley reminisced that "Meadows where frogs were as thick as flies are now silent."* Why were frogs disappearing? What did this mean? Who could save them? Does the earth really need frogs?

The tale of vanishing frogs is by no means unique: more and more, scientists are warning that we are entering the Sixth Great Extinction, with hundreds or more species disappearing every year. In virtually all cases, humans seem culpable, although there is no end of efforts to pin the blame on nature. But there is more! At the same time that many living things appear to be in decline, non-living stuff appears more ubiquitous than ever. Smartphones, iPads, big screen monitors, electric and autonomous vehicles, toxic residues of electronic wastes, and greenhouse gas emissions—all are being generated in ever-increasing quantities, with no end in sight and no quantity of recycling able to take them on. Indeed, things and stuff seem to proliferate beyond all reason, like some alien invader intent on conquering Earth. "The ecosystem collapsed again! Smartphones! We need fewer smartphones!"

---

*Sandra Blakeslee, *New York Times*, February 20, 1990, at http://www.nytimes.com/1990 /02/20/science/scientists-confront-an-alarming-mystery-the-vanishing-frog.html?page wanted=all (accessed December 20, 2017). A comprehensive source on amphibian declines can be found on Amphibiaweb at https://amphibiaweb.org/declines/declines.html.

If we ask "who *can* save the frogs (and all of those other species)?" the answer is not so clear. There are plenty of candidates—governments, scientists, corporations, consumers, control systems, coalitions, social movements—but only limited action. And, of course, there are those who assure us there is no problem at all: as always, this is the best of all possible worlds, and it can only get better. Just in case, solutions are at hand. For the past four decades, for example, the conventional wisdom has come to be that "economic growth" can save everyone: frogs, ecosystems, *and* people. After all, goes the argument, rich countries have cleaned up their environments because their inhabitants could afford to and wanted to. Moreover, relying on growth means that no one would have to act; all that is necessary is to make more, ship more, buy more—and, we might guess—dump more. But does it not seem a bit odd that, confronted by a growing environmental crisis of global proportions, our only response is to do more of exactly that which has brought on the crisis in the first place?

Not everyone agrees that getting the economics right will solve the problem. Like Guindon's scientists, many hope that technology can save us. Fossil fuels are dirty and produce carbon dioxide. We cannot give up energy, and we always need more, so we must shift, as soon as possible, to alternatives. We know how to manufacture solar cells and clean up coal (it is said); wind is plentiful, uranium is cheap. Someday—always in fifty years—we may even have nuclear fusion. That won't solve all our environmental problems—even nuclear fusion will generate radioactive wastes—but it will deal with the increasingly pressing one of climate change. (And if we don't manage to stop climate change, well, we can just geoengineer the planet.) Still, if we have the answers, why hasn't this shift taken place? If we know that it's not nice to fool with Mother Nature, surely we have no choice but to stop our fooling? But we don't. More research is required. It will cost too much. And we would be foolish to act if no one else does (people would laugh at us!).

Some who study global environmental politics (GEP) continue to put their faith in international institutions, politics, and interstate cooperation. Global problems are transboundary and must be handled via collaborative international arrangements. There is no shortage of such institutions and, by now, they number in the hundreds, if not more. Nonetheless, their ability to solve the problems they are intended to address has proved to be limited: the Conference of the Parties to the UN Framework Convention on Climate Change (UNFCCC) just recently concluded its twenty-third meeting in Bonn, with only limited progress (the Rio Summit took place a *quarter of a century* ago!). All evidence suggests that climate change is happening more rapidly and severely than predicted even as recently as 2010; it is only a matter of time before those living in low-lying coastal cities will be standing in water up to their knees.

So what are we to do? The paradox is that we *know* what we must do, and we could begin to do it tomorrow. That we don't act is due to the fact that the problems we face are political and cannot be addressed by markets, technology, or diplomacy and evermore complex and arcane gadgets, schemes, and policies. Only politics can save the environment. By this, we do not mean the politics of liberal markets and democracy, which focus on "dividing the pie." What we need is an ethically based politics, one rooted in an understanding of right relationships between humans and nature and praxis and which seeks to give to nature as much as we take from her, and which is mobilized around protection of nature and elimination of poverty. That is our goal in the second edition of this book.

The research and writing for the first edition of *Global Environmental Politics: Power, Perspectives and Practice*—renamed *Environmental Politics for a Changing World: Power, Perspectives, and Practice* for this second edition—began in the mid-1990s, as a set of notes for a college class. At that time, the "end of the Cold War" still mattered, the 1992 Earth Summit in Rio was more than just a remote memory, and the annual Conference of the Parties to the UNFCCC had not yet convened for the tenth time. Much water has flowed under the bridge (and air overhead) since then. The "war on terrorism" has distracted us for more than fifteen years, with limited success, and climate change has begun to wreak havoc around the world. The Trump administration entered office vowing to "undo" as many environmental regulations and mandates as possible, and is making good on that promise. The twenty-third meeting of the UNFCCC Conference of the Parties tried to instill some substance in the Paris Agreement, a backward evolution from legally binding emission reduction commitments (as in the much-maligned Kyoto Protocol) to voluntary, nationally determined pledges that are widely recognized as insufficient to keep warming to safe levels. Meanwhile, geoengineering technofixes to the excess carbon in the atmosphere are attracting more attention. Beyond climate change, scientists count no fewer than eight other planetary boundaries we risk crossing, or have already exceeded (biodiversity loss and human interference in the nitrogen cycle).* All in all, hope remains but things are looking pretty grim.

In 2003, the field of GEP was still fairly new and limited in terms of size, scope, and scholarship. The second or third generation of graduate students was completing PhDs on increasingly specialized topics. The number of undergraduate texts was in the region of ten, of which two or three offered a somewhat Marxian critical approach. Today, there are dozens of texts and hundreds of scholarly monographs and anthologies. There is an eponymous, highly respected journal, launched in 2000 and now ranked thirteenth out of 163 "political science" journals. But *GEP*, as it is fondly called, has a much broader disciplinary base than political science and international relations, including geography, sociology, economics, anthropology, and even psychology. As the readers of the bibliography to this book will discover, GEP remains

---

*J. Rockström, et al., "A Safe Operating Space for Humanity," *Nature* 461 (2009): 472–75.

a "hot topic." Anything and everything that might be studied and written about relating to some aspect of GEP seems to have been studied and written about. And specializations are becoming more and more narrow.

So why a second edition of this book? Credit must be given where it is due: several years ago, one of us (Doreen Stabinsky) inquired of the other (Ronnie Lipschutz) whether he had ever contemplated a new edition. Thinking about the saturation of the field, Lipschutz said "no." But Stabinsky was eager, and Rowman & Littlefield was willing. So Lipschutz said "yes." Returning to a book after more than fifteen years is a sobering experience, and somewhat embarrassing. The writing is pretentious, the proclamations tendentious, the tone naïve, the content questionable. Can the book be saved? Maybe. This second edition retains some of the material from the first edition, updates a great deal, and adds new content. It is a work of synthesis, not creation, and thus draws on a range of arguments and materials unavailable in 2003. And, of course, there is the internet, which makes it possible to gain access to however much one's university library has subscribed to. Perhaps this is an improvement over the days of searching stacks to see what might be found adjacent to the books one was seeking.

Almost certainly (and perhaps unfortunately), the political, social, and biophysical environments are very propitious for this book, in which we take something of a "new materialist" approach to GEP (see chapter 2). "New materialism" comes in several different flavors; we incorporate it here as a mix of non-Marxist historical materialism, critical sociology, and practice theory, with a dash of Foucauldian genealogy thrown in for good measure. In plain English, this means that (1) we need to work with the material and ideational worlds as they have been handed down to us—ideas are not enough where social change is concerned; (2) society is not an aggregation of individuals and, indeed, there is no individual without society (take that, Hayek and Thatcher!); and (3) we need to find *political* ways to change our *habitus* (in the plural), the embedded and normalized practices that make and remake our worlds and which seem (or are made to appear) so natural as to be impervious to change.

In light of this approach, at the end of the day, the invocations, proclamations, and prescriptions offered in the first edition have not changed very much. We still need to understand the linkages and relationships among space and place, the local and the global, the past, present, and future, if we are to make a new world not only possible but also practical. We must always be aware of these linkages and relationships, even as we act locally, in what Hannah Arendt called "the spaces of appearances." What we do where we are can be a model and inspiration—for better or worse—to people in other places, and indeed we look to those on the frontlines of environmental and social struggles for inspiration in our own work. There are no silver bullets, magical formulas, or new technologies that can, or will, provide a shortcut to the hard, political work of social change. If for no other reason, repeating that proposition seems sufficient rationale for a second edition.

ACKNOWLEDGMENTS AND GRATITUDE

Doreen: First, deep gratitude to Ronnie for agreeing to the opportunity to help revise a book that I have used in my courses for many years and which, years after they've graduated and gone forth into the world, students tell me they still have on their bookshelves and/or still fondly remember. Second, well, I'm extraordinarily fortunate to be able to say that there are many others whom I could thank who have provided me with similar opportunities to push the boundaries of what I knew or understood about the practice and praxis of GEP. I wish I could thank you all by name, but please know that my appreciation runs deep and wide. But there are a few who cannot reside in anonymity. To my colleagues in the Equity and Ambition Group who have taught me almost all that I know about the UNFCCC and how to radically engage in such a complex and challenging space, you have made the last eight years the most politically and intellectually fulfilling of my existence to date. To Monica Moore, Michael Picker, and Angus Wright, you set me on this path with confidence and inspiration, and I'm so pleased to be able to acknowledge you in a book that you have in many ways written through your own work as scholars and practitioners of environmental politics. To Carol, thanks for being my biggest fan since day one. Finally, to Dave, heaps of love and boundless gratitude for keeping the fires burning and the cats fed, and so much more.

Ronnie: My deep appreciation and gratitude to Doreen for instigating this revision, being optimistic that it could be done, and providing so much energy and support in seeing that it was done, especially in taking care of many details.

Ronnie D. Lipschutz   Doreen Stabinsky
Santa Cruz   Mount Desert Island, Maine
March 2018

1

What Are "Global Environmental Politics"?

★ ★ ★

SHOE PHONES AND IPHONES

In the 1960s television spy spoof, "Get Smart!" Secret Agent Maxwell Smart owned a shoe phone. In those days, spies came equipped with all sorts of outlandish devices, and Smart's shoe phone, as ridiculous as it seemed, satirized such espionage fetishism, especially as it appeared in the James Bond films. Today, phones come in all shapes and sizes and, with the advent of smartphones, no one would blink an eye to see someone talking to their shoe or to any other object, for that matter. If you are a college student, it is likely that you own a smartphone. You probably can identify every icon on your smartphone's screen, and every app available in your smartphone's memory. But do you know where your phone came from? Do you know who made it, or where? Where do old smartphones go to die? What is inside your phone, and how were those materials produced? Are the workers on the assembly line exposed to toxic materials as they are putting phones

Photo 1.1   Secret Agent Maxwell Smart

1

together? Do the innards of old smartphones get recycled, or are they burned for precious metals somewhere in Asia? In a word: are smartphones good or bad for the global environment, and why should you care?

Here are some numbers and facts: in 2017, according to best estimates, there were more than two billion smartphones in the world, a number expected to grow to more than six billion by 2020. In the United States, there are more than two hundred million such devices, and about 70 percent of adults own one (in the eighteen to twenty-nine age group, the fraction rises to 86 percent!).[1] According to Statista.com, in 2015, more than 1.4 billion smartphones were sold around the world, two hundred million more than in 2014, which suggests that phones are replaced frequently.[2] The phone you hold in your hand, according to some, has more computing power than was possessed by NASA in 1969, when the first men landed on the moon.[3] And it can meet (or exceed) all of your communication, photography, music, and information needs. But not everything about smartphones is on the upside.

Smartphones contain more than seventy elements, among them as many as sixty-two metals and sixteen rare earth elements (see Table 1.1).[4] These come from all over the world, often from heavily polluting sources with regressive labor practices and poorly paid workers.[5] Environmental impacts occur all along a smartphone's life cycle; one study calculates that the Global Warming Potential of a single smartphone used for three years is fifty kilograms of carbon dioxide equivalent.[6] Multiplying that by two billion and dividing by three years points to an annual addition of one hundred thousand tons of greenhouse gases per year to the Earth's atmosphere!

Table 1.1. Elements found in a smartphone

Component	Elements
Color screen	Yttrium, Lanthanum, Praseodymium, Europium, Gadolinium, Terbium, Dysprosium, Indium, Tin, Oxygen, Aluminum, Potassium, Silicon
Glass polishing	Lanthanum, Cerium, Praseodymium
Phone circuitry	Lanthanum, Praseodymium, Neodymium, Gadolinium, Dysprosium, Copper, Silver, Gold, Tantalum, Nickel, Silicon, Oxygen Antimony, Arsenic, Phosphorus, Tin, Gallium, Lead
Speakers	Praseodymium, Neodymium, Gadolinium, Terbium, Dysprosium, Copper
Vibration unit	Neodymium, Terbium, Dysprosium
Battery	Lithium, Cobalt, Oxygen, Carbon, Aluminum
Casing	Carbon, Magnesium, Bromine, Nickel

Source: Randy Krum, "The Periodic Table of iPhones," February 4, 2013, http://coolinfographics .com/blog/2013/2/4/the-periodic-table-of-iphones.html (accessed December 19, 2017); Compound Interest, "The Chemical Elements of a Smartphone," 2014, http://www.compoundchem .com/wp-content/uploads/2014/02/The-Chemical-Elements-of-a-Smartphone-v2.png (accessed December 19, 2017).

Disposal of old phones also creates risks and hazards for humans and the environment. Like cars, smartphones can be used long after they have gone out of style. Apparently, however, they are used on average for two to three years before being broken or dumped for a newer model (more than 150 million per year in the United States alone).[7] Some phones go into landfills, some to recycling facilities where they are carefully disassembled, but many are shipped abroad, where they end up poisoning children, polluting water and piling up in dumps. As CNET reporter Jay Green discovered on a visit to China,

> When spent circuit boards find their way to Guiyu, villagers heat them over coal fires to recover lead. The ash from the burning of coal gets dumped into the city's streams and canals, turning them black and poisoning the wells and groundwater. It's one reason why Guiyu reportedly has the highest level of cancer-causing dioxins in the world, elevated rates of miscarriages, and children with extremely high levels of lead poisoning.[8]

This is an example not only of health impacts but also of "environmental injustice"—the unfair imposition of health and material impacts on those who have the least capacity to resist them and whose lives and livelihoods depend on further, and sometimes deadly, exposure to those impacts. This is a theme that we will pursue throughout this book, along with several others.

BANANAS: "MOTHER NATURE'S PERFECT FOOD?"[9]

Like smartphones, bananas are pretty ubiquitous in the lives of Americans and Europeans. Everyone knows what a banana looks like. Everyone knows where bananas come from. Everyone has eaten a banana. And that's all anyone needs to know—or is it? Bananas raise difficult questions, too: How are they grown, and where (how come we can't raise bananas in the Global North)? Who takes care of them while they are growing? How much are workers paid? What kinds of pesticides and fumigants are used on bananas? What happens to those chemicals? Do they make the workers sick? Can they make consumers sick? How do bananas get to the United States and Europe? What determines their cost? And why should such things matter to you?

Bananas are a big business. They are mass-produced on plantations in one part of the world—primarily Central and South America—and sold, in large quantities, in markets in a different part of the world (mostly the Global North), which is where we first see them. But between the images of green banana trees swaying in the trees (hello, Carmen Miranda!) and yellow bananas piled high in supermarkets (hello, Foodmart!) there is a void, about which we can ask many questions, not all that different from the ones we asked above about smartphones. Here are some numbers: in 2011, more than one hundred million metric tons of bananas were grown in 130 countries around the world.[10] Of those, only about twenty million metric tons were traded internationally, mostly in the form of the familiar yellow Cavendish variety—which could soon be rendered extinct by "fusarium wilt—from some 15 countries."[11] That still made bananas the fifth largest agricultural product traded in world markets, after cereals, sugar, coffee, and cocoa, with a trade value between seven billion and nine billion dollars and a retail value between twenty billion and twenty-five billion dollars.[12] In 2012, more than half of internationally traded bananas were consumed in the United States and the European Union.[13]

Who makes money from bananas? In Ecuador, for example, almost four thousand farms are thirty hectares or less in area (one hectare equals roughly 2.5 acres); only about 140 are larger than one hundred hectares.[14] Farm size makes a big difference in production costs, although it has hardly any impact on the price of bananas in the supermarket. In 2002, 70 percent of the international banana trade was controlled by five transnational corporations: United Brands (Chiquita), Dole, Del Monte, Fyffes (based in Ireland), and Noboa (based in Ecuador). By 2013, that fraction had dropped to 44 percent, as growing competition from supermarket chains, and changes in shipping methods, ate into the multinationals' markets.[15] Relatively little of retail revenue reaches producing countries, and even less goes to banana farmers and workers. A pound of bananas that sells for a dollar in your local supermarket might be bought for five cents a pound or less in the country where it is grown. According to one study, every dollar of retail revenue generated by bananas from Ecuador—a pretty typical "banana split"—is distributed as shown in Figure 1.1.

A great deal goes on in the banana commodity chain (sometimes called the product "life cycle") between plantation and plate, and much of what goes on has undesirable environmental consequences. Annual applications of chemicals—fungicides, insecticides, herbicides, nematicides, fertilizers, and disinfectants—to each hectare of Central American banana trees amount to forty kilograms (or forty pounds per acre).[16] A lot of those chemicals get shipped to Europe and the United States, along with the bananas. Some of the impacts of banana production are relatively local: fertilizer runs off into nearby streams and kills fish; pesticides poison birds; chemicals injure workers. Others are quite global. Greenhouse gases are generated throughout the bananas' commodity chain, the production and consumption process, from the ships that transport them to the trucks that carry away the peels. Chemicals applied to the bananas end up in breakfasts, lunches, and snacks eaten all over the world. And all these stages can be linked to the complex social organization of contemporary life, of which you are a part.

Much of the food we eat travels along similar paths from farm to market, but it does not always reach those who need it. Food and farmers' markets operate on perilously thin profit margins and make more money on high-end fruit, vegetables, and processed foods than on lower-cost staples, and low-income consumers often cannot afford these higher-cost and often healthier products. As a result, large sections of urban areas, in particular, are devoid of places to buy food—these are often called "food deserts"—and residents have recourse

Figure 1.1   A Bad Value Chain?

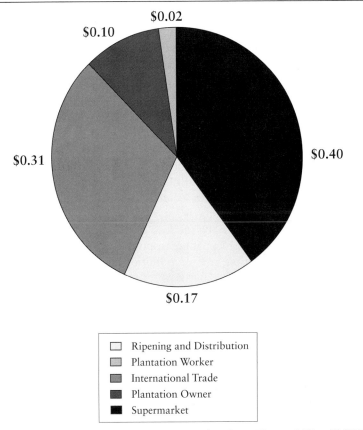

Source: Adapted from Maya Global 2012, "A Founder's Story: Chapter 6," Nov. 17, 2012, http://www.mayaglobal2012.com/a-founders-story-chapter-6/ (accessed Jan. 10, 2017).

only to convenience and liquor stores, whose inventory is largely composed of processed snack foods. Such "food injustice" is of growing concern and has given rise to a dynamic movement seeking to address this problem. Expensive, imported fruits and vegetables are only one contributor to this issue, and there are growing numbers of urban gardening organizations growing fresh foods in cities.[17]

But maybe you don't like bananas. "None of that has anything to do with me; I don't buy 'em, I don't eat 'em. Never have, never will!" But you probably like apples. Or pears. Or hamburgers. Cotton T-shirts. Gasoline. Tablets. Skateboards. Or any one of the hundreds of thousands of things produced somewhere else, shipped across oceans, and offered for sale where you live. Bananas, like all these things, are a *commodity*, a good (a consumable item) produced in large quantities and sold to individuals primarily in retail markets. You buy and consume commodities, which have environmental and social impacts when they are grown, shipped, and eaten and when the skin (or other detritus) is thrown away. Even if you detest bananas, you are part of a complex and complicated global system of politics, economics, and social order organized around the production of commodities, virtually all of which give rise to the environmental impacts along those commodities' life cycles. Indeed, the "commodity chains" associated with the things we produce and consume extend far beyond home and hearth and reach into the lives and well-being of millions of people as well as thousands of plant and animal species, with consequences for health, welfare, and justice. And we cannot ignore the elephant entering the room, either: climate change, with its own health, welfare, and justice problems.*

You, as an individual and as a member of a technological society, are deeply embedded and implicated in the condition and fate of the global environment, primarily through the social system we call "capitalism,"† which, as we shall see, tends to treat all things as having a value and price. This doesn't mean that you alone are responsible for the fate of the global environment, but it does mean that you should accept some responsibility for global warming, the ozone hole, species extinction, marine pollution, deforestation, pesticide poisoning, and other environmental matters. But you might well ask: how has this situation come to pass? How were decisions made? Who asked me? Do I have any say now, when policies and solutions are being discussed and deployed? And what can I do? These are the kind of questions that we seek to address in this book. The answers are not always so evident. Nor is what we can or might do so clear. But as the saying goes, a long journey begins with a single step, and we have a long journey before us. So, let us begin.

THINKING ABOUT BANANAS, PHONES, TABLETS, POLLUTION, AND OTHER SUCH THINGS

Environmental Politics for a Changing World is a book about the politics of the global environment; about the ways in which the global environment is being affected and changed by myriad activities, needs, and desires of human beings and societies; about the ways in which people *think* about the environment; and about the ways in which people, societies, organizations,

---

*As this book was being revised, Hurricanes Harvey, Irma, and Maria, thought by some to be especially intense due to warmer ocean waters, ravaged Texas, Florida, and Puerto Rico, among other places, while wildfires ravaged parts of California, followed by deadly mudslides resulting from intense winter storms. Those who recall Hurricane Katrina in 2005 know, as well, of the health, welfare, and justice issues arising from such climatic disasters.

†Some might recoil at calling capitalism a "social system." It encompasses or reaches into virtually every facet of contemporary life, and shapes lives in ways that are, in long historical terms, unprecedented. See, for example, Ellen Meiksins Wood, *The Origins of Capitalism: A Longer View*, London: Verso Books, 2002, rev. ed.

movements, groups, corporations, and states are responding to their impacts on their environments. It is also a book about the complex connections that constitute the relationships between humans, as individuals and groups, and the biogeophysical world, that is, nature. There are a number of terms that we will use throughout this book, and which we define and discuss in the remainder of this chapter. Table 1.2 provides a listing of these terms and brief definitions and examples of each.

Politics, in this context, refers to the ways in which societies make decisions about the order and functioning of their daily and long-term existence. Inasmuch as there is considerable disagreement over decision-making, order, function, and even the reasons for human existence throughout the world, it behooves us to better understand what we actually mean by the term "politics." But this book is about more than just politics. As suggested above, with our examples of iPhones and bananas, the "global environment" extends from the local to the planetary, and both we and the planet are subject to a broad range of material and social forces, practices, and impacts that involve economics, sociology, culture, psychology, technology, and other disciplines and forces. All of these are relevant to global environmental politics and the global environment.

What, then, do we mean by the term "environment?" *Environment* is frequently used to refer to *nature* (the "biogeophysical" sphere), the worlds of plants, animals, geology, geography, air, and water that are "out there" in forests, plains, oceans, rivers, and so on, and which comprise and live in *ecosystems*. Broadly speaking, an "environment" is a set of material conditions within which an object exists, although, to be wholly accurate, an environment can also include less tangible social and conceptual conditions. Thus, beyond the natural environment, we can also speak of the urban environment, the international environment, and the political environment. Human beings live *in* the environments they construct, but often behave as though they are not themselves part of nature, which is thought to be somewhere "out there" and not part of daily life. Experience indicates this is often not the case. Nature affects us when, for example, we experience tornadoes, hurricanes, and floods; we affect it when we cut down forests, cause global warming, and pollute waterways. But there is a widely held perception that, given enough time, knowledge, and technology, humans could separate themselves entirely from nature. Nothing could be further from the truth. Without nature, humans could not and would not exist; without humans, nature would not exist *as it is today*. There would still be nature, of course, but no natural environment and no one with conceptions of *Nature*.*

---

*A note on terminology. We use *nature* (lowercase) to refer to the physical and biological world. We use *Nature* (capitalized) to refer to the abstract concept of a philosophical realm of life, thought, and action. The first term is about the natural world as it is; the second as we imagine it or would like it to be.

Table 1.2.   Critical concepts for thinking, analyzing, and doing

Concept	Definition	Examples
Basic terms		
Politics	Processes whereby societies make collective decisions	Elections, social movements, parliaments
Environment	Material conditions within which people and nature co-create	Cities, forests, oceans, plants, animals
Value	Monetary or social significance	Value of ecosystem services
Philosophies		
Utilitarianism	Action is valued for its outcomes	Maximization of wealth through free markets
Deontology	Action is valued for its ethics	Distribution of wealth through regulated markets
Anthropocentrism	Actions center on humans	People have rights over nature
Biocentrism	Actions center on nature	Nature has rights and value in itself
Terms of knowledge		
Ontology	Causes and reasons for existence	Humans are nature's caretakers
Epistemology	How we come to know things	Through science or from transcendent sources
Method	Analytical frameworks	Qualitative or quantitative ways of analyzing information and data
Terms of action		
Power	Ability to shape social order	Social forces deposing a government
Agency	Individual freedom to act	Action through social movements
Structure	Social constraints on action	Limits on violence as a tool in politics
Habitus	Normalized social practices	Normative ways of doing things
Terms of analysis		
Political economy	Relations mediating power and wealth	Reasons for rich-poor gap and its maintenance
Political ecology	Relations mediating politics and material conditions	Reasons for control of resources and limits to access

Sometimes, environments are modified by humans and lose all resemblance to what we think of as nature: the centers of New York City or the clear-cuts of forests across the world come to mind in this respect. But the natural features of any particular city or place are a result of people's intentional physical transformation of local habitat over time and their importation of non-indigenous elements of nature not available locally (such as food, plants, and animals). It is not that Manhattan is divorced from nature. Rather, nature in Manhattan is quite different than what existed there before the arrival of Europeans and, for that matter, what is to be found outside of all cities. Human actions have removed large parts of nature from the material city, and those parts that remain must be conserved and protected, while those that have disappeared should, perhaps, be brought back in if humans are to survive and thrive there.[18] Thus, among the brick and asphalt, we also find parks, gardens, cleaner air, and even coyotes and raccoons. What is natural? What is artificial? What came before? What will come in the future?

What is important is that there is no pristine nature anywhere on earth. The conditions that characterize our "environment" are largely of human origin and in some cases have been that way for thousands of years.[19] This might seem to go too far—aren't there "wilderness areas" that are still untouched? Aren't they devoid of human impacts? What about Antarctica? What about the ocean depths? What about the highest mountains? Alas, it turns out that most of the world has been affected by human activities since people began to migrate out of Africa. As ecologists have studied more and more of the Earth's ecosystems, they have generally come to agree that there are few, if any, parts of the world that are untouched or unmodified by human activities. Even those parts that seem remote, such as the North and South Poles, feel the impacts of human activities half a world away.* We should be careful, therefore, about imagining that there exists, somewhere, a pristine nature, untouched by human hands, to which humans might return. Whatever we do in the future, we will live in an environment that has been substantially shaped by our ancestors and us, and that is ours to protect or destroy.[20]

Given this state of affairs, how might we think and act in ways that could foster a better understanding of our relationship to Nature and the natural environment and create a global environmental politics that can protect and sustain them? There is no single way of answering this question, although it is pretty common to say that we should apply the "best" science to those understandings and relationships. Even that might not be enough: while science can provide accounts of how nature functions, evidence of human impacts on nature, predictions of how nature might be transformed in the future, and what we might do to protect and sustain nature, science cannot decide *what* we believe—and there is ample evidence that many people do not believe what science tells them is fact. For better or worse, what we *think* and *believe* has major consequences for what we *do*, and what we *do* has major consequences for the future of nature.[21]

What, then, are the analytical contexts through which we approach global environmental politics? To complicate matters, what does "global environment"

---

*The appearance of the Northwest Passage and collapse of Antarctic ice shelves, for the first time in recorded history, is indicative of this "action at a distance."

mean? What is included in it and what is excluded? Who makes the rules about the environment? How might we understand how the bananas, smartphones, and other commodities we produce, trade, and consume play a role in the global environment? How can we explain the fact that activities that appear to be restricted to one place can have impacts on other places, far away? How long have these conditions existed and what is their history? What has been the role of technology in shaping the global environment and what will be its role in protecting the global environment? Questions are endless—where to begin?

Thinking about Nature and Acting upon It

Let's begin by exploring how the processes of thinking and acting actually work. In chapter 2, we will examine a number of environmental philosophies and worldviews that shape thinking and doing; here, we simply provide some examples of how what we *think* can shape what we *do* and, in turn, how what we *do* shapes what we *think*.[22] Two approaches to thinking about nature are *utilitarianism* and *deontology*. Both focus on "values," but each emphasizes different ways of assessing "value" (of which more in the following). *Utilitarianism* (sometimes called "consequentialism") values thinking and doing in terms of costs and benefits to individuals and groups.[23] We consume resources and benefit from the products and services that result—for example, extracting rare earth minerals so we can have smartphones. The benefits to people (we assume) are far greater than the costs to users and people all along the commodity chain. In other words, nature is a means to an end. By contrast, *deontology* focuses on *means* rather than ends,[24] on "letting nature live" because ecosystems and the species and materials that comprise them have value in *existence*.[25] In other words, if mining rare earths for smartphones would result in the destruction of a tropical forest, it might be of greater intrinsic and moral value to leave nature intact and forego the benefits of smartphones (this, however, is probably a minority view). Some people would label these two positions as "anthropocentrism"—human needs and desires come first—and "biocentrism"—humans must respect nature even if our needs and desires are left unsatisfied.[26]

We could also ask what might happen were humans to pursue a purely utilitarian or a purely deontological approach to thinking about and acting with respect toward nature? If, on the one hand, we were to recklessly consume and destroy nature in pursuit of unlimited consumption and economic growth, the world might soon become a rather barren place[27] and, quite possibly, inhospitable to humanity and other living beings. If, on the other hand, we were to "let nature live" and stop all exploitation of the environment, we might well be reduced to living at a "paleolithic" level (no bananas or smartphones, or even landlines and telegraphs).[28]

Beyond this, our worldviews often seem to imply that we, as human beings, are external to the "environment," that we are "in here," within society, while nature is somewhere "out there." It affects us, we affect it, but little more.

Nothing could be less correct. Without nature, humans could not and would not exist; without humans, the natural environment would not exist. There would still be nature, of course, but no natural environment and no conception of Nature. This might seem like sophistry, but it underlines an important point: unless we recognize that Humanity and Nature are mutually constitutive of each other—indeed, are virtually one and the same—we cannot meaningfully speak of protecting or sustaining the earth's environment. This is not a position often encountered in the literature on nature and environmental politics. To be sure, nature does not exist to serve humankind, but nature's continued well-being does depend on human actions. Similarly, human beings survive only by co-existing with nature (or, as some might put it, *being in nature*) and not destroying it or transforming it out of all recognition.

Between the two poles of utility and deontology, there are many different approaches to thinking and acting with respect to nature and environment (Table 1.3 shows a few of these). Some may represent more realistic approaches, under present-day circumstances, to environmental protection than others. It seems foolish to believe that humans could return to being hunter-gatherers or rely wholly on technological innovation and human inventiveness to free us from all material limits (note that those pursuing a "Paleolithic lifestyle" do not necessarily eschew the technological accoutrements of contemporary life). At the same time, we should not dismiss out of hand ideas that push us beyond the simple reproduction of our current beliefs and practices, and have about them a whiff of utopianism.[29] The challenge is to explore the feasibility of a union between "what is" and what could be, and how to get from here to there.

As suggested in Table 1.3, various forms of utilitarianism largely govern human relationships with nature and environment. Capitalism, which is the dominating social system of twenty-first-century life and which we discuss in detail throughout this book, values money and profits above almost all else. It matters little whether a tree can generate a dollar by being reduced to lumber or by being a tourist attraction to which admission can be charged; the point is to generate that dollar.[30] But not all utilitarian approaches seek to maximize economic value above all else, as Table 1.3 suggests. Which one dominates depends on *politics*, on social choice mechanisms and the institutions and practices that facilitate human activity and endeavors.

Knowing Nature

How do we know what we know about humans and nature, about relationships between humans and nature, and about how we act on nature and nature acts on us? Again, we ordinarily invoke both science and experience to explain such knowledge. Science applies various analytical methods to processes, devises theories and accounts of cause and consequences, and fosters various means to exploit the material world.[31] Experience, by contrast, is what happens to us and it often provides the "raw data" for the explanations we devise for ourselves about what has happened. Science is considered to apply anywhere in

Table 1.3.   Thinking about the value of nature and environment

Beliefs and practices for valuing nature	What it means
Pure extractivism	Nature's value lies in the maximum economic return it can produce when exploited—all nature is vulnerable
Raw material provision	Nature should be protected because it contains valuable chemicals that can be extracted and marketed
Ecosystem services	Nature should be protected because it provides valuable services at relatively low cost
Externalities	Damage to nature arises because it is available for free or at low cost and is not properly valued via price or taxes
Sustainability	Nature must be protected so that future generations are not deprived of valuable services, products, and aesthetics
Intrinsic value	Nature should be left alone because it has value in itself and its existence, exclusive of human needs and desires
Wilderness	Nature should be left alone in bounded areas in which human activities are highly limited or banned
Paleo-life	Nature can thrive only if human impacts are reduced to eliminate extraction and exploitation (including agriculture) beyond basic human needs
Human extinction	Nature can thrive only if humans are entirely eliminated from the Earth and unnatural impacts are reduced to zero

the universe; experience only to individuals and groups in particular places. But both are based on specific foundational assumptions without which explanations would make no sense. These assumptions comprise an *ontology*, which is a basic account of the nature of the world and existence. An ontology asks: Who are we? Why are we here? What is our purpose and the purpose of nature and life? Was Earth created by a god, or is it merely the result of chance? Is there such a thing as "human nature?" Do living things have inherent value? In other words, an ontology explains the purposes and principles for our being in the world, both as individuals and groups, and the way we and the societies we live in have come to be and are now. An ontology is usually based on certain a priori assumptions about human nature, about the world, about power, morality, and justice. For example, to say that humans are on Earth to fulfill certain divine purposes is an ontological proposition. To say that humans are on Earth by random chance is a very different ontological proposition.[32] The first ontology might explain that human pain and suffering are the result of divine displeasure with our corrupt nature; the second that humans are neither good nor evil by nature, even though they engage in both types of practices (as judged by human morality). The first might enjoin us to observe God's commandments; the second, to make of life what we can.

An environmental ontology operates in much the same way, explaining the relationship between nature and humans through propositions about each. As we saw earlier in the discussion of utilitarianism and deontology, one environmental ontology might regard humans as part of nature and its protectors or stewards by virtue of awareness of Nature, another that humans are separate from nature and put here on Earth to use its resources and devise substitutes as natural things become scarce and disappear. While these two propositions are not necessarily mutually exclusive, they posit a different purpose for both humans and the world around them and lead to different practices with respect to nature.

"Human nature" is another ontological concept central to our understanding of the environment and nature: are there natural and essential features to human behavior and existence to which humans are enslaved, so to speak, and which cannot be changed—for example, a propensity to collective violence? Or are humans shaped substantially by the societies in which they grow up, which might imply that odious practices such as war could be controlled and even eliminated? Another ontological proposition relating to human nature addresses selfishness and altruism. Some, such as Thomas Hobbes, regard people as inherently self-interested, violent, fearful of injury and death, and in need of discipline.[33] To others, such as John Locke, human beings are inherently altruistic, peaceful, cooperative, and self-organizing.[34]

To acquire the knowledge that helps to support ontology, we need ways of knowing. *Epistemology* has to do with knowledge and how we know or find out about the world. It relates a set of principles that account for sources of knowledge and how it is acquired (note that this is not the same as "method," which is a set of steps for inductive production of data; see the following). In other words, how do we know what we know, and how can we rely on what we know to be true or correct? For many (tens of) thousands of years, humans and societies relied on *transcendent* sources—gods, spirits, and the heavens—for knowledge about the world. For example, creation stories, frequently explained as having come from divine sources, provide knowledge about how the world came to be, how humans came to be in the world, and how humans and nature should relate to one another. Stories about the constellations provided knowledge about changing seasons and when crops should be planted and harvested. Stories about spirits in nature provided knowledge about how people should relate to animals. Living in secular, scientific society—at least, as of 2018—a common epistemology is to credit the cosmos with functioning according to physical laws that humans can discover and use to manipulate and shape the world. While it might not seem likely that any environmental epistemologies would rely on transcendent sources for knowledge, in fact, several do, as we shall see. For the most part, however, the modern perspective on knowledge about nature is an inductive one, seeking rules that can be discovered through the scientific method.

Finally, we require procedures, or *methods*, that allow collection of knowledge about the world so that the world can be explained. You are probably somewhat familiar with quantitative and qualitative methods, which are frequently counterpoised against each other. The first seeks numbers about phenomena, on the assumption that these can be organized and analyzed to reveal causal-consequential patterns. The second relies on human description and social interpretation of events, often inflected by particular and local cultural systems. In the former case, however, interpretations of experimental results are not a given; they are based on a priori assumptions regarding the nature of the phenomena under study. In the latter case, the researcher must also make such assumptions but relies more on how phenomena are interpreted and understood by research subjects and sources. In this book, we rely largely on qualitative methods of interpretation and explanation and cite data where it is available and appropriate.

Terms of Action

So far, we have focused on aspects of thought about humans, nature, and the environment. But humans *act* and, we assume, make choices about what is possible and how to do what they want or must. To this, we can add that no one decides and acts free of social norms and constraints that specify what is permissible and what is not, what is normal and what is not, and how things are to be done or not done. Our terms of action are *power*, *agency*, *structure*, and *habitus*.

Power is central to politics and is often defined as the capacity of one person ("A") to get another ("B") to do something that B does not want to do.[35] Although it is common to think of such influence being exercised solely through threats and force, in fact, A can accomplish her goals in many other ways: bargaining, bribery, persuasion, or changing the beliefs and values of others. In all these instances, however, power is defined by A doing something to B, which remains a very narrow and impoverished conceptualization of power. It leaves us to assume that everything else that might affect, encourage, or restrict people's behavior, such as morals, norms, laws, and practices, are something other than power.

Power is a far more complex phenomenon than the mere application of force or threats. Moreover, power is not simply something possessed by dominant social groups and exercised against the weak; power constitutes the self in society and makes human society and civilization possible. In this sense, power is *social* and makes it possible for things to happen collectively, in groups. Social power also makes it possible for the "weak" to defy and resist domination by others. Table 1.4 lists several forms of power relevant to this book.

Direct power and compulsion involves the threat or application of force and violence, or the wielding of wealth and sanctions to manipulate or prevent behaviors.

Agenda-setting involves the capacity and right to allow participation, to exclude issues for consideration by those with authority.[36]

Hegemony rests on wide acceptance and naturalization of dominant worldviews as truth through myths and ideologies, and marginalization of other beliefs and views as anathema and threat.

Table 1.4. Forms of power

Form of power	How it happens
Direct power/compulsion	Coercion through application or threats of force, injury, sanctions
Agenda-setting	Exclusion via non-decisions and denial of resources
Hegemony	Shaping of consciousness via myths, ideologies, false consciousness
Discourse	Delineating the boundaries of thought via limits on concepts, expression
Self-discipline	Socialization of obedience through family, schools, religion, etc.
Social power	Constitution of individuals and groups via social relations

Discursive power is the exercise of hegemony, that is, "the power to delineate the boundaries of thought," as Karen Litfin has put it, and render some ideas as unthinkable.[37]

Self-discipline is imposed by the self on the self, in the form of norms, manners, and other behavioral expectations of peers and society in general.

Social power exists everywhere; it is embedded in relationships among individuals, peers, and groups, and constitutes the human. As we use it in this book, it is the practice of local politics and refers to the political activity and activism of people in civil society, outside of the institutionalized politics of representative democracies.

What do these concepts mean in "real life?" Consider the case of the four-billion-dollar Dakota Access Pipeline (DAPL), proposed in 2014 and under construction as this book goes to press.[38] DAPL runs from the Bakken oil shale fields in North Dakota to Illinois, where it links with existing oil pipelines.[39] One segment of the pipeline runs beneath the Missouri River and Lake Oahe at a location from which the Standing Rock and Cheyenne River Sioux tribes get drinking water. These tribes have opposed the Missouri River crossing since learning about the project in 2014; the site was the focus of extended protests and an encampment, the Sacred Stone Camp, in 2016. Toward the end of the Obama administration, the Army Corps of Engineers decided to proceed with a "full environmental review" of the project but, as soon as Donald Trump entered the White House in January 2017, the Corps was ordered to rush construction approvals.[40]

During the 2016 encampment, in a show of *direct* power, local and state governments deployed police forces in order to contain the protesters. When the project was being planned, *agenda-setting* excluded the Native Americans from consideration. *Hegemony* appeared when DAPL supporters argued that Bakken oil was essential to the economy and security of the United States. *Discursive* power appeared when an executive of the company building the pipeline compared those blocking the project to "terrorists."[41] *Self-discipline* was exercised by the protesters in comporting themselves in particular non-violent ways when confronting police forces. *Social power* was evident in the establishment, organization, and operation of the Sacred Stone Camp, and the broad support the protests generated around the world.

What the DAPL case illustrates is that power is not something that is inherently (im)moral or (un)ethical. Moreover, power should be judged in deontological, rather than utilitarian, terms. What matters is *how* power is exercised, by what means and to what ends, and that is what must be judged on moral or ethical grounds. Moreover, power is not something exercised only by the powerful against the weak; to follow Michel Foucault's argument, power is everywhere and it circulates everywhere (a point we return to later).[42] The trick is in discovering how to exercise such power—in the "microspaces" of life, as he put it—to accomplish the political ends we desire.

It can be said, therefore, that when we act, we exercise both power and *agency*. In this instance, "agency" does not mean some organization, such as a travel agency or welfare agency, whose staff arranges things. Rather, "agency" refers to an individual's or group's capacity to act freely under the conditions of constraint imposed by the social institutions and material limits of habitat and society, and to challenge or change such constrained or limited situations and relationships (this is why political scientists often speak of "actors"). Action might seem fairly easy and straightforward, but agency implies more.[43] In any given situation, not all actions are possible, desirable, or permissible. Sometimes, discovering what is politically possible under such limits is critical to exercising agency and influencing outcomes.[44]

The limits on such possibilities of action are often ascribed to *structure*.[45] In this instance, again, structure does not refer to buildings or bridges or roads; rather, structures are limits, restrictions, and prohibitions on the possibilities of action as these boundaries exist among the members of a society. Structures are like manners: you gain more respect from peers by being polite. Structures may also mandate particular roles and actions for people in particular contexts. Structures take the form of rules, sometimes written as law, often as unwritten or implicit norms, manners, and practices; they are known, but often cannot be seen; they must be learned and are not instinctual. Structures may also be contradictory, applying to a specific group or situation and not to another.* For example, university faculty are enjoined to behave in particular ways when they are teaching—they do not, generally speaking, hop up on a table and dance, even though this is not expressly forbidden. It is simply something that is not done—and most of us resist such temptations. Similarly, many students attend

---

*As Anatole France pointed out in *Le Lys Rouge* (*The Red Lily*, Paris: Calmann-Lévy, 1894, chapter 7), "Another reason to be proud, this being a citizen! For the poor it consists in sustaining and preserving the wealthy in their power and their laziness. The poor must work for this, in presence of the majestic quality of the law which prohibits the wealthy as well as the poor from sleeping under the bridges, from begging in the streets, and from stealing bread."

their classes on a regular basis, even though there are no penalties for sleeping in or surfing—going to class is simply what a student does. But structures are rarely determining, that is, they enjoin but do not prevent agency and choice.

Structure and agency are counterpoised to each other as, somehow, opposites: either we are fully free or we are not. Here, we encounter another useful concept: *habitus*. Habitus is a term ascribed to the late French sociologist Pierre Bourdieu.[46] It is not a cognate for "habit," which is a repeated pattern of individual behavior (for example, my habit is to shower every day). Rather, habitus refers to the *practices* in which people engage in social "fields," which are functional environments linked to specific roles. For example, the faculty and students mentioned above are based in a field comprising the university, in terms of both a specific campus and higher education more generally, in which they fill specific roles—educator, researcher, PhD candidate, undergraduate—with which are associated specific practices, or behaviors. As a result of engaging in these practices in particular ways that enhance status and prestige, the occupants of fields generate "social capital." Unlike money capital, however, social capital cannot be spent—at least not in the same way. It consists of recognition for achievements and, possibly, reciprocal favors and obligations.

Let us consider an example of habitus: recycling as normalized practice. Fifty years ago, almost no one recycled used materials, and almost no governments and agencies sponsored programs for this purpose.* Today, at least in the Global North, places without some kind of recycling program are rather uncommon and there is a general expectation that people will recycle cardboard, plastic, newsprint, jars, bottles, and so on. This is a practice inculcated into small children by their parents, teachers, and media. People who ignore recycling strictures tend to be regarded by those who do recycle with disdain and distaste, and the former lose what social capital they might have possessed in this field. Or, consider the professor who is highly lauded by students, thereby building social capital, with the results that her classes are always full, compared to the professor who is derided by students and faces half-full classrooms. The former has accumulated social capital in the teaching field, the latter has spent most, if not all, of his (if he ever had any).

Terms of Analysis

The final set of terms of interest to us are *analytical* frameworks that can illuminate critical causes of environmental impacts. Such frameworks are akin to theories of realism and liberalism (further discussed in chapter 2) but seek to reveal and understand how social structures and relationships are infused with power and wealth and naturalized so that no other outcome seems possible. Such frameworks would help us understand, for example, why most bananas

---

*During World War II, the military need for various materials led to high rates of recycling, but this practice vanished rather quickly as resources and materials were directed back to civilian uses.

are grown on massive plantations, cared for by poorly paid farmworkers who often suffer from ill health, sprayed with massive amounts of chemicals rather than grown organically, and turned into billions of essentially identical and mundane yellow objects with little flavor and limited nutritional value, with a shot of pesticide. Such a framework would explain how power and wealth have been used to create and sustain this particular system, and why it has proven so difficult to change it. It would show us why the global banana trade is mostly in the hands of a few corporations, why growers and workers receive so little for their labors, why the use of chemicals helps to increase profits, how the desire for bananas had to be created, and why consumers demand perfect fruit without any brown spots, even though ripe bananas taste better than greenish ones. And it would identify possibilities and points of action to address these impacts and injustices.

Our two analytical frameworks for answering these questions are *political economy* and *political ecology*, leavened with recognition that *material history* is critical to analysis and understanding. Karl Marx once famously observed that "People make their own history but they do not make it just as they please; they do not make it under circumstances chosen by themselves, but under circumstances directly encountered, given and transmitted from the past."[47] Marx was referring not only to ways of thinking about and doing things passed from one generation to the next but also the material environment as changed by human action. Conestoga wagons followed foot trails to the American West, railroads built over the track left by wagon trains, and highways followed the rails in their turn. Today, even coast-to-coast flights tend to follow the interstate highway system, at least when it is visible. In other words, what was materially constructed in the past—railroads, ports, courts, buildings, orchards, plantations, institutions, corporations, schools, and laws—often continues to exist, having left imprints on and obstacles to what we deal with today and what we might face in the future. This is true even if those things have disappeared in law and practice. Slavery was abolished over 150 years ago in the United States, yet, as evidenced by recent debates, controversies, conflicts, and struggles, its effects have not disappeared. The freeways in Los Angeles follow the routes of long-vanished streetcar and electric train lines, which then fostered and facilitated the urban sprawl and traffic jams that are so problematic now.* A significant fraction of the carbon dioxide produced during the first century of the Industrial Revolution remains in Earth's atmosphere today and, along with annual increases, drives the global climate change of the future.

All of this—people, cities, nature, things, ideas, and practices—are embedded in a *global political economy*. An *economy* is a system of production, exchange, and consumption; a *political economy* is to economy as hardware

---

*To address freeway traffic, Los Angeles is spending tens of billions to build a new light rail system, some of whose lines follow the freeways that followed the old regional rail lines. Go figure.

(machinery) and software (rules and relationship) are to a factory. When we speak of "economy," we refer mostly to markets, supply, demand, and prices. "Political economy," by contrast, addresses the rules and regulations that govern markets and the role of power and interests in shaping those rules and regulations.[48] In many books about the global environment, political economy is raised primarily with respect to environmental *externalities*, such as the impacts of trade on resources or the air and water pollution generated by specific industries or the wastes that result from disposal of worn-out items or the need to incorporate the costs of environmental damage into the price of goods. In this book, we seek to apply a *critical* approach to political economy, which seeks to reveal and critique the often opaque ways in which power shapes the economy and the ways in which the economy, in turn, shapes the application of power.

Political ecology involves the ways in which societies are shaped by struggles over the control and distribution of resources, and access to nature.[49] Political ecology is also a critical approach, in that it does not assume pre-given actors whose behavior and roles are determined by particular interests. When we examine environmental contexts and struggles, it is as important to examine what is not being said or acknowledged as those factors which seem obvious and unquestioned. We tend to associate "politics" with certain institutionalized practices, such as voting, executive power, legislatures, and, maybe, courts. It is misleading to limit an understanding of politics only to that which appears publicly and has been naturalized as "the way things are." Thus, the fact that the U.S. Environmental Protection Agency is required to hold public hearings when developing new administrative rules is widely accepted, even if the procedures are sometimes not applied evenhandedly. But behind this façade of public participation is a system that has been shaped by power to its desires, so to speak. Political ecology seeks to make the parts of this system visible and the subject of struggle.

To illustrate political economy and political ecology as analytical frameworks, let us return to the case of DAPL, described above and shown in Figure 1.2. The first thing to recognize is the importance of material history in this case. First, the presence of the Sioux tribes in the area of the pipeline crossing is a consequence of a treaty signed in 1858, which created the "Great Sioux Reservation," of which the Standing Rock and Cheyenne tribes occupy only a part, bordering the Missouri River. At the time, the interest of the U.S. government was to move Native Americans to as remote a site as possible. Of course, oil played no role in delineating the reservation—that came much later. But the reservations are material and historical facts on the ground, and the tribes have certain sovereign powers granted to them through treaties with the U.S. government. More recently, as the Bakken shale was tapped, oil producers sought a means more efficient than tank cars to move the stuff (tank cars have a propensity to explode and burn in accidents and derailments). But the oil pipeline infrastructure and oil refineries were located so that the proposed

Figure 1.2   Sioux Reservations and DAPL

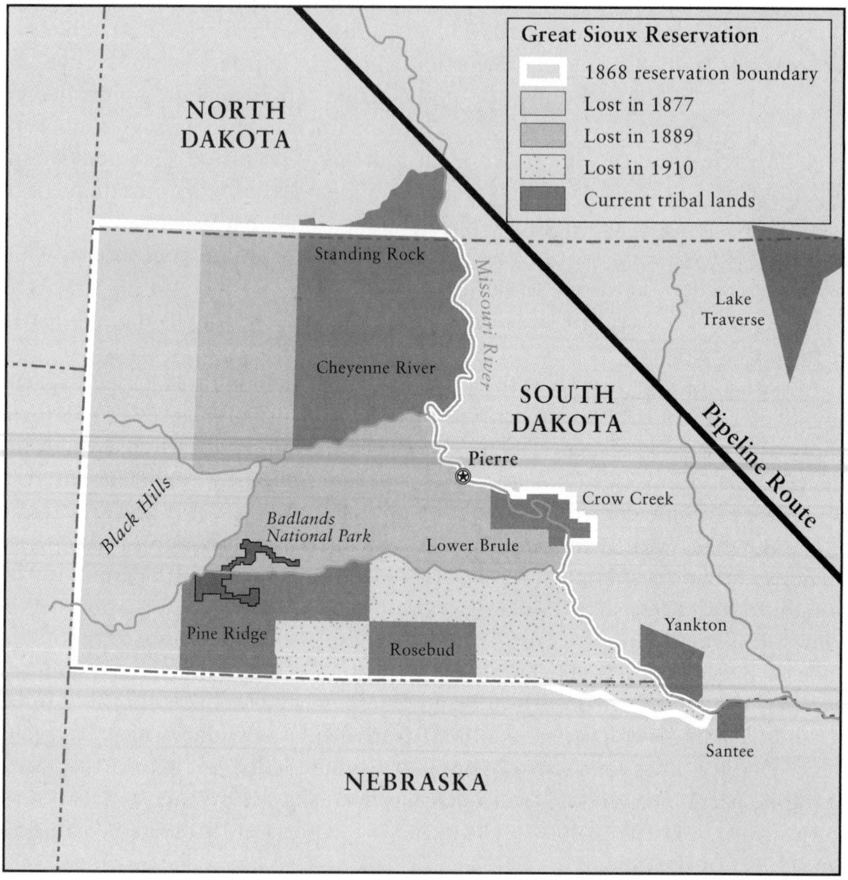

Source: https://commons.wikimedia.org/wiki/File:Siouxreservationmap.png.

pipeline just skirted the northernmost tip of the Standing Rock reservation. It could have been rerouted so as not to cross the Missouri River at exactly that point, but that would increase pipeline costs and impact profits.

That this crossing might impose serious costs on the Sioux did not matter; on that, the law is clear. The political economy of the situation is such that wealth trumps health. The Sioux, finding that ordinary modes of politics failed to get them a hearing, resorted to occupation and blockade of the crossing site, an action that generated immense public interest and support, bringing together Native Americans from all over the United States. Indeed, as a result of this struggle, the Obama administration put a stop to construction, which was subsequently overturned by the Trump administration, which owes more to the oil industry than did its predecessor.[50] Political ecology helps us tease out and make visible practices of power and contested politics around control of natural resources (in this case not merely oil, but more importantly water and land) and impacts on environment and community.

THINKING SOCIALLY

Although the natural environment is a physical and biological system, one that can be studied in whole and in part and understood at least partially in scientific and material terms, the politics that play out around environmental change and degradation are *social*. Politics is a human activity, organized and structured through a specific material history and based in sets of beliefs about individuals, societies, and the world and the practices involved. Both history and many beliefs may have the quality of seeming fixed and immutable, and it is this "naturalized" character that gives them their power and makes them seem "true." "Thinking socially" involves, therefore, recognizing the social forces and structures that play a role in political struggle for and over nature.

Human interventions in nature are, for the most part, intentional, although impacts are often unintended. Given, however, appropriate and just *social institutions*—values, rules, roles, laws, and practices devised by human societies to ensure their survival—it has been possible for people to live in, thrive in, and even modify a variety of natural environments—some of them quite inhospitable—and to do so in a relatively fair and just fashion. But not always. For example, the idea that humans are entitled to exploit nature was hardly questioned in the Global North until a few decades ago;[51] a worldwide movement was required to challenge such ideas, legitimate heterodox questions, and change practices and values. Sometimes, such changes take place without much apparent action or attention. Thirty years ago, people did not have smartphones to stare at while they walked around; today, such behavior is regarded as normal, even if it sometimes leads to pedestrian collisions or unfortunate traffic accidents. And people think nothing of jumping in their cars to drive two thousand feet to a convenience store—mobility is, at least in the United States, both a value and an unreflective practice. Some practices may need to be called out if they are to be changed or eliminated in order to protect nature and the environment.

If well-being and survival are under threat due to defective social institutions, what can we do, especially given the constraints imposed by them on people? What can we do if, for example, water supplies are reduced by drought even as water managers continue to provide large quantities of water to farmers (as was, until recently, the case in California)? What if the atmosphere and oceans can absorb no additional carbon dioxide without significant and potentially dangerous changes to climate and to species with calcium carbonate shells, while corporations and institutions blithely continue their polluting activities? This is where agency becomes crucial. The key to social change is, therefore, to discover what can be done and how to do it.

As you will discover in the chapters that follow, there are serious reasons to doubt that institutionalized systems of law, economics, and practice will suffice to address growing and pressing environmental problems, both local and global. A serious mismatch exists between those institutions assigned the task of dealing with environmental damage, no matter what the scale, and the sources and nature of those problems. Most laws and institutions, whether local, national, or global, originally emerged to serve particular social functions and interests and, perhaps less evidently, to establish, support, and maintain particular relationships of power. As a result, rather than devising effective institutional responses or designing new institutions appropriate to addressing contemporary environmental problems, we have resorted to framing the problems in terms of the institutions and practices already in hand. This is most clearly seen in the generally uncritical application of economic and market-based tools to environmental scarcities and externalities (or excesses, in the case of greenhouse gas emissions). Environmental problems come to be defined in ways that fit existing institutional structures, rather than in ways that could, among other things, clarify who is responsible for problems, who holds power over whom, and to what ends power is being directed by those who hold it. This is a point to which we will return in subsequent chapters.

Indeed, we have often been told that only states (i.e., countries) can address global environmental matters because they are the highest authoritative political institutions on earth, it is their boundaries being transgressed, and they govern (virtually) all inhabited space on the earth. As a result, pollution that originates from one country and has impacts in a neighboring country can be addressed only internationally, for to do otherwise would be to infringe on the sovereign rights of each state and to demand action where there is no government.* A *gedankenexperiment* (thought experiment), offering two slightly different environmental scenarios, will illustrate this obstacle to environmental governance.

Scenario 1—one state: Imagine a river valley with a number of factories running its entire length. These factories were built over a period of decades, during which the entire valley has been under the jurisdiction of a single national government. Culturally and linguistically, the inhabitants of the valley are identical and, indeed, are all descended from a few founding families. One day, a factory at the north end of the valley begins to produce a rather unattractive and mildly toxic plume, which drifts to the south, where it sickens a number of residents. The local health department, empowered to act against such polluters, closes down the factory and requires that the owners fix the problem, which they do. While not much has been addressed beyond the immediate pollution problem, that and its associated hazards have, at least, been eliminated.

---

*This, of course, is the often-invoked condition of international anarchy—about which there are good reasons to be skeptical.

Scenario 2—two states: Now, consider the same scenario with a twist. Our valley is crossed by an international boundary, with one country in control of the northern half, the other country in control of the southern half (think Alsace-Lorraine—now called Alsace-Moselle—which was "traded" back and forth between Germany and France during the nineteenth and twentieth centuries). The inhabitants are still related, speak the same language, and some even cross the border to go to work. The geography of the valley and the factories operating in it remain the same. One day, one of the factories at the northern end of the valley begins to produce a rather unattractive and mildly toxic plume, which sickens a number of residents living at the southern end, which is *now part of a different country.* The health authorities in the affected southern region have no power to shut down the offending factory, because it is under the jurisdiction of a foreign health authority. Instead, the health authorities in the southern end must appeal to their national government to contact the government of the neighboring country, which governs the valley's northern end, to ask that something be done. If the northern state is willing, it can order local authorities in its part of the valley to shut down the factory. But, of course, since each state is sovereign in its own territory, the government that exercises jurisdiction over the polluting factory is under no compulsion to shut it down.[52]

★　　★　　★

The difference between the two scenarios has to do with institutions rather than geography or culture or ecology. Under a single government, residents of the valley can protect their health; under two, they might not be able to, *even if the border is fully open and people cross it all the time.* Moreover, entrenched economic and political interests—factory owners, politicians, municipal authorities, workers—might want to keep the factory open and resist a resolution to this problem. These parties are implicated in the same institutional conditions, as citizens of their particular country. Not to idealize the possibilities, but if the valley were treated as a single habitat or sociocultural ecology, with institutions organized in such a way as to take into account the valley's history and political economy, the obstacles preventing a solution in Scenario 2 might not arise.

It makes some sense to conceptualize the global environment in terms of existing institutions, such as states, borders, international organizations, and capital, but this virtually ensures that most transboundary and international problems will not be addressed effectively. A generation ago, in 1987, the World Commission on Environment and Development (the Brundtland Commission) in *Our Common Future* lamented that "The Earth is one but the world is not."[53] And, notwithstanding hopes during the 1990s that such divisions might be overcome, the current nationalist revivals across the world do not bode well. At any rate, this lament is an empty one. Were the world actually "one," as the commission thought necessary, who would be in charge? Who would make necessary decisions? Who would pay the costs? The past three decades have hardly seen answers to these questions, and it is not very far-fetched to speculate that One World, were it to exist, would operate much to the advantage of the wealthier states (and classes) of the planet and the disadvantage of everyone else. The poor would remain poor, or perhaps become poorer. Moreover, a single unified world might create many more problems than we now face. Indeed, the international approach to global environmental issues might even be considered reductionist. This may seem an odd charge: how can a process that treats the whole world as one be reductionist?

From a technological perspective, a global command economy is becoming more and more feasible. Surveillance, data storage, and computing power have grown so much over the past quarter-century that a system of central planning tracking individual and collective activities and managing negative and positive feedbacks is probably within reach. Who would operate such an arrangement is not at all obvious. Moreover, the bane of centralized management is diversity and difference. Managers—especially global ones—prefer predictability and similarity so that a limited number of operating principles and rules cover all possible contingencies—not that all contingencies can ever be anticipated. (Consider how every significant commercial nuclear power plant mishap or accident over the past fifty years was unforeseen.) Yet it is also evident that the world is an extremely diverse place and that, now more than ever, people are willing to fight in defense of diversity and difference.

The problem of reductionism arises, consequently, in trying to conceive of and deal with the world as a single unit, rather than as many, very particularistic places. No single strategy, no single program, no single directive can apply to every person, every place, every moment, every society. Moreover, because many different types of activities produce the gases that contribute to global warming, are we to treat all of these sources as equal? Do we count emissions of methane from rice paddies in Asia as somehow the equivalent of emissions from rice paddies in California? Is carbon dioxide from household fires in Africa the ethical equivalent of emissions from long-distance leisure travel in Europe? Is it morally legitimate to rank countries by their aggregate emissions, or do we need to be more aware of per capita levels and activities involved?[54] Paradoxically, universalism may be reductionist, while particularism is holistic.

WHAT'S IN THE REST OF THIS BOOK?

Chapter 2, "Deconstructing the Global Environment and 'Global Environmental Politics,'" presents contrasting philosophies and worldviews for understanding and explaining global environmental problems and politics. People subscribe to many different foundational assumptions about life, about nature, about humans. Some assumptions are taken for granted, hardly recognized, and rarely questioned; these become the basis for dominant (or "hegemonic") worldviews. Other assumptions are contested, ridiculed, and marginalized; these might be thought of as "counterhegemonic" worldviews of resistance. Dominant worldviews tend to set the institutional conditions under which social problems, such as environmental damage, are addressed; in a capitalist system, the market comes to be regarded as the universal solution to all difficulties. But there's a catch, as noted earlier in this chapter: the practices that have been institutionalized and naturalized in dominant worldviews are often the cause of the very problems that lead to the environmental impacts of concern. If damage to the global environment is due to the particular structures and practices of capitalism, for example, how can capitalism also be the solution? By examining the ways in which different worldviews and philosophies approach the very concept of "global environment," we may better understand how we have arrived where we are today, and what might be possible in the future.

Chapter 3, "Capitalism, Globalization, and the Environment," examines why the degradation of the natural environment has become a global concern.[55] What processes have led to the "transgression" of national borders by air and water pollution, trade in endangered species and genetic resources, and the global dissemination of toxic wastes? What is new about this? Advocates of some philosophies, especially those focused on economic growth, argue that the environment is being damaged because of the absence of certain property rights, low resource prices, and inefficient allocation of funds for protection. To a growing degree, for example, the solution offered for global warming is tradable emission permits—rights to pollute the atmosphere that can be auctioned to the highest bidders—and "carbon taxes" imposed on goods and energy in proportion to their carbon content. Alternatively, consumers are urged to be "green" and buy only environmentally friendly products carrying appropriate certifying labels.

The specific focus of chapter 3 is, therefore, on the environmental impacts of industrialism, capitalism, and globalization. To be sure, changes in nature as a result of human action did not begin with either industrialism or capitalism. Still, the scale and scope of associated impacts did not reach much beyond regional limits until the extension of markets to all parts of the world and the emergence of industrialism several centuries ago. More to the point, it was not until nature came to be regarded as something that could be exploited for accumulation and profit that the physical and biological world began to be regarded as mere capital, to be bought and sold at will and on a whim.[56] But does reliance on capitalism and markets to protect the environment simply license "rights to pollute?" Should such rights even exist?

Chapter 4, "Civic Politics and Social Activism: Environmental Politics 'On the Ground,'" begins with the aphorism "All politics is local," attributed to the late Democratic Congressional Representative Thomas P. "Tip" O'Neill of Massachusetts. O'Neill, revered and reviled for his staunch liberalism, recognized that he had to respond to the needs of his constituents, who were more concerned about their neighborhoods than the planet. He also saw, however, that it is people who are political, who engage in politics, and who, ultimately, must be persuaded that actions are both good and to their long-term benefit. All environmental politics is, consequently, about the local: about landscapes, land use, people's beliefs and practices, and insults to the environment that originate in and affect specific places and regions. Some of these matters do manifest globally (for example, global warming), but specific places and spaces feel their effects. Other problems have a much more localized character (for example, soil erosion), originating from specific local sources and behaviors and motivating specific local actions in response. This chapter offers a context for understanding social power and an overview of social power as it is deployed both locally and globally.

In recent decades, environmental movements exercising social power have become more and more important in environmental politics and policy all over the world.[57] Social power often emerges in response to the failure of political authorities to address specific public and environmental problems and the dearth of effective mechanisms for influencing policymakers and corporations. Social power also functions outside of conventional political institutions and practices, and may serve to revitalize politics and political activism.[58] Environmental movements are fragmented between and within countries, and between and within groups and organizations themselves.[59] There is a furious debate over the methods used by some environmental groups to lobby policymakers and politicians and over the ethics of corporate sponsorship, a method of fundraising that has transformed some environmental organizations into corporate mouthpieces.[60] Businesses and corporate associations have become prominent both in debates and in policymaking and sometimes have set up "false flag" groups which can have considerable political impact.

Chapter 5, "Domestic Politics and Global Environmental Politics," examines states and the environment. Most books on international environmental politics begin with the state. We don't. But we argue that little attention is paid to the way national policies and practices contribute to international problems of concern, even as those policies and practices are formulated to address those problems. Although national environmental agencies and ministries are of relatively recent provenance, some nation-states have had "environmental policies" for well over two centuries. In earlier times, these were motivated mostly by agricultural and industrial development for the purposes of state building. The practices of one country were frequently adopted by others. One result was similar patterns of environmental degradation around the world.

Over the past few decades, states have instituted policies and established institutions intended to address insults to the natural environment and their effects on the health of human beings. Each country tends to approach and address such matters differently as a result of differing domestic contexts.[61] At the same time, however, certain practices, such as forestry or water management, become "globalized" and set the standard for the replication of policies around the world.

Some policies work in one national setting but not another; conversely, some practices and policies fail in one place but are nonetheless adopted in another, where they fail similarly. Drawing on the literature on comparative environmental politics and policy, this chapter offers capsule studies of several environmental issues and the ways in which they have become (inter)nationalized.

Chapter 6, "Global Environmental Politics, Society and You," asks, "Where do we, both reader and writer, fit into all of this?" As consumers of goods and producers of pollution, we are directly implicated in the damage being done to the earth's environment. We can make careful choices about what to buy and how much stuff to use. If everyone made the same choice, there would be changes but, acting alone, we can never make our choices add up to more than those of a single individual. Experience suggests that large-scale changes in preferences and practices won't emerge simply through changing consumption. Meaningful impact requires collective action and, to repeat the theme of chapter 4, such action involves all kinds of institutions, political and social, ranging from the household to the global.

OUR HOME IS OUR HABITAT, OUR HABITAT IS OUR HOME

Planet Earth is the habitat and home of humanity, plus some two million to one trillion other plant and animal species. We are all familiar with the blue-and-white ball as seen from space; it is probable, notwithstanding dreams of Mars and stars, that Earth is all we will ever have. In one sense, Earth is a single ecosystem, and we really do not know how robust or fragile that system might be under the environmental insults initiated by human societies. In another sense, it is thousands, if not millions, of ecosystems and habitats, and some of those are clearly quite fragile even if others are less so. For each of those species, for each member or group of each of those species, the place where they live is their habitat and home. Most, if not all, of those individual habitats are also part of human habitats. And those habitats are being modified, disturbed, or destroyed by human actions, either deliberate or incidental, individual or collective. Most, if not all, of these actions are a consequence of the operation of networks of social institutions, technologies, practices, and actions that make up the fabric of human existences. Again, "wheels within wheels." Where to begin?

The best place to begin, we believe, is where we are, where we live, and to work "from the ground up." States (countries) are an abstraction. True, we encounter the state every day, through its representatives, money, buildings, roads, rituals, and beliefs. The international system is even more of an abstraction: There is the United Nations building in Manhattan, of course, but what does a building mean? The planet is all around us, yet we can never experience more than a very small part of it, the part where we are: our habitat, if you will. Place is therefore of central importance in determining our identities, defining the meanings of nature, and acting to protect the environment. By caring for the Earth's parts, through politics, we can care for the whole.

Lest this sound hopelessly romantic, it is not. This is not an admonition to stick to your last,* to the exclusion of all else, to pay attention only to your own surroundings and to ignore others. It is essential that we always be aware of the place of our place in the whole, of the relationships, insofar as we can tell, of our part and other parts to the whole. It is essential that we understand the manifestations of international economic forces and international political processes as they appear in the small places that we, ourselves, occupy, individually and collectively. It is essential that we build alliances within places and between them, alliances that are consciously devised to challenge those forces, processes, and practices of power that, in millions of places, degrade or destroy habitat. The "fate of the earth," to use a phrase coined in the 1980s by the late Jonathan Schell and adapted to many other purposes, does not rest on what we think or believe as isolated individuals but on what we do together through and by politics.[62]

FOR FURTHER READING

Kuehls, Thom. *Beyond Sovereign Territory: The Space of Ecopolitics.* Minneapolis: University of Minnesota Press, 1996.

Paterson, Matthew. *Understanding Global Environmental Politics: Domination, Accumulation, Resistance.* Basingstoke, England: Macmillan, 2000.

Polanyi, Karl. *The Great Transformation: The Political and Economic Origins of Our Time.* New York: Farrar & Rinehart, 1944.

Rivoli, Pietra. *The Travels of a T-shirt in the Global Economy.* New Jersey: Wiley, 2015.

Smith, Ted, David A. Sonnenfeld, and David Naguib Pellow, eds. *Challenging the Chip: Labor Rights and Environmental Justice in the Global Electronics Industry.* Philadelphia: Temple University Press, 2006.

Steinberg, Paul F. *Who Rules the Earth? How Social Rules Shape Our Planet and Our Lives.* Oxford: Oxford University Press, 2015.

Wapner, Paul. *Living through the End of Nature: The Future of American Environmentalism.* Cambridge: MIT Press, 2010.

Wright, Angus. *The Death of Ramón González: The Modern Agricultural Dilemma.* Austin: University of Texas Press, 1990.

---

*For those unfamiliar with this term, there is an old saying of Roman origin, *sutor, ne ultra crepidam*, which translates to "shoemaker, not beyond thy shoe." A "last" is the form on which shoes are made, so the meaning is don't pass judgment or criticize that about which you don't know.

2

Deconstructing the Global Environment and "Global Environmental Politics"

★ ★ ★

What does it mean to deconstruct something? Chapter 1 offered a list of descriptive and analytical concepts to highlight different normative perspectives about environment, society, and the relationships between them. Each of these differing normative approaches is based on specific foundational assumptions regarding nature, human nature, the social world, and the structures that comprise them. It might seem that, as perspectives or "worldviews," these different approaches are purely descriptive and explanatory, and have no impact on "real world" conditions and circumstances. But this is not the case; as we noted in chapter 1, problems are often defined in terms of worldviews, which in turn shapes how they are addressed (rather like the old saw, "If all you have is a hammer, everything looks like a nail"*). The results are that what comprises global environmental politics are, often as not, disagreements and struggles over (1) the actual and desirable relationships between nature and society, (2) definitions of causes and effects of environmental impacts, and (3) policies and practices that both exploit nature and seek to address negative impacts. Note also that the particular worldviews held and acted on by various parties reflect not only conflicting normative systems but also differentials in power, knowledge, and wealth. As a consequence, those who govern and benefit from particular

---

*This saying has been attributed to Abraham Kaplan, Bernard Baruch, Mark Twain, and Abraham Maslow, as is often the case with such sayings. See L. Don Wilson, "If All You Have Is a Hammer!" *Dental Economics* 102, #4 (2013), accessed June 29, 2017, http://www.dentaleconomics.com/articles/print/volume-102/issue-4/feature/if-all-you -have-is-a-hammer.html.

social arrangements and practices are likely to endorse existing relationships, whereas those who are poor and weak may dissent from or resist the social order, or do nothing at all.

In order to clarify the role of worldviews in deconstructing and understanding both global environment and global environmental politics, it is useful to return to the discussion, in chapter 1, of *production* and *reproduction*. To survive, human societies must *produce* their needs from the materials provided by nature and they must *reproduce* social organization as well as new human beings. *Production* involves the ways in which technology, capital, and social relations produce the conditions, goods, and services—the *material base* or "hardware"—that allows societies to operate, survive, and reproduce. *Reproduction* is then the rules, roles, and relationships that organize and legitimate the organization of the material base and the distribution of power and wealth that result—in technical terms called the *superstructure* but which can also be thought of as the "software" or "operating system." Production and reproduction are fundamental to all societies, even as the particular methods and rules are often very different from one group to another. These two terms, "material base" and "superstructure," are regarded by some as too deterministic and no longer of much theoretical utility. For one thing, they suggest a certain stasis over time, which is not at all the case for existing societies. For another, it is not so easy to separate the hardware and software of social organization since the two work together and mutually constitute each other.

Some worldviews distinguish between material and "ideational"—between the role of the real world in shaping production and reproduction and the role of ideas and beliefs in directing people and what they do and accomplish. In contemporary society, much of the impetus to social change is thought to arise from ideas rather than the material world, which suggests that the former shapes the latter without much reciprocal influence. This tendency to separate the material and the ideational arises because of the ways in which, since at least the Enlightenment of the eighteenth century, humans have been regarded as separate from nature.[1] Living in built environments—cities, buildings, infrastructures, long-distance transport of commodities—it is easy to forget how dependent urban dwellers are on the stability and reliability of nature*—even though change in the built environment is constant and, sometimes, much for the better. Indeed, the *means* of production of the material base will change over time, as technology and social and economic conditions change, and, especially, if it becomes more difficult to procure materials because nearby sources have been played out. Reproduction of the social superstructure will change, too, if power and wealth are redistributed and if the material base changes radically. None of the structures, practices, or beliefs that constitute

---

*Again, consider the impacts of Hurricanes Harvey, Irma, and Maria; the Mexican earthquakes; and the California fires and floods of 2017.

a particular society's base and superstructure are fixed or "natural" in any fundamental sense, although those who are most advantaged by existing conditions and relations will struggle mightily to prevent or slow changes that appear to their disadvantage.

When we speak of *construction*, we are referring to the ways in which basic, taken-for-granted beliefs, meanings, and practices serve to organize and discipline a society's culture and life, direct the manipulation of nature, and specify rules and roles for the individuals and groups within that society—or, to put this another way, the organization of social power, as defined in chapter 1. When we speak of *deconstruction*, we seek to unpack and examine those beliefs, meanings, structures, and practices, the better to understand why it is that some forms of social being and activity appear more "natural" than others. We would also like to identify the deployment of power, in particular, and ways of changing those ideas, policies, and practices which are most destructive of nature.

As an example of this, consider whales. When whales were seen as terrifying monsters and, later, a source of oil and food, whole fleets of ships were dedicated to seeking them out and killing them. Any suggestion that whales might deserve to be left alone was, simply, unimaginable. When whales came to be seen as intelligent mammals and (some) caricatured as friendly, whole fleets of ships were dedicated to seeking them out and watching them. Any suggestion that they might deserve to be killed for food and oil became unimaginable (except in Iceland, Japan, and Norway). This change did not arise due to the whales themselves—although the imminent depletion of some species did give rise to the International Whaling Commission in the 1940s. What changed was the demand for whale-based products, such as meat, oil, and baleen (for corsets and combs, among other applications), as substitutes were discovered or invented, and the rise of scientific research into whales, in the effort to conserve and protect them (possibly, for future harvesting). Returns to whaling, as species were decimated, declined and made the industry noneconomic. Finally, in the twentieth century, the middle classes were exposed close-up to seemingly friendly orcas ("killer whales"), which probably played a role in attitudes toward marine mammals, including dolphins, porpoises, and whales.*

In this chapter, we will examine and deconstruct four categories of environmental worldviews and consider how they interpret the global environment and global environmental politics in ways that frame, motivate, and justify particular practices and policies. Although we use "global environment" as shorthand for the biological and geophysical systems of the planet Earth, the term is much more political and sociological than it is scientific or descriptive. Furthermore, to the extent that humans shape and manipulate those systems,

---

*Who older than sixty years could forget the 1960s television show *Flipper*, which played such an important role in this change of attitude?

they, too, are critical elements in the "global environment."* As is often the case with language, different terminology conveys or connotes different constructions and actions and, as often as not, specific terms are associated with different philosophies, reflecting contrasting worldviews and epistemologies. To put this another way, societies explain their activities and existence through different narratives which explain their place and role in nature and in relation to it.

DUELING WORLDVIEWS

A *weltanschauung* or *worldview* puts ontology, epistemology, and philosophy together in a coherent package, so to speak. It explains why we live as we do, provides a plan for how we ought to live our lives, and even instructs us how to act in particular circumstances. For example, a religious worldview might posit that God created the world and that humans should obey her rules and dictates as codified in holy texts, or risk punishment and eternal damnation. By contrast, a naturalist worldview might regard the world as coming to be as it is as a consequence of undirected natural forces, for example, geological, evolutionary, chemical, and physical ones, playing out over billions of years but with no particular direction or trajectory. The "laws of nature" (or science)† impose certain constraints or dictates (for example, life must capture energy to survive) but say nothing on an appropriate social order or what is "right" behavior (the rules of morality and ethics).

A more illustrative example is offered by the worldview that dominates in the United States today. Simplifying and idealizing just a bit, this can be described as a form of *liberalism*—sometimes called "neoliberalism"—whose devotees practice representative democracy and seek to foster free market capitalism.‡ Liberalism extols individual freedom, protects both private property and individual rights, and advocates minimal state intervention into the private (family) sphere and civil society. Although it need not be, liberalism also tends to be highly utilitarian, pursuing policies and outcomes that maximize utility and happiness and minimize pain and misery, usually for the greater number. Liberalism is not a religion, per se, although its vision of an improving

---

*This is one of the justifications for naming the contemporary geological era the "Anthropocene." See Working Group on the "Anthropocene," "What is the 'Anthropocene'?"

†The laws of nature are not the same as "natural law," which invokes a transcendent nature to formulate dictates that are then said to direct human behavior. Natural law and God's laws are, for our purposes, very similar to each other. Compare, for example, Celia Deane-Drummond, "Natural Law Revisited: Wild Justice and Human Obligations for Other Animals," *Journal of the Society of Christian Ethics* 35, #2 (2015):159–73, and Michael Shermer, "Scientific Naturalism: A Manifesto for Enlightenment Humanism," *Theology and Science* 15, #3 (2017): 220–30.

‡Note that "liberalism" as used here is not the same as its use in everyday political discussion of liberal versus conservative or left versus right. Liberalism relies on certain features of natural law for legitimation (for example, the priority of the individual over the group).

and orderly world is rooted in Protestantism, in particular.[2] Liberalism regards humans as fallible yet improvable as they acquire knowledge of the world and the self, and human nature as self-interested but also committed to the greater good. How do we know that these are liberalism's purposes and goals? Primarily through deduction and inference. That is, since no one tells us why we are here, and we cannot determine our purposive ends through science, we must use our ability to reason. Surely we are not here to be miserable and unhappy—and if we are miserable, it is our own, individual fault. We are conscious of our individuality and we do not share consciousness or consciences with each other. We dislike being told "No!" and yearn to be free of restrictions imposed on us by others.*

Are these ontological principles true? Is our deductive epistemology correct? Yes and no. To be a liberal and to practice liberalism, one has to treat these principles and dicta as true and the epistemological deductions as correct. Through belief and practice by hundreds of millions of people in Britain, the United States, and elsewhere, over several centuries, a liberal worldview has come to be taken for granted as natural, whereas other social philosophies, such as Communism or fascism, are regarded as simply wrongheaded or evil. Because liberalism so evidently accomplishes what its beliefs proclaim—or so many like to think and argue—not only are its principles and precepts correct and true, how can there be any other valid philosophies? This last question is a bit disingenuous, of course: if everyone behaves as if a belief is true, and it seems to generate the desired outcomes (at least for some), how are we to know whether it is true or not?† If everyone relentlessly pursues their individual self-interest, how are we to know whether this is human nature or simply masses of people pursuing a practice they have been told to believe is true and natural? Ultimately, to paraphrase British Prime Minister Margaret Thatcher (who is supposed to have said "There is no alternative" to liberal capitalism), when a particular philosophy is dominant in a society, it comes to seem as if there are no practical alternatives.[3]

It is possible to analyze other worldviews in a similar fashion: each one incorporates a set of ontological assumptions, based in a particular epistemology. Often these are so deeply embedded in everyday life that they are not recognized as being just that: assumptions whose "truth" cannot be determined

---

*Thomas Hobbes is sometimes credited with being the "creator" of liberalism. Perhaps it is more appropriate to say that he was one of the first English philosophers to describe and document what was emerging as liberalism during the seventeenth century—although it is questionable whether he, himself, ascribed to liberalism. See C. B. Macpherson, *The Political Theory of Possessive Individualism: Hobbes to Locke*, Oxford: Oxford University Press, 1962.

†Note that a philosophical belief is not necessarily the same as a materialist belief. I believe in gravity because experience repeatedly demonstrates its existence—although there are places where our experiential rules of gravity do not seem to apply. I believe in democracy as a good, although there are orderly societies in the world that are only minimally democratic (or not at all).

by any means except reasoning or deduction. Take, for example, the environmental correlate to the liberal philosophy described above. Ontologically, even if the world were not created specifically for humans, as the Bible suggests,[4] the Earth's resources are widely thought to have value only insofar as they contribute to human utility and well-being (as discussed in chapter 1). Hence, humans must "naturally" dominate and exploit nature in order to use it to these ends. How do we know this? Because humans have been dominating and exploiting nature for many thousands of years and nature has not, so far, punished humans or rendered them extinct.* Moreover, because humans, through their ingenuity, continuously seek to improve their lives and living conditions, nature *must* be dominated, controlled, and exploited.† It is as simple as that!

The deconstruction and contrasting of worldviews of "global environment" that follow in this chapter will illustrate how each leads to different and often incompatible or conflicting accounts of both environmental degradation and political solutions. This exercise will also highlight how much of what we call "global environmental politics" is characterized by conflict and struggle among competing philosophies and the way they manifest in people's everyday interactions with the natural world. That many of the poor countries of the world believe global capitalism to be a major cause of their environmental problems, while many of the rich countries believe global capitalism to be the solution to those environmental problems, is not merely a matter of contested data or evidence or even values. Rather, it goes to the core of what people believe, how they act on those beliefs, how their actions inform their beliefs, and, indeed, how human societies have been organized.[5] Let us begin this discussion by offering, in Table 2.1, four philosophical "worldviews" (some of which can be found in textbooks on international relations) as they appear and are applied in the literature on environmental politics (and were previewed in chapter 1): realism, liberalism, political economy, and political ecology. Each of these worldviews has its own set of subsidiary worldviews that share basic premises but differ on important points. These can be seen in Table 2.2.

Realism and Its Subsidiary Worldviews

In international relations theory, realism is based on the premise that nation-states exist in a condition of anarchy and, with nothing to prevent them from

---

*It could be argued, of course, that nature has repeatedly rendered humans extinct or endangered at local scales (Black Plague, Ice Age, Hurricane Katrina, Superstorm Sandy) or that God repeatedly punishes humans for their hubris or even that humans are too smart for their own good. Such arguments have not had an especially notable effect on American society or philosophy.

†Sir Francis Bacon: "Knowledge and human power are synonymous, since the ignorance of the cause frustrates the effect; *for nature is only subdued by submission,* and that which in contemplative philosophy corresponds with the cause in practical science becomes the rule," *Novum Organum,* edited by Joseph Devey, New York: Collier & Son, 1902; Book 1, Aphorism III, p. 11.

Table 2.1. Comparing environmental worldviews and philosophical roots

Worldview	Ontology (nature of the world)	Epistemology (how we know about the world)	Human nature and Nature's "nature"	Characteristic actions
Realism	A real and dangerous world exists, characterized by war and endless conflict	Direct observation of danger, violence, war, death, resource conflicts	Violent and uncertain behavior in the face of social and natural threats and scarcity	Conquest and control of societies and spaces with needed material resources
Liberalism	The world is shaped by competition for scarce and valued resources and goods	Inference from human insatiable desire and inability to fill it; "trucking and bartering"	Exploit nature to overcome scarcity via development of nature as resources	Devise market-based mechanisms for pricing scarce resources for consumption
Political economy	The world is shaped by distribution of power and wealth, and exploitation of labor and the poor	The overall wealth of the world; the few rich and many poor who inhabit it	Exploit people and nature to accumulate capital and foster differential wealth, with little regard to poor	Control resources through markets and rules devised to grant control and ownership to wealthy and powerful
Political ecology/ liberation ecology	The world is shaped by naturalization of domination and power via postcolonial rules and structures	People are unaware of the true nature of domination or how to effectively resist post-colonial structures	Capitalist societies invest in and drain wealth and resources from dominated countries and groups	Resist exploitation of resources by capital and empower local people to control resources and gain social power

Note: A number of authors have created typologies of worldviews based on different configurations of ontology, philosophy, etc. (for example, Jennifer Clapp and Peter Dauvergne, *Paths to a Green World: The Political Economy of the Environment*, Cambridge: MIT Press, 2005). These typologies are incredibly useful to discern and elaborate differences between different sets of ideas, but in that way should be seen and used as heuristic tools rather than fixed categories. Because philosophies and worldviews differ along a multitude of axes, each author may see and frame these typologies in slightly different ways. For example, Clapp and Dauvergne identify a set of four worldview categories rather distinct from and overlapping with the four we identify: market liberals, institutionalists, bioenvironmentalists, and social greens. We leave it to the reader to tease out the differences and overlaps with the categories we define here.

Table 2.2. Subsidiary worldviews considered in the text

Realism	Liberalism	Political economy	Political ecology
Geopolitics	Neoliberal	Redistribution	Ecocentrism
Ecological security	institutionalism	Environmental	Ecoanarchism and
Malthusianism	Sustainable growth	justice	social naturalism
Scientific	and sustainable	Steady-state	Ecomarxism/
naturalism	development	economy	ecosocialism
	Ecomodernization	"Sharing the	Liberation
		Commons"	ecologies/critical
			transhumanism

going to war, they may do so at any time.[6] As used here, realism is based on the proposition that human and national (state) social and political relations are constituted through concrete material conditions and capacities observable in the real world.* Because national survival is a paramount value, realists must be extremely sensitive to those material conditions that might threaten or extinguish the existence of both citizens and states. Such conditions can include a real or perceived denial of access to vital resources such as petroleum, minerals, water, or food. The state must guard against such scarcity through stockpiling and territorial control of domestic resources but, should denial by antagonistic states become a real threat to security, it may become necessary to engage in violence or war.[7]

Within the broad category of realism we find *geopolitics, Malthusianism,* and *ecological security*. Geopolitics views the surface of the earth in terms of geographic features, their occupation or control by nation-states, and the strategic threats and potential consequences if such control is not forthcoming.† Because goods are claimed to be scarce—a view we will question in chapter 3—there is never enough of anything to satisfy everyone or meet all of their needs. What is possessed by one will be jealously sought by another. For our purposes, geopolitics tends to assume that all desirable and necessary natural resources are scarce. That is, whether mineral deposits are limited or supplies are short, the demand for them is always greater than the supply. States and

---

*Realism, as defined here, is not the same as scientific or philosophical realism. The latter takes it as given that there is a really existing, material universe that can be known scientifically and/or experientially. The former invokes claims about "how human social relations really are," which acknowledge material capacities but posit that non-material factors actually determine such social relations. One source for your further exploration of these other realisms is the online Stanford Encyclopedia of Philosophy, https://plato.stanford.edu/.

†The *loci classicus* of geopolitics are: Halford Mackinder, "The Geographical Pivot of History," *The Geographical Journal* 23, #4, (April 1904): 421–37; and Nicholas Spykman, *America's Strategy in World Politics*, New York: Harcourt, Brace, 1942. Today, see Michael Klare, *The Race for What's Left—The Global Scramble for the World's Last Resources*, New York: Metropolitan Books, 2012, which is one contemporary representative of this worldview.

Table 2.3. Realism's subsidiary worldviews

Worldview	Definition
Geopolitics	States struggle to capture and control strategic resources, topography, land, and populations
Ecological security	Scarcity or degradation of vital needs—food, water, land— can provoke conflict, violence, and war
Malthusianism	Population growth outstrips resource availability leading to deprivation, starvation, death, and potential conflict
Scientific naturalism	The world is material and real, operates according to basic physical and biological laws that can be known and cannot (must not) be violated or transgressed

their populations require natural resources to survive. Therefore, competition for access to those resources and goods is seen as a perennial problem.[8] Scarcity, however, is caused not by absolute supply but by distribution. Minerals are unevenly distributed around the world and are not necessarily found within the borders of the states most in want of them. Therefore, it may become necessary for a state to take by force what it cannot acquire through markets or guile.

Ecological security hypothesizes that degradation of the environment and nature can trigger conflict among states and groups. This view can be seen clearly in the "Briefing Book for a New Administration," written by a group of retired military officers, ex-government staffers, and strategic specialists and issued by the Center for Climate and Security in Washington, District of Columbia, in which the authors warn of the "risks climate change poses to the [national] security landscape."[9] There have been many predictions that water supplies will become a focus of conflict and war in the future: "Fresh water has long been a vital and necessary natural resource, and it has long been a source of tension, a military tool, and a target during war."[10] A 2012 report from the U.S. intelligence agencies predicts that "during the next 10 years, water problems will contribute to instability in states important to US national security interests," but that "a water-related state-on-state conflict is unlikely during the next 10 years."[11] And some argue that a 2006 drought played a factor in the onset of the civil war in Syria, as farmers abandoned their fields and moved into the cities, hoping to find work there (not everyone agrees with this claim).[12] Concrete evidence to support such claims and warnings is itself quite scarce: the State Failure Task Force, a Central Intelligence Agency–sponsored study during the 1990s (requested by then Vice President Al Gore), concluded that "Environmental change does not appear to be *directly* linked to state failure" via conflict or war over resources; rather, a complex set of variables must intervene for this connection to emerge.[13]

What about conflict over renewable resources, such as forests? Here, the scarcity involved is of a somewhat different character, involving the erosion of physical, biological, or life support systems. For example, trees play an important role in the global carbon cycle, absorbing carbon dioxide while emitting oxygen and water.* By permitting a forest to be destroyed, the state whose territory it is on could be said to be contributing to global warming conditions that might harm other states. The affected states could argue that the resulting violation of sovereignty justified physical intervention in order to prevent further destruction of the forest.† It would be difficult to sustain such a claim under international law. Any government willing to invade for such nebulous reasons would find little support for either pretext or action and would face difficult, if not impossible, conditions for occupation.

Malthusianism also views the natural world in terms of scarcity, but at the social rather than the state level. Thomas Malthus, an eighteenth-century English cleric (and a contemporary of Adam Smith, the famous Scottish economist), sought to account for the large numbers of destitute people he saw in Britain's cities and towns and wandering around the countryside. At the time, church parishes were required to provide "poor relief" in the form of bread to residents who were in dire need of food but, with so many homeless about, the parishes were hard-pressed to identify who was local. Malthus explained that the poor had too many children and were so numerous because of this relief, which relieved them from the task of finding employment. But this was only a part of the problem: he argued further that, inevitably, population growth would outstrip the country's capacity to grow food, which would expose both the poor and the rich to starvation. There would then follow violent struggle over remaining supplies and the destruction of English society. His solution was to let the poor starve now, rather than in the future, in the belief that only hunger would lead them to have fewer children.[14]

Malthusianism remains a staple of discussions of global environmental politics and problems, and it continues to exercise widespread influence. In many narrative explanations, large and growing populations in poor countries are presented as the "cause" of resource scarcity, such as in situations in which seasonal rains fail to arrive leading to drought and widespread hunger. The poor, it is sometimes said, are forced to rely on local land, food, forests, and water, and, in doing so, destroy their environment. For decades, Malthusianists have argued that providing food aid to poor countries is of no benefit to them, or the world, since it only fosters further growth in populations; better to let

---

*Trees are not, as is often claimed, "the lungs of the planet." During the day they fix carbon (from carbon dioxide) into sugars in the plant through the process of photosynthesis, releasing oxygen in the process. When they break down those sugars for energy they use oxygen and give off CO_2. Oxygen makes up about 21 percent of the atmosphere (nitrogen is most of the rest), whereas CO_2 is less than 1 percent. Hence, trees are much more critical in regard to the CO_2 level in the atmosphere, as they fix more carbon (as wood) than they respire, and have very little effect on oxygen levels.

†Kofi Annan, secretary general of the United Nations, pronounced a "responsibility to protect" to intervene with force in cases where governments are abusing the human rights of their citizens, which was passed as a UN resolution. Why not, then, a similar right with respect to the abuse of nature? See Alex J. Bellamy, "The Responsibility to Protect Turns Ten," *Ethics and International Affairs* 29, #2 (2015): 161–85.

the poor die of hunger.[15] Yet, Malthus's analysis, and those of his successors, was flawed in many ways, not the least in that an important reason that those who are poor tend to have more children is so that the household has a better chance of survival; when their prospects and general health improve, the birth rate declines.*[16] During famines, moreover, food is often available, but the poor often have no money with which to purchase it. As a result, they starve.[17] Finally, over the two centuries since Malthus first penned his treatise, new agricultural technologies have made it possible to produce ever-growing quantities of food and resources—although there are indications that such returns to technology are not growing at the rates associated with the peak of the Green Revolution.[18]

Scientific naturalism (or scientific realism) is not, strictly speaking, a realist worldview in the above sense. Its proponents represent the preponderant view of those who believe that the scientific method should be the basis for social choice and political decisions. The world and universe are really existing, material phenomena, governed by scientific (physical, biological) laws that can be known and cannot (and must not) be violated or transgressed. To do so is to risk well-being and survival. To give a very mundane example, the law of gravity applies everywhere (although diminishing in relation to object mass and distance from said object). To step off a tall building in the absence of a net or flotation device is to court almost certain death. You can't argue with Mother Nature, as it is said! By the same token, states are also bound by these laws—or are they?

The difficulty here is that, notwithstanding universal laws, scientific naturalism is *social*. This means that states and people can deny the veracity of phenomena arising from those laws and proceed on that basis, as we see in the case of climate change. Indeed, the phenomenon we call "climate change" is the result of a long and complex set of causes, factors, and relationships among the various elements of society and nature—what is sometimes called an "assemblage"†—made up of both conscious decisions and deliberate policies, various material appurtenances, and the operation of physical and biological systems. If we insist on a pure causal account, it is relatively easy to point out that one part or another of that long chain is in error and thereby throws

---

*Malthus is only the best known of population doomsayers; he was preceded in his arguments by numerous French philosophers, not all of whom came to quite so gloomy a conclusion. See Joseph J. Spengler, *French Predecessors of Malthus: A Study of Eighteenth-Century Wage and Population Theory*, Durham, NC: Duke University Press, 1942.

†An assemblage is an arrangement constituted by a "multiplicity of heterogeneous objects, whose unity comes solely from the fact that these items function together, that they 'work' together as a functional entity" (K. D. Haggerty and R. V. Ericson, "The Surveillant Assemblage," *British Journal of Sociology* 51, # 4 (2000): 605–22, p. 608, citing P. Patton, "MetamorphoLogic: Bodies and powers in *A Thousand Plateaus*," *Journal of the British Society for Phenomenology* 25, #2 (1994): 157–69, p. 158). The term comes from Gilles Deleuze and Felix Guattari, *A Thousand Plateaus: Capitalism and Schizophrenia*, Minneapolis, MN: University of Minnesota Press, 1987.

the larger claim into doubt (as the Red Queen in *Alice in Wonderland* said, "Why, sometimes I've believed as many as six impossible things before breakfast"). To carry the point a little further, climate change scientists and others may argue that, if we don't change our ways, climate change will change them for us, with devastating consequences. To be sure. But we supposedly have free will to choose what to do, and science cannot *force* us to make decisions that comport with the scientific facts. To act as the science dictates is really an ethical question: is it fair or just to impose extreme costs on those who have not reaped many of the benefits of industrialism? As we shall see, responses to this question are rationalized in many ways, some more compelling and doable than others.

Liberalism and Its Subsidiary Worldviews

In international relations theory, *liberalism* is commonly presented as a philosophy distinct from realism, especially in terms of the contrast between conflict and war, on the one hand, and cooperation through markets and exchange, on the other. In fact, realism and liberalism share a number of common propositions—Thomas Hobbes, who we mentioned in chapter 1, is often invoked as a political realist *and* can also be regarded as one of the founders of political liberalism.[19] Liberalism, like realism, is wary of threats to survival, although it is less inclined to rely on force alone for protection (although many political liberals become realists once they go abroad). Much like some realist worldviews, liberalism tends to view the world in terms of scarce resources and goods. Where the two part ways is that liberalism regards scarcity not as a threat but, rather, as an opportunity for economic gain.

Both realism and liberalism see a world of radically individual humans or states who must be disciplined (or self-disciplined) by a firm hand, whether a domestic sovereign who will punish violators of rules and laws or an international hegemon that can impose order upon anarchy.[20] Liberalism is a theory of individuals or corporations acting competitively in anarchic markets; realism is a theory of states acting competitively in anarchic politics.[21] Both theorize action in settings where there appear to be no governing authorities and few, if any, rules. But where Hobbes imagined a "state of nature" that could be overcome only by a sovereign imposing peace among men, a century or so

Table 2.4. Liberalism's subsidiary worldviews

Worldview	Definition
Neoliberal institutionalism	States can find mutually beneficial cooperative strategies through negotiation, bargaining, and exchange
Sustainable growth	Global poverty and environmental degradation can be solved only through high rates of global economic growth
Sustainable development	Global poverty and environmental degradation can be solved only through economic development and technology transfer
Ecomodernization	The ecological crisis can be addressed through diffusion of green technologies, goods, and services

later, Adam Smith believed that individual self-interest was enough to establish the power of the "Invisible Hand."[22] The uncoordinated actions of individuals would, through markets, result in an aggregate social benefit. But what would keep people from stealing instead of trucking and bartering? Smith thought that religion and fear of God would provide sufficient moral authority to keep social peace and keep buyers and sellers honest.[23] Secularization appears to have rendered this particular sentiment weak, if not wholly moot.

Like many of the worldviews considered in this chapter, liberalism is strongly *idealist* (in comparison with materialism). That is, liberalism is motivated by the notion that ideas make the world, and that superior ideas—*whose* ideas is rarely specified—can change it, if they make "sense" within the hegemonic discourse and are deployed through policy and practice ratified by those in charge (much of the contemporary discourse on innovation, disruption, and entrepreneurship is driven by such idealism). At the extreme, there is a hint of utopianism here in the belief that liberalism could bring the world to a happy and harmonious state via markets, democracy, and individual freedom. That is a proposition that has yet to be demonstrated.

Worldviews associated with "liberalism" focus mostly on economic aspects of environment and nature, and these tend to dominate in both environmental politics and ways of regarding and treating nature, whether local or global. *Neoliberal institutionalism* draws largely on neoclassical economics: markets are the appropriate means of distributing goods and resources—although they may require some regulation. *Sustainable growth* and *sustainable development* regard markets as the best and proper means of bringing the poor out of poverty, through economic growth that is sensitive to environmental conditions and constraints. *Ecological modernization*—which has declined in favor over the past decade—proposes that technological and social innovation, driven by high market returns for new goods and services and, possibly, high costs of carbon, can reduce resource intensity while bringing benefits to the entire world.

Before we explore these three liberal worldviews, however, we need to review some of the basic principles and practices of neoclassical and environmental economics, which underpin those worldviews. According to economic liberalism, everything must have a price if it is to be valued, and it is of no value if it cannot be priced. If something does not have a price, it cannot be bought or sold (that is, no property rights can be assigned to allow transfer or exchange).* By the same token, something that is in near-infinite supply, such as air, cannot be monetized unless access to it can be limited (as is the purpose of carbon emissions permits discussed in the following). By definition, scarcity of goods and resources is desirable—otherwise, no one would want to pay for them—with a price determined—in theory—by what people are willing to pay

---

*This does not cover all exchange, of course. Goods and services might be exchanged as gifts, without attention to their economic value. Or, they may be traded through barter, which implies reciprocal valuation without money prices, but the exchange "value" must be determined through negotiation and bargaining.

in an equilibrium market situation (a condition rarely met under real-world conditions).

Through the magic of markets and individual self-interest, therefore, scarcity makes it possible for goods and resources to have prices, to be alienable, to generate profits, accumulate capital, and, quite possibly, secure society through market capture of high-cost resources. The cost of a good or resource depends on demand for it: we value rare earth metals (whose mining can be quite damaging to the environment) because they are relatively difficult to accumulate and essential for modern electronics—hence, for example, the price of pure scandium has ranged between four thousand and twenty thousand dollars per kilogram over the past decade.[24] We don't value air, because there is so much of it and nobody owns it (yet). You would be hard put to find a price for it. But market-based prices tell only part of the story: we could probably survive without scandium, but not air. One difficulty with monetizing nature is that the benefits of exploitation are usually realized over much shorter time periods than are the social and environmental costs. We return to this matter in chapter 3; suffice it here to say that such costs are rarely, if ever, factored into prices or analyses of costs and benefits of specific projects and actions.

But what about absolute scarcity? Inadequate supply of a good or resource, for whatever reason, normally leads to a rise in price.* If there is not enough of a resource to meet demand, people will be willing to pay more for it; if there is too much, people will only be willing to pay less. People will decide, given their individual economic situations, whether they can afford the good and act accordingly. If a rise in price is large enough and sustained for a period of time, an opportunity emerges to devise or develop substitutes that can provide the same service at the same or lower cost (recall here the rush to develop alternative fuels when the price of oil passed $140 per barrel in 2008). In some instances, the emergence of substitutes might actually lower costs of certain goods and services; in others, "new and improved" may result in higher prices. Those who cannot pay the market price will be denied the good—treated by neoclassical economists as a "rational choice" rather than a forced or constrained decision, or a denial of access. Nevertheless, "freedom of choice" does not mean that everyone can afford to choose from the options on offer.

Monetizing a good or a bad depends not only on supply and demand; it also requires the creation of property rights of some sort, so that the good or bad can be bought and sold. It is easy to see how this applies to goods and resources: one party contracts with the owner—whether public or private—to pay a certain sum, at once or over time, in order to take possession. It is less clear how one can come to own a "bad," such as air pollution. Yet pollution emission permits are precisely such rights: By creating property rights in

---

*Note that "inadequate supply" is a problematic term. For what purposes? In whose view or whose need? An apparent "scarcity" of diamonds designed to support high prices is quite different than a "real" (or apparent) scarcity of food that puts prices out of the reach of the very poor.

"scarce pollution space" and putting those up for sale, the true value of that "pollution space" can be determined. The theory is that high permit prices will motivate polluters to either reduce emissions or come up with lower-cost substitutes (although pollution producers might also go out of business or speculators buy up permits in order to force up prices). Exchange will take place between those who have a surplus of such pollution space and those who need more than they have. This idea may seem more than a little bizarre—after all, who would want to buy pollution? Yet this is how carbon markets work: Parties can buy and sell a scarce resource, in this case, the right to emit greenhouse gases (we return to this paradox in chapter 3).

The real problem here is, according to environmental economics, that the present and future costs of environmental and social damages are not normally included in the prices of resources, goods, and services. Most environmental and neoclassical economists argue that the prices paid to use (and abuse) the environment are too low, allowing those who are responsible for damages to "externalize" them onto others who suffer from those "externalities" while realizing few or no benefits. Were the costs of such damage to be included in the price of resources, goods, and services, through a tax, for example, the responsible agent would be motivated to 'internalize" those costs ("polluter pays"). This argument is articulated in the famous "Coase Theorem,"[25] whose details need not concern us here but which argues, in effect, that whomever feels the greatest damage from an environmental insult should be most willing to pay to have it eliminated (or internalized), *whether that party is responsible or not for the cause of the damage.* Coase's arguments point in several different directions, but the most important one is to recognize that it relies on the existence of markets, prices and cost data, and property rights, if it is to be properly applied.

Let us return for a moment to consider air, which is of considerable value but has no price. Arguably, air pollution in the form of various chemical emissions from factories, cars, and other sources causes visible damage to nature and human health. Polluters can be made to internalize the cost of these damages by direct regulation or taxes. By contrast, the damages resulting from greenhouse gas emissions from the burning of carbon are still largely invisible.* Through access to a limited number of tradable emission permits, a polluter can decide whether to reduce pollution and sell permits to others who might need them, or continue to pollute and buy permits from those who are selling them. Here, internalization is either taken on by the polluter who reduces emissions—and who may pass the cost of reductions on to customers—or distributed over society by those who have purchased societally sanctioned rights to continue their pollution.

---

*This depends, of course, on whether one attributes the intensity and destructiveness of storms such as Katrina, Sandy, Harvey, Irma, and Maria to global warming or random chance.

In effect, this worldview regards the creation of such private property in nature as a net benefit to society. At the limit, some "environmental libertarians" suggest that *all* natural resources—air, water, plants, species, land—should be privatized, on the premise that the owners of property are likely to use their own resources more carefully than those who use public resources (such as parks, game animals, rangeland).* This proposition is generally applied to resources that are "owned" collectively by society or government, for whose use individual members (are said to) pay too little and who usually experience only minor costs and inconveniences as a result of crowding or depletion.[26] For example, summer visitors to Yosemite National Park often find it to be overcrowded and congested, even though each pays an entry fee and relatively high accommodation costs to visit. The problem, according to the libertarian view, is that visitors should pay much more for access, but political pressure prevents the park's managers from setting the entry fee at a level that would sustain revenues but reduce the number of visitors.

As a result, all visitors experience the externalities of crowding and congestion, and the Valley experiences non-optimal degradation of nature.[27] Whether the results of privatization are fair or equitable is beside the point: viewing environmental politics and policies through the lens of markets results in a very narrow view of both means and ends. Everything comes to have a price, and anything whose price does not generate market activity is not worth owning or saving. Those who have the wherewithal to set prices in markets for environmental goods or bads, through either supply or demand, can simply decide the game is not worth playing, and the benefits are not worth the costs. In that case, those without will suffer.

Neoliberal institutionalism derives from liberalism and rests on the creation of "political" markets among states and other social entities.† By "markets," we mean that state parties engage in diplomatic bargaining over interests and make rational tradeoffs among them, ones that are satisficing if not optimal. If bargains cannot be achieved, there is "no sale," and war may follow. Political markets more closely resemble barter in markets than pure exchange, since the monetary value of trades is difficult to calculate,

---

*This is one way to interpret Garrett Hardin's parable of "Tragedy of the Commons" (*Science* 162 [Dec. 13, 1967]: 1243–48). Hardin argues for "mutual coercion mutually agreed upon" as a way of limiting reproduction but, alternatively, a "cap and trade" market in births could have the same effect. That is, based on his argument, too many babies impose an externality on the world that could be internalized if the price of reproductive permits were set in a free market.

†It is important to differentiate between "neoliberal" as the term is used here, and which draws on work from the 1980s by scholars such as Robert Keohane and Stephen Krasner, and "neoliberalism" as it is applied to the global economic system and discussed in chapter 3 in conjunction with globalization. The two are related but not the same. Stephen D. Krasner, ed., *International Regimes*, Ithaca, NY: Cornell University Press, 1983; and Robert O. Keohane, *International Institutions and State Power*, Boulder, CO: Westview, 1989, esp. chapter 1.

and tradeoffs usually affect citizens, who are rarely consulted on a deal and may strongly object to it. At times, of course, money may exchange hands: Napoleon sold "Louisiana" to the United States for fourteen million dollars, while Russia sold Alaska to the United States for a mere two million dollars. What is required for such bargaining and exchange to take place, however, is a framework of rules and regulations that governs what is permissible and what is forbidden. Hence, neoliberal institutionalism, like markets, must be constructed and regulated.

From a pure power perspective of force and influence, such cooperative bargaining via political markets appears anomalous. Because cooperation means compromise, giving up valued things and even suffering a loss of power relative to others, no state should willingly cooperate with any other (unless, as Hobbes argued, it is forced to by an international sovereign). But why is such cooperation thought to be problematic? This is explained by something called the "collective action problem,"[28] based on the propositions of individual self-interest and reluctance to pay the costs of a cooperative deal. Both people (and states) are capable of reason and will seek to maximize benefits and minimize costs, even at the expense of others. Cooperation usually means less than maximum benefits and more than minimum losses. In the absence of information, incentives to cooperate are not always clear.* Furthermore, for a deal or group project to be successful, each participant must have an individual stake in the outcome, and each must be prepared to contribute part of the cost—in time, money, or whatever—toward the project's completion.

In large groups (more than, say, ten people), each individual's contribution to the whole is likely to be small (less than 10 percent) and, if not provided, goes the argument, its absence is unlikely to be noticed or affect provision of the benefit. In other words, it may be possible to realize a benefit from bargaining without having to contribute to the costs. Because each notional participant recognizes this "fact," each will be inclined either to "free ride" and get the benefit of the deal without paying anything or to "defect" and not be bound to the project at all. Therefore, concludes the argument, collective action is difficult to explain because self-interested individuals ought to be loath to cooperate on anything that does not offer clear benefits and minimize costs to them individually. In the international realm, moreover, there is no authority to compel states to contribute to the costs of a joint deal or project. Any such action that does succeed must do so because one or a few parties are willing to pay the costs of group action.[†]

---

*We can see echoes of Coase ("Problem of Social Cost") here: there are externalities arising from non-cooperation and, to internalize them, parties must agree on what each is willing to pay. This, in turn, requires an estimate of costs and benefits from making such payments, although some parties might be willing to suffer some costs in order to derive other benefits (such as domination).

[†]This is also critical to Garrett Hardin's arguments in "The Tragedy of the Commons." In his parable, farmers free ride on each other, thereby destroying their shared pasture.

Given this rather pessimistic outlook, realists find it something of a surprise to see just how much interstate cooperation and dealmaking actually takes place.* Bilateral, multilateral, and international agreements are quite common—by some counts, in the tens of thousands—and the number of these having something to do with the environment are by now estimated to number more than thirty-five hundred.[29] In real life, actors are hardly as antagonistic toward each other as realism might have it. States, corporations, and individuals often share common interests and similar desires and are willing to negotiate, bargain, and collaborate on solving joint problems. How can this be explained?

According to the neoliberal institutionalist view, cooperation arises from the combination of two, somewhat conflicting incentives. Straight exchange through markets, in pure pursuit of self-interest, can result in economic benefits for each of the parties involved.† But the parties may have a broader sense of what they want from a deal or project, one that includes norms, values, principles, and expectations, and they may act on a conception of the public good that involves trades off among these different elements. For many reasons, we cannot know precisely how to behave in order to achieve our goals, and we must be willing to negotiate and bargain in order to find ways of getting to desired outcomes. Because no market is created spontaneously, we are likely to need quite elaborate rules for negotiations, bargaining, and behavior. One example, albeit so far largely unsuccessful, are the long-standing international negotiations over global climate change, ongoing in one form or another since the mid-1980s.‡

Moreover, as conceived by neoliberal institutionalism, even cooperation is often infused with unacknowledged imbalances. In any group project, not only are some participants likely to be wealthier or more powerful than others, they may also be able to control knowledge and language and shape a deal via discourse. The less powerful can find it in their interest to go along with the demands of the more powerful, if only to get a piece of the action. But they may also be pressured, through various sanctions and inducements, to agree to terms and to act in ways that they would reject in other circumstances. There is, however, power in numbers: if enough poor states are willing to support a provision disliked by the rich ones, the latter may find it necessary to compromise or even yield on certain issues. Institutions are designed to "freeze" rules and to obscure a history of struggles; they become the "normal" way of doing things. As a result, power seems to vanish. It is in this gap where competitive

---

*This is not the same as the "art of the deal." Anyone who comes in demanding a maximum share of a deal, or offering a lousy bargain, is not likely to get much, if any, outcome, much less the desired one.

†But this does not mean equal or even equivalent benefits: consider, for example, a used-car dealer selling a junker to an uninformed customer.

‡Paradoxically, in spite of more than a quarter century of climate negotiations, the states involved have been unable to agree on a set of procedural rules, and so the annual meetings continue to function under a consensus rule without any actual voting.

liberalism meets "cooperative" institutionalism, producing both *sustainable growth* and *sustainable development*.

These two worldviews are quite similar to each other, at least insofar as society, nature, and environment are concerned. Sustainable growth posits a constantly expanding world economy as the savior of the global environment, ideally via ecological modernization (see the following). Sustainable development could require major modifications in modes of production and consumption, although it can hardly be said to eschew capitalism or growth as the means through which environment and development can be reconciled. Both are organized around the idea that economic and social improvements in the condition of people's everyday lives will result in greater public concern for and attention to the environment. To put the argument another way, if people are sufficiently well-off, they will be less exploitative of nature and more willing to pay to protect it. Development-based philosophies also hold that people who are better off will be happier and less likely to commit violence against each other or against nature.

According to one common argument, many if not a majority of the world's poor, and especially those in the Global South, engage in subsistence production and are highly dependent for their survival on access to the natural resources around them.[30] Furthermore, it is argued, they are in no position to use these resources carefully (perhaps because they are not familiar with ecology). Even though each poor individual consumes very small quantities of nature's goods, their aggregate number and demand is so great that the total far exceeds supply. The result is damage to forests, soil, water, and species, which endangers everyone. By contrast, if we examine the industrial countries (aka, Global North), we find few people engaged in subsistence production or resource exploitation. Most are deeply linked into capitalist markets, through which they acquire the things they need to live and want to own as well as the services they need or demand. In the industrial countries, we also find high levels of environmental awareness and protection, low rates of population growth, and, when technically feasible and economically rational, substitution of capital for natural resources in the form of new materials and technological substitutes (such as fiber optics for copper).[31]

Proponents of sustainable growth therefore propose to reduce pressure on the environment, especially in the Global South, by moving the poor from subsistence to market-based consumption, making everyone richer via rising incomes and environmentally friendly technologies. As an additional benefit, growth in incomes ties people into networks of exchange through which the monetized value of the environment can be established and commensurate investments made to protect nature (since the well-off are, it has been claimed, more willing to pay for such amenities; we will examine this argument more closely in chapter 3).

To put this another way, if the poor were to more easily gain access to markets, they could purchase their basic needs imported from places in which environmental impacts are more subject to robust management and rely less on heedless exploitation of local resources.* For example, households that use locally produced wood or charcoal from overexploited forests could instead purchase kerosene (or biofuels) and burn them in more energy-efficient stoves.[32] Pressure on local forest resources would diminish and so would pollutants from fires. Such changes in resource use would also offer new market opportunities to producers and suppliers to sell fuel, equipment, parts, and repairs. But many people are too poor to afford kerosene, the stoves in which to burn it, or the spare parts needed to repair them. Moreover, importing such fuels from countries where it is produced requires often-scarce foreign currency in payment. What are people and countries to do? The answer is to be found in export-oriented development strategies (as pursued, for example, by China).

In order to pay the costs of imports in hard currencies, it is necessary for poorer countries to export to hard-currency countries. The more that is exported, the more hard currency can be earned.† But what to export? The global economy is characterized by an international division of labor in which countries, based on their comparative advantages, specialize in the production of certain things—raw materials, manufactured goods, and, in some cases, even technology, information, and workers—which are exported primarily from the Global South to the Global North, where most of the world's well-off consumers are to be found (even so, most of the world's trade still takes place among countries of the Global North). Ideally, the added value from such export and trade accrues to the producing country, which can then use the funds earned to import goods or finance further investment in production to the benefit of the country in which the activity takes place.

Take, for example, a cocoa farmer in West Africa. She could use her export earnings, or borrow funds, to buy more land, plant more trees, and hire more workers. Alternatively, a factory owner could add more machines to the plant, employ more people, and produce larger numbers of shoes or can openers or iPhones. When people are employed, they earn wages that can be used to buy manufactured goods, including imports. They rely less on local resources and move beyond subsistence. As they become richer, people will pay more attention to the aesthetic quality of their surroundings and be more able and willing to pay the costs of keeping air and water clean—or so goes the argument (this is discussed in chapter 3).[33]

---

*The logic here might not be immediately apparent: if we apply rational cost-benefit analysis to exploitative practices, and include near- and long-term social and environmental costs in our calculation, and put an hourly value on the time people can work for wages rather than laboring in fields, it might be more economically efficient to import goods and resources from elsewhere.

†Not that this is always the case: if exporting contributes to a global glut of something, the world price is liable to collapse, leading to a decline in foreign earnings. This is what has happened in most raw materials markets since the 1980s. See, for example, Charis Gresser and Sophia Tickell, *Mugged: Poverty in Your Coffee Cup*, London: Oxfam International, 2002.

One of the basic contradictions with sustainable growth is that it cannot go on forever: high rates of economic growth in Asia have brought hundreds of millions out of deep poverty, but at what environmental cost? China is now plagued with innumerable environmental difficulties—earth, water, air, among others—that will be very costly to address and reduce or eliminate in the short term. Notwithstanding the emergence of a new middle class of three hundred million or so in China who are, presumably, concerned about the many environmental externalities that have arisen from the high growth rates of the past few decades, there remain many hundreds of millions of very poor rural and urban residents more interested in day-to-day than long-term survival.[34] And whether sustainable growth can be "sustained" is as much a matter of the growing burden of externalities as it is the availability of land, water, resources, and even clean air.

What, then, about "sustainable development?" This oxymoronic coinage was made famous by the World Commission on Environment and Development (WCED; aka, the Brundtland Commission) in its 1987 report, *Our Common Future* (although the term was first used by the International Union for the Conservation of Nature in its 1980 *World Conservation Strategy*).[35] The fundamental charge from the United Nations to the WCED was to determine whether economic development and environmental protection involved compatible values and goals and how the two might be reconciled. This conflict between development and environmental protection had been central to the 1972 Stockholm Conference on the Human Environment, and it has remained a point of disagreement between industrialized and developing countries ever since (and it continues today). Developing country governments argued that growth must come first, and the environment could be protected later. Northern governments, under some public pressure, thought that the environment must have precedence and that poverty could best be addressed through family planning and fewer people. If development rested on growth in production and consumption, how could the environment be preserved?

According to *Our Common Future*, however, there was no inherent contradiction between environment and development:

> Sustainable development is development that meets the needs of the present without compromising the ability of future generations to meet their own needs. The concept of sustainable development does imply limits—not absolute limits but limitations imposed by the present state of technology and social organization on environmental resources and by the ability of the biosphere to absorb the effects of human activities.[36]

This is an enormously confusing and opaque definition, which relies largely on technological innovation to meet growing "needs," address the global environmental crisis, and make life better for all. We will examine this notion of relying on technological innovation (that is, "ecological modernization"); what has most troubled critics over the thirty-odd years since *Our Common Future* was published is the apparent failure to recognize the contradiction between "limits" and the economic growth that seems necessary to meet future needs. No one in rich countries will accept a lowered standard of living as the price of providing a higher standard for poor countries, but if the global poor are brought to the income level of the relatively well-off, the environmental burden will grow considerably. Consequently, to achieve sustainable development without igniting a global political firestorm, wrote the commission, economic growth would be essential. By the end of the twenty-first century, proposed the WCED, a five- to ten-fold increase in the size of the world's economy would be necessary to eliminate poverty and preserve the global environment. And only by increasing demand in the Global North for goods and resources from the Global South, reasoned the WCED, would there follow a sufficient transfer of wealth from the former to enable the latter to deal with the problem of poverty and environmental damage.

None of this, it might be noted, really defines sustainable development. Is it, as the WCED proposed and others hope, environmentally friendly economic growth? Does it require a reduction in consumption in richer countries and an increase in consumption in the poorer ones? Does it mean a steady state economy (SSE) where growth, as we know it, is tightly controlled or completely abolished? For the WCED, sustainable development was clearly sustained economic growth and this did not, then or now, seem a likely solution to the global environmental crisis. Many who studied the problem concluded that sustainable development was a concept riddled with problems and contradictions and that *Our Common Future* was a badly flawed document.[37] Growth was probably not sustainable in the long run, and no amount of rhetoric would make it so. Today, as a result, and notwithstanding mighty efforts to square this circle, there is little conceptual difference between sustainable development and sustainable growth.* For the most part, sustainable development remains more a mantra than a strategy, a phrase that bears repeating because it promises a green future and has a calming effect. It seems to imply that, given human ingenuity, we can find our way out of what appears to be an almost intractable dilemma, and we can do so on a global scale.

What the WCED and many others have since prescribed to address this dilemma is *ecological modernization*. This worldview relies on technological and social innovation in environmentally friendly products, services, beliefs, and practices that will "green the economy" by eliminating environmentally damaging competitors. For this process to work, such innovations must substitute for older ones with greater environmental impacts, must be provided by producers who see greater profit opportunities, and must be accepted and

---

*The United Nations' Millennium Development Goals of 2000 and Sustainable Development Goals of 2015 both set very ambitious plans for eliminating "a whole host of social ills by 2030." See United Nations, "Sustainable Development Goals Kick Off with Start of New Year," accessed June 20, 2017, http://www.un.org/sustainabledevelopment /blog/2015/12/sustainable-development-goals-kick-off-with-start-of-new-year/.

adopted by consumers who believe the innovations will save and protect the global environment. If only green products and services are available at reasonable cost, and purchased by enough people, goes the argument, a new set of *social* norms and practices will emerge that make non-green production and consumption inconceivable. Starting this transition is not easy and, under capitalism, is difficult to direct, given the contradictions involved.

Consider, as an example, our old friend the iPhone (or any smartphone, for that matter). As we saw in chapter 1, such devices contain a devil's brew of materials, some of which are highly toxic and difficult, in any event, to extract from discarded electronics. Recycling in the Global North is costly, since a considerable amount of high-cost labor is required to take the phones apart. As a result, many old phones are shipped to poor countries, where entire communities work to take them apart and refine precious metals out of the pieces. In the process, these people are exposed to toxins and, when they are finished, probably dump the carcasses into nearby pits and canals.* A truly "green" phone would not require such disassembly, but that would deprive these poor communities of what is one of the only ways to make a bare living. A green phone would also have to be long-lived—lasting, say, ten years rather than the current two or three before replacement. This would mean fewer new phones, reduced sales volumes and profits, and the end of planned obsolescence in both hardware and software—which is how phone sellers keep up volume and turnover. Consumers will have to get used to the idea that they will not be "out of style" if they own an older phone.†

But what ensures that the necessary technological innovation will take place? The argument is made that ecological modernization may be driven not only (or even primarily) by changes in consumer tastes and opportunities for profit, but also, through internalization of environmental and social bads, by the rising costs of mining, manufacturing, servicing, and consuming. In the context of higher costs resulting from pollution penalties and taxes, producers will seek out technologies and designs that consume fewer resources and emit less pollution. Through a "ratcheting up" process, succeeding generations of technologies will become increasingly greener.

---

*On this very day of writing—in thirty minutes, in fact—Apple will introduce the new "iPhone X" which will sell for $999 and which, we are assured, no iPhone user can do without. See Brian X. Chen, Farhad Manjoo, and Vindu Goel, "Apples New iPhone Release: What to Expect," *New York Times*, Sept. 12, 2017, accessed September 12, 2017, https://www.nytimes.com/2017/09/12/technology/apple-iphone-event.html?_r=0.

†This is not beyond the realm of possibility: the average age of the American auto fleet has been creeping upwards for the past few decades as the cost of new vehicles has gone up. Perhaps smartphones are too cheap! Europeans have the possibility to own a Fairphone, an eco-modern modular phone that is easy to repair and thus extend the life of the phone. The name references the intent of the company to source conflict-free minerals and provide fair working conditions in their factories. Fairphone, "A Better Phone is a Phone Better Made," accessed August 7, 2017, https://www.fairphone.com/en/our-goals/.

For example, after World War II, European and Japanese automakers were confronted with high fuel taxes and designed small, fuel-efficient vehicles (which were also more suitable for the narrow streets found in those parts of the world). When the international price of oil quadrupled during the 1970s, foreign automakers were able to export their small cars into the United States.* Their success, in turn, motivated American automakers to turn out smaller, more fuel-efficient cars. In more recent decades, the U.S. government and the State of California have mandated gradually greater fuel efficiency in the interest of both reducing petroleum consumption and greenhouse gas emissions. Even so, because the real price of gasoline is as low as it has ever has been over the past century, consumers have gravitated back toward larger, more fuel-hungry vehicles. Auto efficiency has risen but, overall, gasoline consumption has not declined very much. It is possible that, if carbon taxes were imposed on fossil fuels, electric vehicles might become more common than gasoline-powered cars. But such a tax would need to be quite substantial to motivate this shift and, for the time being, carbon taxes are a non-starter, at least in the United States (we return to such taxes in later chapters).

Does ecological modernization actually deliver on either its premises or its promises? Yes and no; it is difficult to find unequivocal evidence to answer this question. There are numerous case studies that report on specific sectors that reflect "ecomod's" fundamental premise, but there are few, if any, meta-evaluations that provide an overall empirical evaluation of the concept and its practice. From a conceptual standpoint, however, let's consider the case for electric vehicles, which would allow widespread carsharing and ostensibly eliminate energy-related carbon emissions. Several caveats are required prior to determining whether electric vehicles would foster ecomodernization. First, does carsharing actually reduce the overall number of vehicles on the road? If drivers choose to purchase vehicles for their own private use, the actual number on streets and highways is likely to increase. If more and more drivers use ridesharing to make ends meet, might not more cars end up on the road?[38] Second, does the electricity which charges the electric vehicles come from non-carbon-emitting sources? If coal and natural gas are burned for this purpose, emissions may be displaced from tailpipes, but they might not be reduced overall. For example, a recent study of plug-in hybrid electric vehicles in eight Canadian cities found that, while such vehicles have lower carbon dioxide emission rates than conventional ones, the level of emissions can vary by a factor of two, depending on the electricity source.[39] Third, what are the environmental impacts of the vehicle's total life cycle?† When manufacturing of

---

*One of the authors test drove both small foreign and domestic cars in 1971 and ended up buying a Toyota Corolla for its superior performance.

†As indicated in chapter 1, "life cycle analysis" involves assessment of cradle-to-grave environmental impacts of manufactured goods and extracted resources. See, for example, Kersty Hobson and Nicholas Lynch, "Ecological Modernization, Techno-politics and Social Life Cycle Assessment: A View from Human Geography," *International Journal of Life Cycle Assessment*, 2015, accessed Sept. 12, 2017, https://link.springer.com/article/10.1007/s11367-015-1005-5.

components and end-of-life disposal are taken into account, emissions may be considerably larger—especially due to battery materials—albeit not as great as from comparable gasoline-powered vehicles. Another study of the diffusion of more energy-efficient appliances into the U.S. market suggests that ecological modernization may be dependent on the appropriate public policies and incentives, as well as regulatory pressures by the state. Absent those, it is unlikely to take place.[40]

There are other reasons to be cautious about ecological modernization and its concomitant, "greening the economy." First, greening the economy relies purely on individual consumer choices to shift society away from its current preferences and practices; no political movements or mobilizations, nor government action, are assumed or involved. Moreover, in the absence of clear economic incentives and signals, the "cost-conscious" consumer will frequently select a less-costly, non-green option.[41] Second, producers and consumers must make a conscious and deliberate decision to produce and purchase only green goods and services—and at the moment, there are very few such purists in the market. Furthermore, there is no force or authority or immediate penalty or punishment that compels producers and consumers to go green, or that prevents them from changing their minds and actions in mid-stream.[42] Third, although the externalities from environmentally unfriendly goods and practices might impose costs on consumers, those tend to be distributed over large numbers of people who only experience a small impact individually, today and in the future.* More costly green commodities that, arguably, internalize the future costs of environmental impacts will appear but the higher prices discriminate against the less-wealthy. Finally, greening the economy offers the opportunity to commodify environmental "threats" so that money can be made even on resource degradation, as through sale of and speculation on pollution emission permits. Notwithstanding the efficiencies that may arise from such privatization, the distributional effects are morally unjustified, especially when the poor cannot afford such permits or the higher price of goods that internalize permit costs.

The basic premises of the dominant philosophies of development—that growth is not only necessary but good—are open to serious challenge. However development might be defined, the market by itself cannot generate either growth or sustainability. What is required is some degree of regulation of markets and social policy and, for this to be society- or worldwide, governments and other political actors must become involved.[43] Finally, wealth may well be a greater problem than poverty for nature.[44] Rising rates of production and consumption in both rich and poor countries have led to growing quantities of all kinds of emissions and wastes, particularly in the European Union and the United States and as exemplified by, for example, the disposal of "obsolete"

---

*This is not to say, however, that such impacts are evenly distributed across large numbers of people. Many impacts are very local and very intense locally—an issue addressed by environmental justice, about which we say more in the following.

cellphones, discussed above. This phenomenon is especially visible, for example, in the life cycle of personal computers and a myriad of other electronic devices. The same can be said of cars, CD players, microwave ovens, and all the other paraphernalia of twenty-first-century life.

Political Economy and Its Subsidiary Worldviews

In political science, the term *political economy* has a very specific, depoliticized meaning: analysis of the relationship between power and wealth as it affects states, corporations, individuals, and other groups.[45] Such relationships are, generally, viewed through the lens of neoclassical economics and the role of wealth and power are taken as given, that is, how wealth has been acquired and transformed into power is hardly questioned or critiqued. This has not always been the case: prior to the mid-nineteenth century, and before the rise of marginal utility theory in economics,[46] political economy was understood as an analysis and critique of the ways in which politics and class power were applied to *shape* economic systems, thereby facilitating capital accumulation by those who made the rules. To put this another way, if you *play* by the rules, you may win the game; if you can *devise* the rules of the game, you can guarantee that you will always win the game.[47]

As we use the term in this book, "political economy" is meant in a more *critical* sense, in that related worldviews are largely framed in terms of regulatory (rule) changes to the existing capitalist system—we might say, at the level of scoring points rather than how the game is advantageously structured for some. Critical writers would suggest that the capitalist global political economy must be substantially and substantively reformed so that flows of resources, goods, services, capital, and labor are redirected to the poor and weak. This might require that some version of social economy or redistributive (eco)socialism be deployed, but it does not rely on "ending capitalism." In the following section, we will address *political ecology*, which is based on revealing and understanding how social relations and institutions are shaped by political power in ways that are more mystification than revelation, one result of which is that nature becomes nothing more than an input into capitalism,

Table 2.5. Political economy's subsidiary worldviews

Worldview	Definition
Redistribution	The world's resources are accessible to the rich and not the poor, and redistribution of wealth and access must take place
Environmental justice	Environmental and ecological bads are unfairly distributed to impact poor communities of color and the marginalized in the Global South
Steady-state economy	Inputs and outputs into productive systems must be balanced in order to ameliorate environmental crisis
"Sharing the Commons"	The Earth must not be privatized but related to in terms of access and ownership by both humans and nature

rather than an independent or autonomous realm or subject. Through political ecology, we examine not *how* markets operate in a political context but, rather, *who* shapes markets and makes rules so as to generate benefits for the shaper.[48] This critical approach leads to heterodox understandings of how the global economy is organized and how that organization affects both global environmental politics and the global environment. It also provides some insights into the weaknesses and leverage points in existing political and economic systems that might be exploited to generate social change.

As we have already made clear, there is a gross and growing disparity between the world's wealthy and poor populations captured in the observation that 8.2 percent of the world's population (398 million people) controls 86.2 percent of the world's wealth ($220.5 trillion) while the remaining 91.8 percent (4.44 billion) possess a mere 13.8 percent ($35.2 trillion).[49] As noted earlier, with respect to sustainable development, this gap is believed to be at the heart of the global environmental crisis, and there have been and are innumerable programs and projects intended to redress it. To a growing degree, however, those who are concerned about this disparity and its environmental effects point to *overdevelopment* and *excessive consumption* as the primary causes of the crisis. This includes not only the well-off in countries across the Global North but also the emerging "middle classes" of India, China, and other growing economies in the Global South. It is consumption by those well-off consumers that largely drives the world economy and remains the dominant driver of economic growth. The problem, in other words, is not the six billion poor but the one billion rich, who continue to consume roughly 75 percent of the world's resources.[50]

The magnitude of this excess is reflected in the warnings by the Intergovernmental Panel on Climate Change that carbon emissions and fossil fuel consumption must be reduced to close to zero by the middle of the twenty-first century if there is any possibility of avoiding the worst effects of global warming.[51] Notwithstanding rapid growth across the world in aggregate power and energy provided by wind and solar photovoltaics, the demand for oil and natural gas cannot decline without major transformations in the economies and practices of the world's middle and upper classes. And it remains extremely difficult to see how a decline of this magnitude can be engineered without major and politically problematic regulation not only of the markets in which fossil fuels are bought and sold but also consumption, more generally. Many consider the necessary reductions to be politically unfeasible. Moreover, the emphasis on growth rather than redistribution of the world's wealth and resources and the necessary reduction of consumption among the well-off one billion illustrates the contradiction between trajectory and need.

What would be the impacts on incomes as a result of the required reductions in consumption proposed by the Intergovernmental Panel on Climate Change? The best metric for seeing this remains per capita income, even though it fails to take into account the vast differences between consumption by the well-off and the poor. Median global income is in the range of three to four U.S. dollars per capita per day, although half of the world's people (3.7 billion) live on less than about U.S. $3.50 per day.[52] According to some economists, the absolute minimum income required to belong to the global "middle class" is ten U.S. dollars per capita per day (in purchasing power parity), or U.S. $3,650 per capita per year, which adds up to somewhat less than fifteen thousand dollars annually for a family of four.[53] Compare this sum to the 2016 median household income in the United States of roughly $56,500,[54] with the "middle class" defined as having a median household income between one hundred thousand dollars and $350,000 for a three-person family.[55] Now imagine what might be politically possible in a country in which support for a fifty-cent gasoline or carbon tax is enough to deny election or re-election to virtually all most politicians and candidates for public office (such taxes are one of the "third rails" of American politics, so to speak).

There are powerful ethical arguments in support of *redistribution*, having to do with fairness, justice, and our "responsibility to others." That is, from a humanitarian and cosmopolitan perspective, some argue, the well-off are obligated to give until they are no worse off than anyone else.* This perspective is based on the utilitarian argument that no individual should be allowed to suffer any more than is absolutely necessary, especially if the means are available to relieve such suffering. That the relative impoverishment of one's family in order to save another's might expose loved ones to discomfort and even some deprivation is not an argument against redistribution. Needless to say, while this may be a valid ethical position, it is not one that is widely accepted, especially since it would involve very high degrees of taxation and confiscation of assets. And all of this is, ultimately, premised on the idea that improving the lot of the poor will help to protect and save the global environment—not the argument that overconsumption is the problem.

Environmental justice focuses on the ethics and impacts of the distribution of *bads* (that is, externalities) rather than goods. While environmental justice is not the opposite of redistribution, and not necessarily opposed to capitalism, an environmental justice worldview begins to examine and assess how colonial power, racism, and bias shape the distribution of environmental and social goods and bads to the disadvantage of the poor and weak. It has been widely noted and empirically confirmed that the poor (disproportionately including communities of color and indigenous communities) experience more exposure to environmental toxins and degradation and less access to goods and resources than the well-off and powerful. Moreover, there is often some intentionality in this distribution of externalities, which can be explained by

---

*This was the nub of a debate between Peter Singer and Garrett Hardin in the early 1970s. See Peter Singer, "Famine, Affluence and Morality," *Philosophy and Public Affairs* 1, #3 (Spring, 1972): 229–43; and Garrett Hardin, "Lifeboat Ethics: The Case Against the Poor," *Psychology Today* 8 (September 1974): 38–43, 124–26. Singer argued obligation to all; Hardin, only for the self, family, and co-nationals.

one or both of two processes. The first is what we might call "infrastructural venality," that is, the deliberate siting of polluting industries and infrastructure (for example, highways) in poor neighborhoods and regions, albeit through what are presented as legitimate public processes. The second is "structural venality" (or epistemic violence[56]) resulting from the structures of economics, social relations, and law, which disadvantage those without access to legal resources and normalize their subordinate positions.

Thus, for example, a public authority might propose to construct a road through a poor neighborhood as opposed to a wealthy one. Land is cheaper because of low property values and, often, better located to serve inner cities. Possibly, the route is meant to provide easier access to a business district for middle-class suburban commuters. The authority is required to file an environmental impact statement or its equivalent, which is prepared by in-house specialists and/or outside consultants. An environmental impact statement is only required to evaluate environmental impacts, not to decide whether they can be tolerated in order to build highways—and those who prepare them may well be influenced by the mission of the authority rather than by ethical or political judgments. The authority may then be required to hold public hearings and accept public comments on the environmental impact statement and the construction proposal. Such hearings are more likely to be attended by and responded to by those who actively engage in electoral politics, pay taxes, and want to keep the highway out of their neighborhoods (that is, "Not in My Backyard," or NIMBY). Even though some hearings and public reviews elicit strenuous and concerted opposition to a proposal, the final decision is up to the authority and its funders (and, eventually, the courts), which is all too likely to proceed, notwithstanding the negative impacts on the poor neighborhood. We can identify a similar phenomenon at the international and global levels.

Returning to bananas, we note that they are grown primarily in tropical countries where land and labor are cheap, political and economic elites dominate government, and workers with few or no other employment options are liable to be dosed with a variety of toxic fumigants and herbicides. Organic bananas could be grown in temperate zones, but mass production would require greenhouses and land, higher-cost labor, and substitutes for chemicals, resulting in a high-cost fruit with appeal only to those who can afford it.* The environmental injustice is thus imposed on the poor banana workers who have few other employment alternatives. By the same token, it is cheaper to ship discarded smartphones to Asia, where their toxic after-effects are imposed, again, on the very poor. Finally, the long-term cost of climate change, caused primarily by two hundred years of industrial activity and benefits in the Global

---

*Bananas are being grown near Santa Barbara, California, although no cost information about them is available. See: Kathryn Barnes, "How Tropical Bananas are Grown on Santa Barbara's Mesa," KCRW.com, February 23, 2017, accessed September 12, 2017, https://curious.kcrw.com/2017/02/how-tropical-bananas-are-grown-in-santa-barbaras -mesa.

North, are reckoned to be most severe across the Global South, where people and countries lack the capital and resources to adapt and protect themselves. For the most part, the Global North resists transferring the required resources to compensate.* Needless to say, this North-South conflict has long bedeviled all kinds of international negotiations and agreements (we consider several in later chapters), and continues to do so.[57]

Above, we argued that a primary cause of the global environmental crisis is excessive economic growth and the resulting overconsumption of resources, goods, and services among the world's middle and upper classes. An alternative to growth could be the steady state economy, originally conceptualized by Herman Daly, professor emeritus at the University of Maryland and before that, a senior economist at the World Bank. Daly has published a series of studies and books on this topic, including *Toward a Steady-State Economy* (1973), *Steady-State Economics* (1991), and, most recently, *From Uneconomic Growth to a Steady-State Economy* (2014). Daly argues that the fundamental error in our capitalist economic system is that we rank *flows* of materials and money more highly than stocks, or accumulations. In a steady-state system, by contrast, flows of raw materials into the economy must be approximately balanced by flows of waste out of it, with the rest recycled back into the life cycles of goods and services. It is, of course, potentially possible to design for full recycling and zero waste, which approximates a basic ecological model while still requiring substantial energy inputs, but this model assumes that the costs of full recycling be less than waste disposal (something which has not yet arrived, in any case).† In the SSE, the goal is to minimize waste by reducing the pace or rate of consumption rather than relying wholly on recycling. This means that consumer goods must last a long time, biological and physical resources must be

---

*This discussion would not be complete without reference to the famous World Bank memo co-authored by Harvard professor Lawrence Summers, who was also chief economist for the World Bank and, later, U.S. Treasury secretary for the Clinton administration and director of the National Economic Council during the Obama administration. He wrote: "A given amount of health-impairing pollution should be done in the country with the lowest cost, which will be the country with the lowest wages. . . . I think the economic logic behind dumping a load of toxic waste in the lowest-wage country is impeccable and we should face up to that." Summers later claimed that the memo "was meant to be sarcastic," although, at seven pages, it is difficult to imagine such extended sarcasm from a Harvard economist. See: *New York Times*, "Furor on Memo at World Bank," February 7, 1992, accessed July 12, 2017, http://www.nytimes.com/1992/02/07/business/furor-on-memo-at-world-bank.html.

†For the moment, recycled materials are offered for sale through markets. Unfortunately, manufacturers prefer virgin materials, which are often cheaper than recycled ones, and recycling rates in many parts of the world are sufficiently high that markets are glutted with materials, and recyclers are losing money. China recently announced it would no longer accept plastic for recycling. See Kimiko de Freytas-Tamura, "Plastics Pile Up as China Refuses to Take the West's Recycling," *New York Times*, January 18, 2018, accessed March 21, 2018, https://www.nytimes.com/2018/01/11/world/china-recyclables-ban.html.

minimally exploited, and organic wastes must be composted and returned to the soil. Only if a good has passed its useful and extended lifetime, and cannot be repaired, does it become waste, which must also be recycled to the extent feasible.

Getting to the SSE would seem relatively straightforward, but it is not. Not only is our current economic system based on high rates of flow or throughput, it is also reliant on planned obsolescence to sustain these rates. Again, this process is observable in the frequent release of new versions of any smartphone, which usually offer only marginal improvements or features not available on the preceding generation of phone. But the imminent release of a new model leads people to put off purchases of the older model and to line up in order to buy the new one. It is not that the old one ceases operating; it is that style and status mandate acquisition of the new one.

As an example of an SSE approach, consider the automobile. In an SSE, consumers would select cars based on fuel source and efficiency, greenhouse gas emissions and durability, rather than taste or status (for example, owning a Tesla because it's cool) or commercial indoctrination (that is, targeted or false advertising). An SSE would remain capitalist but it would be heavily regulated and motivated by corporate ends other than the relentless pursuit of profits and growth. A well-built vehicle could last for twenty years or more if properly maintained; the metal parts and much of the rest of the body and chassis could be recycled; with car- and ride-sharing, the size of automobile fleets could decline (subject to the caveats addressed earlier). The flow of resources going into cars and wastes coming out would be reduced considerably. An even more effective strategy would be to shift reliance for transportation to public transit systems (although, in the United States, there would be considerable public opposition to such a transformation).

A successful SSE would also require the normalization and internalization of social practices associated with cars and mobility, and changes in tasks and goals that, today, can be accomplished only by driving. There are indications that some practices are changing: in the United States, for example, a decline has been observed in the percentage of teenagers and young adults getting driver's licenses.[58] Online ordering of various goods, including foods, is becoming much more common, and even if deliveries come in large trucks, the number of automobile trips required declines. Car-sharing companies, such as Zipcar, and ride-sharing ones, such as Uber and Lyft, are seeing growth as people recognize the high costs and inconvenience of owning a car. These do not constitute a step toward an SSE; they are simply examples of the kinds of changing practices that would have to become "normal" for an SSE to work.

Long-lived and reliable automobiles would clearly have an effect on society in other ways, especially one as heavily dependent on the automobile as is that of the United States. Many fewer cars would need to be manufactured each year. Employment in the auto industry would drop (although many assembly-line tasks are now done by robots). The demand for replacement parts and repairs would decline, and many dealerships would not survive (many are already in trouble). Reduced employment might mean lower incomes, although the costs of auto purchase and operation would also decline, freeing up money for other purposes. There would be a host of other consequences, many of which are difficult to anticipate (that is, fewer gas stations, tire companies, insurance policies, etc.). And, even in an SSE, there would still remain major environmental consequences. Long-lived products would mean less-frequent innovation, and less-energy-efficient items might remain in service and continue to pollute for extended periods. The cost of owning and operating a vehicle through high licensing, fuel taxes, and road use fees would need to be made unattractive by comparison with the cost and ease of walking, biking, or using public transportation. The last would require major new investments in mass transit and even reconstruction of cities and suburbs to increase accessibility and reduce the need to travel long distances. None of this is impossible—it has been and is done, to some degree, in western Europe—but it is politically unimaginable in societies like the United States as well as countries like China and India, in which auto ownership is being encouraged as part of an economic growth strategy. And an SSE does not necessarily address the issue of resource distribution or environmental justice.

What might a steady state world look like? What would constitute the economic units of an SSE? In a 1989 book co-authored with John Cobb Jr.,[59] Daly proposed that the proper unit of "development" in his steady state world would be the nation-state.[60] Each country, according to Daly, ought to be prepared to live within the limits of its own natural endowments, possibly in the belief that national self-sufficiency would eliminate conflict over resources and contribute to world peace. Daly provides no real explanation for why the state, which is certainly among the most unnatural of humanity's social institutions, should be the basis for an attempt to implement a steady-state, sustainable economy. And nationalization of the steady state would do nothing to reduce disparities between well-endowed and poorly endowed countries. Realism, in the form of a "Green Leviathan," might well rear its ugly head in Daly's world.*

For a moment, however, let us consider whether the technical capacity exists to enforce an SSE. While Daly sees the SSE as remaining a fundamentally, if heavily regulated, capitalist system, it is much more likely that it would have to be pursued through a centrally planned economy. Only through a centrally planned economy would it be possible to allocate resources to meet the requirements of production and limit waste through regulation of consumption. The centrally planned economy has received a bad name for the manner

---

*The "Green Leviathan" is a state that regulates and enforces ecological and environmental laws and limits through coercion, if necessary. See: Robert L. Heilbroner, *An Inquiry into the Human Prospect: Looked at Again for the 1990s*, New York: Norton, 1991; William Ophuls and A. Stephen Boyan, *Ecology and the Politics of Scarcity Revisited: The Unraveling of the American Dream*, San Francisco: Freeman, 1992.

in which it was implemented in the old Soviet Bloc. Central planners issued production quotas to industrial enterprises but relied on self-reporting by the latter, who had incentives to overstate production and hoard resources for future needs. The result was chronic undersupply of virtually everything except military equipment, production of shoddy goods with abbreviated lifetimes, and careless handling of wastes, since there were few restrictions on and no penalties for improper disposal. A key obstacle to the operation of this system was the absence of timely information, which was often fabricated or not available at all.

If not today, then not far in the future, the computing and communication capacity will exist to monitor resource and waste flows, and both production and consumption, on a real-time basis, and to actually implement centrally planned economies. Producers of goods and services would be allocated resources, labor, and capital in response to demand, and consumption would be regulated in such a way as to foster stable and predictable demand. This will be possible because the world's industries and well-off consumers are deeply linked into globe-spanning electronic communications systems which collect and transmit data to algorithmic systems programmed to monitor, allocate, restrict, or deny all kinds of activities. Imagine, for instance, that the land-based ice sheets of Greenland and Antarctica do melt under the pressures of climate change, leading to a very short-term and high-risk crisis that galvanizes the world's governments into action. It would not be very difficult to identify specific high-emission activities, such as travel by air, and limit access to them by limiting the purchase of tickets. Similarly, the purchase of new or replacement goods could be regulated, limited, or denied. Lest this seem too nightmarish and infeasible, all of the elements to make such a system possible are already available. The Green Leviathan might well emerge not through tyrannical government but, rather, via mundane cybernetic controls.[61]

Could further privatization of nature provide a solution to the environmental crisis? In his book *On the Origin of the Inequality of Mankind* (1754), Jean-Jacques Rousseau famously wrote

> The first man who, having enclosed a piece of ground, bethought himself of saying *This is mine*, and found people simple enough to believe him, was the real founder of civil society. From how many crimes, wars and murders, from how many horrors and misfortunes might not any one have saved mankind, by pulling up the stakes, or filling up the ditch, and crying to his fellows, "Beware of listening to this impostor; you are undone if you once forget that the fruits of the earth belong to us all, and the earth itself to nobody."
> (Part 2, §1)

This is a plaint echoed many times since, from arguments posed by radicals in the English Civil War to the Brundtland Commission's lament that "the Earth is one, but the World is not" to the UN doctrine of the "common heritage of mankind [sic]." For a significant number of observers and scholars, such as Terry Anderson,[62] on the one hand, and Garrett Hardin,[63] on the other, the division and privatization of the "commons" is as things should be if the earth and environment are to be sustained. But, as Nobel Prize winner Elinor Ostrom and her colleagues discovered,[64] there exist many present and historical cases of shared resource systems regulated not by law or title but, rather, common norms and values.

These *common pool resources* (CPR; as opposed to "open access resources," which anyone can exploit) lack the profit and capital accumulation incentives associated with the acquisition and exchange of private property through markets, even as individual self-interest remains important to sustaining the resource. The point here is that group members have a common shared interest in retaining access to the resource and are, therefore, normatively motivated to use no more of the resource than each requires for her own purposes. Moreover, any individual member found violating these normative limits can be sanctioned or even expelled from the group, which can be very costly if said individual relies on the shared resource for basic needs. Such systems have proved remarkably durable, although they can break down under external pressure if not authoritatively recognized by the state.

More broadly, the concept of *sharing the commons* arises especially in the context of environmental justice. The Earth's atmosphere is almost certainly such a commons in terms of its life-supporting properties, notwithstanding claims of state sovereignty over airspace and court decisions that have either forbidden or enabled practices impinging on the rights of states and corporations. Here, however, we enter into tricky territory, because action affecting "the atmosphere" can be disaggregated in a number of different ways. The most common approach is to discuss greenhouse gas emissions as negative externalities that interfere with the ecological services provided by the atmosphere, and to treat this as illegal appropriation of a public commons by private actors. Except in the case of evident pollution and harm, it is difficult to specify either injuries or specific injured parties resulting from greenhouse gas emissions, especially if they are hypothetical injuries whose future distribution is uncertain. This is a problem that continues to confront climate change negotiators, leading to acrimonious and unresolved fights over the area of "loss and damage" and compensation.[65]

As we know, however, the distribution of such entitlements in global commons has favored the states of the Global North, who control the global economy and the making of its laws. The Global South may dominate in terms of absolute numbers of people, but its members remain largely also-rans in terms of relative power and wealth; the best they can do is block legalized privatization (an example can be found in chapter 5 in the discussion on genes and markets). A per capita allocation of title in a common heritage would favor the Global South but result in a potential redistribution of resources and income to the disadvantage of the Global North. As illustrated by the results of distribution of corporate shares in post-socialist countries, where the better-off bought them from those impoverished by the collapse of socialism, even the equitable distribution of shares in the atmosphere or other commons would almost certainly result in the poor selling theirs in order to survive and the rich buying those shares as speculative investments, getting even wealthier in the process.[66]

What is to be done? What if the Earth were treated as a common heritage, as a CPR in which every human being were no longer merely a "stakeholder" but truly a shareholder?[67] What if the "use" of the Earth's ecological resources and services were treated not as private entitlement or subject to unpaid externalities but rather as a common holding, for which any use and damage required compensation to the Earth and its shareholders?[68] What if such compensation were utilized, in turn, to ameliorate the desperate poverty of so many of the world's people and the protection and restoration of environments systematically degraded in the pursuit of both survival and accumulation? What if, in effect, we had to pay interest on loans from so-called Natural Capital? Karl Marx pointed out that capitalism would be hard put to survive were it not for the subsidies to workers' wages provided through women's unpaid labor in the household.[69] By the same token, could capitalism survive if not subsidized by "free" ecosystem goods and services? This is not necessarily an argument for full internalization of negative externalities into the costs of everyday life; it is, however, a proposition that, in a globalized CPR, free lunches from Mother Earth could no longer be the norm.

The creation of an Earth CPR might seem like a radical, if not impossible, proposal (see Box 2.1). What would become of private property, of national sovereignty, of *rights*? Who in the Global North, secure on their little half-acre, would agree to share ownership with billions of people half a world away? And why would anyone pay "rent" to humanity on property and possessions legally and fully paid for? Would not such a project actually commodify the Earth in full, turning it into the object of ever-greater speculation and concentration of wealth? Would this not deny workers the fruits of their labor and capital the benefits of its investments? Yet, at the end of the day, an Earth CPR is no more radical or outlandish than the dividing up of the planet into private property. *Think about that.*

Political Ecology and Its Subsidiary Worldviews
Michael Watts defines *political ecology* as that approach to environmental politics and political economy "which seeks to understand the complex relations between nature and society through a careful analysis of what one might call the forms of access and control over resources and their implications for environmental health and sustainable livelihoods."[70] Whereas political economy primarily examines relationships between power and wealth, politics and markets—or, in the terms we used earlier, the "rules of the game"—political ecology focuses on the structure and organization of the game and how those who make the rules bend them so as to ensure favorable outcomes when games are played. This does not mean, however, that the poor and weak are without

BOX 2.1 | Earthshares®

In the spirit of this section, let us propose something akin to an "independent public offering" on behalf of the Earth. Instead of selling shares to the highest bidder in open auction, however, we would propose to issue twenty billion Earthshares®* each with a nominal face value of five thousand dollars. One Earthshare® would be awarded to every person alive on Earth now through 2050 (we assume here that the Earth's human population will level out at around ten billion in 2050-2100), with the remaining ten billion shares held in trust for Nature. Thereafter, shares would be transferred to newborns following a shareholder's death (shares would not be inheritable). Unlike common and preferred stock and other similar instruments, Earthshares® could not be alienated in equity markets; instead, they would be held in a one-hundred-year trust by an "Earth Commons Bank" (ECB). On the basis of its one hundred trillion dollars in shareholdings, the ECB could issue ten-year bonds for environmental preservation and restoration and social development loans in support of basic needs.

A bond is a promissory note or contract to repay a loan principal with interest, paid at specified intervals. Payment of interest rate and principal are underwritten by the notional flow of future revenues from some reliable source to the bond issuer, whether public authority (taxes), user (fees and rents), or corporation (profits). The interest rate depends on both the cost of money over the length of the loan (maturity) and the long-term reliability of the revenue flow (risk). Hence, a bond considered to be absolutely risk-free, such as a U.S. Treasury bond, may offer an effectively negative interest rate—lenders will pay issuers to ensure the security of their funds—while a junk bond will have a high interest rate, reflecting the possibility of default.†

During the first one hundred years, there would be no active market for trading in Earthshares®—no living person could acquire any material or speculative interest in a bond—although the shares could act as collateral for the bonds issued by the ECB. Revenues to repay bonds and monetize shares would be generated through a 1 percent "user fee" on all recorded economic transactions (in effect, a tax on global gross domestic product). This fee is similar in some ways to a fee on ecosystem services, here to be imposed in recognition of the implicit

---

*These are not really "registered"; the ® just looks more authoritative.

†The interest rate paid on a bond depends on the degree of risk associated with it, which is why junk bonds tend to have much higher interest rates than sovereign bonds issued by states (although markets in Greek sovereign debt illustrate how risky they are considered to be). Rates on short-term U.S. Treasury Bonds, as of March 2018, were about 1.75 percent, since they are certain to be redeemed at face value; rates on thirty-year Treasury bonds were 3 percent, reflecting a long-term risk premium.

(Continued)

negative social and environmental externalities generated by anthropogenic activities (thus, transactions committed explicitly to effective environmental protection and human development would be exempt from the user fee). Given a global gross domestic product on the order of sixty trillion to seventy trillion dollars, and the likely difficulty of collection on low value transactions (say, less than one hundred dollars), this "Earth fee" could generate something on the order of five hundred billion dollars annually.[‡] Assuming a revenue flow of this size, and in the interest of minimizing risks to lenders, the ECB could issue some one trillion dollars in ten-year Earthshare® bonds annually (auction price eight hundred billion dollars). One portion of the incoming revenues would service the bonds (at a 2 percent interest rate), the remainder would be applied to monetize the Earthshares® over the one-hundred-year trust period, at which time they would become fully negotiable and alienable. Finally, in parallel, each individual's Earthshare® could be used as collateral for individual micro-loans to those with annual incomes below a certain local limit (say, five thousand dollars per year), set so as to represent substantial capital for many in the Global South. The ECB, in turn, would make available grants as well as loans at a range of interest rates, which would also be used to service bonds and monetize the bank's capital base.

The issuance of Earthshares® is, at the end of the day, no more absurd than carbon emission permits (further discussed in chapter 3), and the allocation of one share per person no more ridiculous than an annual per capita allotment of emission rights to every living individual. But whereas emission permits can be bought and sold in the short term, Earthshares® represents a commitment to the future sustainability of the planet while providing current liquidity to those most in need of financial resources. Earthshares® are a project that gives *everyone* a concrete stake in the well-being and future of the Earth as well as a means of creating identification with the Earth and all it encompasses. In this sense, over time, each person's Earthshare® will come to represent more than just title to a piece of the planet; it will become a sign of right membership in the Earth's many ecological communities and a commitment to the thriving and well-being of those communities as well as our own long into the future.

Source: This proposal is taken from Ronnie D. Lipschutz, "From Stakeholders to Shareholders: Fostering Global Economic Democracy through Usufruct Rights," presented to the Lund Conference on Earth System Governance: Towards Just and Legitimate Earth System Governance—Addressing Inequalities, Lund University, April 18–20, 2012.

---

‡ Note that total overseas development assistance is on the order of $125 billion a year, while estimates of the cost of restoring and protecting Earth's environment are in the range of two hundred billion to five hundred billion dollars per year. Annual global military spending is about two trillion dollars. See: Ronnie D. Lipschutz and Sarah T. Romano, "The Cupboard is Full: Public Finance for Public Services in the Global South," Municipal Services Project, Queens University, Kingston, Ontario, Occasional Paper 16, March 2012, Table 7.

Table 2.6. Political ecology's subsidiary worldviews

Worldview	Definition
Ecocentrism versus anthropocentrism	Humans must not give priority to their own interests, but approach the world from the position of nature
Ecomarxism and ecosocialism	Exploitation, class struggle, and maldistribution of power are the cause of the environmental crisis
Ecoanarchism and social naturalism	Small, environmentally aware communities in harmony with nature are required to sustain nature and the Earth
Liberation ecologies and critical transhumanism: post-colonialism, feminist political ecology, queer ecologies, new materialism	Domination of people, animals, and nature by the rich and powerful is effected through opaque hegemony, naturalization, mystification, and social practice

influence; if nothing else, they are often able to resist or block unfavorable projects and initiatives. *Critical* political ecology goes beyond analysis to activism, emphasizing not only understanding, but also action.[71] These critical approaches might be called "liberation ecologies"[72] in the sense that they seek the emancipatory goals first articulated by the Frankfurt School[73] as well as strategies and solutions that can liberate those who are being dominated.*

Inherent in the philosophical categories we have examined so far is a matter largely downplayed in all of them: *social power*. This is not power in the traditional military sense, or even as relative wealth, as we argued in chapter 1, but structural and epistemic power, that is, the power embedded in social relations and social institutions, in the ways that they are organized and in the ways that they are reproduced. Power in all its forms normally serves to maintain the status quo or to provide various forms of advantage to some as opposed to others, as we saw with political economy. Therefore, power tends to increase inequities and injustices at the expense of others, who may have little or no ability to resist or act. The result is domination. Although most people regard power in a negative sense, as a tool of manipulation and self-interest, power can also be regarded in a positive light, as something that makes political action possible.[74]

As we saw in chapter 1, there are at least four useful conceptions of power. To repeat, the first is the ability and capacity to get people to do something they would not otherwise do, through persuasion, coercion, or force. The second rests on authority, that is, the right to define which practices are acceptable and which are not, and to set agendas in meetings, to limit discussions in public forums, or to forbid certain behaviors. A third dimension of power is visible in what we can think of as the normalization of practices and beliefs. Finally, a

---

*It is important to understand that critical theories do not deny the existence of a "real world." What they do question is the veracity of language, social relations, and material features that constitute the dominant or hegemonic views of the "real world."

fourth regards power as inherent in social institutions, rules, relationships, and practices, and, indeed, serves to constitute both individuals and society, often via relations between speech and action.

Speech itself can be regarded as a manifestation of power. How is this so? The very structure of language and the way in which it is used can be understood as a form of power—what is often called "discourse"—albeit one that can be challenged in productive ways. Language does not create reality, but it does shape the way we *see* and *act* in the world. Reflect, for example, on the term "resource," as it is commonly used. A resource is something that is deemed to have value for human beings, in that it can be converted from some initial non-utilitarian form into a useful final product or service. When we speak of "natural resources," we treat all of nature as either having value or being valueless. As we have seen, through this language, it is only when something can be turned into a marketable product with a price that it does acquire value. Then the resource is recognized and can be appropriated and turned into private property. Hence, a forest that is used as a common property resource by a specific group is regarded under some legal systems as having no "owner." Unless it becomes state land or is privatized, the forest has no market value and is regarded as "wasteland" (see chapter 5). Denying the legitimacy of common use, especially by the poor, thereby makes the forest vanish and permits authorities to confiscate what is essential to the production and reproduction of the user group. All of this constitutes a "discourse," with power and domination being exercised in often-subtle ways.

As we use it here, *social power* is, by contrast, about being and action. This term should not be confused with *social forces*, a concept frequently deployed in discussions of social movements and political organization. Social forces exercise social power, to be sure, but social power is more complex. To explain social power, we need to refer to the work of Michel Foucault. Although Foucault said many relevant things about both power and nature separately, he never wrote explicitly about the environment. He did, however, write about power and domination and, in particular, the propensity toward management of people through specific forms and rules of social order, which he called "governmentality."[75] Government, in this use, is not the same as that which we associate with state bureaucracies; it is better understood as those internalized and normalized beliefs and practices that constitute governing of "a sort of complex of men and things," both within a state and among states. As Foucault explained it,

> The things with which in this sense government is to be concerned are in fact men, but men in their relations, their links, their imbrication with those other things which are wealth, resources, means of subsistence, the territory with its specific qualities, climate, irrigation, fertility, etc.; lastly, men in their relations to that other kind of things, accidents and misfortunes such as famine, epidemics, death, etc.[76]

In effect, governmentality becomes a means of managing those things that are seen to threaten the welfare of who or what is being "governed."

This is accomplished through what Foucault called "biopolitics." According to Mitchell Dean, a scholar of Foucault and governmentality, biopolitics "is concerned with matters of life and death, with birth and propagation, with health and illness, both physical and mental, and with the processes that sustain or retard the optimization of the life of a population."[77] It is about management of the biological functions and social practices of homogeneous populations. Dean goes on to say:

> Bio-politics must then also concern the social, cultural, environmental, economic and geographic conditions under which humans live, procreate, become ill, maintain health or become healthy, and die. From this perspective bio-politics is concerned with the family, with housing, living and working conditions, with what we call "lifestyle," with public health issues, patterns of migration, levels of economic growth and the standards of living. It is concerned with the bio-sphere in which humans dwell.[78]

To put this another way, all those institutions and practices concerned with exploiting, managing, and protecting the environment, including international environmental regimes and regulated markets, are the governmental expression of biopolitics. They are all concerned with the management of human behavior and populations so as to maintain the material base of life, that is, the global environment.

In Foucault's analysis, however, power is more than a tool of governmentality or something that some people wield over others; power is also something that "induces pleasure, forms knowledge, produces discourse."[79] Power flows between people, it constitutes them, it makes them who they are, and it influences how they behave. This is the case even when power is not visible in its other forms. Social power is not the exercise of power by some people over others and nature but, rather, the result of people acting as they have been produced by power circulating through what Foucault called the "capillaries" of society (we will return to these arguments later, especially in chapters 4 and 6).

Those worldviews subsidiary to political ecology are organized around such notions, especially discourse and social power and, more specifically, in opposition to unjust and normalized (therefore unacknowledged) domination. Generally speaking, these critical approaches have their origins in Marxist theory, although they often arrive at radically different conclusions than did Marx and his followers. Since the first edition of this book, in particular, there has been an efflorescence of theorizing and worldviews around what can be called "transhumanist political ecology,"* encompassing gender, queer ecology, post-

---

* "Transhumanism" and "posthumanism" have been associated more frequently with forecasts of human-computer relations. Here, we use the term in recognition of the essential unity of culture and nature.

colonialism, nonhuman ecology, and the "new materialism." Transhumanism draws on critical political ecology worldviews but extends them to challenge masculinist, anthropocentric, idealist, and heteronormative accounts of nature, ecology, environmentalism, and culture-nature relationships.

All these philosophies posit relations of inequality between two or more groups of people, or beings, or nature. They view such inequality not as a simple matter of differences in individual wealth or coercive power but, rather, as structural and normalized practices. In these philosophies, therefore, the exploitation and degradation of nature is understood to occur because those in power legitimate and reproduce both their dominant positions and the social organization that authorizes their rule through normalized and naturalized actions and beliefs. Moreover, these actions and beliefs are broadly accepted as both legitimate and necessary to the survival of the social order. Domination, in this view, cannot be eliminated by more moderate, work-within-the-system means of reform, such as political participation, or even radical redistribution. What is required is wholesale social change.* That is a tall order for any political philosophy.

Our first subsidiary worldview under political ecology is the binary of *anthropocentrism and bio- or ecocentrism* (already discussed in chapter 1).[80] Realist and liberal worldviews, and utilitarian philosophies, more generally, put human beings and society at the center of their analyses; humans first, then nature. We see this in the monetizing of nature and the prices put on resources via markets. Because these worldviews put humans first, we call them "anthropocentric."† Biocentrism and ecocentrism, by contrast, prioritize nature. Without nature, humans cannot survive; by exploiting nature as they do, humans destroy both nature and themselves, and the future of life on earth. Many ecocentrists are also advocates of "wilderness" and "wildness" (Henry David Thoreau: "In Wildness is the preservation of the World"[81]), on the premise that these are the only places that are truly natural and their elimination would result in both biological and spiritual impoverishment (recall, however, that there are very few, if any, places on Earth untouched by human interventions).

Ecocentrism, then, is best understood as a type of anti-capitalist distributionism. A revolution is therefore necessary to alter the current order of things, a revolution that establishes harmony between people and the natural world, or even gives nature an advantage over people, one in which humans get no more than their fair share of the world's physical resources and biological products. On the one hand, some ecocentrists lean toward a type of Social Darwinian ecologism—a view of "nature red in tooth and claw" related to realism—or Malthusianism, and on the other, some kind of theologically based spirituality. The former bears some similarity to eco-Marxism (see the

---

*Consider one of the resonating slogans of the international climate justice movement: "System change, not climate change."

†We believe that it is difficult, if not impossible, to formulate any worldview in other than anthropocentric terms, since only humans are able to articulate a nature-society distinction.

following) in the belief that domination of the natural world represents a "class struggle" between humans and nature. Capitalist humans exploit and extract the value of the work done by nature, to the point of impoverishment. At some point, if harmony is not achieved, nature will strike back and overcome human domination, to the great detriment of people.

Ecocentrism also bears some resemblance to liberalism and Marxism as a form of romanticism that idealizes nature and regards it as a source of eternal truth and beauty. Some ecocentrists search for transcendental sources of authority, whether natural or spiritual, to legitimate change. In the former case, ecology and science are called upon not only to describe what might happen should human practices remain the same but also as the basis for imposing order on those changes that take place.[82] In the latter instance, a sort of spiritual essence of Nature plays the same role. In both, ecocentrists claim that what should be done must be done, or humanity will inevitably suffer the consequences (disaster, extinction damnation).

Humanity and nature might suffer from our actions, or they might not, but what we see here is an attempt to institute one form of domination to counter another: obedience to an external and infallible authority (much like a Green Leviathan). The command "Repent or die!" is a powerful one, especially if it comes from a source over which humans have no control. Moreover, there is a certain deterministic quality to such warnings: fail to listen and perish; do as you are told and flourish. The Marxist conception of history has a similar teleological quality to it, as do both Western religions and liberalism. All demand certain kinds of prescribed behaviors as the price of salvation and survival, and all prophesy doom should those prescriptions be violated. In other words, ecocentrism proposes replacing human domination with natural domination (whatever that might mean in practice), in order to overturn the existing system. In this instance, a class coalition between nature and bourgeois human environmentalists could overturn what is, in effect, capitalist domination, replacing it with human-nature harmony in which the full potential of both can be realized.

By contrast, *ecoanarchism* and *social naturalism* are based on cooperation, seeing them as central and even "natural" elements of human social organization, based on the premise that humans are social animals who could not have survived outside of small cooperative systems. Both worldviews are focused on relationships *within* society and *between* human culture and nature. They differ from each other in that ecoanarchists argue that humans must not look to nature for lessons in politics and social ordering, while social naturalism argues that nature is "organic" and points to ecology and ecosystems as models for human associations. Consequently, the two lead to rather different conclusions about how to deal with global environmental problems and politics.

Ecoanarchism is a form of leftist anarchism whose best-known advocate was Murray Bookchin.* Hierarchy, the state, capitalism, oppression, and war,

---

*Bookchin called his philosophy "social ecology," but we prefer "ecoanarchism."

all of which are anathema to ecoanarchists, are a consequence of the historical and unjust exercise of power by men over other men, by men over women, by men over nature. In an anarchist society, there is no central government—at most, a kind of confederal association of anarchist communities, or what Bookchin calls "municipal libertarianism."[83] Politics is centered in self-governing, egalitarian, voluntary communes, in which face-to-face interactions are both possible and necessary so that the tendencies and the practices associated with hierarchy and power can be eliminated.*

Bookchin argues that contemporary environmental problems are rooted in an irrational, expansive anti-ecological society and that piecemeal reforms cannot address them. These problems originate in a hierarchical, class-ridden, competitive capitalist system, which regards nature as a mere agglomeration of "resources" for human production and consumption and as a result of which humans are bound to dominate nature. Class is linked to the ownership of private property, which permits individuals to treat nature as they wish. Hierarchy is linked to the exercise of power in ways that maintain the primacy of private property and patriarchy. Finally, capitalism fragments the natural world into units that can be privately owned and exploited. All of these practices are antithetical to the protection of nature. Bookchin, in particular, rejects much of the idealism and romanticism often found in anarchist theorizing. He regards nature in its present form as a product of human action and transformation rather than something that ever existed in, or could be restored to, some antediluvian or pre-human condition.

Ecoanarchism sees human society as rooted in ecological praxis, or practice, without being structured in organic ways dependent on ecological function. The preservation of nature depends on the reorganization of society into small, relatively self-sufficient units in which altruism and mutual aid are practices of central importance. The impacts of such social organization are twofold. First, these communities will not engage in the kind of globalized capitalism that distances production from consumption and externalizes pollution and other forms of environmental degradation onto others, whether near or far away. They will engage in local production and trade with neighbors, but the limited scope and scale of both will result in a much smaller environmental burden on nature. Second, people will be closer to nature, since they will depend on it for their needs and well-being, and will be much more sensitive to the effects of their activities on the world around them. People will transform nature—Bookchin believes this is both necessary and good—but the result will be a rational, technically based culture, refined by spiritual and intellectual insights but not determined by them, and sensitive to and concerned about nonhuman nature.[84]

For some of his critics, Bookchin's ecoanarchism is too humanistic and political, and does not give enough importance to the place of humans in

---

*This is, of course, a crude description of anarchism, which itself is riven by many different approaches. Bookchin's work provides ample evidence of this.

nature. Links to both Enlightenment rationality and socialism are evident in Bookchin's commitment to a science-based understanding of nature and the possibilities of positive transformation of humans and nature together, which some critical scholars question. Bookchin highlights the struggle to achieve human freedom from state and capital, again through a form of class struggle, albeit one that pitches humans and nature against capitalism as a whole. And although conscious of the role of power in fostering domination, Bookchin seems to treat power as something that will not be necessary or even exist in ecoanarchist communes. This seems unlikely.

In contrast to ecoanarchism, social naturalism regards culture and nature as bound together in a kind of pre-existing social community, based on principles of ecology and harmonious cooperation among humans and between humans and nature.* The origins of social naturalism are to be found in nineteenth-century romantic (and often utopian) views, not all of which eventually developed into progressive forms.[85] Proponents of social naturalism have adopted Prince Kropotkin's notion of mutual aid and cooperation between species as the basis for community:

> It is evident that it would be quite contrary to all that we know of nature if men were an exception to so general a rule: if a creature so defenceless as man was at his beginnings should have found his protection and his way to progress, not in mutual support, like other animals, but in a reckless competition for personal advantages, with no regard to the interests of the species.[86]

Social naturalism, however, goes further than mere admonition of human-nature cooperation. John Clark, who uses the term "social ecology,"[87] describes social naturalism as a worldview that seeks to relate all phenomena to the larger direction of evolution and emergence in the universe as a whole, reaching its apex in some kind of organic harmony between society and nature. As Clark puts it, social ecology examines the course of "planetary evolution as a movement toward increasing complexity and diversity and the progressive emergence of value and views, and the realization of social and ecological possibilities as a holistic process, rather than merely as a [Darwinian] mechanism of adaptation."

> This evolution can be understood adequately only by examining the interaction and mutual determination between species and species, between species and ecosystem, and between species, ecosystem and the earth as a whole, and by studying particular communities and ecosystems as complex and developing wholes. Such an examination reveals that the progressive unfolding of the potentiality for freedom . . . depends on the existence of symbiotic cooperation among beings at all levels.[88]

---

*The paradox here is that "cooperation" within an ecosystem means something quite different from the conventional usage of the term.

The goal of social naturalism is, therefore, the teleological creation of a cooperative ecological society "found to be rooted in the most basic levels of being."[89] Clark uses the concept of "community" here in a very expansive sense, to include not only people but also animals, plants, ideas, language, history, ecosystems, and other elements of the material and social world.

In this respect, social naturalism strongly resembles the worldviews and beliefs often attributed to indigenous peoples, many of whom see the world as being of an interconnected piece.[90] No one part of that world can be damaged or destroyed without the rest being affected negatively, but the value of each part to the whole is intrinsic to the whole rather than instrumental to its functioning. More problematically, however, advocates of social naturalism also believe that human practices and institutions must be depoliticized. There can be no harmony so long as there are politics. Clark argues that legislative assemblies "must be purged of the competitive, agonistic, masculinist aspects that have often corrupted them. . . . [T]hey can fulfill their democratic promise only if they are an integral expression of a cooperative community that embodies in its institutions the love of humanity and nature."[91]

In Clark's view, capitalism might still comprise the economic base of such a community, notwithstanding its competitive and individualistic elements: "The dogmatic assertion that in an ecological society only one form of economic organization can exist . . . is incompatible with the affirmation of historical openness and social creativity and imagination that is basic to a social ecology."[92] Clark's vision appears to be one of the harmony of Eden before Eve offered Adam the apple, of society unpolluted by conflict, struggle, or politics, of a world in which the corruption of nature by humans has been redressed.

Curiously, perhaps, some roots of social naturalism are also to be found in G. W. F. Hegel's philosophy of the state.[93] In essence, Hegel argued for the organic nature and teleological destiny of the nation-state, as an entity that is the "natural" collective community for those who are its members. Hegel believed that the state represented the highest and best form of human social and political organization and that human potential could be realized only through the state. For Clark, the "natural" community replaces the "natural" state, and biological potential replaces human potential. All else is very similar. We might doubt, however, that Clark's teleology is any more likely to come to fruition than has Hegel's.

The Gaia Hypothesis,[94] a curious and somewhat idealist combination of social naturalism and scientific naturalism, should be mentioned here. The nub of this worldview is that the Earth is a self-correcting and self-equilibrating system that might even exhibit a form of consciousness (the "noosphere").[95] Gaia emerged out of beliefs having to do with transcendental human consciousness and spirituality, ecological balance, systems analysis and cybernetics, and natural history. The geological record suggests that, even when subject to significant perturbations, the Earth returns to some state that represents a physiological balance among its many components. In this, Gaia comprises a complex system in which the relationships among its components are impossible to fully map but which affect each other, and the whole, through cybernetic-like feedback systems. Finally, transcendental fulfillment of human purposes on Earth requires a greater "spirit" such as Gaia. It is perhaps ironic that, even as Gaia was emerging as a worldview, ecologists were discovering that disequilibrium and change are much more characteristic of ecosystems, with the "balance of nature" being largely a teleological and theologically inspired myth.

Something of a synthesis of ecoanarchism and social naturalism can be found in at least two cases: ecovillages and bioregionalism.[96] The Global Ecovillage Network defines an ecovillage as:

> an intentional, traditional or urban community that is consciously designed through locally owned, participatory processes in all four dimensions of sustainability (social, culture, ecology and economy) to regenerate their social and natural environments. . . . Ecovillages are living laboratories pioneering beautiful alternatives and innovative solutions. They are rural or urban settlements with vibrant social structures, vastly diverse, yet united in their actions towards low-impact, high-quality lifestyles.[97]

Within these parameters, there is considerable variation and no consensual definition, except self-identification. Embedded in these various definitions and conceptions is a notion of diffusion by example: as more and more sustainable communities are established, more and more people will recognize their virtues and join one. Ultimately, perhaps, the ecoanarchist vision will come into being through ecovillages.

Bioregionalism, by contrast, hews more closely to social naturalism, proposing that the "natural" political community must be established within defined ecological boundaries or units, such as the watershed. The author of a recent article arguing for bioregionalism as a post-capitalist, non-Marxist strategy for ecological sustainability argues that

> The economic system that operates within a bioregion arises from the social and ecological culture of a bioregion, and from the inherent logic of bioregionalism: to localise and respect natural limits for ecological and environmental benefit. . . . A bioregional economy would be oriented towards meeting people's needs efficiently so as to enable a high quality of life for all. As such, economic activity would be based on the criteria of satisfying human needs and production would be carried out directly for human use rather than to perpetuate the accumulation of capital.[98]

Inside these units, it is hoped, there can be a high degree of self-sufficiency, within limits imposed by nature: "Bioregions are self-reliant economies, in the sense that imports are treated as a last resort. Bioregions are designed to be largely self-sufficient in basic resources so that only goods that cannot be produced within the bioregion due to their complexity or specific climate requirements are imported."[99] Although the inhabitants of bioregions might share normative perspectives on the preservation of nature, it is an open question as to whether the people's staying within their own "natural" borders would breed harmony or distrust.[100] Indeed, why Malthusian resource geopolitics might not raise its ugly head in a world of bioregions is not at all clear.

However conceived, what the worldviews described above are unlikely to accomplish is any real change in the basic social relations and class structures comprising a global capitalist economy. The rich will remain rich; the poor, it is to be hoped, will become somewhat richer, but certainly not so much as to pose an effective challenge to the rich. And unless the rich prove willing to invest the financial resources necessary to address the environmental crisis, it is unlikely to be dealt with. Indeed, according to neo-Marxist analyses, capitalism requires economic inequality to operate because, otherwise, labor at affordable wages will not be available. Moreover, societies are composed of contending classes—capitalist, worker, bourgeoisie—who struggle for the opportunity to organize modes and relations of production in their particular interests. This is the engine of capitalism.[101]

Although they are not identical, we treat *ecosocialism* and *ecomarxism* as essentially the same. Both argue that capitalism is the root cause of the environmental crisis, and that the domination of nature by the capitalist class is parallel to its domination of labor. Both forms of domination—of nature and labor—emerge out of the material base of society; as David Pepper puts it, "Material production and the exchange of products constitute the basis of all society."[102] Capitalists constantly seek low-cost raw materials and low-cost waste disposal, with both "ends of the pipe" exploiting nature as though there were no costs associated with damage to the environment. Capitalism is inherently anti-ecological and anti-nature. Capitalism also alienates humans from their selves, the things they produce, and the nature that makes possible both them and things they produce. In commodifying everything, capitalism also drains social meaning from work and nature. Finally, capitalism creates "mystification" and "false consciousness" in that things as they are perceived to be is not only right and legitimate, but how things must be, now and always. To protect nature, we must exploit it: that is how it has been, and how it will always be.* To eliminate these conditions requires nothing less than the full overturning of the capitalist system.

Classical Marxism—that is, the historical materialism of Karl Marx—saw capitalism as a necessary step on the road to an eventual workers' revolution, under which the fruits of development and technology would be equitably available to all. But although capitalism might disappear, economic growth would not. An ever-expanding industrial product was central to Marxism, albeit not driven by consumer desires but by some conception of the public good. "From each

---

*This last process is often called "naturalization." Treatment of the market as a "natural" institution is an example of naturalization.

according to his abilities, to each according to his needs."[103] Liberation would come, according to Marx, when each individual's basic needs were provided for, allowing her/him to spend part of the day working and the rest on culture, study, sports, whatever. Marx and his followers were not very concerned about nature and had few illusions about it: resources were there for the taking and provided necessary inputs into the process of industrialization and production. However else "really existing" socialist countries such as the former Soviet Union failed to hew to Marxist principles, they hardly deviated from Marx's attitude toward nature. Resources were treated as a free input to production, and pollution was seen as inevitable and unavoidable when industrial growth was at stake.[104]*

According to ecosocialism and eco-Marxism, consequently, the way in which production takes place—the *mode* of production—matters for the environment because producing things is one way in which human beings interact with nature. More to the point, in making things, humans change the substance of nature into socially useful forms, via the forces of production (labor) and the means of production (technology). The social organization through which things are produced—the *relations* of production—are the most critical part of the mode of production. These relations include those political and legal arrangements that sustain social organization and particular forms of social consciousness, domination, and control. Social change is possible, therefore, only if the material base of society is changed, a process that will lead to the restructuring of its social organization. The protection of nature thus requires nothing less than changes in the material base.

In 1991, James O'Connor provided an elaboration of what he called "the second contradiction of capitalism,"[105] which addresses the damage and waste side of the equation. The first contradiction addresses the tendency of capitalism to arrive at crises of overproduction in the absence of mechanisms that either limit production of commodities or foster high levels of consumption. The second contradiction involves the rising costs of exploiting nature as supplies become more expensive to extract and social and environmental costs increase. As O'Connor puts it,

> The basic cause of the second contradiction is capitalism's economically self-destructive appropriation and use of labor power, urban infrastructure and space, and external nature or environment—"self destructive" because costs of health and education, urban transport, home and commercial rents, and the costs of extracting the elements of capital from nature will rise when private costs are turned into "social costs."[106]

The implications of these contradictions for capitalist societies go rather against the grain of those worldviews that argue that some degree of reform

---

*To be entirely fair, during the early decades of industrial capitalism and well into the twentieth century, smoke-spewing factories were an object of pride and admiration in most industrial towns and cities.

and regulation of markets will be adequate to save the environment—they will not end the continual destruction of nature. What, then, is to be done? Eco-socialists and eco-Marxists regard capitalism as fundamentally unredeemable and their programs rely heavily on revolutionary action. As David Pepper has put it, "Trying to smash capitalism violently will probably not work while capitalists control the state, so the state must be taken and liberated in some way for the service of all."[107] At the global level, and without a world government, this becomes an awesomely difficult proposition. It is a pretty forbidding one at the national level, too.

Consequently, Pepper tends to favor more localized, less complex, and ambitious approaches to social change, to wit,

> It follows that the most potentially fruitful kinds of action are those which emphasise people's collective power as producers, which directly involve local communities (particularly urban) and increase democracy, which enlist the labour movement and which are aimed particularly at economic life. . . . But to be more positive and dialectical, perhaps they ["new" approaches to production and social organization] represent part of that order whom the existing economic and social arrangements do not satisfy and which will eventually, by struggle with the existing order, produce a new socialist synthesis.[108]

The difficulty here, as with all Marxist and socialist theories, is that the state, which is essential to the operation of capitalism and is part of the problem, also becomes central to whatever the new system might be. Marx believed that the state would, eventually, "wither away," as basic needs were met and the sources of class struggle were eliminated; ecosocialists are considerably less sanguine about this prospect, especially since there are no really existing examples of such transformation. Small-scale collective action, to which we return in chapter 4, begins to sound a great deal like the ecoanarchist communities and confederal system advocated by Bookchin. Could these generate the transformation in relations of production necessary to give rise to this "new socialist synthesis" and global social change?

Since the first edition of this book, the field of "consumption studies" has blossomed,[109] and many scholars have shifted their focus from the supply side of environmental destruction to the demand side, that is, consumption. This move has been motivated by recognition that the waste stream from production is, in many ways, more amenable to reduction, since it arises from a limited number of production sites. By contrast, consumption takes place in the household or enterprise, by individuals, and the resulting waste is much more problematic and difficult to reduce. Moreover, it is the demand for goods that drives capitalism and, as consumption has accounted for growing fractions of advanced economies' gross domestic product—more than 75 percent in some cases—not only has end-of-life waste acquired a greater role in environmental destruction, it has also motivated more individualized responses and strategies, such as recycling, green purchasing, and raised consciousness. The more critical aspect of consumption studies focuses not on getting people to waste less but, rather on how capitalism motivates consumption, through forms of false consciousness, in order to increase demand and profits.

Critical transhumanist worldviews encompass a broad range of approaches, all of which share a few basic principles. In particular, the domination of people and nature by other people is a result of unequal power relations among classes, genders, races, social fractions, and humans and animals, which foster and normalize exploitation of and dispossession of land, labor, life, and resources by the stronger. These activities are "naturalized" through various rationales and predominant narratives (for example, "the poor will always be with us"; "animals have no feelings") rather than recognized and acknowledged as features of social structures and relations that the rich and powerful (frequently men) use to protect their property and extract value from culture and nature. Moreover, what you see is not what you get: the normalized and naturalized claims and practices of realist and liberal philosophies are actually forms of internalized bondage and constraint that prevent people from understanding how they are being exploited and oppressed (via false consciousness and hegemony). As a consequence, individuals and social groups support actions and interests that may be contrary to what is best for them, for example, continued reliance of fossil fuels because they are abundant and cheap, even though their burning is leading to increasingly untenable environmental conditions across the planet.

We discuss four critical transhumanist worldviews here: postcolonialism,* feminist political ecology, queer ecologies, and the new materialism. From the environmental and ecological perspective, *postcolonialism* focuses on how domination of people and nature has been propagated and internalized through the legal and social relations devised, originally, by colonial regimes and carried on, since the 1960s, notwithstanding independence of virtually all colonized countries. *Feminist political ecology* represents a critical approach to ecofeminism, in that it focuses on social structures and multiple genders while eschewing the essentialism that continues to characterize much of "women and nature" literature and practice. *Queer ecologies* seek to examine the domination of humans and nature not only through consideration of heteronormative biases and actions against non-masculine genders (women as well as lesbian, gay, bisexual, transgender, and queer) *and* nature but also to disturb and displace the social and material structures and practices that sustain discrimination and domination. Finally, *new materialism* seeks to overcome some of the idealist proclivities of many critical theories by proposing that "things," natural, animal, and human, as well as material and ideational, play a major role in shaping and even determining what humans do. Its advocates argue that

---

* "Postcolonialism" is usually defined somewhat differently than we have used it here, but we regard it as a useful term for denoting unequal distribution, control, and development of property arising from social relations of domination and subordination. The *locus classicus* of postcolonial theory is Edward Said, *Orientalism*, New York: Pantheon Books, 1978; and *Culture and Imperialism*, New York: Vintage, 1994.

the omission of material "actants" from analyses and policies constitutes yet another form of discrimination and domination. According to Dianne Rocheleau, these worldviews recognize that, as she puts it, "The center of gravity is moving from linear or simple vertical hierarchies (chains of explanation) to complex assemblages, webs of relation and 'rooted networks' with hierarchies embedded and entangled in horizontal as well as vertical linkages."[110] Needless to say, this complicates our understanding of the world considerably.

What might be called "green" postcolonialism[111] begins with European extraction of resources from colonial territories for export to the metropole, with little or no regard for either the people or nature in those colonized spaces. These incursions and occupations were justified on the basis of "bringing enlightenment" to non-European peoples even as political, social, and economic systems and institutions perpetuated the domination and oppression of indigenous peoples. Huggan and Tiffin point out that, "The operation of the European empires both initiated and depended upon a globalism that still provides the . . . foundation for the highly diverse interconnections grouped as 'globalization' today."[112] And, they continue,

> The naturalization of uneven development relegated colonized peoples to a stage in the European past, and the geography of difference was reconfigured as a living history of the rise of civilized man, with European development as the natural goal. Different cultures, with very different notions of time, all found themselves on the lower rungs of the ladder of progress, wrenched out of a time of land and ancestry and subjected to the exigencies of Greenwich mean time or, in its modern form, 'corporate time.'[113]

In writing about postcolonialism, Pablo Mukherjee proposes that

> any field purporting to theorise the global conditions of colonialism/imperialism, decolonization and neo-colonialism (let us agree to call it "postcolonial studies") cannot but consider the complex interplay of "environmental" categories such as water, land, energy, habitat, migration, with political or cultural categories such as state, society, conflict, literature, theatre, visual arts. Equally, any field purporting to attach interpretative importance to "environment" (let us call it eco/environmental studies) must be able to trace the social, historical and material co-ordinates of categories such as forests, rivers, bio-regions and species.[114]

Green postcolonialism converges with a number of worldviews already considered in this chapter as well as a variety of disciplines other than political science. Environmental injustice is a consequence of histories, categories, and forces similar to those listed by Mukherjee. New materialism, described in the following, is focused on features of postcolonialism, while "world ecology" studies examine the historical plundering of nature by capital.[115] Green postcolonial scholars come from geography, sociology, anthropology, and literature, among other fields.

Theorizing green postcolonialism is one thing; operationalizing it is quite something else. A key feature of this worldview is the proposition that, only through repossession of control over material goods, including land and natural resources, can unequal power relations be overturned, oppressed societies freed of the constraints imposed upon them, and true emancipation (from capitalism) be achieved. This is seen most clearly in indigenous sovereign movements and campaigns against resource extraction projects, such as Standing Rock, and which are discussed in subsequent chapters.

What was once called "ecofeminism" is today encompassed by feminist political ecology.[116] The change in terminology reflects a more complex understanding of gender and the recognition that masculinity seeks to dominate more than just women. In the past, and in many institutional documents even today, women were characterized essentially in terms of their "links" or "likeness" to nature—these having to do with biology, temperament, patterns of thought and reasoning, and so on—and were, thus, treated by men with as little regard as men treat nature. But the recognition that men treat both women and nature badly points toward a structural problem, rather than one that can be addressed through modification of individual male attitudes and behaviors. For that reason, it is useful to look at the ways in which some ecofeminists, such as Ariel Salleh, have applied Marxist reasoning to their analyses.[117]

Whether abuse of women or nature came first is unclear and probably unimportant, but for all ecofeminists the domination of women and nature are inextricably linked. As Salleh puts the point: "Feminine suffering is universal because wrong done to women and its ongoing denial fuel the psychosexual abuse of all Others—races, children, animals, plants, rocks, water, and air."[118] From the structural perspective, the women-nature linkage is not only characteristic of patriarchy and rationalism, which, together, have generated capitalism, it is also central to the maintenance of contemporary social organization in which men and capital hold power. In other words, reform is not enough; as with ecosocialism and eco-Marxism, change must be foundational and must begin with the overturning of patriarchy.

Salleh's version of this structuralist perspective is an intriguing one, although she tends, perhaps, to present women's suffering under patriarchy as identical to the suffering of nature under the same arrangements. Thus, she argues that women's relationship to nature is not an ontological one that claims women to be, somehow, more sensitive to life and nature for biological reasons—which is a position adopted by some ecofeminists. But Salleh nonetheless does argue that "women North and South tend to arrive quite readily at ecofeminist insights as a result of the conditions they live in and the physical work they do." Indeed, according to Salleh, women's labor is designed to protect life, unlike the work of men, and "Women's ecological commitment is fed by an intimate biocentric understanding of how people's survival links to the future of the planet at large."[119] This "intimate biocentric understanding" is not biological, however; it grows out of the very practices of production and reproduction in which women are constantly engaged. Their very experience of oppression, like that of both the working class and nature (were it to have such awareness), provides women with the insights necessary to resist that oppression.

How did such a state of affairs come about? According to Salleh's analysis, the contemporary ecological crisis can be ascribed directly to a "Eurocentric capitalist patriarchal culture built on the domination of nature and the dom-ination of Women 'as nature.'" This grows out of the Enlightenment project of applying rationalism to the management of nature (a theme prefigured in Hobbes's *Leviathan*). "In the West, feminine and other abject bodies are split off and positioned as dirt, Nature, resource, colonised by masculine energies and sublimated through Economics, Science and the Law." Therefore, Salleh continues, "An ecofeminist response to ecological breakdown means finding ways of meeting human needs that do not further the domination of instru-mental rationality." For Salleh, therefore, ecofeminism offers a comprehensive progressive approach to the ecological crisis: "Ecofeminist politics is a femi-nism in as much as it offers an uncompromising critique of capitalist patriar-chal culture from a womanist perspective; it is a socialism because it honors the wretched of the earth; it is an ecology because it reintegrates humanity with nature; it is a postcolonial discourse because it focuses on deconstructing Eurocentric domination."[120]

Ecofeminism has been charged by some with "essentialism" and gyno-centrism—a charge denied by others[121]—and further criticized for pursuing a "flight from nature"

> relentlessly disentangling "woman" from the supposed ground of essential-ism, reductionism, and stasis. The problem with this [feminist] approach, however, is that the more feminist theories distance themselves from "nature," the more that very nature is implicitly or explicitly reconfirmed as the treach-erous quicksand of misogyny.[122]

Feminist political ecology goes beyond Marxist and even much critical analysis to address *gender* rather than just women, and to critique heteronormativity as a technique of violence, discipline, and domination.[123] It "treats gender as a critical variable in shaping resource access and control, interacting with class, caste, race, culture, and ethnicity to shape processes of ecological change"[124] and analyzes "the connections among racism, sexism, classism, colonialism, speciesism and the environment."[125] We can argue, with considerable justifi-cation, that even the "capillaries" of social power are clogged or infected by the normative discipline imposed by the ways in which the webs of relations among humans and nature are characterized in all walks of life. The grounds for action through feminist political ecology must, therefore, be largely "on the ground," in those movements, oppositions and resistances that identify and practice counter-normativity.

Queer ecologies seek to do much the same as feminist political ecology. They are explained by Catriona Mortimer-Sandilands as

> a loose, interdisciplinary constellation of practices that aim, in different ways, to disrupt prevailing heterosexist discursive and institutional articulations of sexuality and nature, and also to reimagine evolutionary processes, ecological interactions, and environmental politics in light of queer theory.[126]

Mortimer-Sandilands quotes the work of Rachel Stein, who writes that,

> by analyzing how discourses of nature have been used to enforce heteronormativity, to police sexuality, and to punish and exclude those persons who have been deemed sexually transgressive, we can begin to understand the deep, underlying commonalities between struggles against sexual oppression and other struggles for environmental justice.[127]

How are these "institutional articulations of [masculine] sexuality and nature" expressed? Mortimer-Sandilands offers the example of U.S. national parks as sites that embodied a "specifically *masculine* ideal of nature . . . born from a gendered and racialized view of nature . . . [that] literally erased aboriginal peoples from the landscape."[128] Much of the initial impetus for designation of national parks came from widespread white European male beliefs that cities were unhealthy places and morally corrupt as a result of immigrants and non-whites, who undermined the health and well-being of middle- and upper-class men and their heterosexuality. "Aggressive nature recreation—sport hunting, mountaineering—[were described] as a means through which to affirm their virility . . . the parks came to embody a specifically *masculine* idea of nature, one that excluded women, the urban working class, and non-Europeans."[129] This practice was replicated, as well, in Africa, where nature was "protected" in order to maintain the population of big game for colonial-era hunters.

These ideals continue to have effects today. How they are to be resisted, opposed, and overturned is less clear. Mortimer-Sandilands points to the history of intentional communities, in places like Oregon,

> based on the combination of ecological and lesbian separatist politics . . . transforming heterosexual relations of property ownership; . . . withdrawing the land from patriarchal-capitalist agricultural production and reproduction; to symbolically reinscribing the land with lesbian erotic presence.[130]

Later in this book, we will consider the implications of an ecological strategy based on locality, much like the ecovillages and anarchist communities described earlier in this chapter. One of the greatest challenges to such (utopian) communities is the difficulty of becoming "self-sufficient" and minimizing external relations with the global heterosexist division of labor. Time and again, highly committed groups have sought to break away from industrial society only to discover that they are unable to adapt to the lower standard of living this implies. Yet, it may well be that only a "bottom-up" strategy of ecological politics and practice can change the world.

The *new materialism* motivates us to ask: What is materialism? And why is this one "new"? Conventionally, as discussed earlier in this book, materialism is contrasted with idealism, the former concerned with the world as it really is, the latter with the world as we might like it to be. There is a tendency, in much philosophical theorizing, to regard the material world as inert, with no free will or agency, and something that can be manipulated, especially by humans and their technologies. The term also arises in the concept of "historical materialism," through which Marx and others argued that human social organization and development are the dialectical result of a society's mode of production and its social relations of production. This should be contrasted with the widespread idealist view that ideas alone drive history, humans, and the world. We use the term, in chapter 1, in a rather different sense: that the material infrastructures and actions of one generation are strongly shaped by those of preceding generations, and that without knowledge of this past, one cannot understand the present (or contemplate the future).

The new materialism is "new" for several reasons. First, it brings back much of the "old" materialism in asserting the importance of the material world in human choice, action, and understanding. Second, rather than focusing on structure or agency as explanations of social stasis and change, new materialism examines daily and repeated *social practices*, that is, what beliefs about and actions in the material world arise from social normalization of particular ways of doing—what we do matters, rather than who we are or how much stuff each of us has. Third, the new materialism attributes agency, albeit limited, to nonhumans and things—following Bruno Latour,[131] the term "actant" is often applied to this phenomenon—and *relations* among them. Finally, the new materialism is skeptical about the causal claims that underpin explanations of how and why things happen.

Rather than trying to explain each of these points, it might be more helpful to tell a story, about electricity, lamps, and rivers. Let us imagine the prime minister of a country which, for whatever reason, has an unreliable electricity grid. One night, the lights go out in the capital city. Everyone says, "Enough! We must build more power plants!" In this instance, the malfunctioning lights "act" to motivate the construction of more plants (construction is effected through a complex arrangement composed of people, things, and relationships among them). It so happens that a river runs through a valley near the capital, and the river and valley "act" to motivate the engineering design of a hydroelectric dam to be built at the head of the valley. When construction does finally begin, the particular topography and geology of the dam site will "act" on the project and might even force design changes. Design and construction happen through the practices of hydraulic and civil engineers, not because the dam is their idea, but because that is what such engineers do.

Compare this new materialist story to a more conventional, normalized telling. Energy analysts in a country far from our place of concern are working for the World Bank on an assignment to evaluate the power supplies and demands in the Global South through the year 2050. They must acquire data to conduct this analysis and make assumptions about future demand in order to prescribe electricity supply solutions. For our country of concern, they also examine potential energy sources and decide that a hydroelectric dam at the head of the river valley near the capital is an optimal site. Eventually, the ministers of finance and interior of our country of interest will meet with World Bank officials to negotiate a loan that will make dam construction possible. Of course, there will be many others involved before the dam ever sees the light of day, but you get the point. Here, the construction of the dam is explained as the result of a series of ideas and proposals, rather than as a reaction to a concrete, material incident—the blackout; geology—contours of the land; and geography—the river.

What do we gain from the new materialism that is not offered by other worldviews presented in this chapter? For one thing, we begin to recognize that choices and actions can be affected, if not determined, by actants in the material world around us. When we plan for action, failure to take such actants into account may result in failure of our plan. More than that, ignoring the existence of actants is an exercise of power that may well lead to domination and oppression of others. For instance, if that river valley is full of people—as it is likely to be—and the deciders at the World Bank and in the capital ignore this material "fact," there is a strong possibility of protest, resistance, and even violence, not to mention building of alliances with supporters around the world. Indeed, this very approach to the construction of large dams in various parts of the world has intruded on the lives and livelihoods of indigenous peoples who do decide to fight, and sometimes win, in the effort to protect and maintain their material ways of life. In doing so, we might note, these groups are likely to invoke notions of sovereignty and right, as practiced and claimed by states, in order to demonstrate the legitimacy of their cause and the vital nature of their land. Is the "new materialism" really new? Perhaps it is more accurate to say that this particular worldview generalizes what we learn when hitting our thumbs with a hammer: it has a mind of its own and often does not go where you aim it (especially when you are using that hammer on all of those things that look like a nail).

SOME THOUGHTS BEFORE LEADING TO A CONCLUSION

These many different and seemingly incompatible worldviews force us to ask: What is the truth, what is true? What does this mean for the global environment and global environmental politics? Is there no one epistemological trail we can follow to analyze, understand, and address the environmental crisis and its many features? And since we are, presumably, examining the same phenomena and associated data, how is it that disparate worldviews give such different accounts? The way in which a problem or issue is defined and understood has a great deal to do with the way in which it is addressed. Indeed, framing an issue in a particular way often points to particular types of solutions, some of which might lead to unintended or even catastrophic outcomes.[132] But our world is even more complicated than this.

The ontological propositions and beliefs that underpin these worldviews lead to different "solutions" to issues as they are defined through that worldview. Consider the ways in which different worldviews might treat "people." We "know" that there are roughly 7.5 billion people on Earth, and that a major fraction suffer from malnutrition. Malthusianism sees a problem of "too many people" in a world of limited food and resources. Fewer people in the world would alleviate the problem. A liberal might say that the problem is not "too many people" but many people without purchasing power—therefore, development of employment opportunities, income, and food markets is the solution. A redistributionist would argue that the issue is not "how many people" or how much money they have, but how widespread poverty is due to the rich dispossessing the poor. The former will have to yield much of their wealth to improve the lot of the latter. A political ecologist would argue that unequal power relations and stacked institutions create the structural forces that immiserate so many people in the world, and that social change is the answer. You can engage in this exercise for other worldviews.

To return to earlier arguments, we believe that the critical point here has to do with *power*: its forms, its distribution, how it is wielded, how it is normalized and naturalized, how it is deployed to the advantage of some and disadvantage of others. To argue that animals have no feelings and, on that basis, legitimate raising and slaughtering them, is a reflection of power. But the growth of vegetarianism and veganism across the United States is also a reflection of power, albeit of a different kind. We have no robust theories of social change or understanding of why some long-established practices are in decline or have disappeared (for example, social smoking; driving by young adults) while others have emerged (for example, recycling; walking while texting). Some kind of social power is at work here—we often call it peer pressure, but that is not exactly the correct term—having to do with "doing" rather than thinking, choosing, deciding, or acting. And there is no denying that point: to deal with the global environmental crisis, we do have to change what we, individually and collectively, *do*. More on this in the chapters that follow.

FOR FURTHER READING

Clapp, Jennifer, and Peter Dauvergne. *Paths to a Green World: The Political Economy of the Global Environment*. Cambridge: MIT Press, 2005.

Dryzek, John S. *The Politics of the Earth: Environmental Discourses*, third edition. Oxford: Oxford University Press, 2013.

Guha, Ramachandra, and Rukun Advani. *How Much Should a Person Consume?: Environmentalism in India and the United States*. Oakland: University of California Press, 2006.

"International Political Economy of Everyday Life." I-peel.org.

Litfin, Karen. *Ecovillages: Lessons for Sustainable Community*. Cambridge: Polity, 2013.

Mason, Paul. *Postcapitalism: A Guide to Our Future*. New York: Farrar, Straus and Giroux, 2015.

Newell, Peter. *Globalization and the Environment: Capitalism, Ecology and Power*. Cambridge, UK: Polity, 2012.

O'Neill, Kate. *The Environment and International Relations*, second edition. Cambridge: Cambridge University Press, 2017.

Paterson, Matthew. *Automobile Politics: Ecology and Cultural Political Economy*. Cambridge: Cambridge University Press, 2007.

Rothschild, Emma. *Economic Sentiments: Adam Smith, Condorcet, and the Enlightenment*. Cambridge: Harvard University Press, 2001.

3

Capitalism, Globalization, and the Environment

In chapter 2, we examined contrasting, and often conflicting, worldviews of global environment and global environmental problems and politics. In this chapter we will focus on the relationships among nature, society, resources, and markets in a capitalist system and on capitalism and its role in environmental degradation. Although capitalism has long been international—beginning perhaps as long ago as the 1500s[1]—it is only over the past thirty years or so that capitalism has become fully manifest through "globalization."* As we shall see, the links between globalization and global environmental degradation are sufficiently compelling to suggest that the first is causally implicated in the second.[†]

We begin with a discussion of markets as seen through the lens of neoclassical economics. The primary objective of production in contemporary capitalism is the *accumulation* of capital (or money) that can either be reinvested in the production process or be distributed to its owners (capitalists).[‡] The capitalist seeks to minimize the costs of inputs into the process so as to maximize the difference between the cost of production and the price of the good in the

---

*At this writing, with the rise of new nationalisms, some are arguing that the end of globalization is at hand. Inasmuch as capitalism has, if not always, been "global" for centuries, there is good reason to question such claims.

†It is often observed that socialism, as it existed in the old Soviet Union and other communist countries, was much harder on nature and that, by comparison, capitalism does an excellent job of protecting the environment. While this was the case in many countries under communism, it is not a reason to treat capitalism as without fault or impact.

‡Neoclassical economics (and economists) assume that accumulation is a virtuous social goal, since profits can be reinvested to employ workers, generate more profits, and contribute to overall social good. But there seems to be a limit to what can be done with large accumulations of capital: after the first billion or two, capital can either be spent on activities other than consumption (for example, lobbying, bribery), charity (for example, medical research and development), or speculation (for example, bonds, futures, derivatives). For a heterodox view of capital and accumulation, see Nitzan and Bichler, *Capital as Power*.

market, a difference that is called "profit." According to neoclassical economics, which is free market capitalism's analytical and explanatory framework,
environmental problems occur if resource inputs are underpriced and their real
value is understated relative to other input costs. The solution to this conundrum is to put a price on everything, including the environment.

In the second section, we turn to the *consumption* side of the equation (to
which we shall return in later chapters). Under a system of market capitalism,
for goods to sell there must be buyers. As discussed in chapter 2, neoclassical
economists generally assume that goods are "scarce," which is to say that they
are not so plentiful as to have no price. But what happens if there are more goods
than there are buyers willing to consume? Then both profits and production
may collapse, and accumulation will stop. To avoid this dilemma, people are
encouraged to consume. But consumption generates waste, and waste can hurt
the environment as well as people's health. The answer is to include the costs of
environmental damage in the price of the product. But if that price becomes too
high, consumers will stop buying. It is difficult to find a way out of this dilemma.

In the third section of the chapter, we investigate *commodification*, and
the ways in which all things, it would seem, can be turned into goods that
can be sold in markets. In common parlance, a "commodity" is something
that: (1) is produced in high volume by a large number of producers; (2) is
subject to highly competitive pricing; and (3) as a result, generates revenues in
excess of costs (that is, profits). Examples of commodities are mass-produced
cars, bananas, iPhones, and shampoo. The trick is to sell as many of these
near-identical items as possible. "Commodification" is also applied to a growing number of goods and services not previously available through competitive
markets—laser eye surgery, trips to Antarctica, pollution permits—but that
can generate large profits after their initial introduction. Commodification of
"positional goods"[2]—those that convey status and are desired by the well-
off—helps to advance accumulation and generate super-profits. Putting a price
on things gives them monetary value and leads to the assumption is they will
be treated more carefully than things without a price. But it also means that
noneconomic values, such as aesthetics or ethics, don't count for very much
unless they can also be monetized. Once we put a price on an intangible, such
as clean air, we will still have other questions to face. For example, is it ethical
to sell "rights to pollute," especially if that results in environmental injustices
against those who cannot afford to buy emission permits?

In the final section of the chapter, we address the relationship between *globalization* and environmental issues. Globalization involves, in part, the expansion and deepening of transnational capitalism and its extension to things not
previously commodified as well as to places[3] and peoples who have not yet
been deeply integrated into market relations and economies. Although there
is nothing especially new about globalization when defined in these terms,
it has taken on new significance because, as more people are pulled into the
orbit of markets, with their emphasis on consumption, the growing volumes
of waste and water pollution that result have ever-widening environmental

consequences. At the same time, however, globalization might also play a role in "ecological modernization" through the reconstruction and social reorganization of production and consumption under capitalism. If resources can be used much more efficiently, if wastes can be reduced in volume through engineering and recycling, if society can establish a new ethical relationship to nature, perhaps environmental crises can be avoided or ameliorated. But there are many "ifs" in this equation, as we shall see.

Capitalism is central to our understanding of global environmental politics because, at the end of the day, it remains the dominant form of economic belief and practice on the planet. Capitalism is deeply enmeshed in the production and reproduction of virtually all societies on Earth, and its smooth functioning is of central concern to virtually all governments on Earth. Growing demands on the planet's resources come, as we have seen, primarily from wealthy human societies, and if the poor do become wealthier, as proponents of sustainable growth and development hope, their demands and consumption will be added to the world's environmental burden. Whether such a capitalist future is sustainable, let alone survivable, remains very much open to question.

One additional point: The market system does not exist as a "natural" institution, self-organizing and self-realized,[4] no matter what its most ardent believers claim.[5] At the theoretical (and fantasized) extreme, free markets are devoid of regulations imposed by governments or other authorities and governed only by mutual interest and respect. But all markets are organized on the basis of social rules, including those that underpin property rights, so it is not far-fetched to imagine that environmentally friendly markets might be bound up in restrictive rules and regulations or based on an ethical system that imposes social limits on production and consumption.[6] For the moment, however, environmental externalities do not simply happen; they are the result of the ways in which production and consumption are organized under capitalism and the ways in which producers and consumers and the rules and practices they follow disregard those externalities. That environmental degradation is, consequently, not a natural result of markets but rather a result of the ways in which markets are organized and operate, suggesting that it might be possible to organize capitalism in ways that are more environmentally friendly. Yet, it is also possible that an "environmentally friendly" capitalism would not resemble very much the arrangements that are currently in place. Indeed, as seen in the authoritarian capitalist economy of the People's Republic of China, some forms of capitalism might not be characterized by "free" markets as we understand them today.

MONEY, THAT'S WHAT I WANT!*

To fully clarify the relationship of free market capitalism to environmental problems, we need to take a brief analytical detour into neoclassical economics.[7]

---

*An informal introduction to neoclassical economics and critiques of it can be found in: Ronnie D. Lipschutz, *Political Economy, Capitalism and Popular Culture*, Lanham, MD: Rowman & Littlefield, 2010.

Neoclassical economics is derived, in part, from the premises of liberalism as discussed in chapter 2, and is based on assumptions of individualism, self-interest, scarcity, and an innate human desire to exchange goods and services (in Adam Smith's words, "truck and barter"). Markets are composed of individuals (or corporate bodies) who seek to acquire goods that will improve their well-being or status. Such goods are, by definition, scarce—otherwise, no one would desire them and there would be no need for a market—and market participants either trade some goods for other goods (barter) or bid a price for the goods they want (exchange). At a price that supposedly reflects a balance between supply and demand, exchange takes place, and buyer and seller are both satisfied. Adam Smith pointed out that such a system of exchange lacks a central organizer or authority yet is able to greatly increase satisfaction and happiness in society. He called this mechanism the "invisible hand,"* for it seemed as though self-interested individuals, pursuing their own desires, nonetheless behaved in such a way as to produce socially beneficial outcomes.[8]

Neoclassical economics, as applied to capitalism, is also based on the assumption not only that markets will clear—that is, that supply and demand will match at an appropriate price and all the goods produced will be sold—but also that producers can make profits through meeting demand and selling above the cost of production. Clearly, producers who lose money on every unit they sell have only limited incentive to keep supplying that good to consumers—although in the short term it might be better to operate at a loss than to cease production entirely. The drive for profit and accumulation is, therefore, the engine of capitalism, and minimization of the costs of production is central to the process.

Under the postulates of neoclassical economics, it is *scarcity of supply relative to demand* that sets prices in markets and is meant to clear them. People will bid for those things they need or want, and those who can afford to do so will pay more if the object of their desire cannot be bought at a lower price (consider the way house prices are often overbid in a hot property market). A rising price is generally taken to indicate a shortage of the resource or good, as buyers compete for a limited supply (although rising prices may also indicate deliberate efforts to hold goods from the market for hoarding or speculation, which induces apparent scarcity). When a good becomes excessively costly in a market, and buyers are no longer willing or able to pay the market price, one of several things can happen. First, a rising cost may motivate more careful use of the resource, as consumers find they can no longer afford to purchase as much as previously ("conservation"). If, for example the price of gasoline rises from $2.50 to $5.00 per gallon, and your vehicle only gets twenty miles per gallon, chances are that you and others will drive less ("elasticity of demand").†

---

*Although the term is mentioned only once in *Wealth of Nations*.

†As we noted earlier in this book, in most countries in Europe gasoline costs between eight and ten dollars per gallon, mostly because of high taxes. This limits consumption but to a much smaller degree than one might expect. Cars have become more fuel efficient and, combined with inflation, the relative cost of gasoline has not escalated that much.

As consumers use less of a resource, supply will drop back into balance with demand and prices will decline. The market will come back into equilibrium until demand at the lower price begins to outstrip supply again, when the cycle will be repeated.

Second, if supply and demand are more or less unbalanced for a long time, and retail prices remain high, producers will see an opportunity to make money by supplying the unmet demand—at least until new production by competitors kicks in. The potential returns on investment associated with the higher price of a good will motivate the search for or production of new supplies of the resource. If no new sources can be found at a reasonable cost, alternatives may become attractive and provided for in the market ("substitution"). Such alternatives could be replacements for the good or resource in shortage (for example, alternative fuels for vehicles) or technological innovations that use less or none of the good in shortage (for example, hybrid or electric cars).

Third, if demand remains low for an extended period of time, prices for the good or resource will decline. Some producers may go bankrupt or be purchased by other producers, who can then exercise more control over supply.* But market equilibria can also occur when prices are low, especially if more efficient technologies have penetrated markets and reduced overall consumption (for example, more fuel-efficient automobiles). As we will see, there are numerous techniques used by producers to prevent decline in demand and support high levels of consumption.

The logic of the neoclassical perspective on scarcity suggests one peculiar conclusion: one can never "run out" of anything. When resources grow scarce, rising prices will motivate a search for new sources or substitutes that provide the same service.[9] To a point, there is considerable evidence to support this proposition. When oil and natural gas became costly during the first decade of the twenty-first century, fracking actually produced so much in new resources that markets became glutted, prices declined, and the United States began to export oil. When cobalt from Africa used in magnets became very expensive during the late 1970s as a result of conflict in what was then Zaïre (now the Democratic Republic of Congo), new magnetic materials (alnico) were quickly developed. These days, solar electricity is rapidly growing and displacing traditional coal-fired electricity generation. These examples seem to suggest there is no need to worry about "running out."[10]

But scarcity is a tricky thing: during the Great Recession in 2008, the world price of oil climbed to almost $150/barrel, an apparent indication of depleted supplies. Was this the result of the aftermath of peak production, that point at

---

*This is how John D. Rockefeller's original Standard Oil cartel was created: facing a glutted market, small oil producers in Pennsylvania went bankrupt and were purchased by Standard Oil, which was then also in a position to manage supplies and drive the price of oil back up. An interesting gloss on this is Upton Sinclair's *Oil! A Novel*, New York: Grossett & Dunlap, 1927, which later inspired the 2007 film *There Will be Blood*. The same process has happened with fracking in the United States.

which global oil production would begin to decline? Was it a consequence of speculation by oil traders and others, seeking a more reliable store of future value than the dollar which looked shaky? Or were oil-producing countries and companies somehow holding oil off the market in order to force prices up? No one seemed certain. Over the following years, oil prices began to decline and, in May 2018, were around seventy dollars per barrel, reflecting oversupply. In real terms, oil and gasoline are probably as cheap as at any time in U.S. history. Nonetheless, it seems likely that at some point in the future deposits of relatively low-cost minerals will have been fully exploited, and global resource and commodity prices will be likely to rise.[11] Substitutes, or substitute activities, will have to be found for these depleted substances or the world will have to do without.* Such logic clearly applies to nonrenewable resources (at least, so far), although the availability of many resources is as much a matter of ability to pay for them as it is of supply. In countries where gasoline is often scarce or expensive, the rich rarely have to go without. It is the poor who are reduced to walking.[12]

It is less clear that the economic logic of substitutability applies to air, water, or "renewable" resources, such as fisheries, species, and forests. Renewable resources are those that reproduce as part of natural cycles. They are constantly being replenished and appear, in principle, to be limitless. Left alone by humans, the hydrological cycle supplies the water that nature demands, and the natural rate of forest destruction roughly equals the rate of regeneration (although forest cover can change or even disappear over centuries or millennia). But with humans in the equation, such "balance" holds true only so long as the rate of human extraction does not exceed the rate of natural replenishment. Once populations are destroyed or rendered extinct, renewable resources, such as wild fish, are essentially irreplaceable. Furthermore, at some point, the cumulative damage to renewables from mostly ignored social and environmental impacts, such as pollution, soil erosion, or climate change, begins to add up. If climate change makes certain parts of the world inhospitable to plants and animal species that have long been found there, they might well disappear entirely. The limited capacity of renewable resources to absorb such impacts is today being overtaxed, and the eventual effects on human as well as ecosystemic health and welfare are of particular concern.[13]

For the purposes of our neoclassical discussion, however, the problem is that renewable resources have never been regarded either as scarce or as belonging to anyone. There have been lots of water and trees and animals, and "pollution space" has been more than ample for human needs over history. These renewable resources have not been bought and sold at prices reflecting their true value, which, in any event, would be extremely difficult to assess. Consequently, producers and polluters have had little need to use these resources more carefully or in lower quantities. Neoclassical economics has an answer to this problem—see, for example, the Coase theorem discussed in chapter

---

*Predictions of looming scarcity have been made during almost every decade since 1900; see Lipschutz, *When Nations Clash*.

2—but this requires what amounts to active intervention into the operation of markets, something that many free-marketeers abhor. That is why they prefer finding the "proper" value and prices for resources.

Historically, as a result, natural resources have been treated largely as free for the taking with externalities, such as pollution, imposing no costs on extraction, processing, or production.* Why? For Smith and other classical political economists (including Marx), nature and its resources were given by God or Nature. Without human intervention, nature had no economic value until it was transformed into goods that could be exchanged.† No one cared very much about the despoliation of the landscape, or the pollution emitted in the production process. Those were not things that could be exchanged in markets, anyway. A similar logic applied to agriculture, forestry, and any number of other resource-related activities.

In the past, neoclassical economics treated resources and nature only as inputs into production, whose costs include only what it takes to acquire and process those materials (for example, machines, land rent or purchase). These inputs, or *factors of production*,‡ whether raw materials, agricultural commodities, manufactures, or services and information—and which also may include labor, capital, technology, and knowledge—appear in the manufacturer's costs. But what is sometimes called "natural capital," the intrinsic value of nature, does not.§ The cost of a resource as an input to production has, historically, been limited largely to the expenses of paying for access (sometimes called "rent" or a "royalty"), extraction or removing the resource from its original location, transforming it into usable form, and transporting it to sites of further transformation and production. When rents or royalties have been paid on minerals or oil, they have generally been kept low so as to encourage high levels of extraction and maximize short-term revenues.[14]

Externalities: Things for Which No One Pays

How do neoclassical economists calculate the cost of a resource? This quantity is expressed generally as follows:

---

* This is not to say that there have been no environmental laws in the past; in 1306, the English King Edward I banned the burning of sea coal, to no avail. See, for example, Jaz Brisack, "Edward I: Environmentalist by Accident," *PlanetGreen.org*, March 19, 2012, accessed August 1, 2017, http://www.planetgreen.org/2012/03/edward-i-environmentalist -by-a.html.

† This is basically John Locke's argument in *Two Treatises of Government*.

‡ "Factors of production" are the basic elements necessary to create usable things: labor to do the work, capital to pay wages and material costs, technology to facilitate creation, and knowledge about how the process is conducted. For more details, see: "Factors of Production," chapter 2.1, in *Principles of Economics*, Minneapolis: University of Minnesota Libraries Publishing, 2011, accessed August 2, 2017, https://open.lib .umn.edu/principleseconomics/chapter/2-1-factors-of-production/.

§ This is not the same as arguments about the ethical intrinsic value of life or nature; it refers to the value of that which can be provided only by natural processes or "ecosystem services," and are difficult to replace or regenerate.

$$A + B + C = D,$$
where
A = costs of capital (technology and other such inputs);
B = costs of labor;
C = costs of extraction, processing, royalties, licenses;
D = total cost of production.

As noted earlier, what has not appeared historically in such calculations is the monetary value of any environmental damages or social impacts, unless these materially affect the capitalist's investment.[*] Under historical practices, owners of private land could do with it as they saw fit. No one had any incentives to "own" pollution and reduce or eliminate it. Damages to the environment and health of workers were costs imposed on nature and society, appearing as unwanted outputs, dumped in air or water, or manifest as soil erosion or toxic wastes or ill health. The marginally greater value of each unit of resource extracted, due to depletion,[†] was never included. Finally, the cost of disposal of wastes, both toxic and benign, would appear elsewhere in the economy, or not at all.[‡] In other words, in the absence of taxes, fines, or other regulatory costs, the producer did not have to pay for the lost value of the despoiled landscape or polluted water or final disposal, or any negative impacts on those living near production sites (thus producing and reproducing environmental injustices).[§]

The subfields of ecological and environmental economics[¶] seek to incorporate environmental and social impacts in production costs, which requires adding to the above equation the following terms:

---

[*] For example, if an impure metal solution were to trigger an explosion in a refinery and destroy it, or a process required pure rather than polluted water.

[†] Depletion refers to the declining volume of a resource at both an extraction site and in general. In practice, each additional unit extracted costs a bit more than the preceding one, which is called the "marginal cost of production." Since it is difficult, if not impossible, to measure this marginal increase, the cost of extraction is conventionally averaged over all minerals from a single source, which is called "average cost."

[‡] The common phrase for this is "privatization of gains and socialization of costs."

[§] Much of the romantic movement of the 1800s was a response to the impacts of industrialization on people and landscapes, including the work of both Karl Marx and Charles Dickens, and even artists and musicians of the time.

[¶] Ecological economics and environmental economics are not precisely the same, insofar as the former incorporates notional values of nature into its calculations, while the latter is more concerned about the costs of resource supplies and externalities imposed on society. The journal *Ecological Economics* defines its focus as "extending and integrating the study and management of nature's household (ecology) and humankind's household (economics)" (see https://www.journals.elsevier.com/ecological-economics/, bold in original). According to the U.S. Environmental Protection agency, "environmental and natural resource economics is the application of the principles of economics to the study of how environmental and natural resources are developed and managed" (https://www.epa.gov/environmental-economics).

$$A + B + C + [S + P + I + H] = D,$$

where
S = scarcity value due to depletion;
P = costs of pollution and disposal generated by the production process;
I = intergenerational costs of environmental damage and depletion;
H = costs to people's health as a result of environmental damage and depletion.

These costs include not only negative impacts on the environment, such as air and water pollution, but also the monetary and social value of things irreplaceable to the present generation as well as the costs of the absence of resources to future generations.[15] Thus, for example, the logging of an old-growth forest involves not only damage to ecosystems, soil, water, plants, and animals, but also destroys the present value of no longer having those forests and the environmental services they provide (soil retention, food, and shelter for other species). Also excluded are the costs, to future generations, of not having those forests as a result of their elimination. All of these costs are left out of what the producer pays to acquire, cut down, transport, and process a tree. As long as such factors seem not to affect anyone in direct physical or monetary terms, there are presumed to be no costs associated with them. The absence of these costs then shows up as lower expenditures or, conversely, higher returns on investment (profit). Environmental damage fosters accumulation and makes societies wealthier!*

As explained in chapter 2, the social and environmental costs associated with resource use but not included in financial calculations are called *externalities*. A negative externality is an impact or effect borne by those who do not benefit from the processes or activities that cause the environmental damage; a positive externality is a benefit for which the producer does not pay—what is negative for some is positive for others. For example, the sulfur dioxide wafting across the fence of a power plant and drizzling down on nearby houses constitutes a negative externality for those who live in the houses and a positive one for the power plant owners. The fact that air—that is, pollution space—is free and is not incorporated into the costs of power production is a benefit for the plant owners. In this instance, it is commonly said that the market is operating "imperfectly," because of a lack of information about the distribution of costs and benefits and the difficulty of getting someone to pay for those costs.

---

*Paradoxically, perhaps, the costs of recuperating after natural disasters show up as additions to gross domestic product! See Drake Bennett, "Do Natural Disasters Stimulate Economic Growth?" *New York Times*, July 8, 2008, accessed September 14, 2017, http://www.nytimes.com/2008/07/08/business/worldbusiness/08iht-disasters.4.14335899.html; and the Adbusters uncommercial, "The GDP Goes Up," accessed September 19, 2017, https://www.youtube.com/watch?v=0hTtAfjBhyo&list=PLLWWwP1Ombs3_EB0kIm6KdmVrsbAS4anA&index=2.

In a "more perfect" market, such costs would be included in the price of the good or service. As mentioned in chapter 2, the Coase theorem provides, via neoclassical economics, a theoretical solution to this problem that is claimed to be neutral in regard to who pays for those costs.[16] The objective under Coase is to eliminate the externality, not to be fair or just. The logic of his argument is that the costs of pollution from a plant should be incorporated into the costs of production—a process called *internalization*—so that the producer no longer receives a free benefit and the plant's neighbors no longer have to suffer the impacts.

According to neoclassical analysis, there are several ways to internalize such costs. First, working in the public interest, the state can put pressure on plant operators to reduce emissions and community groups can organize boycotts to force internalization. The state is also in a position to set limits on pollution and impose sanctions or fines if limits are exceeded. The state could also impose a tax on each unit of pollution produced.* Incorporating the cost in this fashion reduces profits, and the producer should then seek ways to eliminate the pollution and restore former profit levels. Alternatively, if the plant owners claim they cannot afford to reduce emissions, it might fall to the government or community to pay the cost of cleanup. While this would seem to impose unfair costs on those who did not generate the pollution, the Coase theorem argues that such a solution makes sense: if the cost to the neighbors of internalization is lower than the cost of the damages from the externality, they can realize benefits by paying for internalization, since they will save the difference between the two costs. If the costs of internalization are very high, it might be more cost-effective to shut the plant down; plant owners sometimes threaten to do so rather than pay the costs (and thereby blackmail the communities in which they operate). Clearly, this "solution" could deprive adjacent districts of jobs and tax revenues. But if the benefits to human health and the environment of closure are greater than the costs of those impacts, shutdown may make economic sense. Of course, those who are affected by the pollution could move, especially if the aggregate costs of moving are less than those of reducing emissions. The plant owners could pay for the move, or the homeowners could bear the costs themselves. Finally, the plant owner could simply give money to its polluted neighbors, who could use the funds to move to cleaner climes, or to remain and suffer. In theory, the "optimal" solution depends on which is most cost-effective or leaves no one worse off than he or she was before internalization (this state of affairs is called Pareto Optimality[17]). Deciding which strategy will bring the parties closest to this point is not easy, however. Information about individual

---

*With a few exceptions, most economists believe it necessary to impose a price (with many preferring a tax) on carbon in order to reduce greenhouse gas emissions, with only a few preferring other methods of internalization. See: "Do Economists All Favour a Carbon Tax?" *Economist Free Exchange*, Sept. 19, 2011, accessed August 2, 2017, http://www.economist.com/blogs/freeexchange/2011/09/climate-policy; Dana Nuccitelli, "95% Consensus of Expert Economists: Cut Carbon Pollution," *The Guardian*, January 4, 2016, accessed August 2, 2017, https://www.theguardian.com/environment/climate-consensus-97-per-cent/2016/jan/04/consensus-of-economists-cut-carbon-pollution.

preferences is required, but such knowledge is difficult to come by and often costly to obtain. Moreover, one of the parties may be able to realize a benefit by concealing information from the others.

An alternative to these forms of internalization is to find a means of establishing a cost for the resource that is, as noted above, essentially equal to the marginal cost of controlling the insult to the environment (although even this might not result in the proper pricing of the resource). That point would arrive when the revenue earned from an additional unit of electricity (in this instance) were equal to the cost of eliminating the additional unit of associated pollution. How could this be done? The neoclassical answer is through private property rights linked with a market in "rights to pollute," as described in chapter 2. By privatizing the right to pollute or damage the environment, economic incentives to reduce or eliminate the offending activities can be created, according to proponents of this approach. For example, in the case of a polluting power plant, the owners might be required to buy permits, sold in a state-managed market, to emit carbon. Assume that the plant owners can afford to pay only one dollar per carbon unit; if they pay more, their profits will vanish. The plant's neighbors get together, bid $1.10 per pollution unit, acquire all the available permits, and retire them. The plant is forced to shut down and the air will, as a result, no longer be polluted.

What is worth noting here is that the plant owners are facing a scarcity of "free" pollution space. Recall that "free" inputs are free because they are not scarce and no one sees reason to acquire title to them. How, then, can "free" inputs acquire a price that will reflect their growing scarcity value? According to the neoclassical assumptions described earlier, if such "space" becomes scarce and all else is equal (as it rarely is), market prices for the right to pollute will rise to reflect growing demand for pollution space.* One way to manufacture a market is through creation of new forms of property rights that either create artificial scarcity (through regulation) or raise costs and prices (pollution or emission permits). (That's why oxygen bars exist: clean air becomes so scarce that the owners of full tanks of oxygen can make money by "renting" their tanks to those who can pay for bottled air.)

Creating Property: The Scarcity Factor
A basic principle of capitalist markets is that one must own, or hold title,† to that which is being exchanged or sold. Otherwise, the transaction is considered

---

*This is why marginal, rather than average, costing is so important: only if it is possible to sell at the margin will costs be great enough to incorporate externalities and provide incentives to use less. Of course, sellers may realize windfall profits as a result, but these can be taxed away and used for socially important costs.

†The origins of "title" are not in law but in the relationship between nobility and land. In England, for example, the sovereign owned vast tracts of land which s/he was able to convey to family, friends, and supporters through literal *title*, such as baron or duke. Title was then an inherent characteristic of the land rather than a result of ennoblement. This became something of a problem when land began to be sold through private markets.

illegitimate. Under Western law, ownership is vested through property rights expressed by lawful title. Such title may be a legal document, as in the title to a plot of land, a house, or a car, or it may be a receipt for goods purchased from a seller or previous owner (title can also arise through custom or oral tradition, but these rarely have legal standing). Title confers legal possession even as it helps to create scarcity, which is why stolen goods (lacking title) are usually much less costly than those being sold by their legal owners. The purchaser of stolen goods is taking a risk that they might be confiscated by the authorities, if found out.

If lawful title to a good cannot be demonstrated, and there is no one who holds such title, the good is presumed to have no owner. In regard to land, the Latin term for this state of affairs is *res nullis*. When Europeans established colonies in other parts of the world, they found that the concept of written lawful title generally did not exist as practiced in Europe. Property rights in those colonized places were understood in other ways, related to history of use, occupancy, kinship, and community.[18] When the inhabitants of these areas could not produce written title, the Europeans colonizers declared the resource to have no owner and took possession, issuing written deeds to the state, settlers, and investors at home.[19] Land and its resources—trees, water, minerals—were thus acquired at virtually no cost to the new owner, who could exploit nature and sell resources to the highest bidder (this is sometimes called "primitive accumulation"). In other words, the creation of private ownership rights took away access from local and indigenous users and in the process created scarcity. Users would now have to pay to gain access to those resources, even if they were previously free. Paradoxically, in this instance, the creation of scarcity through titling fosters consumption rather than conservation or careful use.

Various forms of regulatory limits and emissions permitting, both of which convey certain types of property rights, can also be applied to reduce pollution and resource use. As we saw in chapter 2, carbon emissions permits convey a "right to pollute" the atmosphere with a specific mass of carbon (dioxide). As permits grow scarcer and more expensive, emitters should seek ways to reduce their pollution commensurate with the rising cost of pollution. Similarly, state-mandated regulations (sometimes called "command and control") impose specific limits on the freedoms of property owners to use their possessions as they see fit. Lest you think that this constitutes unfair intervention by the state, driver's licenses not only convey a right to move across streets and highways, subject to traffic rules and regulations, they also implicitly limit what a driver is allowed to do with a car. To be sure, one can drive without a license, but the penalties for doing so can be quite severe, especially if an unlicensed driver is involved in an accident.

CONSUMERS 'R' US

So far, we have focused on only one side of the equation: production. Consumption is the other side. There must be customers for the goods being produced, and the more these customers consume, the more goods are sold.[20] Producers and manufacturers who cannot sell their goods to consumers, be they potato chips or microchips, will end up with costly and useless stockpiles and quickly confront bankruptcy. Consumers are key to contemporary capitalism; without them, the system will crash. Capitalism without growth, or with very slow growth, is conceivable, as proposed in chapter 2,* but the social reorganization required to slow down growth in consumption might have untoward political complications, for several reasons.

First, people have been socialized to believe that the accumulation of goods and money is a measure of personal success and that the possession and consumption of more things must be better.[21] If your personal wealth is not increasing, you are doing something wrong; if you are not interested in acquiring more stuff, you are odd, or even a failure. Second, in market-based democracies, there is the implicit promise and widely held belief in progress, and that "things always get better." This generally means growth in income and greater capacity to buy things. "Growing the economy" becomes a political imperative, especially for those holding public office. Finally, because the populations of some market societies (especially the United States) are growing—new consumers are born or are arriving in the country every day—economic growth is the current system's means of providing jobs and income to them. Consequently, constantly growing consumption seems essential to capitalism's social and political viability.[22]

The importance of consumption in contemporary capitalist societies can be seen in economic statistics. What is called "personal consumption" accounted for 69 percent of the U.S. economy in the second quarter of 2017, as seen in Table 3.1 (in current dollars).

Similar numbers apply to both rich and poor countries. In fact, personal consumption constitutes a greater percentage of national income in poor countries, as can be seen in Table 3.2, because most money there goes to the purchase of basic necessities, such as food. The contrast in per capita materials consumption between rich and poor countries is much more striking, however, when we consider minerals, materials, and biomass (Table 3.3). The data indicate the differences between affluent and poor countries, mineral and non-mineral producers, more- and less-industrialized states, and those highly dependent on biomass versus those less so. Australia produces and exports a great deal of metallic minerals, China a considerable quantity of plastics (much of which is exported),

---

*In addition to the sources listed under the discussion of steady state economy in chapter 2, there is also interest in "degrowth," that is, reducing the size of the economy as a path to sustainability. See, for example, Federico Demaria, et al., "What is Degrowth? From an Activist Slogan to a Social Movement," *Environmental Values* 22 (2013): 195–215; Hali Healy, Joan Martinez-Alier, and Giogos Kallis, "From Ecological Modernization to Socially Sustainable Economic Degrowth: Lessons from Ecological Economics," in *The International Handbook of Political Ecology*, edited by Raymond L. Bryant, pp. 577–90, Cheltenham, UK: Edward Elgar, 2015; and Lisa L. Gezon and Susan Paulson, eds. "Degrowth, Culture and Power," Special Section of the *Journal of Political Ecology* 24 (2017): 425–666.

Table 3.1. Components of U.S. gross domestic product

U.S. gross domestic product	$19,227 billion
Personal consumption	$13,297 billion, of which
Durables	$1,453 billion
Non-durables	$2,788 billion
Services	$9,052 billion

Source: U.S. Bureau of Economic Analysis, "Table 1.1.5. Gross Domestic Product," U.S. Dept. of Commerce, July 28, 2017, accessed August 2, 2017, https://www.bea.gov/iTable/iTable.cfm?Req ID=9&step=1#reqid=9&step=3&isuri=1&903=5.

Table 3.2. Personal consumption as a fraction of national gross domestic product

Germany	54%
Portugal	66%
Mexico	66%
Philippines	74%
Cambodia	77%

Source: World Bank, "Household Final Consumption Expenditure, etc. (% of GDP)," 2016, accessed August 2, 2017, http://data.worldbank.org/indicator/NE.CON.PETC.ZS.

and Nigeria a lot of oil and biomass. Even so, the consumption ratios between rich and poor countries are sizable: the average American consumes more than ten times as much in non-metallic minerals as the average Nigerian, and ten to one hundred times as much electricity as the average resident of South Asia.

Comparison of municipal solid waste* production further illustrates the importance of consumption in national economies, as seen in Table 3.4. National differences in consumption and waste reflect a number of different factors. All of these numbers are *averages*, which disguise the shape of the distribution. Rural residents tend to consume less and produce less solid waste, often because they are less affluent. People living in some countries recycle and reuse more than others. Industrialized and industrializing countries that use synthetics such as plastic produce more municipal solid waste than those heavily dependent on biomass. Even among urban residents, there are major differences as a result of gaps between rich and poor. Finally, there are significant differences even among wealthy countries, such as Japan and the United States, perhaps due to the rapidly aging population of the former, or less-consumptive practices.

---

*The Organisation for Economic Co-operation and Development defines municipal solid waste as "waste collected and treated by or for municipalities. It covers waste from households, including bulky waste, similar waste from commerce and trade, office buildings, institutions and small businesses, yard and garden, street sweepings, contents of litter containers, and market cleansing. Waste from municipal sewage networks and treatment, as well as municipal construction and demolition is excluded." Daniel Hoornweg and Perinaz Bhada-Tata, "What a Waste—A Global Review of Solid Waste Management," Urban Development & Local Government Unit, World Bank, March 2012, #15, Annex J, p. 4, accessed August 3, 2017, https://siteresources.worldbank.org/INTURBANDEVELOPMENT /Resources/336387-1334852610766/What_a_Waste2012_Final.pdf.

Table 3.3. Material consumption in eight countries

Country	Domestic material consumption[a]	Material footprint[b]	Domestic metal ore consumption	Domestic non-metallic minerals[c]	Domestic biomass consumption	Ratio of non-metals to biomass	Ratio of metals to biomass
Australia	42	37	22	7	8	0.88	2.75
China	17	14.5	1.2	10.8	2.2	4.9	0.55
Honduras	5.8	4.0	0.25	1.8	3.5	0.51	0.07
India	4.2	3.3	0.2	1.5	1.7	0.88	0.12
Japan	9	20	0.4	3.2	1.75	1.83	0.23
Nigeria	2.6	1.9	0.2	0.4	2.0	0.2	0.1
United States	20	27	2.0	5.5	4.9	1.12	0.41
Viet Nam	8.2	7	0.5	5.1	2.0	2.55	0.25

Source: Heinz Schandl, et al., "Global Material Flows and Resource Productivity," UNEP International Resource Panel, 2016, pp. 99–141, accessed August 3, 2017, http://www.resourcepanel.org/file/423/download?token=Av9xJsGS.

Notes: Data from 2010. All statistics in metric tons per capita. Columns 1 and 2 include non-metallic minerals, metal ores, fossil fuels, and biomass.

[a] Domestic material consumption: This indicator is designed to capture the territorial consumption of primary materials, whether they are sourced from domestic extraction or imported.

[b] Material footprint: This takes into account upstream material flows involved in producing products, rather than just the tonnage of the product itself.

[c] Excludes fossil fuels.

Table 3.4. Urban municipal solid waste generation (2012)

Country	Urban population	Total population[a]	Total municipal solid waste (metric tons/day)	Per capita kilograms of municipal solid waste per day
Australia	16,233,664	22,700,000	36,164	2.23
China	511,722,970	1,350,700,000	520,548	1.02
Honduras	2,832,769	7,900,000	4,110	1.45
India	321,623,271	1,236,700,000	109,589	0.34
Japan	84,330,180	127,600,000	144,466	1.71
Nigeria	73,178,110	168,800,000	40,959	0.56
United States	241,972,393	313,900,000	624,700	2.58
Viet Nam	24,001,081	88,800,000	35,068	1.8

Source: Daniel Hoornweg and Perinaz Bhada-Tata, "What a Waste—A Global Review of Solid Waste Management," Urban Development & Local Government Unit, World Bank, March 2012, #15, Annex J, accessed August 3, 2017, https://siteresources.worldbank.org/INTURBANDEVELOPMENT/Resources/336387-1334852610766/What_a_Waste2012_Final.pdf.

[a]2012 population data from World Bank, "2014 World Development Indicators," Washington, DC, 2014, accessed August 3, 2017, https://openknowledge.worldbank.org/bitstream/handle/10986/18237/9781464801631.pdf.

One popular way of assessing the environmental impacts of consumption is by way of the "ecological footprint," which is the land area, in hectares (1 ha = 2.47 acres), that would be required to provide the resources consumed by a person living in a particular country.[23] It should be immediately clear that the ecological footprint is a heuristic and a rather crude measure of impact,[24] inasmuch as it relies on quantification of land area; many countries are major importers of goods and do not produce much of the resources attributed to them. But the idea is a useful one in that it provides a basis for comparison of the environmental impacts of rich and poor. A second quantity, the "ecological deficit," is the additional area outside the country required to provide imported resources. An individual's total ecological footprint is the sum of these two measures.

A more useful metric emerges from *life cycle assessment* (or analysis),[25] which calculates all energy and resources involved in the production and disposal of a particular good or service ("cradle-to-grave"), including wastes generated in the production of goods; energy, water, and other resources used in production, transportation, and sale of goods; resources consumed and wastes produced while the goods are being used; and the environmental and health impacts of disposal (including the long-distance transportation of toxic materials). But determining either individual or per capita ecological footprints from such data is tricky. Although it is relatively easy to calculate the water content of, say, an imported Polish ham, it would be a good deal more complicated to account for *all* the water that went into raising, slaughtering, and processing the pig from which the ham originated (including the impacts of growing the food for the pig); manufacturing the container in which the ham is packaged; transporting the ham across the Atlantic to the local market; washing up after

Table 3.5. Comparing national ecological footprints (in global hectares per person, 2013)

Country	Gross domestic product per country (2010 dollars)	Production (global ha/per capita)	Consumption (global ha/per capita)	Biocapacity (global ha/per capita)
Australia	$53,704	13.5	8.8	15.7
China	$5,722	3.5	3.6	0.9
Honduras	$2,227	1.5	1.7	1.7
India	$1,555	1.0	1.1	0.4
Japan	$46,288	4.2	5.0	0.7
Nigeria	$2,462	0.9	1.1	0.6
United States	$49,941	8.6	8.6	3.8
Viet Nam	$1,522	1.4	1.7	1.0

Source: Global Footprint Network, "2017 Edition National Footprint Accounts: Ecological Footprint & Biocapacity (2013)," accessed August 4, 2017, data.footprintnetwork.org.

Note: A global hectare (ha) is a biologically productive hectare with world average productivity.

the ham is eaten; disposing of the tin and whatever else is left; and apportioning those impacts among the ham's final consumers.

Are We What We Consume?

Why are the differences in these tables so great among countries (domestic comparisons might show even greater variations, between the super-rich and truly impoverished)? Beyond the acquisition of basic necessities such as food and shelter (which vary among societies), consumption is not given or automatic. There is no human instinct or gene to acquire things or to "shop until you drop." How do you know, therefore, that you need new shoes? Or that your computer is obsolete? That you want to drive through mud in a four-wheel-drive truck? That you must have the latest music download by Beyoncé or Katy Perry? And what would happen if you didn't acquire these new possessions? Evidently, something motivates our acquisitiveness. That something tells you about things you need or want, even if you have never thought about them or wanted those things before. It warns you when one of your possessions is no longer stylish or useful, so you can get rid of the old and bring in the new (as in the latest iPhone). And it admonishes you about what might happen if you do not avail yourself of lots of new stuff. It is difficult to imagine life without all of your possessions. Yet, for thousands of years and for billions around the world, life has gone on without such stuff. What drives consumption?[26]

We can identify three facets of the "will to consume": psychological influence through advertising and peer pressure; engineering of "new" product designs to replace "old" ones (planned obsolescence); and the structural imperatives of capitalist growth. In the first instance, we are constantly being exposed to messages and blandishments encouraging us to look, buy, and consume—estimates vary from 250 to five thousand each day.* Such messages are not merely about the services provided but also about the necessity of acquiring such items if the consumer is to have a complete and fulfilling life. Some products are advertised as "lifestyle" or "identity-oriented" goods; others are touted for their superior performance or particular characteristics. People are not, of course, so gullible as to take at face value the promises made in advertising media (although something must be happening, since global expenditures on advertising were more than five hundred billion dollars in 2016![27]). Children might demand something solely on the strength of the claims made in an ad they hear or see, but adults have learned that the purpose of such messages is merely to get the listener or watcher to recall the product when a choice is to be made, or to notice the brand of the product when a colleague or acquaintance has bought it. The object is to hook a potential consumer to a specific product or brand and to create loyalty where there was none before.[28]

In the third instance—structural imperatives—consumers are made aware of their importance to the vitality of the economy. To a growing degree, as we have seen, personal consumption not only comprises a major fraction of gross domestic product but is also the primary driver behind continual expansion of national economies.[29] No longer is heavy industry a primary source of national power and expansion; today, the consumer has a national obligation to consume. Such admonitions were seen and heard in the days and weeks after the September 11, 2001, attacks on New York and the Pentagon in suburban Washington, DC, when it was feared that people's uncertainty about their safety might carry the American economy into serious recession. Today, we are told that the future of the American economy depends on bringing the production of goods back to the continental United States, in order to restore the economic growth rates and industrial employment levels of the past. Ironically, perhaps, many of the goods purchased as a patriotic duty are not even manufactured in the United States but come, rather, from the People's Republic of China, a key economic competitor.†

---

*There is little agreement on these numbers: the first is roughly how many individual ads are *looked* at during the day, and includes the brand names on the products we use and consume; the second is how many we are *exposed* to, but might not look at—for example, online banner ads. See: Choice Behavior Insights, "The Myth of 5,000 Ads," Hill Holliday, n.d., accessed August 3, 2017, http://cbi.hhcc.com/writing/the-myth -of-5000-ads/; and Sheree Johnson, "New Research Sheds Light on Daily Ad Exposures," *SJ Insights*, September 29, 2014, accessed August 3, 2017, https://sjinsights.net /2014/09/29/new-research-sheds-light-on-daily-ad-exposures/.

†According to an analysis done in 2011, "goods and services from China accounted for only 2.7% of U.S. personal consumption expenditures in 2010." Galina Hale and Bart Hobijn, "The U.S. Content of 'Made in China,'" *FRBSF Economic Letter*, August 8, 2011, accessed August 4, 2017, http://www.frbsf.org/economic-research /publications/economic-letter/2011/august/us-made-in-china/; see also Jay H. Bryson and Erik Nelson, "How Exposed is the U.S. Economy to China?" Wells Fargo Economics Group, August 13, 2015, accessed August 4, 2017, http://www.realclearmarkets .com/docs/2015/08/US%20Exposure%20to%20China%20_%20Aug%202015.pdf.

It is the second instance, planned obsolescence, that is a particularly powerful driver of consumption.[30] Nowadays, many things become useless or undesirable long before they wear out. This is not entirely accidental: increasing the flow of goods through the economy fosters growth. Left to their own devices, consumers would tend not to replace things until they broke down and could not be fixed. Nowadays, many goods are not made to last a long time, as anyone who has recently replaced a toaster can testify.* And given the emphasis on economic value rather than environmental concerns, there seems little point in repairing a two-year-old malfunctioning microwave oven originally priced at fifty to one hundred dollars when it costs fifty dollars merely to have a repairman open the box.[†]

Smartphones offer one of the best contemporary examples of planned obsolescence. Since their first appearance on the consumer market in 2007, new top-end smartphones have not declined appreciably in cost, although their processing capacity has increased considerably. For example, the iPhone 6 has about ten times that of the first iPhone, and more than three times the processing power of a 1985 Cray-2 supercomputer (which cost sixteen million dollars at the time).[31] This does not mean, however, that the latest versions operate more responsively, since new operating systems, software, and applications are constantly being developed and installed. These are things that no user can do without, and to do without is to be left in the dust. Today's phones are almost as powerful (if not more so) than desktop and laptop computers and, as tablets proliferate and smartphone screens grow in size, it is likely that those computers will go the way of the dodo. Paradoxically, perhaps, very few owners of these devices ever avail themselves of the full computing power of their machines (in fact, very few have any idea what that computing power might be or how to use it to its full capacity). For a few basic functions, such as word processing and basic computation, almost any computer from the past twenty years would be perfectly adequate to such tasks.[‡] It is almost impossible, however, to maintain such a computer except through heroic effort, especially as software companies abandon older versions of their programs.

---

*Point of pride: In 1983, one of the authors had to replace his Texas Instruments scientific calculator and purchased a programmable HP 15C. He still owns and uses that device, even though it has almost no memory and is used mostly for cobbling together various project budgets.

†According to Angie's List, "If you have a basic model, which can cost as little as $60, consider a replacement. Replacement also may be the best bet with high-end models, which can run $500 to $1,000. If your broken microwave cost $250 but repairs are $150, you may save more in the long term by getting a new one." Doug Bonderud, "Microwave Repair and What It Costs," accessed August 4, 2017, https://www.angies list.com/articles/microwave-repair-and-what-it-costs.htm.

‡We recognize that one could not surf the Internet with a computer from the early 1990s, but as scholars who studied, worked, and wrote their PhD dissertations before gaining access to the World Wide Web, we know that it is possible to live a perfectly adequate (and ad-free) life without it.

Indeed, almost as soon as you purchase a new computer or device, it is, for all practical purposes, junk—within a couple of years, a one-thousand-dollar machine will be worth no more than one hundred dollars, if that much (old computers are not scarce, even if difficult to maintain). To avoid being blindsided by planned obsolescence, you need to buy a new system every two or three years. The result is that, in the United States and Europe, there is a growing electronics disposal problem. Some old devices can be "handed down" to children or donated to worthy causes. Others end up stashed away in closets or garages, where they gather dust. Disposal of the growing numbers of discarded computers, laced with toxic metals, is no easy proposition. Landfill operators are reluctant to accept them, and the individual components are worthless. Nonetheless, a goodly number of old computers and phones end up in landfills where they can contaminate groundwater or, as we saw in chapter 2, are shipped to poor countries, where valuable metals are extracted while toxic wastes are inhaled, ingested, and dispersed into the surrounding environment.[32] Since the first iPhone hit the market in 2007, more than seven billion smartphones have been produced, although many of those are no longer in use: in the United States, the average replacement cycle is twenty-six months.[*][33]

Wealth: Is It Good for the Environment?

According to the Brundtland Commission's *Our Common Future*, "poverty is a major cause of environmental degradation."[34] According to this argument, the very poor rely heavily on local resources, such as water, soil, and wood, and, via a "tragedy of the commons," overuse them. As people become richer, they have greater recourse to markets which, for a price, offer goods and services with fewer environmental impacts. Advocates of what is called the "environmental Kuznets Curve" argue that as societies get richer and consumers have more discretionary income they also become more interested in aesthetics and the quality and attractiveness of their surroundings, and will demand a cleaner environment and be more willing to spend the money to achieve that end.[35] Proponents claim that empirical evidence for this "effect" is seen in the lower levels of air and water pollution found in industrialized countries compared to developing ones. Early research suggested that higher levels of pollution were to be found in industrializing countries with a per capita gross domestic product in the range from five thousand to eight thousand dollars.[36] Below this level, industrialization has not proceeded far enough to make pollution a serious problem; above this level, society begins to pay for pollution reduction. However, more recent work suggests that the environmental "turning point" in the curve may be might be closer to thirty-five thousand dollars,[37] which is far greater than global median gross domestic product of around fifteen hundred dollars per capita.[38]

---

[*] Not all appliances are subject to this level of planned obsolescence: refrigerators tend to stay in service for fifteen to twenty years or more; stoves even longer. But nowadays, such extended lifetimes may be the exceptions rather than the rule.

The implied calculus is, thus, that it is good to consume, and it is even better to consume more, because higher flows of goods and money through the economy not only lead to a larger pie from which everyone can get a piece, but also foster ecological modernization and environmental improvement as consumers demand and buy new products and services. We might question this, however: if the cost of growth in consumption is also a major increase in the production of wastes all along the commodity chain (or life cycle), wealth can also be seen as a "major cause of environmental degradation." If ecotourism results in an expansion of travel and services, this might well outweigh the overall benefits, as a result of air and ocean travel, and various impacts in host countries. Moreover, in the past, much pollution was generated in the production of things, and a good fraction of both production and disposal took place within the country in which those things were used or consumed. But there is now a dynamic in capitalism driving a shift in this pattern and leading to the transfer of pollution and waste to poorer countries, through offshore production and the lengthening of commodity chains (a phenomenon we return to in the final part of this chapter). The relative wealth of a country thus contributes to its ability not only to eliminate pollution but also to displace it to locations abroad.[39] And while local pollution levels in many places may tend downward as incomes rise, the same cannot be said for energy consumption. Indeed, carbon emissions from the burning of fossil fuels are much higher per capita in industrialized countries.

Growing incomes and wealth might lead to a demand for a cleaner environment, but they also lead to growing wages, expenditures, and consumption, resulting in what has been called the "treadmill of consumption."[40] The material effects of such growth are evident: greater pressure on the environment due to greater extraction from nature in other countries to meet the demand for goods, on the one hand; growing volumes of wastes as products are used and thrown out ever more frequently, on the other. The growing volume of goods produced in developing countries and imported into industrialized ones also encourages consumption by virtue of relatively low prices. Consumers can then buy more goods, even if their incomes increase only very slowly, and feel as though they are better off. And when goods decline in cost, people often buy more of them, more often.* Lower-cost smartphones exemplify this.

Consider, as another example of this phenomenon, a polo shirt with a collar. In the United States and Europe, the cost of manufacture alone runs ten to fifteen dollars, and the final price might be as much as fifty dollars, when other

---

*This is the same phenomenon as happened with gasoline use. The growing efficiency of cars resulted in a decline in the cost-per-mile of fuel, so people tended to drive more and to travel farther distances. See: Cameron K. Murray, "What if Consumers Decided to All 'Go Green'? Environmental Rebound Effects from Consumption Decisions," *Energy Policy* 54 (2013): 240–56.

costs and markups are factored in.* Table 3.6 shows the production and retail cost breakdown for a fourteen-dollar polo shirt produced in Bangladesh and sold in Canada. For the price of one made-in-America polo shirt, a consumer could buy three made in Bangladesh. Now consider the effect of rising wages in the two countries. Assume that the production cost of a U.S.-made polo shirt rises from ten dollars to eleven dollars, due to an increase in the minimum wage. To compensate for this, the retailer might increase the retail price to fifty-five dollars. The consumer sees a five-dollar drop in purchasing power (for only one shirt). But in Bangladesh, wages are likely to remain stable, or rise only a few cents. So there will be no need to raise the retail price in the consuming country. And, insofar as the consumer's income rises during the same period, the foreign-made shirt will actually become less expensive relative to income. (Think about how some people respond to a sign advertising "BIG SALE!" They are liable to buy two items in the belief that they are "saving money," even though they only need one item and the cost for two is greater than for one).

Consumption is also encouraged by appealing to consumers' desire for prestige, or positional, goods associated with higher incomes and the upper class.[41] This is evident with automobiles. Of course, external features on cars are changed every year, even if the mechanical insides remain very similar, in order to emphasize both the aging and declining status of older ones. But auto manufacturers are not above selling two different models with similar mechanical features at different prices. Much of the cost of any automobile is in those parts you don't ordinarily see: the engine, the drive train, the equipment under

Table 3.6. Breakdown of costs of producing a fourteen-dollar polo shirt

Element of production process	Cost	Fraction of total consumer cost
Materials and finishing	$3.69	26%
Labor	$0.12	0.1%
Factory overhead	$0.07	0.05%
Factory margin	$0.58	4%
Freight/insurance/duties	$1.03	7%
Total cost to retailer	$5.67	40%
Consumer markup	$8.33	60%

Source: Report by O'Rourke Group Partners LLC, April 2011, as reported by Rosemary West-wood, "What Does That $14 Shirt Really Cost? *Maclean's*, May 1, 2013, accessed August 8, 2017, http://www.macleans.ca/economy/business/what-does-that-14-shirt-really-cost/.

---

*A more sophisticated discussion of the relationship between labor costs and profits can be found in Robert Pollin, Justine Burns, and James Heintz, "Global Apparel Production and Sweatshop Labor: Can Rising Retail Prices Finance Living Wages," Political Economy Research Institute, University of Massachusetts, Amherst, 2001 (revised 2002), accessed March 20, 2018, https://www.peri.umass.edu/publication/item/161 -global-apparel-production-and-sweatshop-labor-can-raising-retail-prices-finance -living-wages.

the hood, and the frame. Yet even though body and interior features, about which so much fuss is made in advertising and which have so much to do with consumer "identity," constitute the smaller fraction of a vehicle's production cost, in the end they play a significant role in the price of a car. Therefore, why not replace a plain exterior with a fancy one, install leather seats and an expensive sound system, and claim that the result is an entirely different and "higher-class" vehicle? That is, more or less, the difference between the Toyota Avalon and Lexus, even though the former sells for less than the latter.[42] People buy a Lexus as much (or more) for status than transportation.

The consumer is thus encouraged to consume, to consume more, and to consume more costly goods. Few of these items have long lifetimes. Many can be resold, some can be recycled, but a goodly number end up as waste. Reducing the flow of consumables, by making things more durable and de-emphasizing style, would reduce waste but, in a social order predicated on growing incomes and consumer goods, the political consequences of slow growth and limits to consumption could be quite unsettling. What about things advertised as green or sustainable? While the impacts of production might be less, if replacement cycles are equally short, they will generate comparable volumes of waste.

Thus, if poverty is a major source of environmental degradation, as commonly claimed, so is wealth. Poverty, however, does not cast its impacts quite as widely as does wealth. To survive, the poor must generally consume nearby resources, and less of them, and most of the consequent effects are fairly localized (although it is true that the repercussions of soil erosion can show up thousands of miles away, as can the climatic effects of methane from rice production). The rich, by contrast, can afford to export many of the environmental impacts of consumption, which then become invisible to them, what Peter Dauvergne once called "externalizing externalities" through mechanisms discussed above.[43] As a result, the poor experience a double burden: the environmental impacts of their subsistence activities plus those resulting from excess consumption by the rich in other parts of the world. This is the nub of the problem of environmental justice discussed in chapter 2: the inequitable distribution of damage to both nature and health, perhaps most stark in the case of climate change, where those who are responsible for the fewest carbon emissions are most at risk from the livelihood- and life-threatening impacts of more intense storms, rising sea levels, drought, and desertification.[44]

WHAT'S NATURE WORTH?

The central role of commodification and consumption in capitalism suggests that everything has its price and everything can be consumed. In many ways, this is not far from the truth, as we shall see. What about those things that cannot be commodified and do not have a price? What, for example, is the value of a forest left standing, or a species protected from extinction? How much is the atmosphere worth, especially when access to it has always been "free?"* These are not merely questions of philosophy or ethics (although they should be). In a capitalist system, one must make choices about where to invest and what to buy and, perhaps, protect rather than purchase. By conventional measures, the short-term financial return on leaving intact an old-growth forest is a good deal less than what can be earned from the lumber it contains. Over the longer term, income realized from logging the forest can be invested elsewhere to generate much higher returns than those from leaving the forest alone. But the million-dollar questions are: Which is more valuable? Forest or lumber? And how can we know which use is more valuable? Indeed, how can we put a value on ecological necessity and aesthetics?†

There are two quite different answers to these questions. One approach is to establish the market value of the standing timber, impute that value to the forest as a whole, and pay the owner or leaseholder that amount to leave the forest intact (in this case, we are internalizing the damage that would arise from logging). There is a considerable body of historical data on which to base such an assessment: the retail price of a board foot of lumber; the number of board feet that come from a tree; the number of trees in a tract of forest; the cost to log, mill, and transport; the amount of future income that the owner will forego; and so on. A second approach to determining a monetary value for the forest is through a method called "contingent valuation."[45] People are asked, for example, how much they would be willing to pay to prevent the old-growth forest from being logged. This amount is then compared with the market value of the timber. If the market value is less than the contingent value, protection is in order. Otherwise, down with the trees! The problem with the contingent valuation approach is that, because no one is actually asked to pony up funds for forest preservation, we are comparing real revenues with imaginary payments. Moreover, it has been found that, when asked to hand over real money, people are only willing to pay a good deal less than their initial offer.[46] Consequently, efforts to establish alternative monetary values and forge environmental policy on that basis have been largely stillborn.

Another approach to environmental valuation is benefit/cost analysis. In this instance, the benefits of preservation are compared to the value of the economic activities that would not take place as a consequence of preservation. This method is often applied when environmental regulations appear to

---

*Perhaps this is not entirely the case. Nations declare sovereignty over the atmosphere extending from the ground into space; "air rights" can be bought and sold, as was the case when the old Penn Station in New York City was destroyed, with tracks and station moved underground, and air rights sold to build Madison Square Garden aboveground. See, for example, "Air Rights New York," accessed August 8, 2017, http://www.airrightsny.com/.

†The greatest value from the last redwood tree standing might be to cut it into one-inch squares and sell those mounted on plaques (or as Joni Mitchell sings in "Big Yellow Taxi": "They paved paradise and put up a parking lot.")

restrict landowners' use of their property or utilization of contested resources.* The difficulty here is similar to the dilemma posed above: how do humans assess the monetary value of competing projects or uses, and what types of choices do they or should they make as a consequence? Inasmuch as that which is to be preserved is often either on public lands coveted by industries for lumber, minerals, and grazing, or private land where use is incompatible with preservation, pressures to allow exploitation are considerable, especially if non-use results in monetary losses to owners and projects. National, state, and provincial governments possess the authority to create or restrict property rights, as the Obama administration did during its waning months in creating five national monuments on public lands.[47] These were almost immediately declared to be under review by the incoming Trump administration, which was concerned about closing off public lands to oil and gas exploration and mining and subsequently reduced the size of several.[48]

For these and other reasons, a sizable and growing portion of the business of preserving ecologically valuable tracts of land has been taken on by land conservancy and land trust organizations, although the total area of these privately held tracts remains far less extensive than public lands under some form of management (some fifty-six million acres versus 618 million acres).[49] Nonprofit organizations, such as land trusts and nature conservancies, raise funds from private and corporate sources, solicit commitments from public agencies, and often pay something approaching market value for land they wish to protect. Or they try to persuade the owners of environmentally sensitive private property to donate land. Depending on circumstances, ownership of such lands may then be transferred to public entities, which are expected to turn the property into parks. But not all lands become public; some are traded for other tracts of land and some are sold back into private ownership to generate funds for the purchase of more "ecologically valuable" properties. Although the protection of such tracts is a salutary goal, their commodification in this form raises several troubling questions, in particular, what is to prevent them from being put back on the market in the future?[50]

Perhaps the most serious question is whether nature ought to be valued at all in monetary terms. As we suggested in chapter 1, humans value nature in multiple ways, most of which cannot be quantified. Nevertheless, monetary value is the conventional basis on which the benefits of the environment are calculated. In 1997, Robert Costanza estimated the value of the environmental services provided by nature ("ecosystem services" or "natural capital"), such as clean air, water, and so on, suggesting their value amounted to thirty-three

---

*The Trump administration has ordered the U.S. Environmental Protection Agency (EPA) to redo a cost/benefit analysis of the "Waters of the United States" rule that had found net benefits, but which offended various landowners, to show that costs exceed benefits. See Coral Davenport, "Scott Pruitt is Carrying Out His E.P.A. Agenda in Secret, Critics Say," *New York Times*, August 11, 2017, accessed August 14, 2017, https://www.nytimes.com/2017/08/11/us/politics/scott-pruitt-epa.html.

trillion dollars per year, a sum roughly equal to the annual global gross domestic product that year.[51] In a more recent paper, Costanza and his colleagues suggest that annual global ecosystem services in 2011 were as much as $125 trillion (in 2007 U.S. dollars),[52] far greater than the World Bank's estimate of seventy-five trillion dollars for global gross domestic product in 2016.[53] Indeed, following various critiques of Marx's work on labor and reproduction that point out the impossibility of capitalism in the absence of women's unpaid work in the household, it is also possible to argue that, without ecosystem services, capitalism would not be possible, since nature in situ has no monetary value yet is essential to human social reproduction.[54]

But we need to be cautious in monetizing nature, because it may imply that nature and capital are substitutes for each other, that capital can always replace nature, and that survival is possible without nature as such. It also suggests that we can actually calculate the "true" economic value of the things that nature provides, such as air, clean water, habitat, and so on, and in so doing, disregard values that cannot be monetized. Such claims do not stand up under closer inspection, the one on technical, the other on philosophical grounds. Notwithstanding amazing advances in genetic engineering, there is little reason to think that it will become possible to resurrect extinct species or restore old-growth forests to their original condition from fossil deoxyribonucleic acid.* No amount of money can restore that which has disappeared forever. And if an endangered species—say, an amphibian or insect—has no evident "function" in an ecosystem, it will appear to have no monetizable value. Does this mean such a species does not deserve preservation? How can we possibly put a value on a nature we have hardly begun to comprehend and which we have no right to judge?[55]

Nevertheless, neither technical nor philosophical difficulties have stood in the way of attempts to commodify things whose economic value is difficult to assess. The most widely used approach in this regard, as we have already seen, involves the creation of property rights in these things, which then permits their sale in markets. In the case of biological diversity, value is attributed not to ecosystems or species or individuals, per se, but to the genetic resources within the cells of living things, which are appropriated, transformed, patented, and sold.[56] In the case of climate change, permits to emit greenhouse gases are created, apportioned, and traded or sold in a market, or carbon in the form of trees is privatized and priced.[57] And through the creation of conservation offsets, would-be developers can buy credits in forests or species in one location as compensation for destroying habitat in another location.[58] In all these instances, the creation of property rights permits monetary values to be established for parts of nature, and these become crude proxies for those aspects of nature that cannot be valued in other ways. In the following sections, we provide several examples of such misvaluation of nature.

---

*Although there is considerable interest in trying to do this; see Beth Shapiro, *How to Clone a Mammoth*, Princeton, NJ: Princeton University Press, 2015.

Let's Make a Genetic Deal

The rapid development of genetic engineering and biotechnological research over the past few decades has motivated the emergence of a reductionist but increasingly dominant epistemological approach to biological and medical research. Traditional approaches to biology, ecology, and environmental protection tended to focus on species and ecosystems, regarding the whole as more than the sum of its parts.[59] Biotechnology, by contrast, is based on the search for the causes of biological processes in the smallest parts of the organism, the gene, and its place in the genetic sequence, or "genome." Because most of the defining characteristics of both species and individual organisms are to be found in the genome, which is considered a "natural" entity, the implication is that individual deviations from the "norm" in any species have their causes in "errors" in an individual's genome (such as mutations).* This perspective is evident from frequent news reports about the "discovery" of genes linked to specific diseases and conditions. The implication of such findings is that "fixing" the "defective" gene could "cure" the medical problem of concern.[60] The potential economic rewards of such research could be enormous. Imagine, for example, that some way was developed to repair those genes implicated in the development of some chronic or potentially fatal disease, such as leukemia. Imagine further that this repair could be accomplished through the simple ingestion of a pill containing a genetic "repair kit." The company financing such a discovery would be in a position to reap extraordinary profits.

But whether genetic repair and rehabilitation are possible or desirable is less relevant here than the economic and political implications of such research. On the basis of rulings in U.S. and European courts addressing intellectual property rights, any such genetic "inventions"—including synthetic copies of the isolated naturally occurring genes themselves—can be patented and turned into private property.[61] This is the case even if public funds are, somehow, involved in the research (for example, in the education of the researcher, the building of the laboratories, and the research itself) or if the organism harboring the discovery is "public property." Philosophically, the notion of such private title is legitimized on the basis of John Locke's claim that whomever puts labor (whether one's own labor or others' wage labor) into the creation of something is thereby entitled to call it his or her property.[62] We see here how seemingly abstract philosophies and principles come to be embedded in common practice and politics.

Locke was concerned only with physical labor put into the improvement of land and production of crops and argued that, consequently, the fruits of that (agricultural) labor were the laborer's private property. Under the premise

---

*It is interesting to note that human beings do not possess that many more genes than worms and, according to at least some biologists, might be thought of as "mice without tails." Roger Highfield, "Men are Mice: Thereby Hangs a Tail," *The Telegraph*, December 5, 2002, accessed August 14, 2017, http://www.telegraph.co.uk/uk news/1415272/Men-are-mice-thereby-hangs-a-tail.html.

that scientific research, multiplied greatly by capital and technology, constitutes labor, the fruits of such research can be patented and privatized. The holder of a genetic patent is then entitled to commercialize the discovery and sell or license it to whomever is willing to pay a "fair" market price.[63] And what might be the fair market price for such a "product?"[64] Genetic prospecting is not a cheap business; it takes time to gather samples and analyze them in the lab. The research leading to patentable discoveries can be quite expensive—it can cost more than a billion dollars to bring some pharmaceuticals to market.[65] So the patent holder will want to generate sufficient revenues to cover development costs and provide a tidy profit, which means that many new medicines are quite expensive. This is the case even if their biochemistry is based on genes derived from public resources.

Prospecting costs can be reduced. When searching for potential medical products, it helps to have information about medicinally useful plants.[66] Of particular interest, therefore, to corporate and scientific researchers are the remedies and practices of indigenous peoples. Medicines derived from plants known to possess healing properties may have been part of local common lore for centuries or millennia (for example, the active ingredient in aspirin is quite similar to a chemical found in the bark of white willow trees). Such plants are common property resources, in the sense that they belong to no single individual or corporation, and indigenous users hold no recognized legal title to them.[67] As a result, under international and most national law, these plants are assumed to belong to the state, which can sell access rights if it so wishes. Furthermore, under relevant international patent law, traditional uses have legal standing only if these uses have been documented in scientific journals. Indigenous peoples and peasants rarely have the time or interest to conduct such research and publish scientific papers documenting their findings. Needless to say, they hardly have access at all to the relevant scientific journals.[68] In the absence of such documentation, patent holders reap the monetary benefits of access and use, while those whose knowledge leads to medicinal discoveries often share none of the benefits.[69] Traditional users instead may have to pay patent holders to obtain the products manufactured from the very plants they may have used since time immemorial.[70]

There are ongoing efforts to alter this state of affairs. Since 2011, the Intergovernmental Committee on Intellectual Property and Genetic Resources, Traditional Knowledge, and Folklore of the World Intellectual Property Organization (WIPO) has been carrying out negotiations toward a legal instrument that would ensure protection of genetic resources and traditional knowledge. Talks regularly break down for a year or two at a time due to disagreements between developed and developing countries over the mandate of the committee.[71] A new protocol under the UN Convention on Biological Diversity, the Nagoya Protocol on Access and Benefit-sharing, provides a legal framework for countries to demand a share of the monetary benefits from patented resources originating from their territory. However, many states have rather conflictual relationships with indigenous peoples residing within their territory, and there is no obligation under the protocol for states to share benefits with indigenous and local communities.

Meanwhile, what are the implications for plant and animal species of such patent schemes? No longer are such species valued for their contributions to biological diversity and ecosystemic function or their intrinsic value as living things or their roles in indigenous cultures. Instead, they are sought for their "genetic diversity," their potential as sources of medicines and other commodifiable substances, and the value they add to the corporate bottom line and global economy. For countries—especially developing ones strapped for revenues—the sale of genetic prospecting rights to scientists and corporations might be a source of income that could be used for both environment and development.[72] Furthermore, if something valuable is discovered, the owners of the patent rights to the final product ought to be willing to share a portion of their profits, for example by contributing to the preservation of specific ecosystems (this is addressed in the Nagoya Protocol).

But at least two caveats should be stated in regard to this picture. First, the process of establishing private property rights to genetic resources, and offering them for sale in the market, establishes measurable monetary return as the sole criterion for valuing a species. Those in favor of privatizing biodiversity argue that only under private ownership can resources be protected, because the owners of said property will want to maintain a continual flow of revenues from them and will therefore strive to preserve the resources.[73] But value in the market today is no guarantee of protection tomorrow. The specific interest behind genetic prospecting and research is not the container but the contents. If the specific genes and properties extracted from a particular species can be analyzed and reproduced in the laboratory, the supply of raw materials—the species—and its preservation no longer matter for markets.

This is the next frontier of technological transformation for the genetic engineering industry in its 2.0 version, "synthetic biology."[74] With the appropriate chemicals and lab techniques, the substances found in nature can be produced synthetically and manufactured in large quantities, with much greater quality control. Species as sources of genetic material are then no longer important to the process. All that is required is the genetic sequence. A race is on to catalog and contain the genetic sequence information of much of the natural world in huge privately owned databases.[75] Currently, political battles are being fought over privatization, access to, and benefit-sharing from the use of sequences in databases. Once much of the sequencing information of the biological world is contained in databases, the economic rationale for preservation might simply disappear. And as the origin of sequences can be further obscured in the databases, it becomes less and less likely that the original guardians of biodiversity and traditional knowledge will ever be compensated for their knowledge or contribution.

Second, it is sometimes argued that no one knows from where the next great medical or genetic discovery will come and that such uncertainty will induce states and people to protect species.[76] But there are no guarantees here. All species are of great potential value—at least until they have been prospected, analyzed, and their genome digitized and catalogued in a private database. Furthermore, under this accounting scheme, those species that have no special or desirable genetic properties have no value, and they will be protected only if other species in their particular ecosystem do possess such economic value. Inasmuch as the vast majority of species have no evident genetic value, and are unlikely to acquire it in the future, they will not receive direct protection under genetic privatization. In other words, protection via genetic prospecting and private property rights might seem attractive and politically easy, but it is hardly guaranteed. As experience with once-costly commodities, such as eight-track tape players, the Sony Walkman, and flip phones, suggests, current monetary value may be more of a hope than a solid basis on which to base environmental protection and species preservation.

The Air, the Air Is Everywhere!

Climate change offers a somewhat different, but equally problematic, example of environmental commodification. In this instance, as discussed in chapter 2, marketization involves the purchase, sale, and trade of "rights to pollute." The creation of private property rights to unpolluted air is trickier than with property rights to genes, especially since it is so difficult to keep those without such rights from using the atmosphere as a waste dump. By definition, the atmosphere is generally considered an "open access resource."[77] Everyone uses it, but no one owns it. Where there is no owner of a resource, it is often argued, there is no incentive to moderate exploitation. In the absence of strictly enforced regulation or limits, pollution will proceed uncontrolled. As a result, everyone will suffer from bad air even if they gain no benefits from the process generating the pollution. This problem should sound familiar: it is yet another example of an "externality."

One commonly used term for this situation is "the tragedy of the commons," a concept popularized by the biologist Garrett Hardin in a famous and oft-quoted article in *Science*, and already addressed in chapter 2. There, he told a story of a medieval English common on which one hundred villagers grazed their cattle. The self-interested villager, he argued, would like to own two cows, but if everyone grazed two cows on the common it would be ruined. Still, because there were already one hundred cows grazing there, the addition of one more would hardly make a difference (1 percent). So, Hardin's mythical villager buys another cow and sends it out to graze (this is a version of the "free rider" phenomenon as part of the collective action problem discussed in chapter 2).[78] The difficulty arises, according to Hardin, because every villager will come to the same conclusion as the first and each will put a second cow on the commons, leading to two hundred rather than 101. The result will be ruination of all, as the pasture is destroyed.

Hardin offered this scenario of devastation in connection with a then-common fear (in the Global North) of uncontrolled population growth (in the

Global South).[79] He claimed that each additional child brought into the world would make only a very small additional contribution to overall environmental damage, but that the tens of millions of additional children born each year not only constituted a significant burden on global resources but would themselves have children that would multiply the cumulative damage over time. Indeed, Hardin argued, if population growth were not limited, the end result would be ruination of the planet.

Hardin called for "mutual coercion mutually agreed upon" as the solution, that is, a state (or Green Leviathan) that would ruthlessly control fertility and limit childbirth. And, because most of the increase in global population was taking place in developing countries (which continues to be the case today), Hardin's analysis was aimed mostly at them, a point he made even more clearly in "Life Boat Ethics," a 1971 article in *Psychology Today*. There, he compared rich countries to people in lifeboats besieged by the world's poor adrift in the water around them. One or two might be brought aboard without significant risk, but the attempt to save everyone would surely result in the death of those in the lifeboats.[80]

Hardin's formulation was erroneous in several ways. First, the English commons was not open to just anyone with a cow or sheep; villagers possessed limited (or "usufruct") use rights, and violations of the collective restrictions on use would result in the forfeiture of those rights.[81] Second, he presumed that medieval English villagers were selfish individuals who gave no thought to the "public good," an assumption rooted in Hardin's politically liberal and rational choice outlook, with its emphasis on "high individualism," a belief system not necessarily characteristic of the inhabitants of those English villages. Finally, although Hardin didn't know it, being a biologist rather than an anthropologist, there are numerous examples, even today, of such "common property resources" that are shared without being depleted.[82] The gist of Hardin's argument was that the only way to protect an open access resource would be to privatize it. Private owners would restrict the number of cows on their own plots of land and prevent others from putting cows there. To abuse or overexploit the pasture would then impose direct costs, rather than socializing them. Consequently, the plots would not be overgrazed and ruined.*

Thus, along much the same lines as the argument offered about species and the commons, if the atmosphere can be privatized, each of its owners will take much greater care in conserving and protecting it. In chapter 2 and above, we looked at this idea from the perspective of internalization; here, we consider it in relation to the atmosphere as a "commons." Although the notion of privatization of the atmosphere might seem a strange one, as we have seen,

---

*Whether this is actually the case for any or all resources is an empirical, rather than a theoretical, question. Some owners of private land might decide to extract maximum value and then abandon the resource.

it is not an impossible one, through some version of what are called "tradable emission permits" or "allowances," with polluters granted or sold a limited number of allowances to emit greenhouse gases in "carbon dioxide equivalents." Should polluters exceed their permissible limit, or "cap," they would be fined or might even have their right to future permits rescinded. Alternatively, they could "trade" with others who have too many allowances and cannot use all of them. Hence, "cap-and-trade."

How do such schemes work? First, someone—the government, an international agency, a scientific commission—makes an authoritative decision about what constitutes acceptable annual and total levels of pollutants or emissions into the atmosphere*—the cap—and what that level equals in total emission allowances. This quantity is usually based on a scientific assessment of the potential environmental and health consequences of exposure to a specific level of pollution as well as on a political judgment about what constitutes an acceptable social cost (for example, in regard to sea-level rise, ill health, total emissions). The total volume of emissions leading to this level of pollutant in the atmosphere—let us say three hundred billion metric tons—is then divided by the time period of concern—say, one hundred years—to generate the number of allowances available per time period. In this example, the permissible amount would average three billion metric tons per year—but, because the goal of cap-and-trade is to reduce emissions over time, the number of allowances per year could start at six billion metric tons in Year 1 and decline to one billion metric tons or less in Year 100. This quantity would then be divided among the polluters of concern, perhaps as a simple grant, perhaps through public sale or auction (along the lines of an "initial public offering" of shares in a new company). As we saw earlier in this chapter, polluters could keep their units and emit the allowed volume of pollution, or sell them to the highest bidder. The cap theoretically creates scarcity and the necessary conditions for this market to work. This is called "emissions trading."[83] The operating logic here is that each polluter can calculate whether it will cost less to reduce emissions or to use up the pollution units.[†] Through privatization of the atmospheric commons, polluters are induced to conserve their allowances rather than rapidly "burn them up."

Under cap-and-trade, some polluters might find they have more emission permits than they require; others might have fewer than they need. Those who have too many can then sell their permits to those who need more, at prices generated in this emissions market. Alternatively, permit prices in open auction will reflect bidders' costs to clean up their emissions and willingness to

---

*With respect to the global atmosphere, this is called the "carbon budget." See the following and Andrew P. Jones, Stuart A. Thompson, and Jessia Ma, "You Fix It: Can You Stay Within the World's Carbon Budget?" *New York Times*, August 29, 2017, accessed September 14, 2017, https://www.nytimes.com/interactive/2017/08/29/opinion/climate-change-carbon-budget.html.

†As per the Coase theorem ("problem of social cost") mentioned in chapter 2.

pay the auction price. In both cases, the price theoretically reflects producers' calculations of what such a pollution right is worth. Investments in pollution control can be made where they are most cost-effective and offer the highest rate of return. What has been created here, in effect, are property rights to the atmosphere. Each allowance unit gives the holder a right to "use" that volume of air needed to disperse one ton of pollutants. The process is rather like paying an entry fee to dump garbage at a local landfill. A temporary "property right" is purchased from the landfill owner that allows one to enter and get rid of garbage, which "disappears"—in a manner of speaking.

Consider the following example. Let us assume that the cost of pollution reduction by technical retrofitting runs fifty dollars per ton at Plant Victoria, and one hundred dollars per ton at Plant Albert. Let us assume further that the price of a pollution allowance on the open market is seventy-five dollars per ton. It makes more economic sense for the owner of Victoria to retrofit and the owner of Albert to purchase additional allowances. Imagine, further, that each plant has been allocated allowances to emit one hundred tons of polluting gases per year and that each plant emits 125 tons a year. With an investment of $3,750, Victoria's owner can reduce her plant's emissions by seventy-five tons a year, leaving fifty tons of allowances to be sold for $3,750. In other words, she will break even on the deal. Albert's owner, by contrast, would have to invest twenty-five hundred dollars to reduce his plant's emissions to the one-hundred-ton-per-year level. By purchasing twenty-five tons of allowances for $1,875, the Albert plant can continue to pollute and its owner will "save" $625 in the process.

Marketable permit schemes like this have been tried with air pollutants such as sulfur dioxide as well as with carbon dioxide—the main contributor to global warming.[84] The sulfur dioxide trading scheme in the United States is widely seen as successful, with some caveats;[85] carbon trading schemes are a different story, still being written (see Box 3.1 on emissions trading around the world).[86] A fair number of critics see carbon markets as especially problematic, substituting one flawed solution for another (that is, permits in place of taxes;[87] see also Box 3.2 on carbon pricing). In particular, according to Vlachou and Pantelias, emission trading schemes are subject to short-term profit-seeking, speculation, financialization, fraud and speculation, environmental injustice, windfall profits by those who receive free allowances, deployment of cheap modifications and avoidance of innovations, and impeding the systemic changes required to implement a low-carbon economy.[88] Finally, allowance prices to date have been significantly below the estimated social cost of carbon which, depending on one's assumptions, could be between $11 and $105 (in 2015).[89] The price of emission allowances in all EU Emissions Trading System markets has recently been in the range of ten to fifteen dollars (eight to thirteen euros) per metric ton (Box 3.1), but in May 2018 rose as high as eighteen dollars per metric ton.

BOX 3.1 | Emissions trading around the world

The European Union's Emissions Trading System (EU-ETS) was the first compliance emissions trading system, beginning operations in 2005. Sectors involved in the EU-ETS are the most energy intensive: power and heat generation; industries such as iron, steel, cement, and oil refining; and since 2008, civil aviation within the European Union. From its inception, most allowances to emit (EU allowances, or EUAs) have been given to regulated entities at no cost. The bloc is slowly increasing the number of allowances to be auctioned.

Since its creation, however, the EU-ETS has been plagued with oversupply and low prices. An overly generous cap, use of offsets, grandfathering of allowances, low energy prices, and a financial crisis all combined to create the oversupply. The market price of an allowance fell from close to thirty euros (thirty-five U.S. dollars in 2017) to below five euros, hovering between four and ten euros per ton from 2013 to 2018. Perhaps as a result of recent revisions to ETS regulations, prices in 2018 have risen slightly, to between ten and thirteen euros per ton. Unfortunately, this price is far too low to provide incentives for transitioning to renewables and a transformation of the energy sector.

Despite low prices, oversupply, and companies' lack of need to purchase most of their allowances, the volume being traded in the market keeps increasing—some fifty-fold since the opening of the ETS. How can that be? As one set of commentators has noted, the carbon market in the European Union is "mainly a financial market used for hedging and speculation." Compliance transactions in the market account for a smaller and smaller portion of what is traded, challenging "the role of the carbon price whose financial and self-referential evaluation can obviously not reveal installations' marginal abatement costs, the condition of cost-effectiveness expected from carbon trading."

A number of other countries and sub-national jurisdictions—cities, states, provinces—around the world have been setting up their own emissions trading markets. The second largest carbon market after the EU-ETS is California's, created by Assembly Bill 32, and entering into operation in 2013. Regulated entities participating in the new California trading scheme—refineries, power plants, industrial facilities, and from 2015, transportation fuel distributors—are responsible for about 85 percent of the state's emissions. As in the EU-ETS, mostly free allowances and a generous cap have contributed to oversupply and low prices in this market, although a price floor has kept the value of allowances above thirteen dollars per ton for the past few years.

Other jurisdictions that have started or are looking to create an emissions trading market include the Regional Greenhouse Gas Initiative (among nine states in the northeast of the United States), Alberta, China, New Zealand, Quebec, and Tokyo, with the list growing steadily.

(Continued)

EMISSIONS TRADING UNDER THE KYOTO PROTOCOL VERSUS THE PARIS AGREEMENT

Under the UN Framework Convention on Climate Change, the Kyoto Protocol set legally binding caps for developed countries, creating the context and cap for the regional EU-ETS and the possibility of a global trading system. With the adoption of the Paris Agreement under the UN Framework Convention on Climate Change, countries are no longer bound by a globally agreed-on cap, but instead take on completely voluntary emission reduction objectives. The key ingredient for a global *cap*-and-trade system is thus no longer present.

The Paris Agreement does contain language that could set in motion the linking of domestic or regional markets. Caps would be put in place at the sub-national, national, or regional level, while the global architecture would develop rules for linking of markets and trading of emissions between jurisdictions. One essential set of rules to be developed would ensure equivalency of units traded between systems (in the language of the Paris Agreement, internationally transferred mitigation outcomes, or ITMOs).

Sources: Nathalie Berta, Emmanuelle Gautherat, and Ozgur Gun. Transactions in the European Carbon Market: A Bubble of Compliance in a Whirlpool of Speculation, Documents de travail du Centre d'Economie de la Sorbonne 2015.74 - ISSN : 1955-611X. 2015; Environmental Defense Fund, "World's Carbon Markets," 2017, accessed September 30, 2017, https://www.edf.org/worlds-carbon-markets; Chris Buckley, "Xi Jinping is Set for a Big Gamble with China's Carbon Trading Market," *New York Times*, June 23, 2017, accessed September 3, 2017, https://www.nytimes.com/2017/06/23/world/asia/china-cap-trade-carbon-greenhouse.html?_r=0; S.Borghesi, M. Montini, and A. Barreca, "Comparing the EU ETS with Its Followers," in The European Emission Trading System and Its Followers, pp. 71–90, Cham: Springer, 2016.

Regardless of whether these markets work as intended (and in the case of carbon markets, there are many who question their efficacy*), there are at least three ethical problems with such rights to pollute. First, emissions trading is based on the assumption that some level of pollution is acceptable (or necessary), even if this level continues to cause environmental damage or health problems, both widely and in specific locations. Second, wealthier parties will always be able to outbid poorer ones in an open market because of the former's willingness to pay more. The outcome might be an efficient one, but it is not necessarily fair or just.[90] Finally, is it proper or ethical to privatize something that "belongs" to everyone—humans, plants, and animals?[91]

---

*For example, a broad range of indigenous and environmental organizations have issued strident critiques of the carbon market over the past decades, including the Indigenous Environment Network, Friends of the Earth International, FERN, and the World Rainforest Movement.

BOX 3.2 | Putting a price on carbon: taxes and trading

In the United States, on a fairly regular basis, politicians, economists, and business leaders publish editorials in leading newspapers to advocate putting a price on carbon. See, for example, the 2014 *New York Times* editorial by former U.S. Treasury Secretary Henry Paulson or two former chairmen of the Council of Economic Advisors under U.S. Presidents Reagan and Bush, Jr., along with the chair of the conservative Climate Leadership Council.

The economic approach to reducing greenhouse gas emissions internalizes the cost of greenhouse gas pollution in the price of generating those emissions. This is most easily done through increasing the cost of burning fossil fuels that produce the most ubiquitous greenhouse gas, carbon dioxide. There are two main policy approaches to putting a price on carbon: tax carbon directly or create a market in emissions, through a cap-and-trade system, that enables the market to set the price.

A carbon tax is used in at least nineteen national and one sub-national jurisdictions. Sweden has used a carbon tax since 1991, with the value of the tax set at approximately U.S. $131/ton of carbon in 2016. British Columbia, Canada, has had a carbon tax since 2008, with the price currently set at CDN thirty dollars per ton. In British Columbia, the tax is redistributed to families and businesses through a variety of mechanisms, including a low-income tax credit.

Why choose one of these approaches versus the other? We encourage you to explore the various pros and cons of each method in the sources cited below, and just offer some brief analysis here. The most often cited reasons for a tax on carbon: it's much simpler to implement as there is no need to create a whole suite of rules governing issuance and exchange of allowances, whether and what sort of offset credits to allow, etc. Governments just put a tax on sources of carbon emissions, with fuel and power plants the most straightforward products and sources to tax. Why might a government decide against a carbon tax, if it's so simple? An economic reason is that policymakers might not get the price right and decide on a tax that is too high or too low to efficiently incentivize lower consumption to bring about the reductions needed. A political reason is that new taxes are often rather unpopular. British Columbia overcame this reason with what might be called a tax-and-dividend approach, where much of the tax revenue is returned to middle- and lower-income individuals who are taxed. In the United States, a coalition of political and business leaders, The Partnership for Responsible Growth, is looking to sell a carbon tax in the conservative Republican-controlled congress by using half the revenues to offset a reduction in the corporate tax rate.

Sources: Matthew Carr and Anna Hirtenstein, "How Countries Cut Carbon by Putting a Price on It: Quick Take Q&A," *Bloomberg*, April 11, 2017, accessed September 3, 2017, https://

(Continued)

www.bloomberg.com/news/articles/2017-04-11/how-countries-cut-carbon-by-putting-a
-price-on-it-quicktake-q-a; Matthew Carr, "The Cost of Carbon: Putting a Price on Pollution,"
Bloomberg, June 1, 2017, accessed September 3, 2017, https://www.bloomberg.com/quick
take/carbon-markets-2-0; Martin S. Feldstein, Ted Halstead, and N. Gregory Mankiw (Climate Leadership Council), "A Conservative Case for Climate Action," *New York Times*, February 8, 2017, accessed September 3, 2017, https://www.nytimes.com/2017/02/08/opinion/a
-conservative-case-for-climate-action.html?action=click&contentCollection=Science&mod-
ule=RelatedCoverage&region=Marginalia&pgtype=article; Charles Frank, "Pricing Carbon: A
Carbon Tax or Cap-and-Trade," Brookings Institution, August 12, 2014, accessed September
3, 2017, https://www.brookings.edu/blog/planetpolicy/2014/08/12/pricing-carbon-a-carbon
-tax-or-cap-and-trade/; Henry M. Paulson, Jr., "The Coming Climate Crash," *New York Times*,
June 22, 2014, accessed September 3, 2017, https://www.nytimes.com/2014/06/22/opinion
/sunday/lessons-for-climate-change-in-the-2008-recession.html; The Partnership for Responsible Growth, https://www.partnershipforresponsiblegrowth.org/about/; Eduardo Porter,
"Does a Carbon Tax Work? Ask British Columbia," *New York Times*, March 1, 2016, accessed
September 3, 2017, https://www.nytimes.com/2016/03/02/business/does-a-carbon-tax-work
-ask-british-columbia.html?mcubz=0&_r=0; World Bank, "Pricing Carbon," http://www.world
bank.org/en/programs/pricing-carbon; World Bank Group, *State and Trends of Carbon Pricing
2016*, accessed September 3, 2017, https://openknowledge.worldbank.org/bitstreamhandle
/10986/25160/9781464810015.pdf?sequence=7&isAllowed=y.

At the end of the day, the conundrum posed by commodification of the atmosphere is much like that involved with genetic and other biological resources. In this instance, however, marketization is based on the idea that there is a viable and acceptable tradeoff between the economic cost of environmental degradation and the cost of prevention. Beyond a certain point, however, it is almost impossible to calculate the impacts of global warming, especially if the costs are stretched out over decades. By contrast, the expenses of addressing climate change must be paid in current funds, and such expenditures are easily computed. There is a great reluctance to pay these costs, especially in the United States. Ultimately, the hope is that market exchange of pollution permits will generate the resources and technological transfers necessary to moderate the emission of greenhouse gases into the atmosphere. This, unfortunately, is no stronger a reed than that relied on to protect biodiversity.

Saving Nature through Offsetting

There are three ways that regulated firms can meet emission reduction obligations under a cap. They can satisfy those obligations by directly reducing their emissions; or with emission allowances, which can be traded among regulated entities; or they can purchase "offsets": emission reductions by entities that are not regulated but anyway undertake emission reductions that are "bundled" into units that can be bought and sold or traded. These are called offsets because the emission reductions in one place take the place of (offset) emission reductions somewhere else: for example, forest plantations could be grown in

Brazil and take the place of emission reductions from a coal-fired power plant in Montana. Coal keeps being burned in Montana, but somewhere in Brazil, a tree is taking up that carbon. Despite flaws, *The Economist* has argued that, "the idea of carbon offsets is a good one." The same rationale applies here as with putting a general price on carbon through selling permits or allowances: "market forces . . . provide financial incentives for people to find the cheapest ways to reduce or eliminate emissions."[92]

How do offsets work? Let's say a corporation is planning to build a coal-fired power plant, which will be both dirty and carbon intensive. A financial institution like the World Bank comes along and tells the company that it will finance a solar power plant in place of the coal-fired one (for the moment, we will ignore the diurnal nature of solar versus constant output of coal). The emissions avoided by not burning coal over the lifetime of the plant can be calculated and bundled, and distributed between the corporation and the financiers, who can sell them for speculative value (just as with emissions allowances) or to regulated industries that are required to reduce their emissions—the same as with an emissions permit. Forestry projects can generate similar, albeit temporary (we say more about this later), credits through "carbon sequestration," so long as the trees are not replacements for a forest already harvested. Offsets are not solely limited to the use of fulfilling legal emission reduction obligations. Individuals concerned about the emission impacts of air travel can purchase carbon offsets to put their minds at rest.*

One of Martin Luther's biggest complaints about the Catholic Church—about which he posted the Ninety-five Theses in 1517, according to the widely believed narrative—was the common practice of selling indulgences. For a "donation" paid to a priest, one could gain an exemption from punishment for one's sins. Some commentators on carbon offsets see parallels with the purchase of indulgences.[93] In the case of carbon offsetting, the idea is that you can keep flying, driving, consuming—engaging in all of these practices with their attendant carbon dioxide emissions—while in some other part of the world someone else either refrains from similar consumptive activities or actively sequesters carbon into the biosphere—into trees, soils, mangroves. Whether such offsets actually do exist or will exist in the future is more of a bet than a certainty. In the case of the power plant, the comparison pits two hypothetical power plants against each other; in the case of the forest, who can know the fate of the forest decades in the future? And where travel offsets are concerned, *caveat emptor*!

---

*United Airlines, for example, contracts with Sustainable Travel International to offer offsets to its customers and for cargo shipments. See "CarbonChoice Carbon Offset Program," accessed August 16, 2017, https://www.united.com/web/en-US/content/company/globalcitizenship/environment/carbon-offset-program.aspx; and Sustainable Travel International, "Carbon Offsets," accessed August 16, 2017, http://sustainabletravel.org/our-work/our-approach/carbon-offsets/.

Figure 3.1

Source: http://www.geekculture.com/joyoftech/joyarchives/963.html (accessed September 14, 2017).

Most of the carbon-offset market currently is made up of *voluntary* purchases—individual persons and companies buy offsets because they want to, not because they are legally required to. They might be doing so to assuage their guilt or for public relations purposes. A smaller portion of the current offset market is *regulatory* in nature—offsets are purchased in order to fulfill some portion of emission reduction obligations, in regulated jurisdictions such as the European Union and California. Regulated jurisdictions usually limit the quantity or number of offsets that may be used to compensate for emission reductions, in part because of the recognition that the environmental integrity of those emission reductions can be more carefully controlled by requiring the bulk of emission reductions to take place at home.[94]

As more jurisdictions put in place legal requirements for corporate and other large enterprises and companies to reduce their carbon dioxide and equivalent emissions—at supranational, national, or sub-national levels—the demand for offsets is likely to grow. Presently the Clean Development Mechanism (CDM) established under the Kyoto Protocol serves to generate the majority of offsets for the EU-ETS—and the ETS is responsible for purchasing close to 95 percent of offsets generated under the CDM. As we will see in chapter 5, the CDM is facing existential threats due to its legal status as part of the near-moribund Kyoto Protocol. But while the CDM may not continue in its present form, the demand for offsets will continue to grow* and new mechanisms for generating those offsets will be created.†

Let's examine some of the main criticisms of offsetting in a little more detail.

---

*For example, the aviation industry has just introduced a market-based mechanism—the Carbon Offsetting and Reduction Scheme for International Aviation (more commonly known by its acronym, CORSIA)—that after 2020 is expected to generate a huge demand for offsets. Airlines have pledged to cap their emissions levels in 2020 and (while still increasing emissions) will maintain their caps primarily through offsetting, which could generate demand for "1.5 billion offsets in the 2020s alone." Sheldon Zakreski, "Avoiding a Turbulent Ride with ICAO," Climate Trust, May 15, 2017, accessed September 29, 2017, https://climatetrust.org/avoiding-a-turbulent-ride-with-icao-scorcher/; ICAO Environment, "The Carbon Offsetting and Reduction Scheme for International Aviation," accessed September 29, 2017, https://www.icao.int/environmental-protection/Pages/market-based-measures.aspx.

†Another source for offsets (or VERs for "verified emissions reductions") is from REDD+—"reducing emissions from deforestation and forest degradation; and the role of conservation, sustainable management of forests and enhancement of carbon stocks in developing countries"—a global mechanism linked with the UNFCCC and still rather under development. REDD+ offset credits are sold currently only in the voluntary market, and there are numerous technical difficulties still being worked out to enable those credits to be used in regulated markets such as California. In fact, the European Union has already declared it will not accept REDD+ credits in the EU-ETS because of concern over environmental integrity. However, REDD+ is not exclusively an offsetting mechanism, and might be described more generally as a forest protection financing mechanism. For example, the Green Climate Fund will provide "results-based payments" to countries under REDD+ as grants (no carbon credits issued, no offsetting possible), in contrast to the World Bank, which requires countries to privatize and turn over ownership of their forest carbon when receiving "results-based payments" under the Carbon Fund of the Forest Carbon Partnership Facility. Indeed, REDD+ is still very much contested, as you might divine from reading the range of references listed here. See also the further discussion in chapter 5. See Bonneville Environmental Foundation, "Carbon Offsets—International REDD+," accessed August 25, 2017, https://store.b-e-f.org/products/carbon-offsets-international-redd/; The REDD Offset Working Group, "California, Acre and Chiapas," accessed August 25, 2017, https://www.arb.ca.gov/cc/capandtrade/sectorbasedoffsets/row-final-recommendations.pdf; Craig A. Hart, *Climate Change and the Private Sector*, London: Routledge, 2013; Esther Turnhout, et al., "Envisioning REDD+ in a Post-Paris Era: Between Evolving Expectations and Current Practice," *WIRE's Climate Change* 8, no. 1:e425 (2017), accessed October 1, 2017, http://onlinelibrary.wiley.com/doi/10.1002/wcc.425/full.

- **Offsets don't bring about overall emission reductions.** Offsets don't reduce emissions—they offset them. Moreover, they may or may not actually offset them. A study of CDM projects for the European Union found that 85 percent of the claimed emission reductions in the projects analyzed might not actually be additional but would have happened anyway, if they existed at all.[95]
- **Problematic accounting.** Ideally, any offsetting of emissions should be as permanent as the emissions from the activity being offset. If I fly on a plane, those are real emissions that are never going away. Permanent emission reductions should be just that: compensating for the burning of that jet fuel. Many offset projects, however, involve sequestering emissions temporarily, a problem because a ton of carbon taken up by a tree can easily turn into a ton of carbon in the atmosphere when that tree is cut down or burned. Markets address the problem of permanence through discounting—making a future ton of tree carbon worth a lot less than a ton of permanent emission reductions and reducing that tree carbon's future value. Critics say that fossil emissions should not be considered *fungible* with sequestered terrestrial carbon.*
- **Perverse incentives.** What happens when the price someone will pay for offsets is greater than the value of using land for food production? Many offsetting projects involve forest conservation or reforestation. Some "negative emission" sequestration technologies require the burning of trees and other biomass for energy, then capturing carbon as liquefied carbon dioxide or charcoal to bury in the ground. It is not difficult to imagine that the demand for land-based offsets for the wealthy of the world could displace those whose lives and livelihoods are based on small-scale agriculture or forest resources.[96]
- **False assumptions.** Two assumptions are important to interrogate here. The first is that there is an efficiently functioning market that can calculate the price of a ton of carbon, which is directly related to the cost of technologies to replace carbon emissions with renewable alternatives or other means to reduce or eliminate those emissions. It is clear, however, that in the current European market most of the carbon transactions are speculative in nature (see Box 3.2 for further discussion). The carbon price is being set not through the careful calculations of the actual price of emission reductions by market actors but, rather, through the value of carbon futures, options, and asset-backed securities that are being traded on a secondary derivative market.[97] The second assumption is the even more troubling one that we have time and space to trade offsets around the world, postpone actual emission reductions,

---

*The point here is that the emissions from the jet fuel originated in "fossil carbon" long buried in the earth but released through extraction, while the offset coming from a tree might, in the near future, simply disappear. See, for example, Brendan Mackey, "Counting Trees, Carbon and Climate Change," *Significance* 11, #1 (Feb. 2014): 19–23.

and avoid the political difficulties of directly reducing emissions. In the 2015 Paris Agreement on climate change, countries agreed to "holding the increase in the global average temperature to well below 2°C above pre-industrial levels and pursuing efforts to limit the temperature increase to 1.5°C above pre-industrial levels, recognizing that this would significantly reduce the risks and impacts of climate change."[98] What exactly does that mean for what the effort must look like in the coming decades?

In its last assessment report, affectionately titled "AR5," the Intergovernmental Panel on Climate Change (IPCC) published a set of estimates for the total remaining amount of carbon dioxide equivalent—that is, the aggregate carbon budget*—that could be emitted by 2100 while remaining below 2°C or 1.5°C of warming with some calculated probability. Carbon dioxide remains in the atmosphere for hundreds to thousands of years, so scientists look at cumulative concentrations in the atmosphere and correlate those with temperature increases.

According to the IPCC, a two out of three (or 66 percent) chance of remaining below two degrees Celsius requires that cumulative emissions by 2100 be less than twenty-nine hundred metric gigatons (or billions of tons).[99] As of 2011, humans had consumed about nineteen hundred gigatons of that "budget," leaving only one thousand gigatons to be emitted over the next eighty years. At current global emission rates—around forty gigatons of carbon dioxide per year—that remainder will be gone by 2036—or twenty-five years from now. Once this happens, staying below two degrees Celsius will require halting *all* further emissions of carbon dioxide. Avoiding this outcome requires, at a minimum, a rapid phase-out of fossil-fuel burning beginning now with elimination of coal-fired power plants and proceeding from there.[100] To give some idea of how daunting a challenge this actually is, in 2001, IPCC Working Group III (Mitigation) estimated that the economically recoverable fossil fuel supply added up to some five thousand gigatons[101]—this at a time when the cost of a barrel of oil was around thirty dollars (in May 2018, that cost was about seventy dollars, both in current dollars). And as we have noted

---

* According to the IPCC, "cumulative emissions of CO_2 largely determine global mean surface warming by the late 21st century and beyond" (IPCC AR5 SPM p. 62). So while there are several other greenhouse gases emitted that are also responsible for planetary warming, scientists use carbon dioxide concentrations as their reference when talking about remaining atmospheric space. The term "carbon budget" could be a bit misleading. Budgets renew every year. This one doesn't. Some alternatively call the concept *atmospheric space* to convey the idea that carbon dioxide accumulates, we have a finite cap on how much we want to accumulate, and we're rapidly running out of space. Michael Oppenheimer of Princeton uses the visual of a bathtub with a very clogged drain—we only slowly lose carbon dioxide from the atmosphere and we can only fill the bathtub so far.

earlier, factoring in inflation, oil is roughly as cheap as it has ever been over the past 120 years. It will be extremely difficult to give up the habit.

More emissions equals more warming, and offsets don't decrease emissions, they just move them around the planet. When critics of offsets point out that we don't have the time or the space to permit business as usual to continue with just a little less guilt, this is what they mean: offsets cannot solve the problem. Thus, when the International Civil Aviation Organization agreed on a new treaty to continue to allow aviation emissions to increase until 2020 and then merely offset the predicted additional 2.5 gigatons of growth in international aviation emissions through 2035, some activists called it a "plan to burn the planet."[102] If those offsets were based on forest carbon sequestration, as many are advocating—although it is difficult to see how 2.5 gigatons in offsets would be possible—while global temperatures continue to rise, many of those

Figure 3.2 Global CO$_2$ Emissions Past, Present, Future

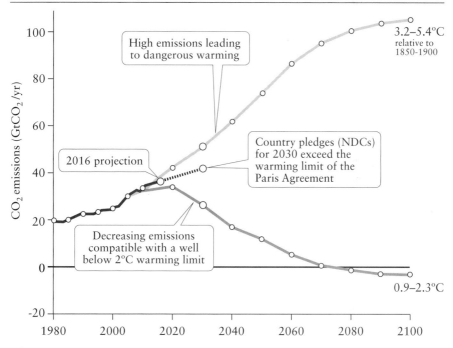

Global CO$_2$ emissions since 1980 (solid black) and country pledges (Nationally Determined Contributions, or NDCs) under the Paris Agreement (dashed) compared to a high emissions scenario (top) and a scenario compatible with limiting warming to 2°C above pre-industrial levels (bottom).

Source: Roz Pidcock, "Analysis: What global emissions in 2016 mean for climate change goals," *CarbonBrief*, Nov. 15, 2016, https://www.carbonbrief.org/what-global-co2-emissions-2016-mean-climate-change (accessed August 16, 2017), based on data from Corinne Le Quéré, et al., "Global Carbon Budget 2016," *Earth Systems Science* Data 8 (2016): 605–49, based on Joeri Rogelj, et al., "Paris Agreement climate proposals need a boost to keep warming well below 2°C," *Nature* 534 (June 10, 2016): 631–39.

forests may become more vulnerable to fires, leading to release of sequestered carbon, and further detrimental positive feedbacks.

Biodiversity Offsets

A more recent twist on the proposition that we can sell nature in order to save it is the concept of "biodiversity offsetting."[103] Working through mechanisms such as wetland mitigation and species and habitat (or "conservation") banking, the basic idea is to conserve a wetland/species/habitat in one place in order to enable the guilt-free demise of similar habitat somewhere else. To create conservation banks, three steps[104] are required:

- Nature has to be turned into recognizable, standardized, countable units that can be valued, bought, and sold, with "complex ecological processes . . . simplified into proxy indicators that can be easily traded,"[105] such as number of breeding pairs per acre.
- Once those units are recognized (or metricized), they must be assigned a monetary value—the commodity must be given a price in order to make it scarce. Differences within species and among sites must be ignored and made irrelevant—the units must be made commensurable and fungible, with a single exchange value.
- Finally, someone—the state—has to write and authorize the rules for all of this, including how markets will operate. These rules are necessary to create both the supply and the demand and bring "orderly" exchange to the process.

This approach to "saving" nature raises ethical issues we have already touched on in our previous discussion of commodification of land and biodiversity. Commentators have also raised at least four other concerns. First, markets in conservation rely on the *destruction* of biodiversity to generate demand for offsets. Without a strong demand through continued development, the price of the offsets won't be high enough to encourage enough species or habitat banking in the first place. This seems like a fairly illogical and irrational system—destroying biodiversity in order to save it—and inevitably leads to net losses over time. A second concern involves equity implications: those with money will be able to direct the uses to which land is put while those without will be pushed off their lands. In some cases, the mitigation value of the land may be greater than productive uses, particularly in developing countries. We examine this concern in more detail in chapter 5, when we look more closely at using land for climate mitigation purposes. Third, market solutions seem easy and straightforward, leading to a preference for simplicity over complex but more substantive solutions that will actually protect species and ecosystems. Finally, a system that turns nature into homogenous units ignores not only the biological and ecological diversity of the system but also social diversity, embedded in local knowledges, practices, and values. All of this diverse value is lost when nature is measured in terms of money alone.[106]

Financialization: The Next Environmental Frontier?

Most of today's trade of carbon credits takes place in secondary markets, as futures, derivatives, and other instruments that bet on higher (or lower) prices in the future or bundle credits (like mortgages*) with potentially high returns.[107] Similar proposals have been made for biodiversity derivatives, for hedging against the risk of species extinction and for green bonds.[108] At the time of writing this book, most of these seem to be still in the realm of academic proposals.† The need of the global financial system for new asset classes seems unstoppable, so we think it is probably more a matter of when, than if, new nature-based derivatives are created. The next phase of commodification is, so it seems, the financialization of nature (see the Earthshares® proposal in chapter 2).[109]

How might such financialization of nature take place? Financialization rests on the proposition that there exist untapped stores of underutilized capital in natural or other assets that can be extracted to provide high-return investments.[110] A simple example would be one hundred acres of grazing land renting for two hundred dollars per year per acre that could be transformed into higher-yielding farmland for rent at, say, five hundred dollars per acre per year. Private equity investors would buy the grazing land outright for its grazing value at the lower price (for example, two hundred dollars per year for twenty years, or four thousand dollars per acre) and rent it at the higher one (five hundred dollars per year for twenty years, or ten thousand dollars per acre), providing an annual return on investment of 250% percent for the term of the lease. This same approach could be applied to nature in terms of carbon sequestered or genes identified.

GLOBALIZATION AND THE ENVIRONMENT

These days, we are more likely to read or hear about the "great unraveling of globalization" than about the wondrous new world it will bring us.[111] For much of the post–World War II era, it was conventional wisdom that "free trade" based on economic liberalism would make the world richer and more peaceful, as innovation and communication brought to the world's people modernization and familiarity with each other. While there have always been those critical of and opposed to unregulated transnational expansion of capitalism and technology,[112] especially as the distributive effects of the uneven development of globalized capitalism have become more apparent, the social

---

*For a primer in speculative financialization of the rental market, see Desiree Fields, "Unwilling Subjects of Financialization," *International Journal of Urban & Regional Research* 41, #4 (2017): 588–603.

†But see also a recent report from Credit Suisse and the McKinsey Center for Business and Environment, "Conservation Finance From Niche to Mainstream: The Building of an Institutional Asset Class," 2016, accessed December 3, 2017, https://portals.iucn org/library/sites/library/files/documents/2016-001.pdf.

externalities arising from globalization have triggered a revival of nationalism, xenophobia, and protectionism. Governments and states, which were once believed to be withering away, have reasserted their power and authority in many areas, seeking to control flows of capital, people, and knowledge. Where this might lead is not so clear.

But what is "globalization"? And why does it matter—even today—for the environment and nature? The term has so many definitions that there is no canonical one; we rely, therefore, on the rather long and detailed definition coming from one of our own books:

> Globalization is a discourse, simultaneously an *idealist* set of beliefs, a *behavioral* set of principles, rules, and activities, and a *material* set of outcomes and infrastructures. Globalization is *idealist* in the sense that it is reified as a complex process that will make the world richer and happier. It is rationalized and naturalized in the name of "efficiency, competition and profit," as an inevitable concomitant of the historical triumph of liberalism. Those who believe, moreover, will surely reap the benefits. Globalization is *behavioral* in the sense that the social reorganization of existing institutions as well as changes compelled in the practices of real, live people alter and disrupt "normal" life. . . . Finally, it is *material* in the sense that capital, technology, goods, and, to a limited degree, labor are able to move rapidly to areas with high returns on investment, without regard to the social or political impacts on countries, communities, and people. Moreover, just as earlier industrial revolutions have left their physical traces on the landscape, altering it forever, so, too, has this latest round. Globalization also has contradictory political effects. It offers numerous opportunities for social movements and other forms of political organization and action even as it disrupts existing beliefs, values, behaviors, and social relations. At times, these disruptions can generate violence and even war.[113]

In many ways, there is nothing new about globalization.[114] Human ideas and practices have been "globalized" for millennia; societies and civilizations have always learned from and traded with one another. The spread of agriculture was an ancient case of globalization; the incorporation of non-European lands into international capitalism was another. In historical terms, therefore, the current phase of globalization is only the latest in a process of international economic extension, expansion, and integration that, by some accounts, began about 1500 and perhaps even earlier. What is new is the scale and volume of capitalist expansion and the commodification of things never before exchanged in markets, such as genes, air pollution, and whale watching. For the purposes of this book, we are most concerned with the alteration of nature and society resulting from the processes described above. Globalization not only involves more complex forms of production, trade, and consumption of material and intellectual goods—it is also very information intensive. It is information intensive in regard to the complexity of production processes, in regard to the commodification of knowledge, and in regard to our understanding of its effects on both social and natural environments. Consequently, there are close links between human impacts on the environment and the so-called Digital Revolution.

One way to gauge the extent of globalization is through the global economy—the combined economic activity of all countries, as measured by aggregated gross domestic product.* This has expanded from less than nineteen trillion dollars in 1970 to more than seventy-five trillion dollars today.[115] World exports have grown tenfold from 1970—from about $2.5 trillion in 1970 to about twenty-two trillion dollars today—even as the global population only doubled, from 3.7 billion to 7.5 billion.[116] And by some estimates, *daily* transactions in international currency and capital markets may total as much as $5.3 trillion (that's almost two thousand *trillion* dollars annually).[117] All of this is rather abstract, but the movement of capital has impacts in terms of where it goes and how it is used. If it is invested in mines and factories, the results may be negative for both environment and workers. Even if such capital is invested in U.S. Treasury or other bonds, the associated wealth can motivate growing consumption.

Another globalization metric is the volume of mobility, via communications and travel. In chapter 1, we noted the extraordinary global growth in the number of smartphones over the past decade. The global number of Internet users has increased from 738 million in 2000 to 3.2 billion in 2013.[118] There are an estimated 4.3 billion email accounts in the world, whose users send more than two hundred billion emails every day.[119] More than fifteen million texts are sent every minute.[120] The volume of airline travel is an indicator of globalization, too, reflecting the growth in both personal and business wealth: the total number of air travelers rose from 310 million in 1970 to 3.7 billion in 2016, who flew more than four billion miles, consuming almost one hundred billion gallons of jet fuel.[121]

Chains, Chains, Chains!

There are several arenas in which the environmental impacts of globalization and the expansion of capitalism over the past few decades can be seen most clearly: commodity and production chains, and transnational flows of money, goods, consumption, and waste disposal. Increasingly freer trade, combined with high labor and social costs in industrialized countries, as we have already noted, has facilitated movement of production offshore from home countries, along with a shift from the mass production of identical products to differentiated production involving goods and services tailored more closely to individual tastes. In the international division of labor and consumption, many (if not most) countries and companies in the Global South send raw materials, commodities, semi-processed materials, and parts to each other and then to the Global North, and receive finished goods in return. The spatialization of

---

*All dollar figures in this paragraph are in 2016 U.S. dollars.

production across many countries has also had the effect of diffusing environmental impacts, and growth in global free trade has made it easier to "export" environmental bads, such as toxic wastes and pesticides banned in industrialized countries, to countries less able to afford and less willing to impose strict regulations.

This process results in four types of environmental externalities. First, as production is moved around and reorganized, environmental impacts are also relocated, with concomitant effects on health and nature in new locales. Second, changing forms of commodification and growing levels of consumption increase the volume, diversity, and, perhaps, toxicity of the resulting waste stream. Third, the drive to increase the efficiency of investment and production results in organizational externalities, especially as rural populations migrate to cities, corporations move leaving behind unemployed workers, and the lower relative costs of goods and services increase consumption and waste. Furthermore, as traditional or familiar forms of social relations and relations of production are altered or destroyed, people may find it necessary to over-exploit their local resources or move to urban areas with inadequate housing and infrastructure. Finally, the rising cost of land in some cities and countries is driving flows of capital to acquire and build on land in other cities and countries, where it is cheaper, hollowing out urban cores, and renting these spaces to foreign travelers on a short-term basis.[122] As tourists seek out lower-cost sites to visit, they impose growing burdens on local environments.

What is less clear, either theoretically or empirically, is how the resulting environmental impacts are distributed across countries and societies—as we have suggested, there is a structural tendency for the wealthy, who can pay for environmental quality, to externalize their environmental impacts—pollution, toxic wastes, slums—onto the poor. Globalization seems to reproduce patterns of environmental injustice (although this view is not shared by everyone.) For example, producers may seek to build factories in countries where environmental and other social standards are lowest, since this will reduce their costs of doing business and allow them to escape the more stringent regulations in the Global North.[123] Moreover, there is only limited consistency in environmental regulation, monitoring, and enforcement among countries, and there has been a general worsening of environmental quality in those countries where production for export has been a driving force behind rapid economic growth (as in southeast Asia and the People's Republic of China).[124]

The extended commodity chains and life cycles that characterize globalization and spread environmental impacts over many countries have three dimensions. First, there is what is called the "input-output structure," consisting of a set of products and services linked together in a sequence of value-adding economic activities, including final consumption. Second, there is a sequence of "territoriality," which includes a spatialized pattern of production and distribution networks made up of enterprises of different sizes and types. Finally, there is a "governance structure," which involves authority and power relationships that determine how financial, material, and human resources are allocated and flow within a chain.[125] There is also a fourth dimension that is not ordinarily considered in discussions of commodity chains: the generation of wastes at each step of the process.

Table 3.7 shows the various steps in a generic commodity chain. The final product could be a car, a T-shirt, a smartphone or tablet computer, or even a banana. Step 1 usually takes place in locations with high levels of intellectual knowledge, where people possess specialized design and analytical skills and work with computers at desks in offices. Even the conceptual design of a product generates various kinds of wastes, as shown in the figure. If we are dealing with a product that requires raw material inputs, step 2 of the chain requires mining, refining, or growing. Here, we find land being transformed, with concomitant soil erosion. Mining leaves behinds piles of spoils (tailings), and the processing of minerals may require not only crushing but also separation with chemicals. What is left after the desired mineral is extracted is of little use and is often simply piled on the ground or dumped into evaporation ponds. Agriculture can also be chemical intensive, polluting land and water and poisoning both workers and other species.

In step 3, the raw materials are further processed. Minerals may be smelted and cast into more tractable forms, such as bars and ingots. Grain may be milled, and fibers woven. Left behind are more unusable tailings, plant wastes, chemical pollutants, dust, and greenhouse gas emissions. Only in steps 4 and 5 does the product begin to acquire a finished form. Here, in a process that may take place in more than one factory or country, bars, ingots, fabrics, and other chunks of material are turned into parts for preliminary or final assembly. Again, all kinds of waste products are left behind, to be thrown out the door, shipped into the countryside, or sent to landfills or for recycling. Some of these wastes may be of value to the poor, but some may also be quite toxic.

Step 6 is shipping. Goods are sent from the host country to the home country, by truck, ship, or plane. Transportation requires energy, and energy generates greenhouse gases and other pollutants. From the port or airport, the goods go to warehouses. More energy is used and more waste is produced in step 7, as the product travels to retailers. Then, the consumer travels to a store to buy the product, and the customer uses the product. Finally, at the end of its useful life, the product is thrown out (step 8). Some parts may be recycled, but not all are either desirable or safe. Others might be burned or buried. "Everything must go somewhere."*

Consider our iPhone (Figure 3.3): It begins its life in the corporate campus of Apple in Cupertino, California, as indicated on the supply chain map. There, Apple engineers decide how the next generation phone should look,

---

*This is one of Barry Commoner's famous "Four Laws of Ecology": (1) everything is connected to everything else; (2) everything must go somewhere; (3) nature knows best; and (4) there is no such thing as a free lunch. See Barry Commoner, *The Closing Circle: Nature, Man, and Technology*, New York: Knopf, 1971.

Table 3.7. Steps in a product commodity chain

Step	Country or location	Process	Steps involved	Wastes produced
1	A	Product design	Concept, market study, schematics, assessment	Office wastes, energy, air and water pollution, land use, greenhouse gases
2	B, C, D, et al.	Primary commodity production	Mining, agriculture, material synthesis	Tailings, land disruption, chemical emissions, air and water pollution, energy, greenhouse gases
3	E, F, G, et al.	Primary and secondary material processing	Smelting, casting, milling, weaving	Excess materials, chemicals, tailings, land change, air and water pollution, energy, greenhouse gases
4	H, I, J, et al.	Component manufacture	Producing parts, cutting and sewing, finishing	Excess materials, energy, air and water pollution, greenhouse gases
5	K, L	Sub- and final assembly	Workers and robots put parts together	Packaging, rejects, energy, greenhouse gases
6	M	Transport	Shipping to consumption site	Energy, greenhouse gases, air and water pollution
7	Many places	Marketing and use	Distribution to retailers, consumer purchases, activities in usage	Energy, greenhouse gases, packaging, sewage, toxics, adhesive bandages
8	N, O	Disposal	Burning or burying, recycling, composting, extraction of metals	Excess materials and packaging, toxics, air and water pollution, greenhouse gases

what it should include, and what will be inside of it. As the map indicates, the various parts are manufactured all over the world, although the sources of materials, such as the rare earths, that go into the iPhone are not shown. The final assembly takes place in China, from whence the phone is shipped to consumers all over the world. Since the iPhone is designed for planned obsolescence, the innards "must go somewhere." Often as not, that somewhere is a country in the Global South, where the phones are disassembled and broken down for the valuable minerals contained in them by the very poor, who are exposed to highly toxic materials as a result (smartphones are too hazardous to go into landfills and labor is too costly to foster disassembly and disposal in the Global North). Figure 3.4 shows a generic (and probably incomplete) map of global electronic waste disposal routes.[126]

Figure 3.3 Supply Chain for Apple's iPhone

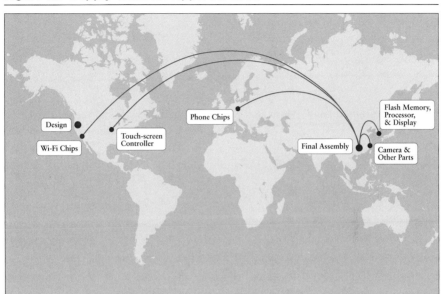

The iPhone is assembled in China from globally sourced components.
Source: Joseph Valacich and Christoph Schneider, *Information Systems Today: Managing in the Digital World*, seventh edition, ©2016. Reprinted by permission of Pearson Education, Inc., New York, New York.

Globalization has had major impacts on food production and food security, in a number of ways that contribute to the export of environmental externalities and their reimportation. Some have called this changing system the "corporate food regime," since it is driven more by profit and accumulation than need.[127] With globalization, agriculture and food systems have changed in several ways. First, with technological changes in farming, refrigeration, and shipping, it is now possible to cool and quickly move fruits, vegetables, and flowers around the world, especially from the Global South to the Global North. Along with diffusion of biotechnologies and expansion of international grain markets, this trend has led to a massive shift from subsistence farming to cash cropping in developing countries.[128] Indeed, cash crops have become a favored generator of both income and hard currency, allowing farmers to purchase imported food and countries to finance the industrialization of agriculture. Specialization in food production has also been fostered by industrialized country trade barriers to certain types of agricultural imports. For example, it is easier to trade in vegetables than grains, as production of the latter is heavily protected in both the European Union and the United States. As a result of the globalization of cash crop farming, during the northern winter it is possible to purchase virtually all kinds of foods imported from the southern hemisphere that were simply not available twenty years ago. Once again, externalities

Figure 3.4 E-Waste Disposal Routes

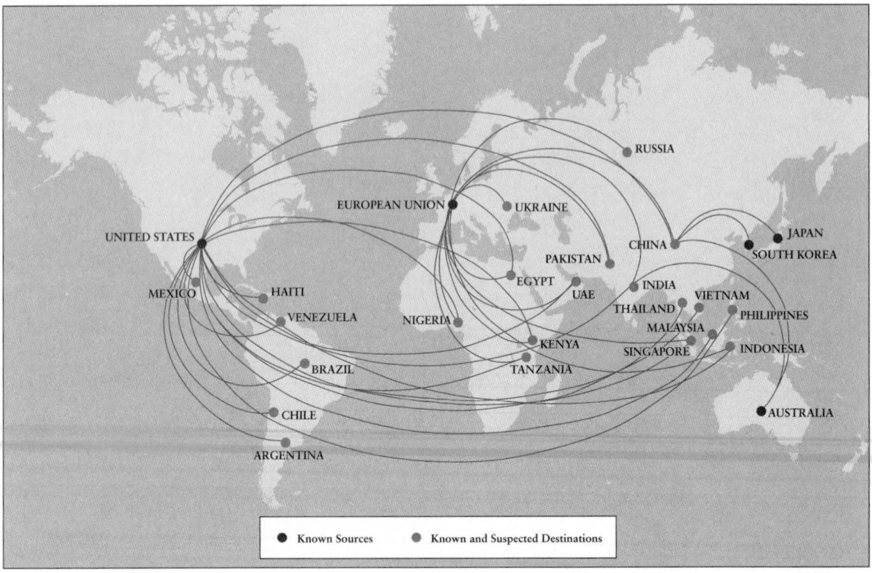

Source: Zak Cheney Rice, "7 Stunning Statistics Show the True Cost of Owning a Smartphone," Mic, April 14, 2014, https://mic.com/articles/87617/7-stunning-statistics-show-the-true-cost-of-owning-a-smartphone#.vLP2W0cdT (accessed August 17, 2017).

remain behind (although some traded food products carry chemical residues with them).

Second, the introduction of patented hybrid and otherwise "improved" seeds into global agriculture has increased some food supplies while imposing severe limits on smallholders. Most export crops are capital, rather than labor, intensive. Farming for export relies on chemicals and high-priced hybrid seed for fruit and vegetable uniformity, machinery for volume, irrigation, and high-quality land for productivity. These tend to favor richer farmers (mostly male) and corporations, who have the capital to expand their small farms and grab vast tracts of land. Owning more land, they are considered more creditworthy than those with less land and money, for such agriculture almost always requires the taking out of loans. Cash cropping often uses more technology and requires less labor than is required for subsistence agriculture; thus, poor farmers are essentially forced to move to lower-quality, often hilly land, and the landless have to migrate to cities. In the former instance, serious land degradation may follow from agriculture on steep slopes; in the latter, growing numbers of impoverished migrants outstrip the capacity of the urban infrastructure, more often than not already in poor condition, to handle the increased burdens that result.[129]

Third, the chemical intensity of much corporate farming exposes those living near such farms to a medley of toxic chemicals that are sprayed into the air and leached into groundwater and which flow downstream in creeks and rivers, creating dead zones at their marine outlets. The overuse of nitrogen fertilizer contributes to this type of water pollution, while the need of improved seeds* for pesticides and herbicides offer ample opportunity for misuse and accidental poisoning (although some deeply indebted farmers have used those toxins to commit suicide).

There is an ongoing and furious debate about whether the globalization of the food system provides net benefits or imposes net costs on the world, and especially what the impact might be on the billions of poor who are undernourished, not because there is a scarcity of food but because they cannot afford to buy it or grow it. Moreover, the epidemic of obesity in some societies arises as much because of this corporate food system as overeating. At the same time, globalization of the food system has provided many with the income needed to move up into the middle class and to ensure household food security. This is not a topic we will address further, but the literature is large and expanding, and well worth exploring.[†]

Data and Dematerialization

Does the global shift to a "knowledge economy" auger a decline in environmental impacts? Yes and no. Clearly, to the extent that practices such as telecommuting, teletravel, and teletourism reduce both the need and emissions arising from cars, buses, and planes, those effects are reduced. This is sometimes called "dematerialization," as information displaces mining and manufacturing.[130] However, the selfsame economy has also led to growing income and consumption by those who are employed in the information industry (beginning computer coders are paid two to three times as much as new teachers) or who benefit from it in other ways. Much of this new wealth goes to goods and activities that still require materials, components, assembly, and disposal. The devices that both make possible the electronics revolution and its exploding growth require significant numbers of servers and electricity to operate them, and their frequent disposal, as we have seen, is a serious waste matter.[131] The source of power also matters: a considerable portion of the world's electricity

---

*This heavy dependence on inputs is one reason why the first time we used this phrase we put the word "improved" in quotes. The choices and direction of technological innovation is always subject to challenge.

†The Springer journal *Food Security* provides a range of articles that offer various analyses and critiques. See also: T. J. Achterbosch, S. van Berkum, and G. W. Meijerink, *Cash Crops and Food Security: Contributions to Income, Livelihood Risk and Agricultural Innovation*, Wageningen, Netherlands: LEI Wageningen University & Research Centre, 2014, accessed August 18, 2017, http://edepot.wur.nl/305638; Ulrike Grote, "Can We Improve Global Food Security? A Socio-economic and Political Perspective," *Food Security* 6 (2014): 187–200; Eric Holt-Giménez and Miguel A. Altieri, "Agroecology, Food Sovereignty, and the New Green Revolution," *Agroecology and Sustainable Food Systems* 37 (2013): 90–102; Kees Jansen, "The Debate on Food Sovereignty Theory: Agrarian Capitalism, Dispossession and Agroecology," *The Journal of Peasant Studies* 42, #1 (2015): 213–32.

generation still comes from the burning of coal, which is the most CO_2-intensive fossil fuel; at the same time, solar photovoltaics and wind electricity produce little, if any, direct pollution, but are hardly clean in either their manufacture or disposal. And many people are concerned about the effects of wind turbine blades on particular species of birds.[132]

The generation of big data through, and stored in, various electronic devices and media is also touted as reducing energy and material use. For example, car-sharing services such as Uber, Lyft, and Zipcar, which depend on rapid communication, are said to have the potential to reduce auto ownership.[133] Cars mostly sit unused during an average day; car sharing means that people don't need to own their own vehicle. Smart buildings are able to sense the presence of occupants, their need for heat, cooling, and light, and the most efficient combination of services to achieve these ends. Autonomous vehicles will make denser traffic possible, as computers and other devices take over the task of driving by unreliable humans; hence, highway and street expansion and congestion will become a thing of the past. All of these innovations will contribute toward better environmental quality, reduced demand for materials and energy, and considerable convenience to consumers. But is this what will really happen? As Zhou Enlai, Mao Zedong's prime minister, is reputed to have said to Henry Kissinger when asked about the significance of the French Revolution, "It is too soon to tell."*

Can Capitalism Save Environment and Nature?

Can today's version of capitalism be mobilized to save the environment and nature? Is the cause of the world's environmental crisis capitalism or industrialism? Can we develop an economy of "sufficiency" that is still capitalist yet not driven by growth? These, too, are questions whose answers "remain to be seen." Capitalism, as it exists today, is dependent on continual growth. Notwithstanding the machinations of Adam Smith's "invisible hand," there appears to be nothing, except crisis (or the collapse of an Antarctic ice shelf that rapidly raises sea levels), that might push the global economic system in an environmentally sustainable direction. But until that happens, most of those with a current interest in the market system as it exists will argue that that crisis is off in the future and can be dealt with then. Once the crisis arrives, of course, it may be too late to do anything. Capitalism, in any event, does not deal well with the far-off future; in fact, the future hardly matters (except, maybe, for those saving up for retirement).†

---

*In fact, an internet search suggests that it is none too clear who asked the question—André Malraux, Archduke Otto von Habsburg, and Kissinger are all cited—or when it was asked—1953, 1971—or whether it was Zhou or Mao Zedong who offered the famous reply.

†This, of course, is of central concern to the fossil fuel divestment movement: continued burning today imperils the future. See: Julie Ayling and Neil Gunningham, "Non-state Governance and Climate Policy: The Fossil Fuel Divestment Movement," *Climate Policy*, 17, #2 (2017): 131–49.

Can capitalism be made environmentally friendly? In principle, there is no reason why not. Environmental and social costs could be internalized into the prices of goods; production systems could be devised that produce much less waste, or none at all; products could be made to last much longer than they do now and they could be fully recycled at the end of their useful lives. The preservation and conservation of land, water, forests, species, and ecosystems could all become central organizing principles of a sustainable economy. The lives of the poor could be improved through education—especially for women and underserved minorities—access to health care services, greater participation in both politics and the economy, and so on. But such a system might be hardly recognizable as the "capitalism" we know today. For the well-off, there would be fewer choices in the marketplace, less opportunity to travel, greater regulation for the public good, and higher taxes to pay for necessary public services, in short, many of those things that contemporary governments are so afraid to contemplate, much less legislate.

Alternatively, ecological modernization, addressed in chapter 2, might provide a means of "saving" capitalism from destroying its material base. Certainly, and subject to limits imposed by the laws of physics, it is possible, in principle, to extract and process raw materials in much more efficient ways, to develop new production processes that use less toxic or nontoxic chemicals under conditions of constant recycling, and to manufacture goods that can be easily recycled and reused. At the same time, however, what is the point of developing ever more fuel-efficient and recyclable automobiles if the result is an ever-greater number of cars driving ever-increasing distances?

What about sustainable development, also discussed in chapter 2? Is it a viable possibility or merely a political oxymoron? As we saw, the World Commission on Environment and Development thought that economic growth in the Global South could be reconciled with the imposition of "limits," albeit those dictated by a more environmentally friendly technology. In 2017, *Our Common Future* celebrated the thirtieth anniversary of its publication, but there is little evidence that it is any more than a statement of what ought to be as opposed to what is achievable. In 2000, the United Nations issued a set of eight "Millennium Development Goals" to be achieved by 2015. Goal 7 was to "Ensure environmental sustainability," including the following:

- Target 7A: Integrate the principles of sustainable development into country policies and programs; reverse loss of environmental resources
- Target 7B: Reduce biodiversity loss, achieving, by 2010, a significant reduction in the rate of loss
- Target 7C: Halve, by 2015, the proportion of the population without sustainable access to safe drinking water and basic sanitation
- Target 7D: By 2020, to have achieved a significant improvement in the lives of at least one hundred million slum-dwellers.[134]

Some progress on these has been achieved, but the goals are still quite far from being fully accomplished (is the glass half full or half empty?).[135] In 2016, the United Nations put forward seventeen more "Sustainable Development Goals," including a number oriented toward environmental protection.[136] Whether these will be achieved also remains to be seen.[137]

So, what are the alternatives? There is considerable evidence that capitalism is less destructive than was "really existing" socialism in the old Soviet Bloc and in the People's Republic of China today. But on its own, capitalism's requirement for continual growth and material turnover is difficult to reconcile with a steady-state economic system, such as that propounded by Herman Daly (and discussed in chapter 2), much less "degrowth,"[138] which would require significant reductions in material throughputs and consumption, or the "sharing economy," whose material intensity remains to be seen.[139]

For others, nothing less than a wholesale cognitive "paradigm shift" away from capitalism and growth is required.[140] This perspective assumes that capitalism is the problem and cannot be reformed. But many advocates of such a shift also believe that, confronted with scientific evidence of the destructiveness of the contemporary economic regime and its implications for the future, thoughtful people will alter their practices and bring a new world into existence. Others argue that we must replace the consumption compulsion with a different version of the "good life" dependent more on relationships with people than things. We can be happy with less. The empirical evidence calls such visions into question and how to get from here to there is not at all clear. There are few credible models of action for social transformation, although it is clear that such changes do take place, albeit not with any deliberate direction.[141] And what do we tell the world's poor, who very probably would like to possess and consume a good deal more than is presently the case?

All of this leaves us with the perennial question: what are we to do? Social change cannot be achieved simply through the market; markets are mechanisms for the exchange and distribution of commodities, not for making vital social choices. The organization of markets reflects social choices already made—that is one reason for their being subject to regulation—possibly via public policy, alternatively as a result of domination, power, and wealth. Without regulation, markets will tend to yield benefits to those who possess resources and impose costs on those who lack them, including both people and nature. To redistribute costs and benefits not only requires political intervention into markets—and not just to internalize social costs. Such regulation also depends on political mobilization to generate public support and motivate directed action. In chapter 4, we will examine how this kind of environmental politics could be created.

FOR FURTHER READING

Beckert, Sven. *Empire of Cotton: A Global History*. New York: Alfred A. Knopf, 2014.

Block, Fred, and Margaret R. Somers. *The Power of Market Fundamentalism: Karl Polanyi's Critique*. Cambridge: Harvard University Press, 2014.

Chang, Ha-Joon. *23 Things They Don't Tell You About Capitalism*. New York: Blooms-bury Press, 2010.

Dauvergne, Peter. *The Shadows of Consumption*. Cambridge: MIT Press, 2008.

Geiser, Ken. *Chemicals without Harm: Policies for a Sustainable World*. Cambridge: MIT Press, 2015.

Hall, Derek, Philip Hirsch, and Tania Murray Li. *Powers of Exclusion: Land Dilemmas in Southeast Asia*. Honolulu: University of Hawai'i Press, 2011.

Heinrich Böll Foundation. *Dossier: New Economy of Nature*. https://www.boell.de/en /dossier-new-economy-nature.

Kloppenburg, Jack Ralph Jr. *First the Seed: The Political Economy of Plant Biotechnology, 1492–2000*. Cambridge: Cambridge University Press, 1988.

Lipschutz, Ronnie D., and James K. Rowe. *Globalization, Governmentality and Global Politics: Regulation for the Rest of Us?* Oxon: Routledge, 2005.

Lohmann, Larry. *Carbon Trading: A Critical Conversation on Climate Change, Privatization and Power*. Development Dialogue 48, September 2006.

Newell, Peter, and Matthew Paterson. *Climate Capitalism: Global Warming and the Transformation of the Global Economy*. Cambridge: Cambridge University Press, 2010.

4

Civic Politics
and Social Activism

Environmental Politics "On the Ground"

ARE ALL ENVIRONMENTAL POLITICS LOCAL?

In chapter 3, we saw that markets have been offered as the "solution" to all environmental problems and seem to offer the possibility of protecting nature with little pain or politics. Throughout this book, however, we take the position that markets are really no solution at all; at best, they displace or replace problems and often create unanticipated and wholly new ones. Moreover, while markets might encourage "efficiency," they do not foster equity or justice. So, if not markets, what? Apparently easy solutions often keep us from asking whether there are other ways to achieve the ends we seek, ways that might even avoid those intractable environmental problems in the first place. The search for alternative means of environmental protection takes us into the realm of politics, for two reasons: first, we may find that much greater changes in our beliefs, practices, and institutions are necessary than might be achievable through market-based methods; and second, that the types of decisions required for such changes to happen require broad-based, collective participation, which cannot happen in markets.* Such changes require us to examine how power is used or is present, and to decide how it ought to be deployed in order to accomplish the goals we set in the search for environmental protection and conservation. We should expect a considerable amount of debate, conflict, and resistance to proposals and actions in the process.[1] After all, societies seek consistency and stability, not change.

How would such an environmental politics be created or constituted? It cannot be simply a politics of reason or economic calculation, not because

---

*To be sure, markets engage literally billions of people, many influenced by the choices made by others. But each decision in a market is an individual one and does not take the community into account.

people are unable to reason or calculate their own interests; rather, we are all strongly and emotionally committed to customs, habits, beliefs, and expectations, as recent research has revealed.[2] An environmental politics cannot be one purely of emotion and impulsiveness, for people would soon discover themselves worse off than before (as illustrated by the thirty-year failure of international climate change policies). Nor can it be a politics of universal global change, for there is too much difference and diversity among societies and across the world for a single program to address. Finally, it cannot be a politics of individual change, either through appropriate choice or raised consciousness, for politics happens only through collective and concerted action.[3] The environmental politics we seek must be one of appropriate scale, with opportunities for productive contact, conflict, and collaboration. And that, it would seem, requires some manner of face-to-face politics. If "all politics is local," as the late Representative Tip O'Neill (D-MA) once put it, then environmental politics must be local in some fashion, too.

This chapter is about local environmental politics, or the deployment of what we called "social power" in chapter 1.[4] In the first section of the chapter, we explore the claim that "all environmental politics are local." They are "local" in the sense that we experience them directly in our own locales and we have the greatest opportunity to act effectively where we live and work with others. To be sure, it is possible to travel to other places in the world and to see firsthand both similar and different problems and politics, but it is the rare individual who is truly "at home" in many widely separated places. This does not mean that problems in those other places are irrelevant to us, in our place and space; rather, it is to argue that, to have broader impact, politics must be exercised locally.

Next, we examine how environmental politics and political activism have manifested in recent decades. Neither is, strictly speaking, a new phenomenon: concerns about the transformation of nature and responses to those changes can be traced back hundreds of years.[5] But most historians of environmentalism believe that the 1960s marked some kind of watershed in the public and political perception and treatment of the environment, one growing as much out of that decade's tumultuous politics as from the increasingly evident damage to nature.[6] Of the many social movements that emerged at that time, environmentalism has become the most institutionalized and bureaucratized, the most normalized and mainstreamed, the most connected with "business as usual."* Indeed, the absorption of the environmental movement into the body politic has become, in a number of ways, something of an obstacle to the practice of environmental politics.

In the third part of the chapter, we take a look at several forms of social power—combinations of civic politics and political activism—that engage with

---

*Contrast the look of today's environmental movement with that of Occupy Wall Street, Black Lives Matter, or #MeToo.

the environment. We use three terms—social power, civic politics, and political activism—to denote, first, the deployment of power through social capillaries, as discussed in chapters 1 and 2; and, second, to describe engagement with local institutionalized politics as well as involvement in movements with a largely local focus. All of these can, we would like to think, cultivate small "d" democracy through Hannah Arendt's "spaces of appearance."[7]

These myriad forms of social action include the projects of guerilla gardeners, pipeline activists, political parties, and even market-based and corporate social responsibility. These do not cover the entire universe of activism, but do they serve to illustrate some important strategic and tactical distinctions among approaches, and underlying theories of change held by scholars and activists, as well as revealing some shortcomings from which several of them suffer. In particular, we may discover that those activities that rely on mainstream methods to accomplish their goals have done little to change the institutions and practices that are the cause of environmental problems in the first place.

The chapter concludes with a brief discussion of social power as a specifically political strategy. This argument, developed in greater detail in chapter 6, is based on the proposition that social change requires deep political engagement with the conditions and arrangements that are the source of environmental degradation. It is not enough to make reasoned arguments that such activities and the damages they cause run contrary to the interests of the offenders and the affected, in the hope that both will see the light and change their behaviors. It is necessary to shift the collective *weltanshauung* through acting and changing business and politics as usual. Environmental politics must therefore involve praxis. *Praxis* is a Greek word meaning "action," but it is action guided by purpose, values, science, and, most of all, ethics applied in a political context.

WHY ALL ENVIRONMENTAL POLITICS MIGHT BE LOCAL

Let us begin with a few definitions: "Social power," as noted above, is the practice of politics that emerges from people acting together in pursuit of shared ethical and just goals. Some of this action takes place through "civic politics," via local institutionalized practices, including voting, lobbying, and volunteering. "Political activism," by contrast, refers to the political activity of people via various aspects of civil society, often external to the institutionalized politics of representative democracies.* Civil society comprises what is sometimes called the "third sector," or those groups and associations that are neither part of the state nor part of the market.[8] But we must be cautious in attributing virtue to civil society as a whole, since it is comprised of both political and apolitical activities, including bowling, economically oriented civic improvement associations, and even reactionary and sometimes violent organizations. Some of

---

* As with Black Lives Matter, Occupy, and other movements.

these—for example, bowling leagues—have no political content at all;[9] some, such as civic improvement associations, have major economic objectives. Others may be part of movements, local reflections of political mobilization all over a country or the world. Many such groups see no farther than their neighborhood or city, whereas others are informed by ideas, practices, and contacts that link them to similar organizations and movements elsewhere around the nation and the world. It is these locally active but globally connected groups that are the practitioners of social power as it most concerns us here.*

In the environmental arena, social power spans a broad range of activities and behaviors, usually informed by social movements, including lobbying of municipal authorities, efforts to affect and pass local legislation, nongovernmental organizations (NGOs) with a local focus, and even so-called NIMBY ("not in my back yard") and "wise use" groups (in keeping with the notion that not all movements are progressive ones). Those engaged in such activities have local concerns and objectives but are frequently informed and inflamed by ideas and practices they have learned about, or know of, in other places.[10] This does not mean that local action cannot have broader consequences but, rather, that political engagement among people in the places where they live, love, work, and play is at the root of social power, civic politics, and political activism.

The argument that all environmental politics are local proceeds, therefore, from two premises: a materialist one based in political economy, and a social one growing out of collective practices, habits, and customs.[11] The first is rooted in the historical development of local and regional political economies, which stand as background structures to environmental change as discussed in chapters 1 and 2. These relatively fixed arrangements, which are visible in the distributional features of and human impacts on landscapes, are nonetheless being constantly transformed through actions and processes linked to the global political economy. Think, for example, of the ways in which the disposition of roads and the distribution of traffic is linked to the production of petroleum half a world away or to the building of factory zones throughout Asia to supply the cars that fill those roads. Consider how slums and food deserts are related to unemployment and lack of money resulting from closing or moving of industry to cheaper locations. And remember that cycles of gentrification, displacement, and regeneration often take a long time to alter landscape and habitations.

---

*A further caveat is warranted here: some understandings of "civil society" (including the authors of this book) regard it as a necessary, if not sufficient, element of capitalist-liberal society, especially as the guardian of private property. It is this role that differentiates civil society in liberal democracies from state-sponsored movements in autocratic societies. See: John Ehrenberg, "The History of Civil Society Ideas," in *The Oxford Handbook of Civil Society*, edited by Michael Edwards, Oxford: Oxford University Press, 2011, accessed September 25, 2017, http://www.oxfordhandbooks.com/view/10.1093/oxfordhb/9780195398571.001.0001/oxfordhb-9780195398571-e-2?print=pdf.

The second premise has to do with the ways in which we imagine the places in which we live, work, and play, and the ways in which these social imaginaries become the models for the changes we then impose on nature.* As Karl Marx put it in *Capital*, "what distinguishes the worst architect from the best of bees is this, that the architect raises his structure in imagination before he erects it in reality. At the end of every labour-process, we get a result that already existed in the imagination of the labourer at its commencement."[12] (But this is surely an insufficient condition. Architects may design that which cannot be built, but even they must work with what is possible—materials, technology, skills—and what is there—terrain, soil conditions, people—if their design is to be realized.)

In other words, we confront human habitats, or *landscapes*, structured by dynamic and ever-changing social relations and ecologies. Those landscapes reflect decades or centuries of patterned and organized human activity and provide the "tracks," as it were, over which society moves today.[13] But those landscapes can be changed—are frequently changed—either deliberately or accidentally, and those changes usually (but not always) are made with some imagined goals in mind. Our goals might be romantic, practical, self-interested, or even destructive, but we envision them before we execute them. Together, these two factors—what is given and what is imagined—comprise the material and conceptual places and spaces in which environmental change and damage are most clearly evident to people on the ground.

A developer might decide, on her own, to build a housing tract that will make major changes to a local landscape. Although she would need to receive all kinds of planning permission to proceed, and might even solicit "public input" into the design, the idea and project would be hers alone. She would need to take into account the historical, natural, and social features of the site but, in the end, would probably be driven most strongly by economic-related considerations. Can permission and permits be obtained without too much trouble? Will the project turn a profit? Will adapting to the site and to the future impose costs that cannot be recouped? Are there potential buyers for the houses at the project price? And will she bear future liability if something should go wrong (for example, flooding that might have been anticipated but was not addressed)? Success depends not only on community approval and nature's benevolence but also on the developer's commitment and will.

---

*There is a clear dialectic at work here, in that it is a combination of what we dare to imagine, what is possible, and what is available (or there) that generates changes in people and landscapes. On "social imaginaries," see: Suzi Adams, et al., "Social Imaginaries in Debate," *Social Imaginaries* 1, #1 (2015): 15–52; Astrida Neimanis, Cecilia Åsberg, and Suzi Hayes, "Posthumanist Imaginaries," in *Research Handbook on Climate Governance*, edited by Karin Bäckstrand and Eva Lövbrand, pp. 480–90, Cheltenham: Elgar, 2015; Lorenzo Pellegrini, "Imaginaries of Development through Extraction: The 'History of Bolivian Petroleum' and the Present View of the Future," *Geoforum* 90 (2018): 130–41.

By contrast, imagine that a group of people want to restore a marsh in a low-lying area of a city that has been flooded periodically and abandoned to nature. While the initial idea might come from an individual, the overall project would require collective collaboration to be designed, developed, and deployed, drawing not only on city departments but also activists, (former) residents, scientists, conservation specialists, restoration biologists, and others. Cost-benefit issues might enter into the process, but it would be essential to compare the near-term project cost to future savings from avoided flooding, since restoration would not generate a stream of reliable revenues. And, presumably, the actual physical process of reconstructing a marsh would require a pool of volunteers willing to commit considerable time to the project, now and in the future. In this instance, power rests on the very idea of environmental praxis—of directed, ethical action—and involves an approach to politics that depends on bringing like-minded people together to discuss, debate, and design a specific environmental project, and then to develop it themselves. This is the essence of social power as praxis.

History, Political Economy, and Place
In *The 18th Brumaire of Louis Napoleon*, Marx wrote that,

> Men make their own history, but they do not make it as they please; they do not make it under self-selected circumstances, but under circumstances existing already, given and transmitted from the past. The tradition of all dead generations weighs like an Alp on the brains of the living.[14]

To this, we might add, as well, that the material artifacts of that history also weigh on "the brains [and bodies] of the living." Human activities are, to a large extent, patterned and habitual. Not only do we fall into rather fixed schedules in our everyday lives, we also tend to repeat behaviors and duplicate social practices that result in the ends we imagine and desire (and many of these practices are normative and internalized).[15] For more than a century, for example, railroads were the most efficient ways of accessing continental interiors and moving around people, armies, and goods.* They were also quite profitable for those who financed and built them (often due to government land grants to entice construction and development). Railroads transformed landscapes through their construction and through the various activities that they made practical and accessible. They affected the success or failure of agriculture, they contributed to the growth and shrinking of cities, they moved mountains of raw materials and commodities over thousands of miles. For the most part, those railroads remain where they were originally laid down, still moving commodities, goods, and people.[16]

---

*Actually, transport by water was much cheaper but only possible to destinations on navigable waterways. The interiors of continents were difficult to reach.

But during the two-hundred-year history of railroads (between about 1820 and 2020), other forms of transportation were imagined and developed, consonant with changing technologies. The development of the internal combustion engine ultimately led to the production of more than one billion automobiles and trucks and to the construction of vast road networks, often on top of or next to already-existing rails (some roads were built on old covered-wagon trails, too). New transformations, new patterns, new landscapes followed. Neither the old landscapes nor the continuing impacts on human consciousness and practice vanished, as new layers were built over old ones.[17] As on a palimpsest, whose earlier writings have been overwritten but not entirely obliterated, older landscapes overwritten by later human practices never vanish completely.*

In other words, the environments that people confront in their daily lives are not easily changed, abandoned, or erased. Those environments are understood best not as a collection of resources or a hierarchy of ecosystems but, rather, as landscapes created over time through complex interactions between patterned human practices and nature. The history of those practices is etched, so to speak, in political economy, in both institutions and artifacts. From this perspective, nature is never "pristine" or untouched by human action, and may even be the result of millennia of human intervention and modification.[18] The landscapes we transform are already part of "second nature."†

But such landscapes are constructed and transformed not only through human methods of production, which is how we often think of environmental change and degradation, but also as a consequence of social reproduction, a process we rarely consider. To understand more fully how imagination is turned into material reality, however, we need to examine how and where such visions originate. What are the sources of "imagined landscapes"? How do these visions mirror the inner life of human societies, beliefs, and behaviors? Why are some transformations accomplished and others not? The antediluvian biological and geophysical world ("first nature") offers opportunities for and imposes constraints on what changes can be made to it—mountains are difficult to move. The same is true of second nature, altered and molded as it is by people in ways that meet human needs and desires. In other words, the transformation of nature involves not only the physical process of modification (and degradation) but also cognitive processes of imagination and design. These visions of nature transformed are implemented locally, no matter how

---

*The best examples of this process can be observed in cities like Rome, where new construction must always (now by law) take into account protection of the ruins built by and terrains shaped by previous generations.

†Recall that "second nature" is nature as transformed by human action; for example, apparently primeval forests were managed by indigenous peoples through controlled use of fire, or consider the nature we find in New York's Central Park. O'Connor, "Three Ways to Look at the Ecological History."

far away the visionaries, architects, and financiers might live. Their politics are local, not global.

There are at least four important types of vision and action involved in such "reconstruction" of nature. First, people arriving in a new place tend to reproduce what is familiar and customary, recreating old abodes and material practices in their new homes. That is why parts of the American Midwest often appear similar to landscapes in northern and central Europe and invasive species appear in so many suburban gardens. But such efforts to reproduce the natal homeland sometimes fail. Nature does not permit all changes, and people already living in such places—whether indigenous or not—often resist new arrivals and their grand ideas. Still, traces of "the old country" can be found in landscapes everywhere.[19]

Second, people's perceptions of the physical environment are mobilized as cognitive elements in the transformation of nature. That is, some might see a wetland as a "wasted space" while others regard it as a marine "nursery." Those who want to "develop" the wetland must convince a host of people that it is useless or inefficient as it is and will be more benign if converted into commodities and capital (as in turning it into housing with canals instead of driveways). If the developers succeed in their project, the conversion of one landscape into another will generate capital, foster accumulation, and enable people to seek out other opportunities for further transformation of nature and accumulation of capital.[20] The altered landscapes can be returned to their former condition only at great economic and political cost, and efforts to protect what remains unexploited often encounter stiff resistance from those who benefit—or think they do—from continuing transformation, modification, and destruction.

Third, there is what is called "primitive accumulation," whereby those with capital and technology steal land, resources, and minerals from those who live on the land, and imagine and build vast plantations and industrial complexes to exploit what nature has created. This point was discussed in chapter 3: while those who take wealth from the poor and weak never imagine what will be left behind after their projects are implemented, they rarely care that what is left behind will continue to exist as ruin and devastation.

A final form of transformation arises from what we could call "environmental idealism": the effort to restore altered or degraded landscapes to some approximation of a previous condition, a historical nature as it might once have been.[21] Nature, in these schemes, becomes not only something to be manipulated in the pursuit of material ends but also a reflection of who people are or whom they would like to become. Restored wetlands and wildernesses are the material realization of such environmental "idealism," expressions of notions of humans as stewards or guardians of nature.[22] As we noted in chapter 3, these may even be commodified in the form of conservation offsets.

Such transformative processes are central to our proposition that "all environmental politics are local." It is not that the causal chains of damage necessarily arise locally or that the effects of our local activities do not extend outward into the rest of the world. Rather, it is that the greatest impacts of environmental action, both destructive and restorative, are primarily local, and it is the pattern or form of praxis in one locale that is reproduced elsewhere. A stream or river that might be dying or dead because of the pollution dumped by factories along its banks is there for all to see. National or state law might mandate cleanup and compel factory owners to reduce or eliminate pollution, but actual action may require legal or even police intervention. People are likely to resent outside interference in their municipal affairs and to resist it. Politics must focus, therefore, not only on changing the material conditions whereby environmental degradation takes place but also on the social (imagined, ideological, and cognitive) conditions that rationalize and support the continuation of that degradation. These are conditions that, if not local in origin, can only be changed through local action, through political activism and civic action. We will return to this point later in this chapter.

The Structures of Contemporary Political Economy

In chapter 3, we saw how much contemporary production and consumption has been organized in the form of long-distance, transnational "commodity chains." Just as goods are produced in places far away from where they are consumed, environmental problems tend to be treated as originating in other, far-removed places, thereby requiring action far away. Thus, it is not uncommon to read that population growth and poverty in developing countries are the "real threats" to the global environment and, therefore, it is in the Global South where population policies must be put into place.[23] The story continues: here in the Global North, these problems have already been addressed. But things are neither so simple nor straightforward: One only need look at the figures linking consumption in the developed world with greenhouse gas emissions in China to know that things are a bit more complicated than that particular narrative makes them out to be.*

The commodity chain model illuminates two aspects of the difficulties associated with managing impacts at a distance. First, the impacts of production and consumption happen, for the most part, in specific places. Second, the arrival of things from "far away" leads to *mystification* of both that process and the linkages between global and local economies and environments. An entire production and consumption chain may be spread over many different locations or countries (see Figure 3.1), but each step of the process occurs in a

---

*The literature on this topic is rich and continues to grow. As just one example, these numbers from Norway are startling—emissions embodied in imports are 67 percent of Norway's domestic emissions. Glen P. Peters and Edgar G. Hertwich, "Pollution Embodied in Trade: The Norwegian Case," *Global Environmental Change* 16 (2006): 379–87. See also: Christina Prell and Kuishuang Feng, "Unequal Carbon Exchanges— The Environmental and Economic Impacts of Iconic U.S. Consumption Items," *Journal of Industrial Ecology* 20, #3 (2015): 537–46.

specific place or country. Raw materials and commodities are extracted from the environment, and parts are shipped to the production, processing, and consumption sites. Various types of wastes and pollution are emitted at each stage, too. Thus, while greenhouse gas emissions from the production and use of an automobile are rapidly discharged into the "global" atmosphere, they originate from very specific and localized sources. Their complex organization means that centralized coordination of behaviors throughout such myriad chains is virtually impossible.

Mystification is rather more complicated, for we are encouraged "not to know" about where things come from, how they were produced, or the kinds of social and environmental problems with which they are associated.* A conventional perspective on the distinction between local and global relies solely on the spatial character of the resource in question and, as with the carbon example above, there is a tendency to treat causes as wholly local, rather than as a complex "externality chain." For example, soil erosion is conventionally treated as a local matter, even though it occurs all over the world. It is usually attributed to the specific practices of farmers, loggers, land developers, and others who modify the landscape in ways that degrade soil retention mechanisms and permit water to carry soil away.[24] Why they engage in such destructive practices, even when they "know better," is rarely considered. It is not merely (if at all) the fault of the individual (that is, "operator error"). Structural factors, such as economic pressures and legal incentives to increase agricultural production, to build on undeveloped land, and to cut down forests, are as important in motivating these behaviors as are individual choices and decisions, but these structural forces are not given much attention, regarded as part of the social background and "business as usual." Responsibility for soil loss is therefore attributed to the uninformed or ignorant individual or group; these individuals or groups must be "educated" and "trained" in less harmful practices and, if they fail to learn the lessons, criticized and penalized.

Within the global political economy, the conceptual and political distinction between individual practices and structural effects arises from the liberal individualist ontology and legal principles that infuse capitalism, and present the connections largely as happening in a black box.† Under contemporary legal systems around the world, most of which are based on either

---

*Indeed, with a few exceptions, the rules of international trade prohibit discrimination against imported goods based on how they are produced, so long as they are "essential equivalents"—more or less identical in form and function—of goods already for sale in the importing country.

†Bruno Latour writes that blackboxing is "the way scientific and technical work is made invisible by its own success. When a machine runs efficiently, when a matter of fact is settled, one need focus only on its inputs and outputs and not on its internal complexity. Thus, paradoxically, the more science and technology succeed, the more opaque and obscure they become." *Pandora's Hope: Essays on the Reality of Science Studies, p. 304*, Cambridge, MA: Harvard University Press, 2009.

Anglo-American or European law, liability for actions is attributed to a specific actor, be it an individual, a corporation, or a state; structural conditions never enter the debate. This point has been made forcefully by Johan Galtung in his discussions of the relationship between Western legal systems and structural violence.* He argues that the legal organization of industrialized, capitalist societies rests on the principles of actor culpability and liability. At the same time, Western law renders invisible the role of structures, such as the organization of and incentives in markets, in engendering material impacts and violations of norms, ethics, and laws. To put this another way, no one has ever been indicted, convicted, and fined or imprisoned for engaging in "business as usual," even though it is arguably the cause of much, if not most, environmental damage.

A structural analysis of environmental impacts in a commodity's externality chain reveals a sequence of processes that concentrate culpable actions at specific points, even as the logic operating throughout the chain is one over which individual actors have little control. The factory owner is compelled to dump wastes illegally because he is a subcontractor to an exporter operating under strict contractual conditions laid down by the foreign contractor. The owner could act otherwise, but the result would be higher costs, loss of business, and, perhaps, plant closure. This emphasis on specific actor liability serves to mystify the causes of environmental degradation and to displace environmental (and other) politics from one arena of action to another. Thus, to return to soil erosion, we can see it as a consequence of an individual's necessary response to compelling factors inherent in market organization and relations. The individual must respond to these compulsions (cost cutting) or suffer the consequences (hunger or bankruptcy). The displacement of responsibility onto the individual obscures the role of power in the political decisions about market rules and organization made and enforced in other places, by other people (as discussed in chapters 2 and 3).

Or we can see the international treatment of climate change as a way to shift responsibility for high levels of greenhouse gas emissions out of the arena of production and consumption within societies and onto other parties whose actual responsibility is much less (as will be discussed in chapter 5). This move transfers an environmental problem from the realm of local and national politics into the arena of interstate market relations, where politics are muted, if not entirely absent. The result is to displace the focus of climate change policies from structural arrangements and practices that encourage continued and growing emissions to questions of how much each individual state should be permitted to emit now and in the future, whether in aggregate or per capita. What is never addressed politically are the ethics of maintaining

---

*Structural violence involves injuries, such as hunger and ill health due to endemic poverty and injustice, that arise not from individual action but, rather, structural conditions, such as the operation of markets. Johan Galtung, *Human Rights in Another Key*, chapter 2, Cambridge, MA: Blackwell, Polity Press, 1995.

greenhouse gas emission rates, linked to production and consumption, at such disparate levels among countries or people, or the unethical imposition of impacts on those whose emission levels are vanishingly small. Again, the role of power in the structuring of these market relations is partially, if not wholly, obscured.

To put this another way, the claim that all environmental politics are "local" is based on what has sometimes been called the "agent-structure problem," which can be understood, in this case, as a disconnect between the individualist orientation of capitalism as a social order and the structural features of capitalist political economy.[25] We make our choices in the market, and in our everyday lives, on an individual basis, often with little thought to how they might affect or be linked to other individuals, groups, and societies nearby and far away. Such choices, as we saw in the discussions of advertising in chapter 3, are not simply the consequence of individual desires or needs. They are connected to structures, and those structures affect the environment and have global reach. These structures are not particularly amenable to change through action by individual agents; acting alone, each person can hardly have an impact (as described in chapters 2 and 3). Collective political action through social power is required to modify or transform such structures, and such action may best be initiated where agents are offered the greatest opportunity to act, in localized settings where praxis is possible and practical, rather than in national or international ones, where no one is very interested in structural change, especially since it is so difficult to accomplish.

THE POWER OF SOCIAL POWER

We are often told that local efforts to engage in structural change make sense only if all other localities act similarly (that old collective action problem again). After all, why should people in your town deliberately and with intent drive fewer miles if no one else is doing it? The difference in national or global emissions would be minuscule, and it would be foolish to suffer the economic losses that would surely accompany such a project.* This argument is valid, however, only if we accept individualist market logic as defining what is "rational." It is the inversion of the old question often posed by parents to their children when the latter want to do something in which their friends are engaged: "So! If everyone else jumped off a cliff, would you do it, too?"

Must rationality be defined solely on an individual basis and the logic of cost-benefit? Would it not be equally rational to define a shared political ethic whereby each locality would recognize, resist, and reorganize the mystifying incentives imposed by globalized structures of production and consumption?

---

*Even so, we are frequently encouraged to "drive less" in order to save energy and reduce emissions (and "save money," as well). This poses a particular difficulty to those who don't have access to alternative modes of transportation.

Such an ethic would not treat the local as one, self-interested individualist point among many, bound to lose in an eternal competitive race if it gives up even one step. Instead, this ethic would see the local in terms of the people who live, work, play, and dream there, a group with collective principles and goals that give due consideration to others, whether human or not. Moreover, those who pursue such an ethic would through their praxis recognize both the social power inherent in collective action and the ways in which praxis in one place could inform similar social power in other places. Civic action and political activism are, potentially, expressions of such praxis.

Is Collective Action a Problem?

But what about the parent's question to the child regarding peer pressure and competition? Is it so easily dismissed? The frequent displacement of politics to higher levels of collective behavior is often justified by a popular ontological argument that problematizes political action outside certain limited settings. As discussed in chapters 2 and 3, this is called the "collective action problem."[26] The collective action problem is a contemporary version of both the Hobbesian "state of nature" (which we first mentioned in chapter 1) and the parable of the stag hunt as related by Jean-Jacques Rousseau, the eighteenth-century French political philosopher. Both are arguments about the individual temptation to defect from collective action and the necessity for some means to prevent defection.

In Rousseau's story, a group of hunters is pursuing a deer, which can be killed only if everyone in the group cooperates in the hunt.[27] If the deer is killed, everyone will eat; if it escapes, everyone will go hungry. One hunter spies a rabbit. He realizes that the group might fail in its pursuit of the deer, but that he will have no trouble snaring the rabbit on his own and satisfying his hunger. What should he do? Men are selfish and self-interested, according to Hobbes, Rousseau, and other theorists of the collective action problem.[28] Our selfish hunter chases the rabbit, catches it, and eats. The deer escapes, however, and everyone else goes hungry. Self-interest mandates defection. Knowing this, how can a hunt even take place to begin with?*

Although Rousseau was interested primarily in the origins of government, his parable presents a purely economistic picture of human motivations.[29] It implies that only through some sort of coercion—custom, law, pressure, threat of loss, or ostracism—will people forego their individually harmful acts and work together for the social good. Adam Smith subsequently introduced the "invisible hand" into this calculus in order to explain how, in markets, individual action can serve the collective good and is, therefore, not a problem.[30] But Smith, as we have already noted, also believed that a shared morality, rooted in

---

*Indeed, if we ponder this question, it seems that the group hunt is a fool's errand in the first place. Yet, the hunters' experience has taught them they must work together, especially if they have families to feed. Which action is more rational?

religious belief and practice, would moderate individual appetite and prevent purely selfish, injurious behavior.[31]

Rousseau's argument, as well as Hobbes's, is much the same as that made by Garrett Hardin in "The Tragedy of the Commons," discussed in chapter 3. As we saw there, Hardin could not imagine a group of herders cooperating to maintain a field owned in common, thinking that each would seek to maximize his individual take, with the consequence that the field would be destroyed. Hardin did not think that a local politics of pasture could exist in a world of self-interested men. He believed that either privatization or police power (that is, a sovereign) would be necessary to keep the pasture from ruin.[32] Hardin's arguments have had an enormous impact on environmental politics and environmentalism, for they seem logical within their own suppositions and assumptions (liberalism and individual self-interest) and frightening in their implications.

In fact, Hardin appears to have been wrong in both his parable and his fears for the future.[33] As noted in chapter 3, there are numerous documented cases of cooperation in the protection of commons without the presence of overt coercion.[34] To be sure, in most of these instances, the users of a resource held in common realize individual material benefits from it and pay costs to sustain it. But the users are disciplined in their use of the resource, not by a sovereign but by bonds of trust and obligation, mutual reliance, and some sense of a common good. Free riding or defection may result in a violator's expulsion from the user group and loss of livelihood and sustenance. Awareness of this ultimate sanction keeps users from engaging in wholly self-interested and, ultimately, destructive behavior. Moreover, observing the rules of the commons is (as Pierre Bourdieu might have it) simply "what is done."

The collective action problem is not without an inherent politics of its own, which is obscured by its compelling economistic logic. Not only is our experience in a capitalist society one of constant and never-satiated needs, but our socialization has conferred on us the foundational assumptions of material envy and self-interested behavior.[35] "Wanting more" is naturalized through belief and practice; to feel and act otherwise makes one a fool. If everyone else is selfish, we are foolish not to be selfish, too. And how could it be otherwise? In effect, by making the claim that collective action is rational only when each individual realizes a clearly defined individual benefit, the reasoning behind the collective action problem makes any other form of collective action (or social power) appear foolish and even inexplicable. Consequently, social activists who engage in politics outside established institutions, and who appear to realize small or nonexistent benefits from their activities, are made to seem both irrational and even dangerous to business as usual.[36] Indeed, if too many people were to act in such a fashion, business as usual might be threatened.[37]

This attitude regarding collective action is evident, once again, in the Trump administration's position on global climate change—at least among those parts of the administration that admit that humans are responsible for some amount of planetary warming. The costs of action will be concentrated over a small group (rich countries) while the benefits will be dispersed over many people or groups (both rich and poor countries). Why not spend that money protecting ourselves and our fossil fuel industries, and continue to benefit from our environmentally harmful practices? And, in the long run, others cannot be trusted to take action under the voluntary emission reduction provisions of the Paris Agreement, so why should we? That we might all perish together as a result of climate change does not seem to have much effect on either practice or consciousness.

Collective Action Is Not a Problem

Does widespread belief in the truth of the collective action problem mean that collective praxis is impossible? Not at all! We see people acting together all the time, apparently in defiance of the logic of the collective action problem. But why does this happen? Are people irrational or do they know something? Indeed, we can construct counterarguments to the collective action problem that invoke collective interests as well as ethical principles, and it is these arguments that help to illuminate both social activism and civic politics. The first argument involves intergenerational concerns, that is, the interests of future generations who are not in a position to make decisions today.[38] The second is based on *intragenerational* equity and justice and the fact that many of those most affected by environmental degradation have little or no power over the processes that cause it (including both people and nature).[39] These are not the only counterarguments available, but they may be the most important ones.

The intergenerational argument rests on the assumption that we would wish to leave our descendants a world that is, at a minimum, in no worse environmental condition than the world in which we live today. This would mean, for example, avoiding further drastic climate modification or a world laced with toxic wastes. In such a world, some are likely to be better off, some will be no better or worse off than we are today, and many others are likely to be worse off. But we do not know who these different groups of people will be and whether they will be kin or not, co-nationals or not.[40] Such uncertainty ought to make us cautious, and for the benefit of everyone in the future, we should act today so as to minimize the harms that we do to the environment, even if the costs are high for the present generation. This is sometimes called the "precautionary principle."*[41] In essence, if we have concerns about the potential for negative effects of an action, we should do all we can to avoid that action.

The intragenerational argument has to do with environmental justice and rests on the claim that environmental impacts resulting from highly distorted

---

*The precautionary principle states, in essence, that if there is doubt about the safety of a practice or product, it is better to avoid it, rather than to take the risk and have to ameliorate the effects. Erik Persson, "What are the Core Ideas behind the Precautionary Principle?" *Science of the Total Environment* 557-558 (2016): 134–41.

and unequal patterns of consumption and wealth cannot be justified under any set of ethical principles that recognizes all humans (and nature) as deserving of equal consideration. This is especially the case if the actions leading to those impacts serve to improve the condition of the better-off while worsening that of the poor. Inasmuch as simple reliance on markets and economic growth to address these concerns appears unlikely to alter greatly the existing distribution of impacts, we must act in a way that does change that distribution.[42] Therefore, on the principle that we should try to minimize such suffering as we might impose on others, we are ethically bound to moderate and modify our current desires as well as the practices causing environmental degradation. While we will experience material costs from such changes—and they may be quite substantial—the benefits to those who are worse off will compensate us not only through the ethical satisfaction we realize from our changed behavior but also through reductions in the environmental impacts of both wealth and poverty.*

The logic of the critique of the collective action problem implicit in these two arguments, while more difficult to articulate, is nonetheless compelling. Ultimately, as conscious beings we do not act only on the basis of self-interests ("interests" that are, as often as not, associated with the interests of capital). There are social norms and rules that we follow in pursuit of a common good, even if not always in a consistent or conscious fashion. Norms are principles that express a society's sense of what is proper and good, while rules dictate appropriate practices and actions in a particular situation or context. Both can be the subject of political debate and action.[43] Much of the collective action we observe does take place on the basis of such norms and rules. For example, the preservation of an old-growth forest, for both ecological and aesthetic reasons, might mean foregoing its conversion into lumber, capital, collective enrichment for a few and "affordable" housing for others.[44] But even if trees can be replanted and forests grown, as is often argued, a world without old-growth forests is one impoverished in many nonmonetary ways. Those who suffer from such actions can be compensated for their losses, but forests cannot defend themselves.[45] There are good reasons not to cut off your nose, even if everyone else is doing it. But such changed awareness and practice does not come easily, especially if we rely on individuals acting by themselves. People may act as individuals, but they do so on the basis of social beliefs, habits, norms, and rules. And societies do not, as a rule, easily or quickly change these foundational practices. Except during periods of crisis, arguments in favor of change are regarded as odd, irrational, or even subversive. Moreover, people tend not only to resist such arguments but to insist that the ends sought by those who make these claims cannot be achieved and would be too costly, if not suicidal.

---

*This is a utilitarian argument, but it also arises from the principle that humans (and nature) deserve respect and should be treated as ends, and not means.

This is the essence of realism as both philosophy and practice: we must deal with the world as it is, and not as we might wish it, with the "is" and not the "ought." Were such realism really the case, social and political change would never take place. That such change does, sometimes, occur indicates why "realism" is given that name and generally deployed to protect those interests and structures that benefit the powerful. Under these circumstances, realism is no more than the ideology of those who dominate society.[46]

Social Power: Against "Realism"

Civic politics and political activism—what we call here "social power"—are responses to the ideology of realism. Institutionalized political practices in democracies—voting, lobbying, representation—are, by contrast, mechanisms that acknowledge realism and seek to reproduce a society's foundational principles and relations in a stable and predictable fashion, without the appearance of coercion or constraint.[47] At best, however, such practices normally generate only small variations in the status quo, in response either to the moods of the electorate or the wishes of elites.* Social power is a means of countering the conservative tendencies inherent in democratic political systems.

In order to understand better the concept of social power, civic politics, and political activism, and associated practices, we need to delve a bit into liberal political theory. Most readers of this book are likely to be living in politically liberal, market-based countries.† Although the details will differ from one to the next—as between the United States, Germany, and Japan, for example—the social world in which individuals find themselves is composed, for the most part, of two arenas of institutions and action: the public and the private. The public sphere is generally associated with the state and comprises what we conventionally understand as "politics." The private sphere is generally associated with everyday life and is comprised of family, civic associations, businesses, and markets.‡ This distinction between public and private is not entirely straightforward but, in market democracies, it is a means of differentiating between "legitimate" political activities that do not challenge the social order, on the one hand, and those activities that support the social and economic order on a daily basis but appear to play no overt role in its political organization, institutions, and relations of power, on the other (ignoring, for

---

*Does the election of Donald Trump belie this claim? Not if the mere act of voting is an enabler of democratic domination. See Herbert Marcuse, *One Dimensional Man— Studies in the Ideology of Advanced Industrial Society*, Boston: Beacon Press, 1964.

†The first edition of this book was translated into Chinese, so this might not apply to those readers.

‡Sometimes, a distinction is made between the private (family or civil society), the state, and the market. As we shall see, the private-public distinction is most germane here. Stanley I. Benn and Gerald F. Gaus, "The Liberal Conception of the Public and the Private," in *Public and Private in Social Life*, edited by Stanley I. Benn and Gerald F. Gaus, pp. 31–66, London: Croom Helm, 1983.

a moment, the relations of power inherent in political economy).[48] To put this another way, even though the private sphere is ostensibly distinct from the public sphere and institutionalized politics, the private sphere remains central to the maintenance of power exercised through established political institutions that support the public sphere, because the social order, or "civil society" defined as "nonpolitical," is the basis for this power.* Because of this separation, "normal" action within the private sphere is structurally unable to challenge or change the organization and institutions of the public sphere. The privileging of the autonomy of the individual and his or her self-interests characteristic of market-based democracies reifies individual action and further obscures the possibilities for social change.[49] A little reflection on the public-private distinction illuminates more clearly its function in the containment of non-institutionalized political activity.

The family is conventionally regarded as central to the private sphere and an arena from which the state should be excluded.[50] Yet, ironically and paradoxically, the state is routinely involved in regulating both forms of and behaviors within the family and throughout the private sphere. Both conservatives and liberals complain about such regulation—for different reasons—but the intrusion of the state into both commerce and civil society illustrates just how artificial the public-private divide really is. In the United States, continuing resistance to non-heterosexual marriage, support for the nuclear family (as opposed to other familial forms) and alteration of the tax code to favor nuclear families, encouragement of sexual abstinence among teenagers, even the requirement that the name of a child's father and mother both appear on a birth certificate, all represent the extension of the public hand into the private sphere (justified as necessary for the state to maintain a "proper" social order). Conversely, media exposure of the sexual peccadilloes of public figures and the illegal activities of their children, as well as efforts to legislate morality through constraints on abortion and the availability of contraception, among other things, show us how the private lies at the heart of the public.

Still, all human activity involves politics. Setting boundaries between public and private disciplines the private even as it serves to depoliticize many so-called private issues that are political and public at heart. This notional separation of public and private becomes a means of stabilizing, reproducing, and naturalizing certain hegemonic social relations, such as the "nuclear family" and heterosexism. Challenges to these relations and their inherent politics, such as those arising from civic politics and political activism, are thereby rendered highly controversial and threatening. There are no officially approved political channels—except, perhaps, courts—for contesting such naturalized arrangements and the politics they carry in everyday life.

---

*This is why Robert Putnam (*Bowling Alone*) bemoaned the decline in bowling leagues and similar private sphere activities: apolitical group activities rely on public rules and regulations but do not seek to shape or change them.

Taken together, civic politics and political activism constitute one arena of civil society. Some political philosophers and theorists characterize civil society as a realm of private, selfish, bourgeois activity; others argue that it is central to the constitution of public politics and the state.[51] This point of dispute is relevant, however, only in liberal democratic societies, in which political action outside of approved channels is regarded as inappropriate, illegitimate, or even threatening to the social order (as expressed, for example, in the application of the term "environmental extremism" to relatively benign activities).[52] The so-called public-private distinctions, and disputes over them, also apply to what has been called "global civil society."[53]

Social power, which includes both civic politics and political activism, originates in the private realm but seeks to intervene in and reshape the public sphere. The dividing line between public and private is rather vague in the case of civic politics, which often involves the activities of professional and bureaucratic associations that deal largely in the formulation, implementation, and modification of public policies. Civic politics is also practiced by both NGOs and corporations, which rely on lobbying, influence, and expertise to accomplish their objectives. Some of these groups have their origins in movements of political activism but have become institutionalized as they have become well established, with trained, salaried staffs, lobbyists, and fundraisers. Strictly speaking, such aggregations of social power violate the tenets of representative democracy, which is organized primarily around the individual expression of preferences through the vote and secondarily on the basis of individual efforts to communicate preferences to their duly elected representatives.

Political activism as we define it here is, by contrast, less attentive to the separation between public and private. The exercise of this form of social power involves provocations and resistance to the public order and takes place outside of the constraints of institutionalized politics.[54] Action occurs when political activists, composed of a broad range of individuals, organizations, and movement associations, sometimes organized into coalitions, seek to extend the private into the public, thereby pressuring governments and institutions to change their policies or even reject them. At times, these movements are able to mount major challenges to established power and come to be treated as threats to the institutionalized public sphere, as noted above.[55]

From Changing Minds to Changing Societies

The conditions that give rise to expressions of social power must arise from changes within society that, somehow, trigger popular imaginaries and mobilization and engender high levels of commitment to shared or common issues and problems. Because we are interested specifically in the environment, these conditions must be related to the arguments about local environmental politics made earlier in this chapter. And drawing on the historical-material-practice framework presented earlier in this book, we can generally identify four conditions necessary for mobilization of social power:

- **Material change:** Changes to the local political economy are a central element in the mobilization of social actors seeking to protect the environment, for they shape the material base that people see and experience around them. Changes in this material base—as in a shift from heavy industry to services—are tied to changes in perceptions and experiences, as well as to alterations in class structures and relations. Impacts on health and well-being are part of such changes in the material base, too, and they alter the ways in which people are linked to their localities and perceive their relations to others. An example of this might be the construction of a road through a local forest or wetland, which results in greater traffic and housing densities, with concomitant effects on air and water quality.
- **Challenges to shared meanings of nature:** People living in a particular place often tend to hold similar perceptions of nature in that place (even when they are at odds ideologically). Certain biological (forests, wetlands) and physical (rivers, mountains) features are marked as signifiers and symbols of place, and their presence acts as a basis for a kind of local identity When, however, these signifiers are threatened physically by "normal" activities undertaken through the local political economy, a growing collective consciousness can develop in opposition. For example, city leaders might urge further development of nearby "underutilized" agricultural lands in order to foster economic growth, generating concern among those who see this as a threat to local environmental integrity (such concerns are often met by reassurances that every precaution will be taken to protect local nature).
- **Common epistemological framing:** Shared, broad knowledge of the environment, and how it ought to be treated, is also important to social power (and here is where ideological differences become critical: preservation or exploitation?). As the material base changes—for example, from direct use via industrial production to indirect utilization through recreation and ecotourism—the epistemological relationship between nature and people is reframed.[56] Such reframings are motivated by processes both external and internal to locality, informed by what people know locally and what scholars, experts, and leaders tell them "globally." Does the endangered status of a local fish species not matter very much or is it critical to particular ecosystems? And what are the broader political and ecological implications of species extinction?
- **Normative shifts:** Finally, changes in the way people value their relationships to nature—a consequence of material shifts around them and epistemological shifts in their understanding of the natural world and even valorization of nature—play a central role in the mobilization of social power. Ultimately, it is the sense of what "ought" to be, rather than what is, that motivates many people to act. Like Marx's architect, they imagine that "ought" and seek to turn it into an "is," and they do so in concert with others of like mind and capacities.[57]

In this process of mobilization, material and cognitive (social) elements prompt and inform each other, encouraging people to recognize material change, comprehend its material impacts, formulate an understanding of its significance, and mobilize around alternatives. Theorists of social power disagree, however, on the mechanisms whereby social and political change ultimately happen. How does action within the private realm of liberal societies lead to political impacts and outcomes? How might change happen? Four approaches to answering these questions—*theories of social change*—are discussed in the following: consciousness-raising, resource mobilization, party organization, and "praxis," that is, active engagement with structures through changing practices.

Consciousness-raising

There is broad agreement that, as a political and social phenomenon, contemporary environmentalism is not much more than forty years old.* As noted earlier, the publication of Rachel Carson's *Silent Spring* in 1962 is often regarded as the starting point, if not the trigger, of the movement.[58] But it is often difficult to separate out environmentalism from the other expressions of social power, civic politics, and political activism that, in many parts of the world, characterized much of the 1960s.[59] What all social movements of that decade did share was a growing skepticism of the authority of political, economic, social, and scientific institutions. In particular, throughout much of the period following the end of World War II, between 1945 and 1960, science was regarded as possessed of awesome powers of explanation, prediction, and control. Although it had produced the atomic bomb and the threat of a nuclear World War III, science also promised a glossy future, absent of the demons of hunger, illness, and poverty that had plagued humanity for so long.[60] *Silent Spring* was only one of several critiques of this optimistic view that appeared during the 1950s and 1960s.[61] Carson's arguments about the effects of DDT on nature were strongly resisted, and even attacked as subversive, by both industry and government.[62] Nonetheless, her critique resonated strongly with that segment of the American public becoming disillusioned with both science and the state (a disillusionment that also contributed to growing support for the Civil Rights Movement and the rise of opposition to the Vietnam War).

Whatever the trigger, there is no denying that the 1960s saw considerable expression of social power around the world, certainly more than there had been for several decades.[63] The anti-war, civil rights, anti–nuclear weapons, and free speech movements in the United States had their parallels in other countries. The events of 1968—in Czechoslovakia, France, Mexico, the United States, and elsewhere—marked it as a watershed year for political and social protest and resistance, which left its imprint on many Western countries.[64] Authority has never fully recovered.[65]

---

*Prior to the late 1960s, "environmentalism" was known by different terms and contexts. For example, Progressive conservationists around the turn of the twentieth century, such as Gifford Pinchot, sought to protect nature for its utility; German and English romantics during the nineteenth century, such as Alexander von Humboldt, focused on the potential of extinction of species, including humans; before that, a central trope was land management. See Richard H. Grove, "Origins of Western Environmentalism," *Scientific American* 267, #1 (July 1992): 42–47.

Environmentalism can be seen, in part, as an outgrowth of its predecessors as well as these other movements and the skepticism of government and expertise that gave rise to them. In the United States, the environmental movement encompassed, in particular, a segment of American society—generally young, white, and well-off—many of whose members had not been deeply engaged by other, more obviously political issues. In its early days, at least, environmentalism was not regarded as the province of conservatives or liberals—some laud Richard Nixon's environmental policies, if nothing else about his administration.[66] In subsequent decades, environmental activism spread throughout the world, expanding to include people and groups left out at the creation, including racial minorities, indigenous peoples, and even nationalist movements.[67] Environmental groups also built transnational coalitions with movements focused on other pressing issues, such as peace and human rights.[68] The emotions and actions associated with the movements of the 1960s fed into what were later called "post-material" and, more recently, "post-humanist" values.[69] Some people were able to be less concerned with livelihood and more concerned about the norms, values, and practices with and through which they lived. Across the Global North, this change was also driven by an ongoing economic shift from heavy "metal-smashing" and highly polluting industries to services and knowledge-based production.[70]

Among other things, "post-material" values included a growing aesthetic appreciation of nature and concern about individual health (now called "wellness"), both of which require more attention to pollution and other forms of environmental damage. Some argue that, while material factors were central to this change, the key element was a shift in individual awareness and reflection upon satisfaction of one's material needs, a process sometimes called "consciousness-raising."[71] Because people find meaning in their lives on the basis of their ideas about life and living, a change in ideas can result in a change in meanings and consequent action. If enough people alter their ideas and practices, it is argued, they can effect social change.

But consciousness-raising has its limits, as can be seen in the example of "green consumerism."[72] A growing number of commercially available products are certified, in one form or another, as environmentally friendly (although the consumer often has to take such guarantees on faith; see the discussion of certification later in this chapter).[73] If enough people shop with an environmental consciousness, goes the argument, producers will see the growing demand as a profit opportunity and will supply appropriate goods. Eventually, the preponderance of products will be environmentally friendly, a result brought about wholly through directed changes in consumer preferences.[74]

Certainly, demand for such goods has been growing, and more and more companies are producing them, even though green products tend to be more expensive than their non-green counterparts.[75] Organic farming has grown by leaps and bounds over the past few decades, largely as a result of consumers' desire for foods free of agricultural chemicals.[76] Yet, the act of buying a particular product, based on preferences, arises almost entirely out of individual choice and not collective decision-making or political activism.[77] Individual action of this type cannot be the basis for social power. Consumer preferences change, often under an onslaught of advertising or peer pressure or economic instability. Manufacturers alter their product lines and supermarkets the products on their shelves, especially if profit margins on particular items drop. If politics is about collective action, green consumerism and consciousness-raising are not politics.

Resource Mobilization

Although there are competing explanations for what are called the "new social movements" and "new" new social movements,[78] and disputes over whether they are really "new," resource mobilization theories are generally based on three elements.[79] First, people are frustrated by certain conditions that are not being addressed through institutionalized political processes. Second, some people possess the resources—skills, knowledge, and money—to support collective action that agitates for something to be done. Finally, entrepreneurial movement leaders espy certain kinds of political opportunities that enable them not only to organize frustrated people but also to demonstrate that frustration publicly.[80] If enough people join together to act, their "contentious politics" may pressure political authorities to respond and change those conditions.[81]

One example of resource mobilization has been visible in the Global Justice Movement, the series of very large demonstrations and protests held during various international economic and trade meetings (in Davos; Genoa; Gleneagles, Scotland; Göteborg, Sweden; Prague; Québec City; Seattle; Washington, District of Columbia; and other cities) beginning in the 1990s.[82] According to the framework of new social movement theory, growing numbers of people are angry about the negative social and environmental effects that result from globalization (see also chapter 3). They are seeking outlets for their anger and ideals, and want to pressure governments and institutions to reform. These repeated demonstrations have sensitized political and economic elites to growing dissatisfaction with the world's economic system, but they also have led governments to rely much more on police power to limit or prevent such activities.[83]

More broadly, numerous local, national, and transnational organizations—in the tens of thousands or more[84]—are dedicated to a variety of postmaterial issues, including environmental protection, enforcement of health and labor regulations, sustainable development, fair trade, human rights and the rights of indigenous peoples, and many others.[85] These groups have their own projects, sources of funding, and supporters, and they are always concerned about bringing more attention to their particular issue area. As we have seen in earlier chapters, it is a relatively straightforward matter to highlight the many linkages among "business as usual" and environmental externalities, and there are many opportunities to bring these connections to public attention or into the political arena. In most countries, however, governments remain committed to neoliberalism and globalization either by choice or necessity. When confronted by activists' demands, individual governments and officials may acknowledge problems but they will also respond that the necessities of remaining internationally competitive do not permit them to change their policies so as to address the concerns of the protesters. To do so would be to cave in to anarchy and the street. Besides, such policies are the result of "market forces," argue governments and defenders of the market, over which no one really has any control.[86]

One consequence of the transnationalization of environmental activity is a shift of responsibility for problems of concern out of the national sphere and into the international one. Various "Groups" (of seven, twenty, etc.), miscellaneous intergovernmental institutions (UN Environment Programme; International Monetary Fund [IMF]; World Trade Organization [WTO]; World Bank), and other regimes and financial entities become the notional sites of regulation and disbursements of funds, even if they do not actually run the global economic system.* A similar logic is applied to transnational corporations.[87] Governments, businesses, and international institutions then become the target of social power and street demonstrations. (In the past private financial institutions and private capital were rarely targeted, even though they accounted for the bulk of international loans and investments; more recently activists have taken up campaigns against private financial players, particularly for their fossil fuel investments or role in tropical deforestation.†)

Given the many projects, campaigns, alliances, networks, and coalitions on offer, in combination with rapid communication and relatively cheap international travel, it has become a fairly straightforward task to mobilize people, "create" a movement, and encourage those people to show up wherever meetings are taking place: some twenty-four hundred NGOs and seventeen thousand individuals attended the 1992 Earth Summit in Brazil; more than forty thousand people, including twenty-five thousand delegates ("government figures along with representatives from intergovernmental organizations, UN agencies, NGOs, civil society" and business) were at COP 21 in Paris.[88] (Authoritarian governments, such as that of Qatar, can deny visas to protesters and keep them far away; democratic ones, such as Canada and Italy, have more difficulty but try nonetheless—there were eleven thousand police deployed in

---

*The "Group of Seven" (G-7) consists of Canada, France, Germany, Italy, Japan, the United Kingdom, and the United States. From 1997 to 2014, the Russian Federation was also a member of the group, which was then known as the G-8. After Russia's invasion in 2014 of the Crimean region of Ukraine, its membership was suspended.

†For example, the movement for universities to divest from fossil fuel investments (http://powershift.org/campaigns/divest) or for individuals to move their money from banks investing in the Dakota Access pipeline (http://www.defunddapl.org/defund). See also Rainforest Action Network, et al., *Banking on Climate Change: Fossil Fuel Finance Report Card 2017*, accessed August 22, 2017, https://d3n8a8pro7vhmx.cloudfront .net/rainforestactionnetwork/pages/17879/attachments/original/1499895600/RAN _Banking_On_Climate_Change_2017.pdf?1499895600.

Paris.[89]) International conclaves have high profiles but the proceedings are relatively boring, so the opportunity for media attention and publicity is quite high. Resources are mobilized and opportunities seized.

The trouble, as is apparent in the case of the global justice movement, is double: first, these international institutions (or regimes) are not quite so autonomous or powerful as they might seem at first appearance (see chapter 5).[90] Although they are classical bureaucracies, and sometimes act independently of the wishes of individual governments, they are nonetheless very much constrained by their member states, especially the powerful ones. The U.S. Treasury continues to exercise an inordinate amount of influence over both the World Bank and the International Monetary Fund, which is not surprising given that the headquarters of both organizations are located in Washington, DC, not far from the White House and the U.S. Congress. The second is that such movements often mobilize and disperse, and are not heard from again. Decentralization has its costs. Resources may have been mobilized for giant demonstrations, but are activists tilting, perhaps, at the wrong windmills?

Are Party Politics the Answer?

Constraints on mass demonstrations suggest, perhaps, that social power ought to be aimed at national politics.[91] Why not transfer the energy of social movements into political parties, which offer greater engagement with existing political institutions and processes, and provide opportunities for longer-term strategic thinking? Might this not bring about the kinds of changes demanded by social movements? If a party can find sufficient popular support, it could end up in the national legislature, or even participate in a governing coalition (sometimes, new parties even become the new government). There, it would be able not only to pressure the authorities to implement environmentally friendly policies and programs, it could even be involved in their formulation and enabling legislation. An instructive, and sobering, case to consider here is that of the German Green Party (originally called Die Grünen, now Bündnis 90/Die Grünen). This was an alliance of Greens from East and West Germany established just before the first postunification election in 1990; in the most recent German federal election, in September 2017, the Greens received slightly less than 9 percent of the national vote.[92]

The German Greens have not always been a political party, nor were they the world's first Green Party, an honor that belongs to the Values Party in New Zealand.[93] The German Greens were, however, one of the first such parties to see members elected to local, state, and national legislatures. Established in the late 1970s as a leftist environmental and peace movement, the Greens began to gain large numbers of supporters during the Euromissile debates of the early 1980s.[94] Getting involved in institutionalized politics was not a simple matter. Even when putting candidates up for office, the German Greens were not a political party in any conventional sense of the word. Its members held diverse views, ranging from Marxist to liberal and even, in a few cases, conservative. Indeed, some of its members already belonged to established political parties (and later dropped out of the Greens). But, as an extra-parliamentary movement, the German Greens were able to organize, educate, mobilize, and demonstrate, and to find growing numbers of supporters who eventually voted for them in elections.

Difficulties arose, however, when the party finally did manage to win seats in the German Bundestag (parliament), and it split into two factions, each with a different idea of how the party should operate. The *fundis*, or "fundamentalists," felt that it was imperative to maintain the Greens' (then) radical social and political positions. They did not want to fall into line with conventional parliamentary practices and behavior (in fact, the *fundis* had serious doubts about joining the Bundestag at all). The *realos*, or "realists," believed that, in order to establish the party's legitimacy, its members of parliament had not only to look and act like those members from older parties but also to cooperate and collaborate with those parties, especially the leftish Social Democrats. Initially, party members agreed that, to prevent the Greens from becoming too bureaucratized, its members of parliament would hold their seats for only one year and then pass them on to colleagues. The rule remained operative for a very short time, as it became apparent that one year was insufficient time to learn the parliamentary ropes. Today, members of the German Greens elected to the Bundestag and other positions in state and local governments serve their full term. It was this decision not to rotate positions that finally split the party and drove the *fundis* out. They returned to their social movement activities.[95]

As a result, the German Green Party today is a largely middle-class one, stripped of much of its early radicalism. Following the national elections of 1998, the party was asked to join a governing coalition. After sixteen years in power, the joint Christian Democrat/Christian Social Union party, headed by Helmut Kohl, were outpolled by the Social Democrats, which offered the Greens a junior role in the new government, in a Red-Green alliance, which held power from 1998 to 2005.* This was quite an achievement. But the party's members found themselves having to compromise and go along with many Social Democrat decisions that they once would have opposed, for example, the bombing of Serbia by the North American Treaty Organization [NATO] in 1999.[96] Remaining in power requires party discipline. If the Greens refused to support the coalition on critical issues, new elections could be called and they might find themselves back in the opposition (since 2005 they have been in opposition, anyway). In other words, political participation is not cost-free, and it can undermine much of the rationale for the original social movement.[97]

---

*In parliamentary systems, governments are formed by whichever party or parties command a majority of the vote in the legislature. These majorities select the prime minister, who holds executive power.

In countries like Germany, with systems of proportional representation, it is possible for social movement parties to gain access to political institutions. By 2014, Green parties held fifty seats in the European Parliament, representing eighteen member states of the European Union*; in 2015 and 2016, Greens won presidential elections in Latvia and Austria respectively.[98] In 2017, Green parties were part of ruling coalitions in Greece, Latvia, Luxembourg, and Sweden.[99] In places with "first past the post" systems, however, reaching public office is much more challenging.[†] In 2010, Caroline Lucas of the United Kingdom became the first Green elected in a "first past the post" parliamentary system; she is still currently serving as the only Green member of parliament in the British House of Commons (representing Brighton Pavilion).[‡]

In the United States, there is a national Green Party as well as state and local chapters. In 2017, at least 135 Greens held elected offices in eighteen U.S. states, most being local posts such as commissioners of soil, water, or fire protection districts, or members of the school board. In the United States, the Green Party has been most successful in California, where three mayors, three city council persons, and one county supervisor are among the sixty-five Green elected officials statewide.[100] As might be imagined, a Green Party political agenda in one country could differ radically from another's, as could the political platform of the party differ from one major candidate (Cynthia McKinney, who ran as a presidential candidate in the United States in the 2008 election) to another (Jill Stein, who ran for U.S. president in 2012 and 2016).

As we saw earlier in this chapter, as for other matters, a complex system of civic politics has developed around environmental issues. Such activities tend to focus on distributive questions—that is, how much funding will go to environmental protection, who will be affected by environmental policies, etc.—rather than more fundamental structural ones. In other words, they reproduce and legitimate those political, economic, and social arrangements that are the sources of environmental degradation in the first place. The German Greens have had major impacts on the country's environmental policies (especially with respect to nuclear power generation), but they have managed to effect few, if any, real structural changes in the economy. Is this all there is, or are there alternatives to demonstrations and reformism?

---

*According to Global Greens, there are now almost one hundred Green parties around the world, although most of them are not represented in national legislatures; see "Member Parties," accessed August 30, 2017, https://www.globalgreens.org/member-parties.

[†]In a "first past the post" system, a candidate must win the most votes to be elected (however many people are running for the office). The United States and the United Kingdom both employ this arrangement. In some places, the two top vote-getters must participate in a run-off, with the winner getting more than 50 percent of the run-off vote.

[‡]Lucas was a Green member of the European Parliament from the United Kingdom from 1999 to 2010.

Praxis: Agents Coming to Grips with Structure

Praxis as a form of social power presents a rather more complicated picture than the actions offered above. Praxis draws on both consciousness-raising and resource mobilization, to be sure, but it also recognizes that *agency-structure* relations, articulated through practices, are important to action and social change.[101] What this means is that individual and group behavior are strongly constrained and limited both by material conditions and by social factors, the latter including laws, norms, customs, perceptions, beliefs, values and what is called *habitus*.*

Briefly, *habitus* is the repertoire of normalized social behaviors, or "practices," in which people in a particular society engage on a daily and almost unselfconscious basis. As discussed in chapter 1, the practices associated with habitus take place in a "field," which is a collective social space associated with a practitioner's particular social role. To repeat, a practice is a "routinized type of behaviour which consists of several elements, interconnected to one other: forms of bodily activities, forms of mental activities, 'things' and their use, a background knowledge in the form of understanding, know-how, states of emotion and motivational knowledge."[102] By contrast with a rational actor approach, which focuses on the "ABC," or "Attitudes, Behavior, Choices," approach,[103] social practice theory shifts the focus "from individual and autonomous agents, or self-directive and purposive technologies, and onto assemblages of common understandings, material infrastructures, practical knowledge and rules, which are reproduced through daily routines."[104] Moreover, the social world emerges from practice; subjective structures such as consciousness, discourse, or ideas do not exist prior to practice.[105] One consequence of this approach is that objective structures and agents do not really exist; they emerge out of the relations and networks produced and reproduced through social practice.

Consider the following example of changes in social practices: the recently observed global decline in the rate at which adolescents and young adults are acquiring driver's licenses (Figure 4.1).† Across the United States, the fraction

---

*Pierre Bourdieu is the name most closely associated with *habitus*. See, for example, *The Logic of Practice*, Cambridge: Polity, 1990. Portions of the following paragraphs come from Ronnie D. Lipschutz, Dominique de Wit, and Kevin Bel, "Practicing Energy, or Energy Consumption as a Social Practice," paper presented to the 2015 Behavior, Energy and Climate Change Conference, Sacramento, CA, October 18–21, accessed August 20, 2015, http://escholarship.org/uc/item/1vs503px.

†There appears to be a secular decline in car usage across the industrialized world; see, for example, Kurt Van Dender and Martin Clever, "Recent Trends in Car Usage in Advanced Economies—Slower Growth Ahead?" International Transport Forum, Organisation for Economic Co-operation and Development, No. 2013-9, April, 2013, accessed July 22, 2015, http://www.internationaltransportforum.org/jtrc/Discussion Papers/DP201309.pdf; Michael Sivak and Brandon Schoettle, "Recent Changes in the Age Composition of Drivers in 15 Countries," Transportation Research Institute, University of Michigan, Report # UMTRI-2011-43, October 2011, accessed July 22, 2015, http://deepblue.lib.umich.edu/bitstream/handle/2027.42/86680/102764.pdf.

Figure 4.1 Licensed Drivers in the United States by Age Cohort

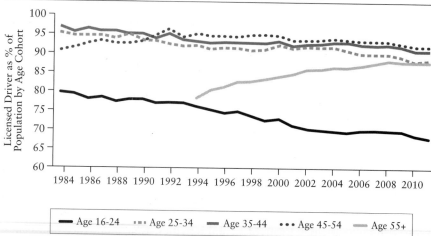

Source: Derek Thompson, "The dubious future of the American car business—in 14 charts," *The Atlantic*, Sep. 6, 2013, http://www.theatlantic.com/business/archive/2013/09/the-dubious-future-of-the-american-car-business-in-14-charts/279422/ (accessed December 28, 2014).

of all age cohorts holding licenses is in decline, with the greatest drop among those sixteen to twenty-four years old. This would certainly appear to be a counterintuitive trend in a society so greatly shaped by and dependent on private automobiles. How might this change be explained? Certainly, there has not been a commensurate increase in accessibility to alternative modes of transportation, and young adults, at least, tend not to rely on their parents to get around. The cost of owning, insuring, operating, and repairing a vehicle is hardly negligible and is rising—which could be a factor behind the decline, although the majority of people in the sixteen- to twenty-four-year age bracket do have driver's licenses and cars to drive. Perhaps the bloody-minded accident videos traditionally shown in high school and during new driver training are finally scaring people out of cars. But these same factors were in place prior to the 1980s, so it is unlikely that they matter very much more now. Note also that this trend might have begun even earlier than 1984, although it has only been remarked on during the last decade or so.

One broadly offered explanation for this decline is the rise of social media, the new communication and social technologies and platforms: smartphones, tablets, Facebook, Twitter, etc.[106] According to this line of thought, cars used to be the "social medium" of choice via which adolescents and young adults interacted and communicated with each other—recall (for those of you who can), for example, the importance of "cruising" and drive-in restaurants in the film *American Graffiti* (or in *Rebel without a Cause* and many others). Today, such social interaction can take place more easily and cheaply via communications technologies and social media, and there is less need (and desire?) for face-to-face contact.* To be sure, not everyone buys into this proposition,[107] especially because the decline began long before the advent of smart technologies. Nonetheless, this example highlights a change in social practice and habitus.[108]

Social practice theory would seem to pose serious problems for praxis and politics. After all, if social change takes place in some kind of response to relational changes among people and the material world, where is active social and political intervention possible?† What is left out of this analysis is the role of the social "imaginaries" discussed earlier in this chapter. The material world and subjective social structures offer both constraints and opportunities to practitioners in fields of habitus, especially in terms of the ways in which people understand the world and their place within it, for example, with respect to nature or climate change. This is the point at which opportunities for politics and praxis arise, via information, knowledge, literature, film, protests, movements, and other forms of action. The point is not so much to "inform" as to reshape imaginaries and offer notional ways of realizing them (as with, for example, Marx's architect) within material histories and constraints.

Social movements represent the coalescence of frustration with political and economic systems, and fulfillment of movement goals requires changes in the way people see and understand the world around them, that is, their imaginaries. At the same time, history and materialism—linked to local political economy, as discussed in chapters 1 and 3—play an important role in the specific form that movements take, as well as their success, failure, or co-optation by existing institutions and parties. The argument presented here draws on various strands of Marxist and neo-Marxist thought (including that of Antonio Gramsci), and is as follows.[109]

First, as argued above, most institutionalized political systems in democratic liberal states foster individualist approaches to politics in that they rely on the aggregated choice of individual voters' self-interests. Outside of running for public office, such systems limit the degree to which any particular individual can become involved, beyond such ritual practices as town meetings, petitioning, periodic voting, communicating with representatives, and lobbying (all forms of habitus). Election results are presented as the articulation of some kind of "general will," even though they are mostly the additive consequence

---

*On the perils of such mediated communication, see E. M. Forester, "The Machine Stops," *The Oxford and Cambridge Review*, November 1909, accessed September 26, 2017, http://www.ele.uri.edu/faculty/vetter/Other-stuff/The-Machine-Stops.pdf.

†Indeed, Elizabeth Shove argues that theories of active intervention move us back to the propositions of the ABC approach; personal communication. See also Elizabeth Shove, "Beyond the ABC: Climate Change Policy and Theories of Social Change," *Environment and Planning A* 42 (2010): 1273–85; Tim Kurz, et al., "Habitual Behaviors or Patterns of Practice? Explaining and Changing Repetitive Climate-relevant Actions," *WIREs Climate Change* 6 (2015): 113–28.

of individual preferences.* Other political activities—which can, of course, be defined so broadly as to include everything that involves more than one person—are limited or constrained not only in the form of explicit laws, but also as norms ("no decent person would do that"), perceptions ("you're a traitor if you don't support the president"), beliefs ("this is the best of all possible countries, and if you don't like it here, go somewhere else"), and values ("God made us great!"). And don't forget money ("it's a free country; anyone can buy the opportunity to express his or her beliefs"). These limits are assumed to have been internalized, even if supposed offenders need reminding.[†] The limits also appear in the form of prescribed "practices" (stand during the playing of the National Anthem, etc.), because the counterpractices are regarded as illegitimate.[110]

Second, no human institution, including well-established political systems, is without its flaws and contradictions. That is, there is no reason to expect material and ideational conditions and belief systems to be fully in concordance with each other or, for that matter, internally coherent (such incoherences are called "contradictions").[‡] Despite considerable evidence to the contrary, it is remains an abiding principle of devotees of economic liberalism that capitalist markets will make everyone richer ("a rising tide lifts all boats"), although, as we observed in earlier chapters, while capitalist economic growth does increase the wealth of societies as a whole, that income is differentially distributed. Some people get wealthier, some face stagnating incomes, and others become poorer, either in relative or absolute terms.[111] But another central principle of contemporary neoliberalism is utilitarianism, which implies that it is not only good but also fair that people be better off (and presumably happy). Therefore, the gradual impoverishment of some people and groups exposes a contradiction between the material conditions and the beliefs of liberal politics and economics. This contradiction can be exploited for political and social purposes (as seen, most, recently, by the Trump administration, but also by earlier administrations and governments around the world).

Let us examine the case of climate change to illustrate these points. The causes and processes driving such changes are many, and their interactions

---

*With the stickiness of incumbency and the status quo, elections become more a forum or plebiscite for ratifying things as they are, rather than as they could be.

†As we write, President Trump has been engaged in invective against African American (and other) football players engaging in various sorts of protests during playing of the National Anthem at sports events, suggesting that they are unpatriotic and disrespectful, among other things. He also criticized Congressional Democrats as "un-American" and "treasonous" for their failure to stand and applaud during his 2018 State of the Union address.

‡A "contradiction," according to Lawrence Crocker, takes the following form: "A and B are contradictory if and only if: (1) A and B are both processes; (2) A and B have natural paths of development; [and] (3) The natural path of development of A and the natural path of development of B cannot be jointly realized." "Marx's Use of Contradiction," *Philosophy and Phenomenological Research* 40, #4 (June 1980): 558–63, p. 560.

are very complex. But the broad response of elite supporters of the global economic system and all that it includes is to seek solutions in the very system that is experiencing the contradiction between endless growth and its limits.* Notwithstanding the need for radical changes, such as 80 percent reductions in greenhouse gas emissions, there are none forthcoming, since the required changes would be in contradiction to the demand for continued growth. These supporters of the current economic and political systems include political leaders and legislators, public intellectuals, news reporters and pundits, academics, businessmen, corporate executives, and the staff of various think tanks and public associations. They are, for the most part, defenders of the "conventional wisdom" regarding what constitutes "proper" social arrangements and behaviors, even in the face of impending disruption and disaster. Their primary concern is to ensure that the material and ideational arrangements supporting capitalism and globalization remain largely intact and, if possible, unchallenged. To accomplish this end, these elites need only engage in business as usual, with tweaks here and there, and justify the system as both necessary and natural.[112] If the consequences of business as usual are negative for some, it is not the fault of the social arrangements. Growth will provide a fix. At any rate, these consequences occur because certain people, such as the poor and unemployed, are not behaving in the manner required to avoid (or be resilient) in the face of flooding (building walls, buying flood insurance, moving out of the flood-prone areas). There is no need for structural change because specific individuals or groups, and their practices, are to blame for their own misfortunes.[†]

Those who are involved in laying the groundwork for the case against business as usual argue that the contemporary global economic regime generates unfair and unjust conditions and outcomes for many people, especially the poor in both the Global South and the Global North.[113] It is not by choice that, for example, formerly wealthy and white parts of southwest Houston are, after decades of flooding, minority-dominated today—that is where lower-cost housing is to be found. It is not the fault of the poor and middle class that one cannot live in Houston without having to drive on flood-prone highways, since the public transportation system is so wretched. And it is not their fault that unbridled growth that has paved 30 percent of the county and excessive

---

*This is not about belief in or denial of climate change; it is about commitment to existing systems as being able to provide solutions to social problems.

†The example of Houston and Hurricane Harvey, discussed in Box 4.1, is instructive here: despite decades of awareness and experience of the growing number and impact of floods on that city, and even possible links to climate change, its political and business class responded only minimally and looked to non-disruptive means of channeling floodwaters. In this instance (with respect to contradictions) it appears, "B" has won, even if only temporarily. See: Manny Fernandez and Richard Fausset, "A Storm Forces Houston, the Limitless City, to Consider Its Limits," *New York Times*, August 30, 2017, accessed August 31, 2017, https://nyti.ms/2wpnVPq.

groundwater pumping has left floodwater much less (marshy) land into which it can percolate or drain. But they bear the brunt of the elite's disregard for the fundamental contradiction that played a major role in the disaster.[114] The flood of capital into Houston over the past fifty years, at least in part the result of oil and globalization, helped create the conditions for flooding across Houston, rationalizing the absence of zoning and ignoring the implications of the growing number of one-hundred- and five-hundred-year and even one-thousand-year floods (see Box 4.1).

BOX 4.1	**History, nature, society—the case of Houston**

As this chapter was being revised and edited (in August 2017), Houston, Texas, the fourth most populous city in the United States (whose metro region comprises some ten thousand square miles, more than the area of New Jersey), was hit by unprecedented flooding due to Hurricane Harvey. Like many cities, Houston was established, in 1836, to take advantage of several natural features, especially access by water. It was a center of the slave trade, and many slaves worked on cotton and sugar plantations across the region. At that time, no one gave much thought to the fact that the highest point in the area was 125 feet above sea level, or that most of the future city, built on a semi-tropical coastal plain, was much lower. Over the past fifty years, the city grew by leaps and bounds, from six hundred thousand in 1950 to 6.7 million in the metropolitan statistical area in 2016.

For many decades, the city consumed groundwater, which led to subsidence, while the constant building and horizontal growth covered over the marshes and swamps that absorbed floodwaters. Moreover, because of its general flatness, whatever water flowed through the city's creeks and bayous moved slowly and, during times of heavy precipitation, floodwaters began to back up into streets and surrounding neighborhoods, and into belowground freeways and underpasses. Houston has been inundated more and more frequently by floods of varying magnitudes over the past few decades, with disastrous consequences,* and climate change is only likely to make "one-thousand-year storms," such as Harvey, come much more frequently. None of this was intentional, of course; merely the convergence of geography, weather, and economy. Yet it seems quite improbable that Houston will be abandoned (many flooded-out residents are already rebuilding).†

---

*There are no natural disasters, only people in the wrong place at the wrong time.

†The population of the City of New Orleans has been in decline since the 1970s, but between 2000 and 2010, it declined by 29 percent (from 485,000 to 344,000) as a result of displacement by Hurricane Katrina in 2005. The city has recovered somewhat, with an increase of about fifty thousand people since 2010.

(Continued)

What, then, of the future? We can speculate about several points. First, periodic flooding has already driven a change in neighborhood demographics, as the wealthy move to higher ground and the less wealthy move into the now-cheaper once-flooded real estate, which they cannot sell because of the risk of future flooding. As a result, minorities and the poor are more subject to being flooded out, as seen during Harvey. Second, the city will make every effort, within its budget, to construct new bulwarks in wealthier areas against rising waters and may leave some less-prosperous neighborhoods to their own devices (as happened in New Orleans, during and after Katrina). Because those living in these areas cannot afford to move to higher and more desirable elevations, where real estate is more costly, they will face the choice of riding out future floods or leaving the city. Third, depending on sea levels and climate change, growing portions of the city may be triaged and abandoned. City planners, developers, and others must, before they approve any new buildings and developments, imagine alternative material futures before they make any changes to the landscape.

Sources: Nadja Popovich and Claire O'Neill, "A '500-Year Flood' Could Happen Again Sooner than You Think. Here's Why," *New York Times*, August 28, 2017, accessed August 29, 2017, https://www.nytimes.com/interactive/2017/08/28/climate/500-year-flood-hurricane-harvey-houston.html; David Leonhardt, "Harvey, the Storm That Humans Helped Cause," *New York Times*, August 29, 2017, accessed August 29, 2017, https://www.nytimes.com/2017/08/29/opinion/harvey-the-storm-that-humans-helped-cause.html; Greg Bankoff, "No Such Thing as Natural Disasters," *Harvard International Review*, August 23, 2010, accessed March 27, 2018, http://hir.harvard.edu/article/?a=2694.

To put this another way, the outcomes promised by globalization's advocates have not been, and cannot be, delivered under the contemporary economic and political structures.[115] If the negative impacts affect enough people, they can be mobilized to demand structural change, and their expressions of doubt will expose and weaken the underlying belief systems that support globalization. Movements such as Occupy as well as the rise of nationalism around the world appear to validate this proposition, although the associated movements are by no means always progressive ones, as noted earlier. One result of these movements might well be a reconfiguration of the global economic system and its politics.[116] The precise form of such change is not determined and cannot be predicted. We might imagine reforms in the operating rules of the WTO, international financial institutions, and private funders that would incorporate both social and environmental considerations. The World Bank has made extensive efforts to do this, but with limited success.[117] Organizations usually change radically only in the face of crisis and, even then, most do not.[118]

Alternatively, we might imagine new international organizations, dedicated to social and environmental regulation, with authority and power commensurate with that of the WTO. Organizations that oversee these two issue areas do exist, in the International Labour Organization and the UN Environment Programme (now known as UN Environment), but they do not have much in the way of power, authority, or financial resources. And there is a continuing debate about creation of a new global agency for the environment.[119] A more powerful organization would represent mostly costs to countries, corporations, and citizens, as against the benefits of free trade accruing largely to capital. Such an organization would not find it easy to contend with the WTO and its unequivocal supporters.

A third possibility is that social movements and organizations might be able to pressure national governments to enforce both international and domestic laws applicable to social and environmental matters. If they were to do so, some states might find themselves at a competitive disadvantage with respect to others.[120] Enforcement is not, therefore, a step to be taken lightly. The likelihood of each of these outcomes would be a function of the effectiveness of political alliances and coalitions assembled by different parties as well as a compelling presentation of the real costs of business as usual. As Margaret Keck and Kathryn Sikkink have pointed out, governments and international organizations can be influenced in many different ways, some of which might, at first glance, appear quite improbable.[121] While it is somewhat dated, a good example of such politics, and coalition building, can be seen in the fate of the Multilateral Agreement on Investment (MAI).

The MAI was originally negotiated by the twenty-five-odd industrialized and developing countries belonging to the Paris-based Organisation for Economic Co-operation and Development. It was meant to establish uniform rules governing both domestic and foreign investment in the organization's industrialized countries, as well as in its two developing ones. The members intended to formulate the agreement, sign and ratify it, and then open it to nonmembers, which were mostly developing countries. These states would have to join or miss out on investment from signatory countries. But the MAI was derailed just as it was about to be finalized. A coalition of NGOs and social movements was able to raise doubts about the MAI in strategic places. Both the French and Canadian governments came to agree with the NGOs and social movement groups: the MAI would be disadvantageous and unfair to developing countries, the vast majority of which had no say in its terms. As a result of the ability of activists to influence key states, the agreement was withdrawn from consideration.[122]

Certainly, although the collapse of the MAI constituted a defeat for international capital, it did not do much to alter the structure of the global economic system. Developed countries brought negotiations on investment into the WTO and at the same time pursued bilateral and regional investment treaties, such as the Trans-Pacific Partnership, with developing country partners with the same or more onerous terms as the MAI. What the collapse of the MAI did demonstrate, however, was that social power, especially in joint action with doubtful governments, can have an impact on the form and operation of global economic structure. At a minimum, those who favor business as usual must take into account the sustained resistance to globalization and the inequities it fosters. Those elites may respond peacefully and agree to the creation of new institutions and rules. Or they may decide that repression is called for, as was the case when police clashed with demonstrators in Genoa in August 2001 and Hamburg in 2017.[123] Outcomes are not determined; they are the result of strategy, tactics, planning, and action.

SOCIAL POWER IN ACTION

These largely theoretical explanations of civic politics and social power—consciousness-raising, resource mobilization, party politics, and agency-structure praxis—are embodied in a broad range of activities and actions relevant to global environmental politics. But what does social power actually look like? By focusing on "real life" examples, we can begin to see how history, political economy, and local action fit together as essential parts of environmental praxis. Recall that it is the history of a place that accounts for the shape of the landscape and its social organization and that establishes many of the patterns of interaction (including habitus) among people and within societies. It is political economy that accounts for the ways in which the local environment has been and is being used and abused, and for the roles of individuals, groups, and business in that process. Finally, it is in the face-to-face encounters of people that debate and action—politics—takes place. This is not to say that all environmental praxis is or must be limited solely to place; it is to argue that a global environmental politics must be rooted in the politics and practices of specific times and places.

Two examples of social power in action are given here: climate justice and market-based activism. Each has a somewhat different character and politics. Led by indigenous groups defending their lands and water, a climate justice movement in the United States is engaging in institutionalized politics and practicing a mix of political activism and civic politics, using a broad spectrum of strategies to bring about an energy transformation away from fossil fuels. Environmental and social issues have acquired such prominent positions around the world that even corporations have gotten into the act, through market-based *corporate social responsibility*. We offer here both description and critique, especially of this last mode of social power in action.

Climate Justice and the Dakota Access Pipeline

We return again to the example of the Dakota Access Pipeline to illustrate several of the ideas discussed earlier in this chapter, in particular, political activism and praxis—agents coming to grips with structure. You recall the basic details of the case: the proposal to build a pipeline from Bakken oil shale fields of North Dakota to Illinois, with part of the pipeline running under Lake Oahe and the Missouri River where the Sioux of the region get their drinking water. After failing to get the project stopped and in response to imminent construction, in April 2016 the Standing Rock Sioux and their allies set up the Sacred Stone Camp to call attention to the conflict, prevent further construction, and push the Obama administration to halt the project completely.

The occupation at Standing Rock has links to two important contemporary precursor movements. First, is the "Idle No More" movement of First Nations in Canada and indigenous groups in the United States, which found its origins in protests against Saskatchewan provincial legislation that would diminish indigenous territorial sovereignty and environmental protections.[124] Begun in 2012, the movement swelled over the next several years to hold large teach-ins, sit-ins, rallies, and marches across Canada and the United States. The mobilization of indigenous North Americans is further linked to growing activism of indigenous nations across the world in their own movements for sovereignty and control of resources in their territories. The resistance movements of indigenous peoples are of course centuries old, though they have seen a powerful resurgence in many parts of the globe over the past few decades.[125] Many of the issues these movements take on as their own also include climate justice, related to fossil fuel exploitation and impacts, which threaten livelihoods, ways of existence, homes, and lands.[126] But the movements also struggle against mining, deforestation, industrial agriculture, development, and denial of access to traditional lands.

The second important sister movement to Standing Rock comes out of mainstream U.S. environmentalism. Frustrated with the Obama administration's slow to no movement to transition the U.S. economy away from dependence on fossil fuels for energy production, climate activists identified a large-scale project that could serve as an iconic mobilization focal point: the Keystone XL pipeline.[127] A long series of conventional protests and civic politics from letter-writing to legislation to legal approaches built toward a demonstration of civil disobedience and non-violent direct action, with hundreds arrested during a sit-in protest in front of the White House.[128] The Obama administration eventually denied approval to construct the pipeline (however, the Trump administration reversed course and approved it; the pipeline is now tied up in court litigation).[129]

The Standing Rock Sioux began their opposition against the Dakota Access Pipeline when it was first announced in 2014, arguing that it threatened their drinking water and cultural sites. Since then, they have used a wide variety of strategies and tactics, civic politics, and political activism, to mobilize social power against the pipeline and put public pressure on the Obama administration. Environmental and climate activists became important allies because of the clear link with climate change; "Keep it in the Ground" was an important rallying cry reflecting the fact that most fossil reserves cannot be burnt if the rise in global temperature is to be kept below two degrees Celsius. A legal strategy was a critical element of the campaign to stop the pipeline. On July 27, 2016, the tribes filed their first lawsuit seeking an injunction against the project, with support of Earthjustice, a nonprofit environmental law organization in San Francisco.[130]

The encampment at Standing Rock was the most visible of the tactics used to stop the Dakota Access Pipeline. Over the months after the first lawsuit was filed, the camp attracted thousands of protestors, including celebrities and political leaders. At several points during the fall of 2016, the Standing Rock protestors put out a call for veterans of armed forces to come help them defend their land and water. Thousands of former servicemen answered that call, some of them twice. For the first "deployment," the group Veterans Stand raised over one million dollars on the website GoFundMe for transportation and supplies to support the veterans.[131] Meanwhile, back in communities across the country, students and other climate activists were challenging banks to stop their financing of the project. A victory was achieved in December 2016, as the Army Corps of Engineers decided to seek another route for the pipeline, turned into a further protracted fight in 2017 as the new president, Donald Trump, reversed the decision to stop the project. The court battles continue.[132] (Paradoxically, the pipeline's operator is having difficulty signing up companies to ship crude from Canada to the Gulf Coast.[133])

The actions and activists at Standing Rock fit squarely within what is broadly called the climate justice movement, which itself finds at least some of its roots growing out of the environmental justice movement.* The origins of the environmental justice movement in the United States are often traced to protests in 1982 against a landfill in Warren County, North Carolina, at which toxic PCB-tainted soil had been dumped.[134] The community around the landfill was poor and majority African American. The first environmental justice protests called attention to decisions to locate toxic waste in communities with little political power and which were often mostly non-white. Another key moment in defining the environmental justice movement was the First National People of Color Environmental Leadership Summit, held in 1991 in Washington, DC. The NAACP now links environmental and climate justice in their work, noting that "environmental injustices, including climate change, have a disproportionate impact on communities of color and low income communities in the United States and around the world."[135] As noted by two academic historians of the environmental justice movement, "climate justice focuses on local impacts and experience, inequitable vulnerabilities, the importance of community voice, and demands for community sovereignty and functioning."[136]

The fight to stop the Dakota Access Pipeline from crossing the Missouri River at the Standing Rock reservation touched the imagination and spirit of

---

*We already mentioned the very important roots in the movement for sovereignty and indigenous rights.

many across the United States and the world. It is at the same time a very local fight to protect water, land, and cultural sites and a global fight of agents against the seemingly immovable structures of a fossil-fuel-dominated economy. Of course, a local fight against a global industry will require building alliances, at least some of them appearing quite unlikely. Whatever its successes or failures, Standing Rock has become a model for indigenous peoples' struggles, not only in the United States but elsewhere in the world (just as many of those struggles no doubt inspired the activists at Standing Rock). Most important about such movements are that they are focused on the *local* manifestations of *global* systems. Capital must have access to resources; denied that access, capital must either negotiate or leave.

"Doing It" in the Market?

Over the past couple of decades, social power has been appropriated by capital through the *corporate social responsibility* movement.[137] As we saw in chapter 3 and the first section of this chapter, the growing complexity of commodity chains and associated consumption have relocated environmental and social externalities from "home" countries of corporations to the widely separated sites where production and consumption take place. The ability and willingness of local authorities in those places to regulate these externalities is highly variable: some countries have much stronger laws and enforcement than others, and the costs associated with production are, therefore, higher in some places than others. Even though host countries may have strong laws on the books, weak enforcement permits capital to reduce the cost of doing business and increase profit margins. All else being equal, capital seems likely to invest and produce in places with lower levels of regulatory attainment.[138] Although it is not clear how large this effect actually is, it appears that some regulatory arbitrage does take place between developing and industrialized countries.[139] Moreover, pollution havens exist not only in the Global South but the North, too.[140]

In an effort to combat egregious violations of laws and norms, a growing number of social activists and transnational coalitions have launched campaigns directed toward the behaviors of specific corporations and industrial sectors. Using publicity, social media, education, and the threat of consumer boycotts, activists try to pressure companies to change their practices within the confines of their specific commodity chains in ways that are both environmentally friendly and more supportive of labor. By engaging in these actions, activists try to induce greater public and corporate attention to social issues.[141] All else being equal, however, corporations prefer to set their own terms for social and environmental regulation. The corporate responsibility "movement" has, subsequently, emerged as capital's response to market activism.[142] In the belief that it is better, for reasons of both cost and public relations, individual companies have adopted codes of conduct and performance standards.[143] Codes and standards are intended to preempt external pressures for responsible behavior even as they tend to be less rigorous than those mandated by local law or desired by market activists. In addition, national and international business associations and corporate groups have developed environmental management standards for various production practices (in chapter 5 we shall see how such standards have been developed and applied to sustainable forestry management).[144] The United Nations has promulgated something called the "Global Compact," which commits signatory corporations to "good global citizenship."[145] Some of the standards in these codes are very general; others quite specific.

Codes of conduct are sets of rules established within factories or company facilities. They are intended to establish minimum wages, working hours, health and safety conditions, grievance procedures, environmental quality, and so on.[146] Companies agree to post the codes in visible places, to inform workers of their existence and significance, and to ensure that management obeys them. Workers, in turn, are expected to perform the duties assigned to them and to accept directions issued by their supervisors. These codes have no legal standing in local courts, however, and there is no obligation on the part of local authorities to see that they are enforced. Thus, although some plants in some places handle their wastes carefully, others are not subject to such pressures. Under these circumstances, it is difficult not to continue a regulatory race to the bottom.[147]

Best practices are somewhat different. They involve the systems that companies use to produce goods and to reduce the pollutants associated with those goods. It is possible to reduce the environmental impacts of production in three ways: end-of-pipe reduction, ecological modernization, and sustainable supply chain management. The first entails the capture of pollutants from smokestacks, wastewater pipes, and other outlets. The second (discussed in chapter 3) focuses on changes in manufacturing processes and technology so that less waste is produced and fewer pollutants require disposal. The third—sometimes called "sustainability performance measurement"—addresses reduction or elimination of environmental impacts at each step of the production process.[148] Producers will, we can assume, select among alternatives so as to minimize their costs of retooling (recall the discussion of greenhouse gas permits in chapter 3). In some instances, the more efficient use of inputs provides enough savings to more than compensate for the greater cost of implementing ecological modernization. Ensuring sustainability at each step in the supply chain requires, however, careful analysis and inventory and implementation of "best practices"—and who pays for such changes is not always clear (generally, it is assumed to be the buyer, in the form of higher prices).* Again, however, the commitment by a corporation to best practices has no domestic or international legal status, and there are generally no political or legal mechanisms to compel good performance.

---

*Nor are empirical outcomes clear. In the absence of standards against which to measure performance, and detailed accounting of performance, it is very difficult to determine to what extent sustainability has been achieved. The International Organization for Standardization has issued ISO 14001, an "environmental management system," discussed in the following.

In these examples, companies are engaged in *self-regulation*. That is, corporate boards and management voluntarily agree to alter environmentally unfriendly company practices, sometimes with the hope that purchasers and consumers will preferentially choose their products (see the discussion on certification). The problem with this approach is evident: if there is no legally binding requirement that the company change its practices, how will anyone know for certain that it is observing its commitments to do so? Under law and legal regulatory systems created and supported by domestic governments, violations of both should be detected and violators punished. Presumably, the threat of punishment acts as a deterrent—if the threat is a real one and the costs of violation sufficiently high*—and individuals and companies alike will obey the law. Under self-regulation, however, there is no system of monitoring for violations, there is no threat of punishment, and there is no incentive to observe the commitments made. There is only good faith and the possibility of exposure by activists. Experience often shows this to be a weak reed. Indeed, Prakash and Potoski conclude that

> analyses of pollution levels for a panel of 159 countries (73 for water pollution) from 1991 to 2005 indicate that ISO 14001 certifications reduce air (SO_2) emissions in countries with less stringent environmental regulations but have no effect on air emissions in countries with stringent environmental regulations. We also find that ISO membership levels are not associated with reductions in water pollution levels . . . irrespective of stringency of domestic law.[149]

One market-based response to the enforcement problem is certification or "eco-labeling,"[150] which relies on the modification of consumer choice or preferences to pressure alterations in corporate behavior. An eco-label is a claim placed on a product having to do with its production or performance. The label is intended to enhance the item's social or market value by conveying its environmentally advantageous features, making the product more attractive to the environmentally conscious consumer. Three categories of eco-labels are widely recognized.[151] *First-party* certification is the most common and simplistic approach. It entails producer claims about a product, for instance, that it is "recyclable," "ozone-friendly," "nontoxic," or "biodegradable." But these are claims made without external or disinterested verification. Therefore, the only guarantee to the purchaser that the product performs according to those claims is trust in the producer's brand reputation.

Second-party certification is conducted by industry-related entities, such as trade associations. These groups establish guidelines or criteria that must be met if producers are to make environmental claims about their products. The group offering the label may or may not actually monitor and verify the

---

*Actual experience shows that, quite often, it is less costly for companies to pay fines than retrofit their operations.

producer's adherence to the guidelines and criteria. Once the standards are met or the guidelines followed, an industry-approved label is placed on the product stating or verifying the product's environmentally friendly qualities. In this instance, the certifying organization will seek to ensure the label's value and to mandate its use, so that no single producer will gain an advantage over any other. An example of second-party labeling is ISO 14001, sponsored by the International Organization for Standardization (ISO). The ISO is an association of national standards groups, some of which are private and others public.[152] ISO 14001 comprises a set of environmental management standards that companies apply within their plants and to other operations in the commodity chain under their control. Each individual company is responsible for certifying that it has followed the standards and, if it has, it is permitted to put the ISO logo on its products.

Third-party, or independent, certification is performed by government agencies, nonprofit groups, for-profit companies, or organizations representing some combination of these three. As with second-party labels, third-party eco-labeling programs set guidelines that products must meet in order to use the label. Sponsoring organizations also generally conduct audits in order to ensure producer compliance with the guidelines. As the name implies, third-party certifiers are not affiliated with the products they label. An example of third-party labeling can be found in the various sustainable forestry management standards and certification programs of various organizations, such as the Forest Stewardship Council (FSC).[153] The FSC has developed and adopted a set of global guidelines called Principles and Criteria for Forest Stewardship, and it accredits certifying organizations that agree to abide by these principles, criteria, and standards. The certifying groups, in turn, inspect and monitor timber operations seeking the FSC's stamp of approval. The final product appears on the retailers' shelves with the FSC logo affixed.

First-party certification is not considered very reliable, inasmuch as a company's claims are restricted only to the extent that relevant law requires "truth-in-advertising." Second-party labeling rests on corporate assurances to the relevant trade association that it has fulfilled the minimum standards required for certification; there is frequently no mechanism to verify such claims. Only third-party certification, which relies on outside inspection and monitoring, may provide a means of checking a company's claims of environmentally friendly behavior. But even approval by third-party organizations is fraught with problems. Who certifies the certifiers? Who has vetted the procedures whereby certifiers conduct their inspections? What, besides consumer disapproval, motivates a company to rectify violations detected by certifiers? What prevents a company from shopping around for the most accommodating certifying organization? And to what extent do consumers actually choose their purchases on the basis of certification, especially if such goods are more costly than uncertified ones? All these questions remain largely unanswered, and viable empirical data that might help to answer them remain sparse.[154]

At the end of the day, the value of certification is very much connected to brand reputation. Many consumers have a preference for food labeled "organic." But whereas brands are the private property of specific companies, certification is not. In the absence of effective regulation and enforcement, anyone could call their food products "organic." This is why certifying organizations are established. For organic foods, farmers and producers have created associations and formulated rules defining how to obtain organic certification. There exists an international association for this purpose as well as local, state, and national groups and laws.[155] But as the market for organic foods has expanded, so have opportunities for both profit and fraud. After all, confronted with multiple certifiers, how is the consumer to know which ones are legitimate? In the United States, the Department of Agriculture has formulated regulations specifying requirements for organic labeling, although these did not, initially, gain the wholehearted support of organic farmers. To a growing degree, the force of law stands behind various systems of self-regulation, in effect, "certifying the certifiers." International organic certification also covers foods being shipped across national borders. In some instances, certifiers use U.S. standards as the basis for their assessments, but other bodies and countries, such as the European Union, the ISO, and Australia, have their own sets of standards.

The path of such standards has not been smooth, however. As the demand for organic food has grown, a number of incidents of illness associated with their production have occurred in the United States and elsewhere.[156] This has taken place, notwithstanding certification, safety precautions, and branding,[157] for several reasons. First, the organic fertilizer applied to some fields may carry traces of dangerous bacteria. Second, feces from both domesticated and wild animals may contaminate the fields via rain and irrigation. Third, contamination might occur in the handling and packaging process. In 2006, contaminated spinach from Earthbound Farms, one of the largest producers of organic greens in the United States, killed three people and sickened two hundred others. The culprit was *Escherichia coli* O157:H7, matched to cow manure from a nearby pasture.[158] Needless to say, this bad publicity did not bode well for Earthbound or other organic growers, who issued assurances that this would never happen again, and that contaminants would be kept far away from farms, fields, and production lines. But who tells the cattle, pigs, and bacteria to stay away?

We must ask one final question: At the end of the day, what are the political impacts of actions such as corporate social responsibility? This question brings us back to a problem encountered in chapter 3 regarding the use of market mechanisms to control pollution. What remains unchanged in the arrangements deployed by producers and certified by third parties is the fundamental structure of the production process. The use of markets as the means of allocating rights to pollute relies on economic tools to achieve political ends. Whether it is right to pollute is never asked. Even the "precautionary principle," which advises erring on the side of safety when undertaking potentially damaging actions, does not inquire about the ethics of the action itself.

A similar critique can be applied to market activism, corporate responsibility, and consumer consciousness. Campaigners argue that companies ought to reduce their polluting activities because it is the right thing to do, and they use the threat of a reduction in market share to goad companies into implementing best environmental practices, but changes take place only in those factories and facilities where their goods are produced. Other enterprises remain unaffected and no change in national law or enforcement results. Responsible corporate executives—recognizing that there might be some commercial advantage in advertising their company's good behavior—try to ensure that these practices are being followed by plant managers and others.[159] But none of this has much impact on more general political conditions or regulatory arrangements within countries. That sweatshops continue to exist throughout the United States, even though they are illegal, simply illustrates the real and potential shortcomings of self-regulation through market mechanisms.[160]

LOCAL POLITICS, GLOBAL POLITICS

The transnationalization of social power, civic politics, and political activism, through networks, coalitions, mobilization, and action, has been made more feasible by the physical infrastructure of global communication and transportation, as well as the diffusion of ideas and practices, and it is this transnationalization that has captured so much attention in recent years.[161] But whether global or local, in order to accomplish the kinds of social changes discussed in this chapter, activists must still affect the beliefs and behaviors of real human beings, whose social relations are, for the most part, highly localized. Ideas do not fall from heaven or appear as light bulbs; they must resonate with conditions as experienced and understood by those real human beings, in the places that they live, work, and play.[162] Moreover, it is in those local places that social power is most intense and engages people most strongly. How such environmental praxis articulates with the national and international will be the focus of chapter 6. In order to arrive at that point, however, we must first examine national and international environment politics more closely. This we will do in chapter 5.

FOR FURTHER READING

AWAKE, A Dream from Standing Rock. Bullfrog Films, 2017.

Bebbington, Anthony (ed.). *Social Conflict, Economic Development and Extractive Industry: Evidence from South America.* London: Routledge, 2012.

Gaventa, John. *Power and Powerlessness: Quiescence and Rebellion in an Appalachian Valley.* Urbana-Champaign: University of Illinois Press, 1980.

Hall, Derek. *Land.* Cambridge: Polity Press, 2013.

Holt-Gimenez, Eric, and Raj Patel. *Food Rebellions!: Crisis and the Hunger for Justice.* Oakland: Food First Books, 2009.

Hou, Jeffrey, ed. *Insurgent Public Space: Guerilla Urbanism and the Remaking of Contemporary Cities*. London: Routledge, 2010.

Jobin-Leeds, Greg, and AgitArte. *When We Fight We Win!: Twenty-First-Century Social Movements and the Activists that are Transforming Our World*. New York: The New Press, 2016.

Magnusson, Warren, and Karena Shaw, eds. *A Political Space: Reading the Global through Clayoquot Sound*. Minneapolis: University of Minnesota Press, 2003.

Reclaim the Streets: The Film. Tactical Media Files, 1998. http://www.tacticalmediafiles.net/videos/4484/Reclaim-the-Streets_-The-Film.

Toxic Bios: A Guerilla Narrative Project. http://www.toxicbios.eu/#/stories.

Walton, John. *Western Times and Water Wars*. Berkeley: University of California Press, 1992.

5

Domestic Politics and Global Environmental Politics

THE STATE AND THE ENVIRONMENT

As we have seen throughout this book, the standard account of global environmental politics begins with the borders between countries: Because many environmental phenomena, such as air pollution, cross national borders with alacrity, they escape the jurisdictional control of individual governments. Because no state is permitted to violate the sovereignty of another—at least in theory and international law—such problems can be addressed only through negotiation and joint action among governments—or so it is commonly said. Through application of scientific knowledge to the give-and-take of bargaining among countries, agreements can be developed that deal with such transboundary issues and serve the interests of all concerned signatories. Although there may be some degree of residual unfairness in such *international environmental regimes* (IERs), their objectives are said to serve the common global good.[1]

But is this really the whole story? From where do these agreements and regimes emerge? Are they the product of uninterested technicians of diplomacy, who seek to ensure that the resulting "balance" of duties and responsibilities is equitable and fair? To what degree do they reflect science as opposed to politics? Do these regimes accurately represent the interests of all states, especially those that are most severely affected by the problem? And who decides what principles, tools, and practices are to be applied to the problem? Where, exactly, do those principles, tools, and practices originate?[2] There are very few (if any) international institutions that do not have their sources (or origins) in the principles, tools, and practices of particular states, especially the wealthier and more powerful ones.[3]

The story we offer in this chapter is somewhat different from the conventional one about global environmental politics. The difficulty with the standard account is not that cooperation or conflict do not occur (both a focus for the "neoliberal institutionalist" view) or that international agreements and regimes are mere epiphenomena of state power (the "realist" view) or even that struggle over the "distribution of the world product" is unimportant. Rather, it is that an approach that puts states and the borders between them at its center is bound to ignore history, political economy, and, most important, social forces, many of which also "cross borders." Indeed, as we saw in chapter 1, those very borders are a product of history, political economy, and social forces, and not given by geography. Their locations have been constituted historically and socially, in response to both natural conditions and human activity and as a result of specific exercises of power over time and space. What might appear straightforward, therefore, is not at all simple.

We begin this chapter with a brief review of the concept of an IER. A "regime," according to the now-standard definition, is an international institution established by states to deal with shared or common problems.* We then turn to the national origins of these IERs, or what may be stylized as "environmental (inter)nationalism." (This is not the same as the "transnationalism" discussed in chapter 4, associated with social power operating across national borders.) The parentheses around the prefix "inter" are intentional and are meant to connote the extension (and projection) of particular forms of domestic action, nationalism, and national practice into the international realm. Those states with greater resources and power have, both historically and today, managed to impose their own ways of "doing things" on others. Thus, for example, international projects for sustainable forestry management depend heavily on approaches developed within specific countries. The design of the Paris Agreement and the resort to market mechanisms in the climate regime is a direct result of American political context, preferences, and domestic practices. And so on.

Consequently, the point of origin of international environmental practice is to be found in the historical (materialist) relationship between the sovereign state and nature. The environmental policies and practices of specific countries cannot be explained without understanding the way in which national development requires "natural development."[4] A second point is also largely national in origin but global in objective. As we have seen throughout this book, nature has been commodified and turned into a good to be exchanged and consumed within the global economy, rather than standing as a focus of ethical concern requiring political attention. The result—which is common to all international endeavors—is a constant tension between the national and the global, between expressions of sovereignty and alterations to that sovereignty, between what states might wish to do and what they find themselves compelled to do. These tensions and constraints are illustrated through several of the case studies that run through this book. We will examine forestry and

---

*The *locus classicus* for this definition is found in Krasner, *International Regimes*.

water policies in historical terms, as a set of practices originating largely in one country but diffused to others. We will then turn to the contemporary version of this diffusion process, evident in the commodification of goods, in the form of forests and biodiversity, and bads, in the form of greenhouse gas emissions and other wastes. Along the way we briefly shine spotlights on civil society contributions to shifting narratives and altering the political dynamics for regime formation.

Finally, we shall conclude with a brief inquiry into the implications of this tension for the roles, rules, and practices of IERs. Inasmuch as IERs constitute the focus of much work on global environmental politics, they can hardly be ignored.[5] But analyses that argue that IERs are, on the one hand, simply the handmaidens of powerful states or, on the other hand, the product of complex bargaining among states, miss their historical and normative content, reflected particularly in the current trend toward market-based approaches to environmental governance and management as well as opposition and resistance to them.[6]

INTERNATIONAL ENVIRONMENTAL REGIMES: THE STANDARD ACCOUNT AND SOME CAVEATS

An international regime is an interstate solution to the transboundary scenario described in chapter 1 and the collective action problem discussed in chapter 2. Recall our discussion in chapter 2 of the realist and liberal perspectives on relations among states. Both realists and liberals view states as operating in a Hobbesian "state of nature," lacking any social institutions and governed by only one rule: save thyself. This particular injunction can be read to mean that, even if your activities harm another, when those activities are central to survival, they are not only justified but necessary (liberals call this "rational egoism").[7] It follows, therefore, that cooperation among states is similar to a herd of cats: neither exists. Unlike cats, however, the leaders and governments of states can be persuaded, by power, wealth, and knowledge, that their interests may be better served through collective action bound by rules and fulfilling interests. Some animals, such as caribou, find safety and survival in herds. Are states more like cats or caribou?

An international regime is the answer to this dilemma: it facilitates collective action among states without sacrificing rational egoism (at least, in theory). According to the now widely accepted terminology, a regime is a set of "principles, norms, rules, and decisionmaking procedures around which actor expectations converge in a given issue area."[8] This unusually opaque definition masks what is, in fact, a common activity: patterned relationships and interactions among individuals, in which behaviors and actions are governed by rules and whose results are intended to be fairly predictable (see chapter 1). We all participate in such practices, or institutions, every day of our lives, even though we think of ourselves as cats rather than caribou (think of the deprecatory implication of the phrase "running with the herd").[9] In other words, regimes are no more than institutions, except that the central actors are states rather than people.*

Why, then, are regimes not called "interstate institutions" and thereby recognized as a common form of human relationship, albeit among, rather than within, political communities? Aside from the many ways in which we use the term "institution"—an organization, a building, a research center, a long-established custom or practice—the term "regime" serves two particular functions. First, it suggests an apparent uniqueness in international relations: states, the term seems to imply, really are different. Second, it obscures the extent to which interstate institutions are actually quite common and nothing new. Even war, as uncontrolled as it may be, is hedged about by all kinds of written agreements, implied understandings, and customary practices.[10]

An IER is, therefore, no more than an institution intended to facilitate specified patterns of behavior with respect to some contested or shared resource or problem. The "transboundary" concept recognizes that, within states, there are usually many institutions to address such matters, whereas, among states, there are many fewer, if any. And such interstate institutions often need to be created from scratch, as it were, especially if a new issue or problem has been identified as being of shared concern. At the same time, no regime simply emerges from a social vacuum: the "principles, rules, norms, and decisionmaking procedures" that constitute any regime emerge out of very specific social and historical contexts and circumstances. More often than not, these contexts are associated with particular countries and periods of time.

We might usefully compare two different IERs to see how context affects form. The International Whaling Commission (IWC) is one of the older IERs in existence, established in 1949 when the very survival of virtually all species of whales was in doubt.[11] Although IWC membership was historically composed of countries with whaling fleets, this was not a requirement for membership. Every year, the IWC met to set whaling quotas, which specified how many of each type of whale could be taken and which allocated shares among countries. The quotas were set, more or less, according to the concept of the "maximum sustainable yield" that would maintain the number of each species at a relatively constant level—the hope being that the individuals would reproduce at a stable replacement rate.[12] Maximum sustainable yield was also applied to various bilateral and multilateral fishery regimes, although it did not, in the end, prevent serious depletion of many fish stocks around the world.†

---

*In keeping with this parallel, we can also regard state cooperation via international regimes as approaching *habitus*, something that states *do* as states.

†There are problems with the notion of maximum sustainable yield, not the least of which is getting accurate counts, which, among some fish species, can vary drastically from one year to the next. The idea has since been replaced by other concepts and practices. See, for example, Lesley Evans Ogden, "The Complex Business of Sustainable Exploitation of Wildlife," *BioScience* 67, #8 (August 2017): 691–97.

The point here is that the very concept of a quota involves rationing of access to the resource (as we have seen with respect to emission allowances). Rationing is a *political* function requiring centralized knowledge and decision-making, for it presumes that someone knows how much of a particular good is available, and how much should be kept in reserve for emergencies at a later time. Having this number in hand, a duly appointed authority can then decide how much to allocate to each individual or participant.* It is not surprising that the IWC was based on what is, essentially, a form of rationing.[13] During World War II, all of the combatants found it necessary to ration civilian access to food, fuel, and other goods, and some countries, such as Great Britain, continued rationing into the 1950s. Although rationing is usually criticized as costly and inefficient, the presumption is that it is fair, if not always equitable. But it also means that some people will gain access to less than they need, whereas those who are willing to pay more than the going rate can usually satisfy their desires through the black market. The IWC has not altered its basic approach to the regulation of whaling—although in 1982 it put in place a moratorium on whaling that is still strongly contested by several countries and groups.[14] Other fisheries have found, however, that their quota systems are not only subject to cheating but have not prevented decline and collapse in many parts of the world.[15] The solution that has been settled on is to create regimes that commodify access to fisheries, that is, to let the market "decide" who will be able to fish.[16]

Many of the world's fisheries are both overfished and overcapitalized. That is, there are too many boats with too much fish-capturing capacity. The straightforward answer would be for governments to pay fishers to destroy their equipment. This, however, raises all kinds of technical problems: who should be paid, what will they do if they cannot fish, where will the resources come from to fund such a project? An alternative is to have the fishers pay for the *right* to fish, creating a property right in the fish catch and setting up a cap-and-trade system for the fishery. If permits to fish are sold by auction, the marginal cost of a permit will rise to the level of the marginal benefits expected from fishing. Larger, more efficient fishers will likely acquire more permits, while less efficient, often smaller fishers, will leave the industry. The cap theoretically reduces pressure on the fishery and stocks can grow back to sustainable levels.[17] The approach allows decision makers to push more of the politically painful decisions about who stays and who leaves the fishery to the market.

A number of developed countries now operate many of their fisheries with a version of this fish cap-and-trade system, first introduced in New Zealand in the 1980s. A catch limit (cap) is set for the season and property rights to

---

*This is the same process as what is often called "command-and-control regulation," through which the state imposes limits or requirement on permitted activities, products, and services.

a proportion of the catch are in some way distributed among or auctioned to fishers (confusingly for our story, the tradeable units are called individual transferable quotas [ITQs] or individual fishing quotas [IFQs], not permits). However, in contrast to a true free market system, at least at present, many countries restrict the sale of permits/quotas to some degree to prevent industrial concentration and/or to preserve size diversity so small fishers have some chance to remain in the fishery if they choose.[18]

The permit/quota method is touted as being more "efficient" as well as more "effective than rationing," but it marks a major shift in the operating principles of IERs, a shift that can be ascribed directly to discursive and material changes—the shift from the welfare state to a neoliberal one—in the United States and Great Britain between the 1970s and 1990s.[19] In effect, talking about the benefits of ITQs convinced relevant parties that they would work, even in the absence of empirical confirmation. Fixed quotas and rationing are difficult for governments to implement and enforce, and cheating is always a problem, but these approaches are motivated by some sense of equity and the common sharing of burdens. People may not get as much as they want under rationing but, for the most part, they do receive a fair share of what is available. By contrast, marketable permitting systems are complicated to set up but, once in operation, are relatively easy to monitor.* They generate up-front revenues for the authority that creates the system, and operating costs can be recovered through a small tax on transactions. But, as noted above, access is directly related to the individual's ability to pay for the permit. This approach is philosophically linked to the American and British shift to neoliberalism† and a greater reliance on markets, rather than politics, as allocational mechanisms.[20]

To put this point another way, IERs whose procedures are rooted in market-based transactions provide structural advantages to the rich, whether individuals or countries. There is nothing inevitable about such an outcome, of course; the negotiations and bargaining that are the staple and basis of all IERs are intended to address such structural inequities. Poor countries may be granted certain exemptions or provided with access to special funds or technologies or allocated a certain number of permits at concessional prices. All these mechanisms are designed to "level the playing field," to some degree. Yet, they do not alter the basic structure of the IER or the discursive imprint put on it by dominant countries, such as the United States. This pattern—that is, the national origins of international environmental practices—is apparent in other IERs, as well.

---

*These systems do require reporting on catches, either by monitors on the docks or self-reporting by fishers.

†Here, neoliberalism refers to a more market-based approach to resource allocation and includes reduced state welfare spending and privatization of public services. Ramesh Mishra, *Globalization and the Welfare State*, Cheltenham, England: Elgar, 1999; Philip G. Cerny, "Structuring the Political Arena: Public Goods, States, and Governance in a Globalizing World," in *Global Political Economy: Contemporary Theories*, edited by Ronen Palan, pp. 21–35, London: Routledge, 1999.

THE HISTORICAL DOMESTIC ORIGINS OF
ENVIRONMENTAL INTERNATIONALISM

These types of discursive imprints on the structural design and reorganization of IERs cannot be explained simply by recourse to the changing short-term interests of states or corporations. If we apply the analytical approaches laid out in chapter 1—political economy, political ecology, and material history—it is evident that we need to look more deeply into the history of today's political economy to account for the shape and operation of IERs. For our purposes, the significant starting point of this discussion occurs around the beginning of the nineteenth century, a period during which two related trends, capitalist industrialization and nationalism, were joined together in order to exploit nature.[21] This union— and some argue that the two processes were of a piece and never separate—was mediated by cultural, economic, and social forces that differed among individual countries.[22] The ways in which forestry was practiced in Germany differed from that in Great Britain, although, over time, the two influenced each other's practices, as well as those imported into India and other colonies.[23]

Most studies of nationalism focus on one of two of its aspects. The first involves the rhetorical articulation of the nationalist vision and the mass mobilization that is presumed to be required to support nationalism.[24] The second addresses the "protection" of a society's cultural attributes from intrusions by others through various defensive or offensive measures.[25] This second aspect is often identified with the industrial requirements necessary to the accumulation and application of military power and, more recently, with "new nationalist" opposition to immigrants.[26] But what is often ignored in studies of nationalism are the ways in which the general concept overlooks the great differences among individual countries' specific institutions and practices. There is no single way, for instance, of organizing industrial production or educating people or growing food. Yet development of these institutions and practices, and their protection, is central to the nationalist project, for they provide the material goods and socialization that underpin both public support for the nation-state and its ability to secure itself relative to other nation-states.

Nature fits into this scheme because it is one of the two raw materials—the other is people—out of which the nation-state is constructed. The "taming" of nature involves the transformation of an "undeveloped" environment through the application of capitalist and industrial methods in order to increase the production of commodities and goods, especially military ones. Dams are constructed, swamps are drained, and rivers channeled in order to control floods and expand arable land; canals and railroads are laid out to facilitate the movement of minerals, grains, animals, and armies; water supply systems are built to irrigate farms and provision cities; and so on. Landscapes are transformed in pursuit of the nationalist project.[27] These projects not only bring many of nature's hazards under human control, but they also facilitate extraction of material inputs and wealth for state building.

Of course, nature was exploited long before nationalism emerged as a pan-European phenomenon during the nineteenth century, as that continent's countries ranged abroad, creating colonies, mines, and farms, and bringing back raw materials, foods, and plants.[28] Until then, the exploitation of nature was driven by the sovereign's desire to accumulate wealth and power, but in the results we can begin to see the stirrings of what later became nationalism. As the historian E. L. Jones put it,

> Cultural homogeneity seemed desirable because it confirmed loyalty to the crown, simplified administration, taxation and trade. . . . Policies successful in one province might more easily be extended to all corners if the state were unified. . . . A motive [for internal expansion] . . . was to secure internal supplies of raw materials. Scientific ventures of the day . . . were often just as much resource appraisals as science. . . . A further motive, distinct from the aims of earlier periods, was to keep faith with a sense of cultural and political nationalism, and unify and develop state and market. . . . Motives and processes were therefore mixed . . . but they all drove towards the settled, occupied and unified nation-state.[29]

In England, the rapidly growing use of coal as an energy source and the appearance of factories for the manufacture of standardized goods began, during the 1700s, to transform landscape, economy, and nation. But it was not until the following century that industrialization was put fully into the service of state power. As a result of the Napoleonic Wars, in particular, nationalism arose as a mass social phenomenon, spreading from one country to another.[30] Geographically, Europe was too compact a landmass for the practices and experiences of one country to be ignored by others. Indeed, one state's failure to pay attention could result in occupation and conquest by another.

It became clear to elites that the security of state and sovereign was now intimately tied to national industrialization. To resist future invasions and maintain balances of power among them, states had to maximize their military capabilities through the mobilization of both industry and labor. And, even though few countries actually followed the French example of mass mobilization until much later in the nineteenth century, the state meant to muster its citizens for both work and war.[31] In other words, the project of state building launched by intellectuals and engineered by economic elites had as its goal not only the creation of a national consciousness but also a nation-state that could hold its own against the other existing or emerging nation-states within Europe.

Under the new regimes of nationalism, nature came to fill a double role. First, the construction of a nation-state and its nationalist "civil religion"[32] demanded a historical account of the attachment between people and land, a link that, in some places and in later decades, might be further legitimated through archeological exploration and discovery (although, as has been noted, many of the resulting histories were pure fiction).[33] Second, as argued earlier, nature and people came to constitute the raw materials out of which the nation-state was built. As the control and "improvement" of nature came to be one of the primary responsibilities of the newly bureaucratizing, nationalist state, techniques of nature management became the focus of scientists, and practices of nature management became the focus of administrators. One could say that, around the world, lessons were learned from England, France, and, what would, eventually, become Germany (Prussia). The strategies and programs of one country were carefully studied and reproduced by the scientists and administrators of others, although not always with complete success.[34] What worked well in one part of the world might be an utter failure elsewhere. Whole ecologies could be destroyed, as they were in any number of places.[35] We shall see, later in this chapter, how these processes affected the control and supply of water and the management of forests.

THE CONTEMPORARY DOMESTIC ORIGINS OF ENVIRONMENTAL (INTER)NATIONALISM

Notwithstanding expectations to the contrary, nationalism has not vanished with the arrival of the twenty-first century, and the transformation of nature remains central to the nation-state, as seen in Donald Trump's linkage of coal, jobs, and American "greatness." Today, however, the state has become less an agent for its own national projects than a facilitator of a global project of economic growth via capital and corporations.* In place of the nation-state as the object of popular veneration, globalization and liberalism encourage individuals to seek their own goals through consumption and identity (as seen in chapter 3). In chapter 2, we read about sustainable development and its corollary, sustainable growth. Here, we revisit these two concepts to see how, as forms of "environmental (inter)nationalism," they have their origins in what are, for the most part, the national beliefs and practices of the United States.

To repeat the point made above, the global shift toward market-based approaches to addressing environmental problems has its roots in ideological and material shifts in the United States and the United Kingdom, beginning in the late 1970s.[36] As "free-market capitalism" came to be regarded as the solution to the economic stagnation and inflation of the 1970s, state regulation came under increasing attack.[37] The turn to the market for environmental protection was, if not inevitable, strongly determined by this change, and the trend toward commodification of both nature and environmental externalities initiated in the United States and diffused into the international realm both discursively and through diplomacy and the global economy.[38] Recall the discussion of externalities in chapters 2 and 3. In effect, both taxes and privatization are

---

*This remains the case, notwithstanding the apparent revival of the "war-making" state and the likely temporary nationalist project of the Trump administration. See: Hurt and Lipschutz, eds., *Hybrid Rule & State Formation*.

meant to raise the marginal cost of an activity to the point where it equals the marginal benefit, with individual producers then deciding whether to pay—which seems only fair. But the distributional effects are quite different.

A pollution tax, for example, applies equally to all offending activities above a certain set level, regardless of the economic condition of the polluter. Both the poor polluter and the rich one are subject to the same penalty, regardless of ability to pay. The rich polluter will, however, be in a better position to pay the costs of pollution reduction; the poor polluter may go broke. Privatization of pollution rights, as discussed in chapter 3, responds to the polluter's ability to pay and helps to direct economic resources where they can be used with greatest effect. In this instance, the rich polluter can outbid the poor polluter, or buy the latter's permits. This time, the poor polluter will not necessarily go broke, but she will most certainly have to go out of business. The outcome will be efficient, if not necessarily fair.

Why, then, privatize? First, a market-based approach to protection and internalization "lets the market decide" about winners and losers. This may offer a means of avoiding difficult political struggles, some of which might threaten powerful states and their interests.* By arguing that the market is an impartial institution in which decisions are based only on individual willingness to buy and sell, and ability to pay, the problem of distributive justice and remediation of unexpected outcomes are largely ignored or attributed to individual causes.[39] Second, markets do require rules to function, and those rules are usually ones favored by and favorable to the wealthy and powerful.[40] The poor and weak often find it difficult to participate both in rulemaking and in exchange, and as a result are excluded and further disadvantaged.[41] Finally, we must note that privatization is not actually popular with everyone. Some might choose not to participate at all in such markets (or institutions), regarding them as unfair or disruptive. A lack of interest in, or even hostility to, privatization may be regarded as a threat, and those states who take such a position may find themselves labeled "rogues" or "defectors," cut off from access to other international institutions (Nicaragua, for example, found itself severely marginalized for its initial refusal to join the Paris Agreement, in part because of its objection to inclusion of market mechanisms in the agreement).[42]

This privatization and marketization trend also takes a leaf from the standard narrative of the U.S. economic and political experience. The relative stability of the American political system is often attributed to the expansiveness of its economy. Many political scientists are wont to repeat Harold Lasswell's claim that "politics is about who gets what, when, and how."[43] Because financial and other resources are never unlimited—by the strict economic definition, they are "scarce"—difficult and contentious decisions must be made about

---

*President Trump's imposition of tariffs on steel, aluminum, and other goods imported into the United States reflects a move away from "letting the market decide" and toward political attempts to avoid "losing" (at least, in his terms).

how they will be distributed. In this view, the primary purpose of politics in a democratic system such as that of the United States is the distribution of resources, an activity subject to a great deal of pushing and pulling, lobbying and logrolling. An economy that grows too slowly or not at all, combined with the redistribution of resources toward the middle classes that is typical of democratic systems, may intensify struggles over shares of the pie, triggering class conflict and the instability so feared by political and economic elites as a threat to their interests, status, and position.[44] There is virtue, therefore, in ensuring a constantly growing "pie" or pool of resources, for, even if relative shares of the pie remain the same, absolute shares will increase and people will have a sense that their situation is getting better.[45]

Since the first edition of this book, however, this distributional calculus has changed. Today, markets and rules are formulated so as to advantage the rich to the detriment of the middle class, which has seen its income grow slowly, if at all, even as the wealthy have become much richer. Part of this can be put down to slow economic growth but it may also be caused by offshore production of manufactured goods and the associated decline in blue collar employment in home countries, declining tax rates on the very rich, the massive shift to speculative investments, and even low interest rates. Today, the economic pie is growing very slowly, and not at all for some, and middle-class resentment of both rich and poor is on the rise, as seen especially in Europe and the United States.

Thus, the need and demand for environmental protection may threaten this formula for social stability. Such programs insert a new competitor into the struggle over the division of resources. Intensive environmental protection might require "limits to growth" in some or all sectors of the economy, which could reduce the flow of goods and money through the economy and cause the pie to stagnate or even shrink.[46] Moreover, the demand for high environmental quality has long been seen as a middle-class desire, on the proposition that the poor have more pressing matters about which to worry. Well-off neighborhoods may become cleaner while poorer ones get the garbage.[47] Serendipitously, perhaps, a rapidly growing economy appears to address these difficulties, providing both the growing pie and the resources necessary to protect the environment.

Indeed, as discussed in chapters 2 and 3, proponents of the "green economy" and ecological modernization go so far as to claim that environmental protection is entirely consonant with high rates of economic growth, if that growth comes primarily in environmentally friendly sectors, such as services and information.[48] This is nothing other than an argument for sustainable development and growth, a principle has been exported by the United States into the international arena and offered as the way to pay for global environmental protection. But wait! There is more!

Growing the global economic pie requires appropriate forms of investment and development in places and sectors where it will produce the greatest economic returns (as opposed to the greatest environmental justice). In line with "letting the market do it," governments are not considered to be competent to make such decisions, especially given the dismal record of development programs over the past fifty years.[49] The external (international debt) of the world's low- and middle-income countries (that is, the Global South) exceeds six trillion dollars (total external debt levels across the Global North are, however, much greater, at around sixty-five trillion dollars, while government debt stands at around forty-one trillion dollars).[50] Since international loans are disbursed on the basis of a country's ability to repay, which is linked to revenues not allocated to domestic social and environmental needs, financing such protection is increasingly difficult. A growing amount of environmentally linked financial flows have to do with projects that generate carbon or other types of potentially fungible and tradable credits.

TALES OF PRIVATIZATION

From where, then, will the necessary environmental investments come? The answer appears to be greater corporate involvement, that is, more privatization, a strategy evident through the string of UN-sponsored sustainable development conferences, from the World Summit on Sustainable Development held in Johannesburg in 2002 to the UN Conference on Sustainable Development held in Rio in 2012 (see Box 5.1).[51] Even as governments have deliberately and consciously supported the shift of various aspects of environmental regulatory authority out of the national domain and into the international arena, business has come to be viewed as a potential protector of nature (especially via "corporate social responsibility").[52] This is another trend we have tracked throughout this book. Only corporations, it is argued, have sufficient capital to finance "green" industry and, so long as such enterprises generate adequate profits, only they have the incentive to make such investments. As discussed in other chapters (see, especially, chapter 4), there are good reasons to doubt that corporations can "save" the environment. This will not prevent them from trying, especially with encouragement from the World Bank, the U.S. government, and the European Union, and especially if they can make a nice profit from the effort, as we shall see from the stories.*

---

* As another example of the discursive shift toward the private sector, in the years since the publication of the first edition of this book, international financial institutions have greatly increased resources made available for private-sector lending. For example, between 1990 and 2010, lending from multilateral development banks to the private sector increased tenfold—from four billion to forty billion dollars a year. As a continuation of the trend, the recently established Green Climate Fund created a separate private sector facility to funnel money directly to businesses. See Bretton Woods Project, et al., "Bottom Lines, Better Lives?" April 22, 2010, accessed August 12, 2017, http://www .brettonwoodsproject.org/2010/04/art-566197/; Green Climate Fund, accessed September 25, 2017, https://www.greenclimate.fund/home.

BOX 5.1

Just words? Rio+20 and defining "green economy"

Two worlds collided in Rio de Janeiro in June of 2012. Twenty years after the first Earth Summit produced the Rio Declaration, Agenda 21, and two important environmental treaties—the UN Framework Convention on Climate Change and the UN Convention on Biological Diversity—the international community came back to Rio to celebrate Rio+20.

Government representatives gathered to do what environmental diplomats do at these sorts of meetings. They agreed to a 283-paragraph declaration of intent: *The Future We Want*. They also launched a multi-year process to define sustainable development goals, which were adopted by the UN General Assembly in 2015. In advance of Rio+20, to introduce one of the two themes of the conference, the United Nations launched its preparatory report: *Towards a Green Economy*.

Thus, in the lead-up to Rio+20 and after at least two decades of defining and redefining the phrase "sustainable development," a new narrative battle began over what a green economy might be. Some interpreted the term "green economy" to mean "green growth." The UN Environment Programme said it was an economy "that results in improved human well-being and social equity, while significantly reducing environmental risks and ecological scarcities. It is low carbon, resource efficient, and socially inclusive." Grassroots organizations, indigenous peoples movements, labor unions, and social and popular movements, holding a parallel summit that they named the Peoples' Summit at Rio+20 for Social and Environmental Justice, spun another narrative:

> The current financial phase of capitalism is expressed through the so-called "green economy" and in the old and new mechanisms such as the deepening of the public-private debt, the super-stimulus to consumption, the appropriation and concentration of new technologies, the carbon markets and biodiversity, land grabbing and take over by foreign companies of the lands and public-private partnerships, among other things.

What lessons to take home from this return to Rio and retrospective of twenty years of environmental summits, declarations, and unmet goals? From the seemingly constant definitional battles over terms that often sound platitudinous and empty? We put forward just two here, but feel free to dive further into this debate with the sources below.

First, as much as global meetings like Rio+20 leave us wondering why tens of millions of dollars are spent to convene thousands of people from all over the world to write a rather inconsequential 283-paragraph document, many, like the participants at the Peoples' Summit, consider such meetings to be important sites of contestation. To be sure, they are high-profile opportunities for showcasing the environmental narratives

(Continued)

of neoliberalism, such as green growth, natural capital accounting, ecosystem valuation, and payment for environmental services. At the same time, they create opportunities for a high-profile collision of worldviews, a clash of narratives and counternarratives, and in so doing create possibilities to chip away at the hegemony of neoliberal approaches to valuing nature.

Second, on words. In "The Sustainability Story," Peter Dauvergne reviews the decades of contestation around the term "sustainability." In particular he spotlights the way corporations and big business have appropriated the term, "rewriting the narrative of sustainability, turning what was once a powerful critical discourse of environmentalism into a strategy to expand business (sales, profits, and stores) as well as gain more control over suppliers worldwide." He notes that this has led some environmentalists to give up on the concept, which he says "is a mistake. The narrative of sustainability is too valuable to surrender to business," given the power the term has "when infused with principles of ecology, equity, and social justice."

Narrative battles such as these are contested terrain, where worldviews collide and possibilities to imagine a different world, the future we want, are given life.

Sources: United Nations, "The Future We Want," Outcome document of the United Nations Conference on Sustainable Development, Rio de Janeiro, Brazil, June 20-22, 2012, accessed September 3, 2017, https://sustainabledevelopment.un.org/content/documents /733FutureWeWant.pdf; "Final Declaration of the Peoples' Summit at Rio+20," http://www .mstbrazil.org/news/final-declaration-peoples%E2%80%99-summit-rio20 (accessed September 28, 2017); Peter Dauvergne, "The Sustainability Story: Exposing Truths, Half-Truths, and Illusions," in *New Earth Politics: Essays from the Anthropocene*, edited by Simon Nicholson and Sikina Jinnah, Cambridge: MIT Press, 2016; Thomas Fatheuer, Lili Fuhr, and Barbara Unmüßig, *Inside the Green Economy*, Munich: oekom, 2016; Bruce Rich, *Mortgaging the Earth: the World Bank, Environmental Impoverishment, and the Crisis of Development*, Boston: Beacon Press, 1994; UN Department of Economic and Social Affairs, "Sustainable Development Goals," accessed September 3, 2017, https://sustainabledevelopment.un.org/?menu=1300.

Making Scarce Water Even Scarcer

Urban water supply is one sector in which large-scale privatization efforts have taken place. Water is essential to life. Water is, by many accounts, also growing scarce.[53] More than six hundred million people lack access to clean drinking water, while 2.4 billion lack access to proper sanitation.[54] Many urban distribution systems around the world are in a state of chronic disrepair, and massive quantities of water are lost through old, cracked pipes and other sites of leakage. Building new supply systems and fixing old ones is an expensive proposition that governments claim they are unable to afford due to limited public resources. By one estimate, it could cost one trillion dollars over the next twenty-five years simply to repair existing water systems in the United States alone.[55] But where public authorities see nothing but problems, private entrepreneurs and corporations, encouraged by a range of public and private agencies and advisors, see nothing but opportunity. Private parties can acquire part or all of public water systems, and by charging customers a higher price for water, they can generate the revenues necessary for repairs and earn a tidy profit! That, at least, is the theory; the results of practice are much more uncertain. Nevertheless, the private water supply business—not counting bottled water—earns revenues of $360 billion per year serving several hundred million people in more than 130 countries.[56] Since everyone needs water, the revenue potential appears almost limitless—if customers are able and willing to pay for this vital necessity.

There is some disagreement, even within the United Nations, about the basic need for water. According to the World Health Organization

> Based on estimates of requirements of lactating women who engage in moderate physical activity in above-average temperatures, a minimum of 7.5 litres per capita per day will meet the requirements of most people under most conditions. This water needs to be of a quality that represents a tolerable level of risk. However, in an emergency situation, a minimum of 15 litres is required. A higher quantity of about 20 litres per capita per day should be assured to take care of basic hygiene needs and basic food hygiene.[57]

However, the United Nation's "International Decade for Action 'Water for Life' 2005–2015" suggests that (again, according to the World Health Organization), "between 50 and 100 litres [about 12.5 to 25 gallons) of water per person per day are needed to ensure that most basic needs are met and few health concerns arise."[58] Again, depending on the information source, the average American uses somewhere between fifty and two hundred gallons (roughly two hundred to eight hundred liters) a day, much of that for activities beyond a basic need for water.[59] The global divide in per capita water use is stark, as can be seen in figure 5.1.

Yet should water be treated as a commodity, to be bought and sold according to one's ability to pay for it? This is not only a question of human rights to water; it is also an ethical dilemma. And it continues to divide those who have faith in the market's ability to distribute fairly and those who believe that water should not have a price. The former argue, as we have seen, that water is "scarce"—although what that means, exactly, is not wholly clear, since "water scarcity" is often premised not on basic human needs but, rather, on what is required for a functioning economy, which includes industry and agriculture. As is the case with all declarations of a "human right to water" (see Box 5.2), there is no enforcement mechanism—the goal is aspirational. Nevertheless, there is an ethical challenge in the fact that so many of the world's people lack access to clean water and sanitation, largely because of funding constraints

Figure 5.1   Worlds Apart: The Global Water Gap

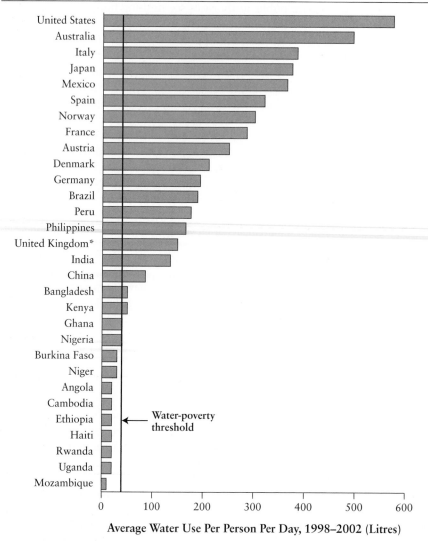

Average Water Use Per Person Per Day, 1998–2002 (Litres)

Source: UN Development Programme, *Human Development Report 2006*, New York: Palgrave Macmillan, 2006, p. 34, http://hdr.undp.org/sites/default/files/reports/267/hdr06-complete.pdf (accessed August 25, 2017).

and the absence of profit-making possibilities. On this basis, those who favor commodification argue that it is more ethical to provide some water at a cost than little or no water for free.[60]

So long as water seemed plentiful, it was virtually free for the taking; now that farmers, industrialists, and consumers are demanding more water than

| BOX 5.2 | An ongoing debate: the human right to water and sanitation |

On July 28, 2010, the UN General Assembly passed resolution 64/92 on the human right to water and sanitation, as a result of a diplomatic effort by the government of Evo Morales in Bolivia. The resolution *"recognizes the right to safe and clean drinking water and sanitation as a human right that is essential for the full enjoyment of life and all human rights"* and *"calls upon States and international organizations to provide financial resources, capacity-building and technology transfer, through international assistance and cooperation, in particular to developing countries, in order to scale up efforts to provide safe, clean, accessible and affordable drinking water and sanitation for all."*

As with most UN General Assembly resolutions, it was not adopted by consensus. 122 countries voted in its favor, zero opposed, and forty-one abstained. In its statement on why it would abstain from the resolution, the United States reiterated its opinion that there is "no right to water and sanitation in an international legal sense as described by the draft resolution."

Sources: UN General Assembly Resolution 664/292, "The Human Right to Water and Sanitation," http://www.un.org/es/comun/docs/?symbol=A/RES/64/292&lang=E; UN General Assembly 64th session, 108th plenary meeting, July 28, 2010, A/64/PV.108, official records, http://www.un.org/en/ga/search/view_doc.asp?symbol=A/64/PV.108; "Our Right to Water," Food and Water Watch, May 2012, accessed August 12, 2017, https://www.foodandwaterwatch.org/sites/default/files/our_right_to_water_report_may_2012.pdf.

seems to be available in many places, how is the supply to be divided? Here, we see, once again, the political costs of equitable rationing, discussed above. Allocation through markets is one means of dividing this scarce resource. This, at least, is the logic that is being propagated globally, to a growing degree, by those individuals and institutions who pay for and play a role in financing and building water delivery systems.[61] But there is something of a paradox here: the growing scarcity of water is, in many ways, a direct consequence of the development of large-scale water projects. Dams, reservoirs, and long-distance distributions systems have, as seen in the example of California, made agricultural, industrial, and urban growth possible, drawing in people and investment. In the early days of such projects, water supply usually far exceeds demand, so consumption is encouraged to increase revenues and subsidies are paid by governments to big users so as to ensure that the water is so cheap as to make it seem free.[62] This makes the deserts bloom and puts cheap food on the table but also discourages judicious use of water. Supply creates demand, and then demand outruns supply. Suddenly, water is "scarce." How could this have happened?

Dam History

To better understand the perverse logic of scarce water, we need a bit of hydrological and hydraulic history. The earliest large-scale water projects were constructed in the Middle East and Asia for irrigation of agricultural lands during dry seasons. We might speculate that the first such projects were quite small and crude and were washed away every year by seasonal flooding. With the rise of city-states, kingdoms, and empires, and the emergence of divisions of labor between leaders, workers, and peasants, irrigation systems became essential to ensuring an adequate and reliable food supply for non-agricultural populations. Thus, the early systems were extended in scope and came under more centralized control. They also began to have serious and widespread environmental impacts, most notably salinization and waterlogging of soils. Some have argued that these problems led to the declines in agricultural productivity that played a major role in the collapse of the civilizations that built those irrigation systems.[63]

Storage and irrigation projects covering tens or hundreds of square miles did not appear again until a little more than a century ago. The dams of the nineteenth and twentieth centuries were built to control floods and generate electricity as well as to irrigate farmland and supply industry. Although the first dams were rather small and sometimes privately built, because of high construction costs, later ones were constructed by governments, at increasingly greater scales and with consequent environmental effects over greater areas. Ultimately, only states were able to mobilize the capital, expertise, labor, and land to build big dams, and only states had the power to force people to move out of the dams' way.[64] In virtually every instance, the goal of hydraulic projects was not only to "manage" the resource but also to transform the landscape and people. The rivers of Africa, the American West, Brazil, China, Europe, India, Israel, the Soviet Union, Turkey, and elsewhere were tamed and their waters distributed widely. All big dams followed a few large-scale variations on a standard hydraulic technology. When completed, these projects were regarded as great national accomplishments.* And those who worked on these projects and on the land were lionized as heroes of the nation. The future seemed bright. It was not.

The gradual failure of large-scale hydraulic technology has become apparent only over time. On the one hand, dams and irrigation systems have been a great success in their specific objectives: vast areas were transformed into productive farmland, cities appeared where none had been possible before, new industries arose only because cheap water was available. On the other hand, those same dams often displaced thousands and even millions of people—who

---

*Prime Minister Nehru once famously lauded dams as "the temples of modern India." Few states look any longer to large-scale hydraulic projects as a means of developing the nation. See: Special Correspondent, "When the Big Dams Came Up," *The Hindu*, March 20, 2015, accessed September 28, 2017, http://www.thehindu.com/todays-paper /tp-national/when-the-big-dams-came-up/article7013509.

most often benefitted neither from the stored drinking and irrigation water nor the hydroelectricity produced—and gradually led to the ruination of lands and ecosystems in some locales, and slowly filled up with silt to become almost useless in others.[65] Large-scale water storage and irrigation systems constructed in dry and near-desert areas created much the same problems with waterlogging and salinization that plagued the early civilizations of the Middle East. Large dams still have their defenders, for they remain central to the development strategies of countries such as Brazil, China, and India. Because hydroelectricity is "renewable"—at least for the lifetime of the dam—and emission-free, the Green Climate Fund[66] and the World Bank continue to lend large sums to countries in the Global South for big dams, now seen as a climate solution. Global environmental activism, and the enormous cost of such structures, have slowed but not stopped large dam projects across the developing world, and activists continue to pay with their lives for defending their rivers from mega-dam projects (see Box 5.3).[67] By contrast, there is a growing view, emerging from the United States, that floods cannot and should not be controlled, but that people should,

---

BOX 5.3 | **Defenders of the earth**

By midyear in 2017, ninety-eight land and environmental defenders around the world had been killed for defending their lands and resources from industries such as agribusiness, logging, and mining. The deadliest year on record so far was 2016, with over two hundred killings recorded in twenty-four countries. The killing that year of the Goldman-Prize-winning anti-dam activist Berta Cáceres in Honduras drew worldwide attention to the rapidly growing threats to environmental activists—primarily but not exclusively in the developing world—including death threats, sexual assault, arrests, aggressive legal attacks, and murder. In 2016, Brazil had the highest number of activist murders, while Honduras had the highest number per capita. Indigenous people are disproportionately represented among the victims—almost 40 percent of those killed were indigenous defenders. Mining and oil are the most dangerous sectors. Killings are carried out by the police and military as well as paramilitaries, private security forces, and paid hit men. Campaigners are demanding that governments guarantee that communities are active participants in decisions affecting their land, with free, prior and informed consent; support and protect the rights of defenders; and hold accountable those responsible for aggression and killings, as well as the international actors supporting the aggressors.

Sources: "Berta Cáceres," The Goldman Environmental Prize, accessed August 8, 2017, http://www.goldmanprize.org/recipient/berta-caceres/; Global Witness, *Defenders of the Earth*, July 13, 2017, accessed August 10, 2017, https://www.globalwitness.org/en/campaigns/environmental-activists/defenders-earth/.

insofar as is possible, move out of riverine flood plains. Further, a growing number of older dams are being decommissioned due to siltation.[68]

With growing global food demand and changing weather patterns due to climate change, irrigation for agriculture has become more essential than ever. New types of genetically engineered and hybrid crops require reliable supplies of water, as do the burgeoning populations of the world's cities. What to do? It appears that the growing need for water must be achieved through demand management rather than new supply: water will have to be used more efficiently, and many people may have to make do with less. Conservation, however, does not make much of a contribution to "state-building" in the traditional, heroic sense—no one is ever lionized for using less water—so other conservation methods must be identified and implemented. This is why privatization and sale of water through markets is popular: when water becomes more costly, farms, industries, and consumers will use less. But raising the price of heretofore cheap water makes few people very happy.

Selling the Springs of Life

The water supply problem is, therefore, twofold. To repeat, first, in many parts of the world the infrastructure of water storage and delivery is not only breaking down, it is also unable to meet the demands of increased food production and growing populations, especially in cities but also in rural areas. Second, agriculture, industry, and cities are all competitors for the water that is provided through distribution systems, and each believes it has the greatest entitlement to that water. Because of the enormous monetary and environmental costs of making more water available through new capture, storage, and distribution facilities, it makes economic and political sense to manage demand by those with ready access, especially because it usually costs less to save water through reducing demand than to capture new water.[69] But demand management appears to come in only two forms—rationing or higher prices—and neither can address the global water divide. As we saw earlier, rationing is fairer, but lends itself to all kinds of problems, including theft, corruption, and black markets (an excellent illustration of this can be seen in the film *Chinatown*). Higher prices should persuade people to value water and use it more carefully.* We know that this approach will reduce demand because the very poor in developing countries use very little water, and they often must buy it from water vendors who offer their product at extortionate prices, or they must carry water long distances from source to the site of use. And, unlike rationing, no complex allocation system is required: people will respond directly to the

---

*Another paradox that arises when water use declines is rising rates. Even though reductions in use should make more supply available, water companies often raise their prices to make up for revenue lost as a result of selling less water! See, for example, Bradley J. Fikes and Morgan Cook, "Saving Water Adds Up to Rate Hikes," *San Diego Union-Tribune*, July 27, 2015, accessed September 28, 2017, http://www.sandiegounion tribune.com/news/drought/sdut-drought-water-prices-rise-2015jul27-story.html.

price signal. But, whereas the poor may have to pay five to ten U.S. dollars for a cubic meter of water (two to four U.S. cents per gallon), those with domestic water taps pay only about thirty-five U.S. cents per cubic meter (U.S. $0.0013 per gallon).[70] Doubling or tripling that cost would hardly make water much more expensive, although it would cause bills to rise and consumers to scream.

How is demand management to be accomplished? It would seem fairly simple to raise water prices, but often it is not. For one thing, many water systems are publicly owned and governments are loath to offend voters whose bills could skyrocket. It is easier to raise funds by selling municipal bonds to pay the costs of repairs, putting off the cost. For another thing, water is still not metered in numerous places throughout the world;[71] in those situations, public water providers have no way of measuring anything but aggregate consumption and charging a flat fee, regardless of use. A private company, by contrast, will value the resource for its profit potential and will find ways to squeeze the greatest return out of its investment. This means repairing existing systems to reduce water loss, improving water quality, and carefully measuring what consumers use. All of this requires capital and the new owners have little choice but to raise prices. When that happens, people will use less water, making the remainder available for "more productive" uses.

It is an interesting exercise to imagine how much water would have to cost in order for demand management to work effectively. In the United States, water from a public supply typically costs a few dollars per thousand gallons.[72] A gallon bottle of distilled water runs around one dollar. Twelve liters of bottled drinking water can be purchased for ten to twenty dollars, whereas a single small bottle might run as high as three dollars a bottle or, roughly, ten dollars per gallon. The typical American family of four uses between three hundred and one thousand gallons of water per day, so an average monthly water bill runs anywhere from ten to thirty dollars. It is unlikely that a doubling, or even a tripling, of this sum would have very much effect on demand. But at three hundred dollars per month—a roughly ten- to thirtyfold increase in price—many people would begin to conserve, cutting back on watering their lawns, washing their cars and clothing, and even bathing daily. It is difficult, however, to imagine any water supply company being allowed to raise rates so precipitously: the public outcry would cow any public utilities commission that authorized such an increase.

At any rate, water privatization has taken the world by storm (although there are a growing number of cases of re-municipalization of water systems).[73] International financial institutions, such as the World Bank and International Monetary Fund, actively encourage privatization of water systems.[74] Some governments sell public water systems in order to generate one-time revenues to reduce budget deficits and avoid future expenditures on infrastructure. Others argue that privatization will lead to greater efficiencies. As noted earlier, private water companies see the opportunities for significant profits, so long as the price of water is set sufficiently high and local politics are excluded (not so easy, as we shall see).[75] In the past, even some nongovernmental organizations (NGOs), such as Environmental Defense Fund in the United States, supported privatization of urban supply (although that is no longer the case).[76] In some settings, especially where wasteful users are well-off or producing high-value products, there is some sense in raising the price of water to that which the market will bear. In others, however, it may be nothing short of a social disaster.

A Bolivian Tale

The story of the water system in Cochabamba, Bolivia, is an instructive one, which has been repeated many times since.[77] During the 1990s, Bolivia was pressured by the World Bank to privatize numerous public corporations, such as the national airline, the train system, and the electric utility. In September 1999, the Bolivian government gave a forty-year concession to Aguas del Tunari to operate the water and sanitation systems of the city of Cochabamba. At the time, the owners of Aguas (including engineering giant Bechtel) were based not only in Bolivia but also in the United States, Italy, and Spain.[78] The logic of the deal was fairly straightforward. With financing provided by international lenders, Aguas paid the Bolivian government for the existing utility. The company was slated to build new water storage and distribution lines and to rehabilitate the municipal delivery system. In return, it would be permitted to raise water tariffs to a level that would provide a nice profit to the company and its shareholders. But there was a catch.

The catch was that the financing could not go forward without the water rates first being increased, to ensure that repayment of loans would begin immediately. And the rates were raised—more than 200 percent, by some accounts.[79] One resident reported that his water bill went from twelve dollars in December 1999 to thirty dollars in January 2000; this in a town where many families earned only about one hundred dollars a month.[80] A municipal uprising followed, along with a four-day general strike, marches, and, eventually, violent confrontations with police. Almost two hundred people were injured. The president of Bolivia at the time, Hugo Banzer, declared a state of siege in the entire country, which led to more violence. At least one person, and perhaps more, were killed.[81] At that point, the government revoked the concession to Aguas and returned the water system to the municipality. Water rates were reduced.

Subsequently, Aguas sued the Bolivian government for fifty million dollars for recovery of investments and profits lost as a result of the concession's termination, under the investor-state dispute settlement provisions[82] of a bilateral investment treaty between Bolivia and the Netherlands (Aguas changed its location of business in 1999 to Amsterdam). The case was heard, in secret, by a tribunal of the World Bank's International Centre for Settlement of Investment Disputes beginning in 2002. In 2005, the International Centre for Settlement of Investment Disputes ruled that it had jurisdiction. But, in early 2006, the main shareholders in Aguas decided to drop the case and made a token payment to the Bolivian government in order to avoid further conflict and controversy.[83]

Not all privatizations fall victim to these kinds of problems, to be sure, but the basic argument continues to be repeated. The commodification of water and the attendant higher prices are judged to be necessary in order to reduce "waste." Why is water being "wasted"? Not because people are inherently wastrels. Rather, in order to encourage economic growth and development, water has been heavily subsidized. And where there was water, people would come, build, and work (the case of California is exemplary in this respect).[84] It might appear as though households constitute the bulk of global water use, but this is not true. The world over, about 69 percent of stored water is dedicated to irrigation; the rest is left for industry (19 percent) and people (12 percent).[85] Why, then, does it seem as though households bear the brunt of water privatization?

In wealthy countries, with many water-thirsty appliances, consumers pay relatively low water rates but can afford relatively large price increases. In poor countries, there are many fewer such appliances, and household water use is relatively limited. But households are not the main source of economic growth; farms and industries are. If farmers have to pay market rates for their water, they might go out of business—in California, farmers have been enormously resistant to higher water costs or even to selling their water in markets. Agricultural conservation technologies are expensive, too, and make sense only for high-value crops whose water requirements are fairly limited. Industries make similar claims about their water needs and, because they are the source of much economic growth, governments are loath to impose high rates on them. Only household consumers are left to be squeezed. Making them conserve will save water, but it will have an impact on only one-fifth of the total demand in most places around the world.

Not Seeing the Forest for the Trees

A similar logic of privatization is being applied to forests, although here the methods of "enclosing the commons" differ considerably. Forests, too, are growing "scarce," less for the timber they provide than for their ecological role in local, regional, and planetary biological and physical systems, whose economic value is rarely taken into account.[86] Aside from the intrinsic value of the various species of trees themselves, forests provide habitat for other plant and animal species and environmental services such as water purification, soil retention, local climate moderation, and reservoirs of genetic diversity. Forests provide food and materials to those who live in and around them, and sequester carbon that would be released if they are burned or degraded. Indeed, for many indigenous groups, the destruction of forests means their destruction as well. With the inclusion in the Kyoto Protocol, REDD+, and the Paris Agreement of provisions allowing for the protection and planting of trees as a means of carbon storage and growing calls for trees to be used in large-scale carbon dioxide removal technologies, more attention than ever is being paid to the "fate of the forest(s)."[87]

The exploitation, preservation, and restoration of forests—tropical, temperate, and boreal—are therefore of growing international concern. Wherever there are trees, however, trade in timber competes with preservation, livelihood, ecosystem functions, and other uses of forest resources. Many countries and companies earn significant revenues from lumber production, and they do not wish to see restrictions imposed on logging.[88] As a result, forests across the world are being replaced by tree or industrial agricultural plantations of soy or oil palm, while others are being left in ruins (see figures 5.2a and 5.2b).[89]

Each country has its own regulations governing the protection and exploitation of forests, but those that export are especially driven to log heavily by competitive pressures and revenue needs. While there exists a nonbinding forest "instrument," negotiated over the course of meetings of the United Nations Forum on Forests in 2007, there does not exist any significant international regulation of forest practices comparable to those for climate and biodiversity, even with the increasing pressures to save forests for climate mitigation.[90] Efforts to craft an international forestry convention have failed, usually because of resistance by governments, who fear intrusion on their sovereign prerogatives, and by corporations, who fear having to pay greater costs for access to timber and sustainable forest management. In lieu of an effective international system, a large number of competing national and private, market-oriented regulatory initiatives have sprung up, seeking to foster forms of "sustainable forest management" and to become the (inter)national standard for the future.[91] The result has been something of a convergence around certain principles of sustainable forestry management without, as noted above, any single, authoritative convention. Here, we see an example of the failure of the collective action problem, on the one hand, and the competition to retain market share driving convergence, on the other. We will return to this point later in this chapter.

The Forests of Nations

Virtually all contemporary forest management systems the world over have been derived from principles and practices of a few countries in response to a growing shortage of wood, especially Prussia and Saxony in the eighteenth century.[92] These management systems were subsequently adopted by Britain, France, the United States, and other countries, revised or altered in various ways, and diffused throughout European colonial territories.[93] In all instances, national practices were implemented as the "best available approach" to forest management (although, in retrospect, it is not always clear that such practices were really the "best"). Much like maximum sustainable yield, "scientific forestry" was based on the precise measurement of the distribution and volume of wood in a given parcel, the systematic felling of trees, and their replacement by standard, carefully aligned trees in rows of monocultural plantations

Figure 5.2   Annual Change in Forest Areas, 1990–2015 (top), and Planted Forest Area Change, 1990–2015 (bottom)

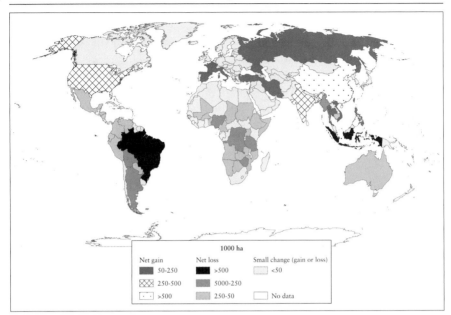

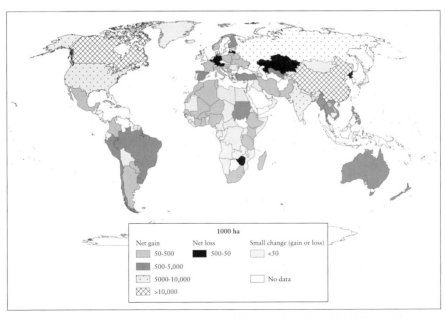

Source: Food and Agriculture Organization of the United Nations, *Global Forests Resources Assessment, Maps and Figures*, 2015, http://www.fao.org/forest-resources-assessment/current-assessment/maps-and-figures/en/. Reproduced with permission.

that could be harvested at scheduled times.[94] As James Scott points out, this approach succeeded beyond expectations during the first growing cycle of eighty years or so.[95] It began to fall short, however, during the second cycle as a result of unanticipated ecosystem damage and destruction. No matter. By then, the model had been adopted around the world and the principles and practices had become the law of many lands.

What is noteworthy about scientific management is that its goal was not preservation of forests, or even "sustainable forest management," in the sense that those are understood today. Rather, as Scott has observed, the objective was entirely economic: increasing the wealth of seventeenth- and eighteenth-century kings and queens.

> The early modern European state, even before the development of scientific forestry, viewed its forests primarily through the fiscal lens of revenue needs. To be sure, other concerns—such as timber for shipping, state construction, and fuel for the economic security of its subjects—were not entirely absent from official management. These concerns also had heavy implications for state revenue and security. Exaggerating only slightly, one might say that the crown's interest in forests was resolved through its fiscal lens into a single number: the revenue yield of the timber that might be extracted annually.[96]

Little changed over the following centuries. Forest management was overseen by state agencies, intent on maximizing production in the national "interest" and for national objectives. That forests might be used by others who were less interested in timber was hardly relevant. Exclusion of peasants from forest commons was one more means of forcing them off the land, an example of land grabs and primitive accumulation.

Actual practices differed from one country to the next. Even though most forest land in the United States and Canada was and is privately owned, for example, a considerable amount is held by the state as a "public good." In the interest of revenue generation, portions of these national forests are systematically leased to private timber producers, who do not always harvest carefully.[97] In India, the British Imperial government took ownership of virtually all forests, declaring them to be "wasteland" and, therefore, no one's property (see later in this chapter).[98] Again, one result was less-than-careful logging combined with denial of access to villagers who relied on these commons for wood and other basic needs. In Indonesia, forests are legally state-owned but in practice are treated as private property. In Brazil, a lack of national government capacity, especially in remote areas, almost literally renders Amazonia's forests open-access commons, available to anyone with an ax.

In all cases, these public forests are viewed as a national resource, that is, the sovereign property of the state, even when mediated through private ownership. In this role, the conservation of forests is tightly linked to the production of timber and other commodities that generate both capital and jobs. The economies of large regions have come to be almost wholly dependent on natural resource production from those public forests, and they suffer greatly when the trees begin to run out or cutting is restricted.[99] In the domestic scheme of things, timber producers are politically influential, both for the jobs they offer and for the campaign and other funds they can sometimes divert to elected officials and policymakers. The timber industry often gets its way when it comes to legislation governing public forests and is able to find ways of getting around any restrictions that might be imposed. Although industries in some countries often argue that they plant more trees than they cut, the results are rarely the equivalents of the forests that have been destroyed.[100] Production generally trumps conservation (pun not intended in the first edition, but relevant now).

Forests are not, of course, only for timber. Beginning in the nineteenth century, with the creation of national parks, and continuing today with various forms of ecotourism, forests have come to fulfill recreational and observational roles—ideally generating some sort of revenue. These uses compete with timber extraction. The growing importance of ecotourists' expenditures in formerly timber-oriented regions and creation of private land trusts and public and private reserves for forests and wildlife are placing more and more areas out of reach of timber companies. Automation of many of the steps involved in logging has also had major impacts on industry employment. While unemployed loggers complain about "tree huggers," others see new opportunities. Forests are thus a subject of fierce contention between economic and environmental interests—no surprise there.

A Tale of India

One example of the way state-based appropriation and development processes play out can be seen in India's forest policy, which is especially complex as a result of its history, geography, and demography. The British colonizers bestowed on India a system of government whose task it was to control a large, fragmented territory of high ethnic and cultural diversity. While the Raj (the colonial British government) tended toward a form of "divide and rule," leaving many administrative tasks to local leaders and villages, after 1948 the government of India pursued strategies intended to unite the disparate parts into a whole. The consequent tension between localism and union has never been wholly resolved and can be seen with particular clarity in ongoing struggles over the environment.

The first postindependence government of Jawaharlal Nehru pursued a program of state planning, industrial centralism, import substitution, protectionism, and urbanization. At the same time, India's growing population required land, water, and food, at any cost. Together, these demands led to exploitation of the country's natural resources as critical inputs into the development process and in general disregard for the protection of nature.[101] Perhaps of greater importance was (and continues to be) the fact that there is almost no "primordial" nature within India. The region has been inhabited for so many millennia, located at the crossroads of so many civilizations and migratory flows, and farmed for so long, that most of the land has been worked and transformed many times.[102] Without romanticizing the care and capabilities of rural villagers, we might nonetheless say that, for the most part, the inhabitants of the subcontinent were able to maximize their usage of resources while also managing to sustain them. This is not to imply that there was any equity in access to nature, of course, since the actual structure of village life is a good deal more complex than implied here, and class and caste differences have always complicated the use of resources and their maintenance.[103]

Nevertheless, what "balance" there was, was upset, first, by the Raj and, to a much larger degree, by postindependence governments. As N. Patrick Peritore puts it,

> The environment is in the hands of complex and elephantine bureaucracies, which apply contradictory policies, and of local elites connected to national parties, contractors, and mafias. Outmoded colonial laws treat villagers as strangers in their own land and criminalize actions creating autonomy. Government permission, requiring years of legal battles and large bribes, is required for villagers to reforest their own catchment watershed or commons or to build small dams and modify village water tanks.[104]

The effects of such legal regimes and their implementation are evident in India's forestry policy. During the Raj, timber was considered essential to British interests and was largely exploited for export. To this end, the British sought to conserve forest commons usually shared by neighboring villages, declaring that they were not owned by anyone and were, therefore, state property.* An 1878 law defined three classes of forests: restricted and closed to the public, protected and open to the public within certain limits, and village forests open to all. The 1927 Indian Forestry Act gave complete control of all forests, except those within the princely states, to the Raj.[105] Imperial foresters took it as their mandate to prevent villagers from gaining access to these now-public lands. Following independence, the Union government of India pursued the same policy, first in order to give state and private corporations access to timber and, later, to conserve dwindling forests. The Indian Forest Conservation Act of 1980, designed with this purpose in mind, further centralized control of forests. It required the permission of the central government for all changes in uses of forest lands by anyone, including government agencies.[106]

As a result, locals continued to be excluded from access to essential forest products, whereas logging went on, sometimes under state sanction, sometimes illegally.[107] Poverty increased, and forests disappeared. These circumstances

---

*Declaring natural resources to be the (initial) property of the state has been common practice, especially where written private title did not exist (see chapter 3). Even in the Global North, vast swaths of land were declared to be "public" and under stewardship of the state.

gave rise to movements such as Chipko Andolan ("Hug the Trees Movement"), which, during the 1970s and 1980s, became well known for its efforts to halt logging in the Himalayan foothills.[108] Chipko has since largely vanished, but similar movements have taken its place.[109] These have achieved some successes. India's Forest Department has concluded joint management agreements with tens of thousands of village councils, establishing committees that care for lands formerly under the strict control of the state. But even this innovation has not completely freed village forests from the development plans and programs of the government.[110]

In 2006, the Indian government passed the "Forest Rights Act: Scheduled Tribes and Other Traditional Forest Dwellers (Recognition of Forest Rights) Act," which recognized both community and individual forest rights of those indigenous people historically dispossessed by both colonial and postindependence regimes. The law stipulates that members of scheduled tribes can claim rights if they were "residing in or dependent on forests prior to 13 December, 2005." All other "traditional forest dwellers" can make claims only if their families have been resident for at least seventy-five years prior to December 13, 2005.[111] Others are not covered by the Forest Rights Act. Those qualified under the act have been using the law to assert rights in the face of state and corporate claims, with several successes. At the same time, the law has not prevented numerous state-approved projects in violation of villages' rights; the state continues to assert control, a pattern which is found the world over.[112] So long as timber remains a valuable commodity and there are forests and lands to be exploited, governments and lumber companies will seek ways to exploit them. Trees, it would seem, are too valuable to be left to the people.

Is Global Forest Governance Possible?

Although the many forest functions enumerated earlier are extremely important to the local and global environments, as we have seen, none is as central to the economies of many countries as timber production. Moreover, many of the secondary benefits provided by forests, especially ecological ones, have much less to do with sovereignty and much more to do with the global good. These might be thought of as positive externalities for which no one pays but from which everyone benefits. In political terms, however, the concentrated interests in and maintenance of *national* control over forests—and, by extension, over biodiversity and genetic resources—far outweigh the diffuse and scattered stake that the world could claim in these secondary benefits. This is evident in the history of efforts to craft international forestry law.[113]

In 1992, representatives of 180 of the world's nations met at the UN Conference on Environment and Development [UNCED], the "Earth Summit," in Rio de Janeiro to consider, among other things, the adoption of an agreement on forestry principles. The document bore the unwieldy title "Non-legally Binding Authoritative Statement of Principles for a Global Consensus on the Management, Conservation and Sustainable Development of All Types of

Forests." That statement was the result of several years of sustained, intensive negotiation and controversy, a product of growing concern during the 1980s and early 1990s about the future of the world's remaining tropical forests. That this meeting was taking place in Brazil was especially apposite for two reasons. On the one hand, the burning forests of Amazonia had, during the late 1980s, served to focus global attention on their survival as well as their role in supporting the global environment. On the other hand, the Brazilian government had expressed strong opposition to any hint of internationalization of its sovereign resources and territory.[114] At the time, national governments were leery of being bound to a single set of rules, and many environmental NGOs believed that an agreement would only foster increased international trade in timber and even higher rates of deforestation than had been taking place.

Unwilling to lose the momentum generated by the Earth Summit's Forest Principles, however, in 1993, the Canadian and Malaysian governments launched a joint initiative to create the Intergovernmental Working Group on Global Forests (the word "global" was later dropped). In April 1994, the Intergovernmental Working Group on Global Forests held a meeting of experts and officials from fifteen key forest countries and several NGOs to facilitate dialogue about and consolidation of approaches to the management, conservation, and sustainable development of the world's forests. By the second meeting, in October 1994, attendance had expanded to include technical and policy experts from thirty-two countries, including Brazil, Finland, Gabon, Indonesia, Japan, the Russian Federation, Sweden, the United States, five intergovernmental organizations, and eleven NGOs.[115] At the end of 1994, the final report of the Intergovernmental Working Group on Global Forests was presented to the UN Commission on Sustainable Development, which, at its third meeting in 1995, proposed to establish an ad hoc Intergovernmental Panel on Forests (IPF) to further examine issues and develop proposals and recommendations. The IPF held four subsequent meetings through 1997, when its final report was submitted to the commission.[116] As a follow-up to the work of the IPF, in 1997 the UN Economic and Social Council established the Intergovernmental Forum on Forests, which pursued the work of the IPF and developed additional action proposals. Ultimately, the IPF and the Intergovernmental Forum on Forests together issued 220 proposals for action.[117] In 2000 the Economic and Social Council established a permanent entity, the UN Forum on Forests, to build on the work of its predecessors.[118] Between 2000 and 2007, the UN Forum on Forests met seven times, ultimately adopting a second non-binding instrument.[119] None of these initiatives has led to a global forestry agreement, and therein lies a tale.

Initially, the United States was a strong supporter of such an agreement, in the view that tropical deforestation represented a major contributor to global warming. Although the United States preferred to see other countries, especially developing ones, reduce their greenhouse gas emissions by emphasizing forest protection, the UNCED Forest Principles were the most to which the developing countries would agree. After UNCED, however, many governments, including European and Canadian and those of developing countries, came to favor a global agreement. By 1996, however, the U.S. position had changed completely, as domestic industry opposition grew because of the inclusion of boreal and temperate forests in the agendas of the various panels and forums addressing deforestation. Environmental organizations, too, continued to be opposed to a global forest agreement and wished, instead, to see forest conservation addressed through the UN Convention on Biological Diversity (CBD).[120] The nail in the coffin, as it were, occurred when countries under the UN Framework Convention on Climate Change (UNFCCC) sought to govern forests under that agreement as valuable sinks and stocks of carbon (first with LULUCF [land use, land-use change, and forestry] accounting and then with REDD+).[121] With this new climate change mitigation* lens, the United States and several other countries began to see in forests the possibility of sequestering carbon in existing forests and new plantations, and thereby avoiding the need to reduce greenhouse gas emissions in other sectors across the Global North, such as transportation and industry. This is an approach that has strengthened considerably over time. Through various projects and mechanisms, carbon emissions in the form of standing trees would be traded (as discussed in chapter 3), and sustainable forestry would then become something quite different from what was originally envisioned.[122]

Privatizing Forest Governance

After so many frustrated efforts to establish an international legal regime for forest protection it is not altogether surprising, then, that environmental actors began experimenting with alternative governance strategies.[123] One element of the evolving global forest governance system is an effort to regulate national logging through markets, especially consumer markets in industrialized countries. Today, there are as many as fifty voluntary forestry-standard-setting programs in existence around the world, including the United Nations Global Compact, ISO 14000, the Global Reporting Initiative, and one we profile here—the Forest Stewardship Council (FSC).[124]

With international government processes in stalemate, the FSC and other *private environmental governance* initiatives have been seen by many as a "magic bullet," a market-driven mechanism able to fill a critical niche in the effort toward achieving sustainable forest management where governments cannot. The FSC was launched in 1993 in Washington, DC, by environmental groups, the timber industry, foresters, indigenous peoples, and community

---

*In the climate change context, the word "mitigation" has a very specific meaning. According to the UN Framework Convention on Climate Change (UNFCCC), "Mitigation involves human interventions to reduce the emissions of greenhouse gases by sources or enhance their removal from the atmosphere by 'sinks.' A 'sink' refers to forests, vegetation or soils that can reabsorb CO_2." From UNFCCC, "Fact Sheet— The Need for Mitigation," accessed August 10, 2017, https://unfccc.int/files/press/back grounders/application/pdf/press_factsh_mitigation.pdf.

groups from twenty-five countries, with initial funding provided primarily by the Worldwide Fund for Nature/World Wildlife Fund. An interim board was elected, a mission statement adopted, and a draft of the guideline "Principles and Criteria for Forest Management" formulated soon thereafter. The FSC is a membership organization comprised of three equally weighted chambers— environmental, social, and economic—and membership within each chamber is also equally weighted between North and South.[125] Each chamber represents 33 percent of the vote at annual meetings, and the board of directors has rotating members reflecting these interests. The FSC currently has around 850 members, consisting of individuals, businesses, and groups.[126] According to its mission statement,

1. The Forest Stewardship Council A.C. (FSC) shall promote environmentally appropriate, socially beneficial, and economically viable management of the world's forests.
2. Environmentally appropriate forest management ensures that the harvest of timber and non-timber products maintains the forest's biodiversity, productivity, and ecological processes.
3. Socially beneficial forest management helps both local people and society at large to enjoy long term benefits and also provides strong incentives to local people to sustain the forest resources and adhere to long-term management plans.
4. Economically viable forest management means that forest operations are structured and managed so as to be sufficiently profitable, without generating financial profit at the expense of the forest resource, the ecosystem, or affected communities. The tension between the need to generate adequate financial returns and the principles of responsible forest operations can be reduced through efforts to market forest products for their best value.[127]

As part of its program, the FSC has developed and adopted guidelines for forest management, and it accredits certifying organizations that agree to abide by them. Purportedly, the FSC also monitors the operations and portfolios of such certifying groups on an annual basis. As of August 2017, the FSC had granted 1,513 "forest management certificates" in eighty-four countries, covering almost 199 million hectares, and 32,802 "chain of custody" certificates, in 121 countries.[128] These certifications are designed to assure producers and consumers that no materials or goods incorporating non-sustainable forest products enter anywhere in the "cradle to consumer chain." Thus, a certification on a chair or table guarantees that the FSC's guidelines have been rigorously followed from tree to retailer.[129]

However, despite its twenty-plus-year existence, the actual ecological and social outcomes triggered by the FSC system are not entirely clear.[130] According to the "Global Forest Atlas" of the Yale School of Forestry & Environmental Studies,

440.3 million hectares of forestland around the world have been certified by . . . FSC and . . . PEFC.* This number represents 10.7% of the total global forest area of 4.03 billion hectares. Annually, these certified forests account for an estimated 523.4 million m³ of industrial roundwood production, 29.6% of the world total.[131]

These numbers must be treated with caution, since there is considerable forest production that escapes industry statistics. Moreover, some indications suggest that, in some locations, the system is not leading to ecological or social outcomes that exceed those already required by existing government policies. In other instances, FSC (or other) standards may not actually be implemented by producers, because of the weak institutional base of the FSC and lack of enforcement. Funding and personnel to monitor implementation are scarce, and penalties for failing to observe the rules are few.[132]

Paradoxically, an additional challenge to the FSC's success may be the broader trend toward green labeling in the forest sector that it has inspired. Its forest product certification program triggered numerous corporate and government responses, and considerable alarm. The large financial stakes involved led forest products companies to become actively involved in standard setting and implementation activities in countries, such as Indonesia, Malaysia, and Sweden. A growing number of business organizations, including the American Forest Products Association and the Canadian Pulp and Paper Association in conjunction with the International Organization for Standardization, have developed their own certification programs.[133] The largest competitor of the FSC, the Programme for the Endorsement of Forest Certification, has developed a "mutual recognition" system among different national standards programs in order to preempt both the FSC's growing domination and the possibility of intervention by national governments.[134] And most recently, a number of large big box retailers, such as Walmart, Home Depot, and IKEA, in addition to carrying FSC-certified products, have created their own sustainability metrics.† Taken all together, this competition from a wide array of producer-backed schemes has likely cut into the reach of the FSC.[135]

Does the fragmentation of sustainable forest management certification matter? After all, if *everyone* were to follow some set of acceptable standards, wouldn't that effectively amount to the same thing as an international convention? Here, we might consider national patent systems as something of a parallel to the sustainable forest management situation. Every country has its own patent

---

*The PEFC (Programme for the Endorsement of Forest Certification) is an international certification organization composed of national forestry associations, national and international companies, non-governmental organizations, and other entities; see PEFC, "Membership," accessed March 26, 2018, https://www.pefc.org/about-pefc/membership.

†These include product life cycle and corporate footprint analysis, carbon accounting, supplier tracing, procurement requirements and audits, and sustainable performance tracking and reporting. Dauvergne and Lister, *Timber*, p. 24.

system, subject to certain international standards required by international trade agreements and organizations. The Trade-related Intellectual Property Rights (TRIPS) agreement of the World Trade Organization specifies what a patent system must do, but it does not require or detail exactly a specific approach to patenting. As a result, national access and enforcement vary from one country to another, and the pirating of patents and property is not uncommon. Still, national authorities are notionally able to enforce the rules and regulations.

This is not the case with market-based certification, for several reasons. First, because such agreements are "private" (that is, non-public), signatories follow rules only so long as it is in their individual interests. Second, it is difficult to ensure that different standards are actually commensurate, which might lead to forum shopping (that is, seeking out the least rigorous set of standards). Third, these systems are purely voluntary, and there is nothing—except, possibly, consumer disapproval—to prevent defection. Finally, as discussed, such private certifications are not *political*, which may be a fatal flaw in such market-based schemes.[136]

Effects and Effectiveness of Certification

Does privatized forestry regulation work? As we have suggested, the answer is not evident either in theory or practice. Individual forestry product companies, groups, and organizations will be attracted to such approaches *only* if environmentally conscious consumers choose their individual environmentally friendly, certified products. Moreover, certification adds to the cost of the product, and it is hoped that consumers will pay a premium for certified lumber, but how much more will they pay? It is one thing to tack a 10 percent green surcharge on a piece of furniture that may cost between one hundred and one thousand dollars; it is quite another to charge an extra 10 percent on a twenty-thousand-dollar remodeling job or a three-hundred-thousand-dollar house.[137] At present there does not seem to be significant demand in developed countries for certified wood, yet despite the absence of consumer demand, large building supply retailers such as Home Depot and B&Q remain motivated to carry FSC-certified products—a likely result of NGO campaigning, or the threat of future NGO campaigns.[138]

It is very difficult to find hard data on the effects and the effectiveness of FSC and other third-party certification on the health of those forests that have been certified. Less than 1 percent of tropical forests are certified, and half of those certified are actually plantations.[139] FSC has certified approximately 7.5 percent of forests globally, and the Programme for Endorsement of Forest Certification 11 percent, but the vast majority of certified forests are in industrialized countries, and it appears that most of those forests were already being managed close to certifier standards.[140] Overall, as noted earlier, certified forests globally account for less than 10 percent of the global timber trade.[141]

Meanwhile, the global rate of deforestation has persisted. From 1990 to 2005, Brazil lost 9 percent of primary forest land; Indonesia lost 31 percent;

and Gabon, 32 percent between 1990 and 2010.[142] Would an increase in certification make a difference, especially where tropical deforestation is concerned? To answer this question requires us to look more carefully at the drivers of deforestation and examine the changing global geography of production and consumption of timber and other uses of forests as well as the growing concentration and power in forest and forest product industries. Global commodity chains of timber have been evolving quickly in the years since the first edition of this book was published. In particular, a set of growing giants has gained significant control over the structure and rules of timber supply chains: big box retailers such as Walmart and IKEA, whose primary market is northern consumers, who already consume three-quarters of wood products and two-thirds of the global supply of paper.* Discount consumption is driving deforestation, as global retailers reinforce a global culture of "disposable consumerism" with products that have relatively short lifetimes and are cheaper to replace than to repair.[143] Other important drivers of deforestation include the clearing of lands for cattle ranches and plantations for timber and fuel pellets, soy, and palm oil.† "High consumption in the First World" is casting "deeper and longer ecological and social shadows . . . onto poor peoples and fragile environments in the Third World."[144]

Commentators find some room for hope in the private governance of timber supply chains. Arguably, Walmart has much greater power in the supply chain than the FSC (on its own, Walmart is China's sixth largest trading partner). When a supplier cannot meet the standards set by a big box retailer, that retailer has many more suppliers to choose from; rates of compliance with retailer standards may indeed be much higher than can be achieved through international law or a voluntary agreement. So we probably should not downplay potential impact of the sustainability goals set by huge retailers like Walmart‡ or IKEA. Still, the retailers put two contravening pressures on suppliers, for the bottom line still reigns supreme—low-cost supply is essential to a business model built on low-cost consumption with a high volume of sales. It should come as no surprise that Walmart and IKEA are having a hard time meeting their lofty environmental goals.§

---

*Dauvergne and Lister (*Timber*) identify two main aspects of that control: (1) companies can shape the means of interaction with and range of options available for suppliers, and (2) they have power to frame policies and discourse—including through crafting voluntary standards and defining what is corporate social responsibility.

†Indeed, many of the main drivers of deforestation may be out of reach of certification programs.

‡Walmart's three aspirational goals: "to create zero waste, operate with 100% renewable energy and sell products that sustain our resources and the environment." See: "Sustainability," accessed August 14, 2017, http://walmartstores.com/sustainability.

§Among other sustainability goals, IKEA set a target of a "30 percent share of the wood used in IKEA products coming from forests certified as responsibly managed—as of mid-2010 it was at just 7 percent." Dauvergne and Lister, *Timber*, p. 158.

Is forestry regulation through the market an adequate substitute for an international convention or national law? For the moment, there is no such agreement stipulating standards for sustainable forestry that could be the basis for policies in many different countries. Although a few governments might prefer to have such regulations in place, timber producers and associations have made clear their preference for "self-regulation." Under ideal conditions, private regulation might be a "second-best" solution to the problem, but the proliferation of certification schemes and, especially, fragmentation among national standards suggests that we are far from even a common private standard. It would be unreasonable, of course, to expect completely effective certification and sustainable forestry even under a global convention, but there would, at least, be a single framework from which to develop the required policy tools.

There is, moreover, an unrecognized trap hiding in the market-based approach to regulation of sustainable forestry, and that is, as we have seen, that markets are particularly weak arenas in which to seek political goals. Politics is, by definition, a public, collective endeavor, whereas markets involve private exchange between individuals. Politics is based on the visible aggregation of power, which markets eschew. Politics through market-based methods, which is what private certification amounts to, rests primarily on attempts to alter the individual preferences of large numbers of consumers in order to put pressure on producers (as opposed, for example, to regulating on a collective basis or altering social practices).[145]

Because consumer preferences are not political and are strongly influenced, if not determined, by the very system of production and consumption that motivates the social disruption and externalities of concern, there is a certain tautological process at work here. If capital is able to acquire political power,[146] it operates more as a form of displacement than an alternative: the "corporate citizen" becomes, in a sense, a franchisee able to cast a vote using its dollars.* The relevant question here, then, is not about "best" or "second-best" solutions to what is presented as a largely technical problem about maintaining best practices in forestry or any other environmental sector. If sustainable forestry is as critical to the sustenance of biological diversity and the environment as is often claimed, institutions that emerge from and through the market are unlikely to provide the necessary normative and legal structures or guarantee that these will last through much more than a few of the required harvesting cycles.

Some authors think that the answer lies in some combination of state-led mechanisms, voluntary certification, civil society advocacy, and global commodity chain market strategies, with private forest governance seen as a tool in a more comprehensive conservation strategy.[147] Twenty years of trying to

---

*The *Citizens United v. Federal Election Commission* case of 2010, lifting corporate campaign spending limits in the United States, illustrates this proposition. See: U.S. Supreme Court, "Citizens United v. Federal Election Commission, (2010)," *FindLaw*, January 10, 2010, accessed September 28, 2017, http://caselaw.findlaw.com/us-supreme-court/08-205.html.

come up with global forest governance illustrate the difficulties ahead. And a new and future driver of demand for forests that we alluded to earlier, the use of trees for climate mitigation—either as intact, biodiverse forests, plantations, or burned in bioenergy facilities—throws another complication into the mix.

Storing Forest Carbon and Selling It

As we noted earlier in this chapter, one of the barriers to negotiating a legal regime of global forest governance has been a discursive shift recasting global forests not as important in their own right but, rather, as technologies of climate mitigation, with the possibility that the carbon they contain can be counted, traded, and/or sold. That shift requires forests to be simplified and their carbon content standardized, as the vast ecological and social complexities of forest ecosystems disappear under the global gaze of a carbon-counting technical elite.* Under this new dispensation, forest value is based on the quantity of carbon that might be pulled out of the atmosphere (sequestered) and stored for some aspirational period of time in the trees and soil. The LULUCF rules under the Kyoto Protocol, which required parties to account for both emissions and removals of carbon, created these two new forest attributes and ways of measuring value: sequestration and storage. In the terminology of the agreement, forests could be *sources* of carbon dioxide in the atmosphere when they were harvested, burned, or otherwise deforested or degraded. They could also be carbon *sinks* and *stocks* if replanted or protected as intact forests.

Under Kyoto Protocol accounting of greenhouse gas emissions, developed countries must account for both emissions by sources and removals by sinks in each economic sector. Of all the sectors involved—energy, transport, industry, waste management, forestry, and agriculture—only the land sector (forestry and agriculture together) can claim removals by sinks, making it particularly valuable. With a large enough land sector, and what some view as rather creative accounting,† countries can minimize the aggregate net emissions they report to the world.

---

*In other words, not seeing the forests for the carbon. See Cathleen Fogel, "The Local, the Global, and the Kyoto Protocol," in *Earthly Politics: Local and Global in Environmental Governance*, edited by Sheila Jasanoff and Marybeth Long Martello. Cambridge, MA: MIT Press, 2003; Felicia A. Peck, "Carbon Chains: An Elemental Ethnography," PhD dissertation, Politics Department, University of California, Santa Cruz, 2016; Constance L. McDermott, "REDDuced: From Sustainability to Legality to Units of Carbon: the Search for Common Interests in International Forest Governance," *Environmental Science & Policy* 35 (2014): 12–19; Ester Turnhout, Margaret M. Skutsch, and Jessica de Koning, "Carbon Accounting," in *Research Handbook in Climate Governance*, edited by Karen Bäckstrand and Eva Lövbrand, Cheltenham, UK: Edward Elgar, 2015; Thomas Sikor, et al., "Global Land Governance: From Territory to Flow?" *Current Opinion in Environmental Sustainability* 5 (2013): 522–27.

†While it is relatively straightforward to calculate carbon dioxide emissions from the amount of coal or oil consumed, it is far more difficult to estimate the amount of carbon taken up or emitted as forests grow, or burn, or degrade. Environmental NGOs have

The Kyoto Protocol, which established legally binding emission reduction obligations for the rich countries of the world, also created a new mechanism to provide developing countries with incentives to reduce their emissions. The Clean Development Mechanism (CDM) is an offsetting mechanism, whereby developing countries would voluntarily host emission reduction projects paid for by developed countries who would then be able to claim those emission reductions as partial fulfillment of their *domestic* legal obligations. As one might imagine, while rules for the CDM were being developed, there was substantial political debate over which sorts of projects would qualify as CDM projects. Developing countries with substantial forest resources were eager to sell as offsets the carbon sequestered in already-existing trees standing in their forests—what was termed "avoided deforestation."[148] Many developed countries and environmental organizations were—at least ostensibly—concerned about the lack of "environmental integrity" in such a scheme. Carbon that has already been sequestered makes no *additional* contribution to the main problem to be solved: the growing quantity of greenhouse gases in the atmosphere, primarily from the burning of fossil fuels for energy production. Negotiators eventually decided that afforestation and reforestation projects would be the only types of forest-related projects that could receive CDM credits. In addition, the sequestered carbon would be significantly discounted relative to projects that permanently avoided emissions from fossil fuels.

Permanence matters, particularly in the context of offsetting (we discussed this briefly in chapter 3). Stored forest carbon is by its very nature temporary, because trees don't live forever, whether they are deliberately harvested or turned into atmospheric carbon by fire, beetles, or another cause of demise. Contrast the carbon sequestered in a forest with the impact of a decision to build a solar farm instead of a coal-fired power plant. The emissions that are avoided when the coal plant is not built will never happen—the emission reductions are permanent.* Forests, by contrast, rely heavily on future decisions by those with the power or capability to cut them down, or climate impacts that increase pest populations, drought, and fire.[149] Laws change, practices change, climate changes.

---

campaigned for years against what they call LULUCF loopholes, energetically challenging the environmental integrity of land-sector accounting rules despite the highly technical arena of debate. See as just one example, Alistair Graham, "The Hypocrisy of Land-use Accounting," *Ecosystem Marketplace*, April 5, 2011, accessed August 15, 2017, http://www.ecosystemmarketplace.com/articles/the-hypocrisy-of-land-use-accounting/.

*Credits from afforestation and reforestation projects are called temporary certified emission reductions and have been priced at around 20 to 25 percent of the price of permanent CERs. Investors in carbon credits apparently care about the longevity of their investment. At writing, the market price of permanent CERs is around twenty cents per ton—the authors were unable to find a market quote for the current price of temporary CERs. Avril David, "Can Carbon Sequestration Be Leased?" *Ecosystem Marketplace*, July 24, 2009, accessed August 15, 2017, http://www.ecosystemmarketplace.com/articles/can-carbon-sequestration-be-leased/.

At present, the climate regime principally privatizes forests by turning their carbon into tradeable commodities, such as CDM credits or voluntary emission reductions (VERs) from REDD+ projects (as we described in a footnote in chapter 3), which are both used as offsets. As emission reductions lag and global temperatures rise, more and more attention is turning to forests for their broader mitigation values, whether or not those are commodified for trading and sale.* In particular, as industrialized economies lag substantially in retooling toward low carbon and carbon neutrality, there is a real and growing need for mitigation to happen *somewhere*. Forests—natural or plantations— are being enlisted to serve that discursive and material purpose.[150] Mitigation becomes a new, powerful reason for enclosure and control of forest resources, to soak up the excesses of the high-carbon-consuming class.

Forests as Our Last, Best Hope?[†]

As we noted in chapter 3, under the Paris Agreement, countries have agreed to limit warming to "well below 2°C above pre-industrial levels and to pursue efforts to limit the temperature increase to 1.5°C."[151] That is a tough order, given the likelihood that the world will pass the threshold of 1.5°C sometime between 2025 and 2030, if not sooner.[152] Recall that, as of 2011, the remaining carbon budget for a 66 percent chance of staying below 2°C is about one thousand gigatons, and that global emissions are currently about forty gigatons per year. According to the Intergovernmental Panel on Climate Change (IPCC), the carbon budget available to stay below 1.5°C, with the same probability, is a mere four hundred gigatons, as of 2011.[153] The odds are not good that we will completely cease emitting greenhouse gases by 2021.[154] We rather urgently need to develop and deploy mitigation approaches that can generate "negative emissions," drawing more carbon out of the atmosphere than is currently entering it.[155]

As noted earlier, there has long been considerable interest in the utility of forests, as well as soils and grasslands, for carbon sequestration and storage. Indeed, sequestration by natural systems is one of the only viable means to create negative emissions (see Table 5.1 for examples).[‡] There are at least three ways that forests (in a rather broad sense) might be configured to generate

---

*Recall the discussion in chapter 3 where we made the point that offsets do not contribute to overall emission reduction (mitigation); they just move emissions around the planet.

†There is still significant lack of scientific clarity about the extent to which natural forests currently serve as a carbon sink. See, for example, A. Baccini, et al., "Tropical Forests are a Net Carbon Source Based on Aboveground Measurements of Gain and Loss," *Science* (2017), accessed October 4, 2017, http://science.sciencemag.org/content /early/2017/09/27/science.aam5962.

‡"Forests are one of the most promising natural places to engineer negative carbon emissions." Fred Pearce, *Going Negative: How Carbon Sinks Could Cost the Earth*, Brussels: FERN, 2016, p. 11.

Table 5.1.   Carbon dioxide removal methods

Technique	How it works
Bioenergy with carbon capture and storage (BECCS)	Crops grown for the purpose are burnt in power stations (providing energy), and the resulting CO_2 is captured for secure long-term storage.
Afforestation and reforestation	Large-scale tree plantations increase natural storage of carbon in biomass and forest soil.
"Blue carbon" habitat restoration	The recovery of degraded or overexploited coastal ecosystems that have a high potential for carbon storage, such as salt marshes or mangroves.
Biochar	Carbon from partly burned biomass is added to soil, with potential for agricultural benefits.
Enhanced ocean productivity	Marine photosynthesis and CO_2 drawdown from the atmosphere is increased, either by adding nutrients to promote phytoplankton growth in the open ocean or through seaweed cultivation in shallow seas.
Enhanced weathering (using silicate rock)	Crushed olivine or other silicate rocks are added to soil surfaces or the ocean for chemical absorption of CO_2. (Could help to reduce ocean acidification.)
Direct air capture	Chemicals (or possibly low temperatures) are used to extract CO_2 from ambient air. Safe CO_2 transport and storage are subsequently required.
Cloud treatment to increase alkalinity	Alkaline rain resulting from cloud treatments reacts with, and removes, atmospheric CO_2.
Building with biomass	A massive increase in the use of biomass (straw and timber) as a building material removes carbon for decades or centuries.

Source: Phil Williamson, "Scrutinize CO_2 Removal Methods," *Nature* 530 (2016): 153–55. Adapted by permission from Macmillan Publishers Ltd: *Nature* 530: 153–55, copyright 2016.

negative emissions: (1) enhancing natural forests, reforestation of degraded forests, and through forest ecosystem restoration;[156] (2) plantation forestry, with extensive monocultures of fast-growing trees, possibly for harvesting as building material, which would store carbon for decades after the trees were cut down; and (3) plantations used as feedstock for bioenergy power plants with associated carbon capture and storage facilities (BECCS—bioenergy with carbon capture and storage).* All these options will require a much greater degree of state and corporate control to deliver mitigation at the required scale. Furthermore, through REDD+, climate change is already serving as a new rationale for control of forests and for exclusion of communities and indigenous peoples from forested territories, the latest form of what academics and activists have termed "carbon colonialism" (which has also been applied to the

---

*BECCS technology actually has not been proven at scale, though significant hope is being placed on what Kevin Anderson calls the "carbon-sucking fairy godmother" scenarios of large-scale negative emissions through BECCS. Kevin Anderson, "Talks in the City of Light Generate More Heat," *Nature* 528 (2015), accessed October 4, 2017, http://www.nature.com/news/talks-in-the-city-of-light-generate-more-heat-1.19074.

CDM).[157] Options that rely on plantation "forestry" to store carbon will likely also be a potential driver of large-scale deforestation in the name of climate mitigation.

Researchers and policymakers are currently exploring "a broad set of methods and technologies operating on a large scale that aim to deliberately alter the climate system in order to alleviate the impacts of climate change," or in a single word, geoengineering.[158] Geoengineering technologies are sorted into two main categories: carbon dioxide removal (CDR) and solar radiation management.[159] Solar radiation management includes technologies such as putting sulfate particles into the atmosphere to reduce the amount of solar radiation, and thus heat, reaching the earth. Carbon dioxide removal technologies include iron fertilization of oceans, chemical weathering of rocks, and a range of technologies associated with forests.[160] BECCS is a forest-related geo-engineering technology that has attracted significant attention. Briefly the idea is to grow a biofuel feedstock—usually wood—to use for energy production, replacing fossil fuels. The feedstock is burned in a power plant equipped with technologies to capture the carbon dioxide emitted. The carbon dioxide is then put under extreme pressure and more or less liquefied, enabling it to be piped to storage caverns underground. Fingers crossed that the carbon dioxide stays where it is put.

The IPCC states the obvious: "most terrestrial [carbon dioxide removal] techniques would involve competing demands for land."[161] Indeed, at the scale of mitigation and negative emissions being considered by some proponents, serious environmental and social impacts will accompany the large-scale deployment of these technologies.[162] A brief example should vividly show the challenge of reconciling mitigation with other valued uses of land, such as food production and intangible social and environmental benefits of intact, diverse forests. As with all climate projections, there are many assumptions hidden behind the numbers offered in our example, too many and too technical to recount here, so view this example as a heuristic to help visualize scale and implications. Estimates of the future potential for negative emissions from BECCS are between 2.4 and ten gigatons of carbon dioxide stored per year—providing a quantity of negative emissions equivalent to up to one-quarter of current emissions (which you should recall is currently on the order of forty gigatons of carbon dioxide per year).[163] Remember, storing carbon through BECCS technologies requires first growing trees or other biomass to burn to generate that carbon.

> Deployment of BECCS could raise food prices, and/or displace agricultural production, in ways that could also imperil food security and violate the right to food. One striking feature of BECCS is the potential amount of land that may need to be diverted from other uses, including food production and live-lihood-related activities, to provide bioenergy feedstocks. Delivery of a rela-tively modest 3 Gt of CO_2 equivalent (CO_2-eq) negative emissions annually would require a land area of approximately 380–700 million ha in 2100,

translating into 7–25% of agricultural land and 25–46% of arable and permanent crop area. . . . This level of emissions removal would be equivalent to
a startling 21% of total current human appropriate net primary productivity.
While it might be possible to reduce these impacts by more of an emphasis on
the use of agricultural residue and waste feedstocks, this option could prove
to be extremely limited.[164]

There are numerous questions, many based on equity considerations that
might be going through your mind right now. Whose land would be used to
grow those bioenergy crops? How is that land presently used? If the choice is
between climate mitigation and food production seriously affected by climate
change, who gets to decide? The distribution and justice issues, now and in the
future, are rather apparent—how much land would be needed in the future is
directly related to the efforts made to reduce emissions now.[165] Emitters and
those whose land would be used to compensate for their emissions are not the
same people, nor are they likely to live in the same country.

The complex difficulties of transforming energy systems—social, cultural,
political, economic—land us in a very troubling conversation about overshoot
and negative emissions.[166] There are clear and compelling reasons to keep warming to below 1.5°C; by the time you read this book, the IPCC will have published a special report on that very number that you can consult for a detailed
scientific explanation. It is hard to imagine we will not pass that threshold,
leaving us in a situation of overshoot, and wondering about the possible recovery of human and biological systems and whether global temperatures might
someday settle at less than 1.5°C above historical norms. Negative emissions
technologies such as BECCS and more benign approaches to enhance natural
land sinks are among the limited options we have going forward.

The confluence of consumption habits—really the consumption habits of
the privileged of the world; power over who gets to decide on national and
global carbon budgets; and a seemingly boundless faith in technological solutions are pushing us down a dangerous path, one that threatens, for the sole
purpose of climate mitigation, not only to privatize but to literally enclose
huge expanses of forests, preventing those who rely on forests from access
to them, much as the British Raj and Indian states did to their peasants and
indigenous peoples.[167] The future does not look particularly hopeful for those
concerned about climate change. And yet, as authors, and like Antonio Gramsci, we remain pessimists of the intellect but optimists of the will. We can put
ourselves on a trajectory that does not require manipulation of incoming solar
radiation or planting an equivalent of half the total current global cropland
area to biofuels. We begin to answer the question "how?" in the next chapter.

Don't Get Caught with Your Genes Down

Not all natural resources have as long a history of state manipulation and
management, combined with a more recent transformation into commodities,

as do water and forests. Today, the struggle for control at the molecular genetic level is becoming as intense as those historical cases. On the one hand, the very concept of genetic resources is, as we have seen, a relatively recent one that has emerged as a result of innovations in biotechnology and gene manipulation—some of this through military research.[168] On the other hand, humans have been engaged in manipulating and trading the equivalent of genes for thousands of years. Almost all of the fruits, grains, and vegetables we eat on a daily basis originate from cultivars that were the result of millennia of collective breeding by farmers to change crop characteristics to suit human purposes: seed pods that don't shatter, better flavor or storability, increased yield.[169] Domesticated animals, too, are a product of selective breeding for favored characteristics, which depend on particular combinations of genes.

As a result of this very long history of genetic interchange and their immense collective value to humanity, plant and animal genetic resources have generally been treated in practice as a common heritage of humankind.[170] Yet beneath this understanding of a global genetic commons is a wealth of historical, political, ecological, and economic relationships reflected in the history of often-colonial exchange among societies. For example, corn (known as maize outside of the United States) was domesticated thousands of years ago in Mesoamerica, and today is a staple foodstuff for billions all over the world—displacing traditional staples such as millets and other small grains. How did that happen? For millennia, humanity's most loved foods were traded, carried, and shipped, by land and sea, across continents, globalizing a food supply and making tea, coffee, sugar, and spices common in kitchens around the world. That process was speeded up in the age of European exploration and colonization, with the "common heritage" of humankind acquiring a particular imperial character. The major botanical gardens and zoos of Europe are largely the result of the acquisitional habits of colonial era biologist/explorers—as often as not, sent out by monarchs and global trading companies. So are the monocultures of plantation crops across the world, transplanted from one colonial territory to another, most often with the labor of slaves.

The modern politics of genetic resources have embedded within them these historical roots—colonial relationships, slavery, movement of crops around the world through trading among economies and cultures. That the vast majority of our major crops, fruits, and vegetables have their biological origins in what is now known as the Global South—much of which was colonized by European states—is another important element of ecological history relevant to understanding current debates.[171] The most important question in these debates is one that, until the twentieth century, hardly mattered: who owns these resources?[172]

The 1972 declaration by the Stockholm UN Conference on the Human Environment affirmed the sovereignty of states over their natural resources, in Principle 21: "States have, in accordance with the Charter of the United Nations and the principles of international law, the sovereign right to exploit their own resources pursuant to their own environmental policies."[173] Through negotiations leading to the 1992 CBD, the understanding of sovereign control over natural resources was updated to include biological and genetic resources, at which point the idea that plant and animal genetic resources, including seeds, might remain a "common heritage of humankind"[174] rather evaporated.

The CBD has three objectives: (1) the conservation of biological diversity, (2) the sustainable use of its components, and (3) "the fair and equitable sharing of the benefits arising out of the utilization of genetic resources, including by appropriate access to genetic resources and by appropriate transfer of relevant technologies, taking into account all rights over those resources and to technologies, and by appropriate funding."[175] The according of sovereignty over biological resources to states lays the basis for the central quid pro quo of the convention: developing countries take action to save their biodiversity for the sake of humanity *and* provide access to their biodiversity to the pharmaceutical and biotechnology companies of the developed world, who in return will share benefits from their inventions with the now "owners" of the genetic resources from which the inventions are derived.[176] The according of sovereignty and ownership over what was in the past common heritage is essential for the market transactions that are central to the bargain of the CBD—with markets, once more, playing a mediating role between agents. And there is another reason why developing countries were keen to claim sovereignty over these resources—the according of ownership also serves to disrupt historically unequal trading relationships that look quite a bit like outright theft—as, for example, when seed companies of the developed world take farmers' seed, privatize it through a system of plant breeders' rights, and sell the seed back to farmers as varieties protected by intellectual property rights (IPRs).[177] The biological diversity of the developing world was a vast resource for imperial accumulation, as noted above. Consequently, as the CBD was being negotiated, developing countries were quite aware that many of their plants and animals, such as the rosy periwinkle of Madagascar, had already been "bioprospected" and were yielding millions in revenues for the pharmaceutical industry.[178] Those countries wanted a share of those profits.

But such sharing of benefits, so it is argued, cannot happen unless countries allow the privatization of genes and genetic resources. When genes and inventions based on those genes (for example, genetically engineered crop plants or pharmaceuticals) can be patented, companies may then extract monopoly rents on the use of their inventions. Only when companies make large profits on the sale of these inventions by way of these monopolies are there benefits to be shared. So as designed under the terms of this international environmental regime, saving biodiversity and sharing its benefits *requires* first, privatization, and second, alienation through markets.[179]

The devil remains in the details of the access and benefit-sharing arrangements, a fact that developing countries were well aware of. They pushed for negotiations on a subsidiary protocol under the CBD—the Nagoya Protocol on Access and Benefit-Sharing. Negotiations concluded in 2010 and the agreement entered into force in 2014. The access provisions in the protocol are, not surprisingly, more developed than the benefit-sharing ones found there, as developed countries hold significant knowledge and power in that intergovernmental negotiating sphere.*

The contest over access to and sharing of benefits from genetic resources continues under both the CBD and the Nagoya Protocol. As we mentioned in chapter 3, the new technologies of synthetic biology raise a host of new political questions about privatization of genetic information: for example, if a company never touches a physical sample of a plant or animal or microorganism, but merely accesses a sequence of its deoxyribonucleic acid from a database, does it have the obligation to share benefits from the invention that results?[180] Does the definition of "genetic resource" include digital copies of its sequence information residing in private (pirated?) or open-access (common heritage?) databases? Developed countries with pharmaceutical and biotechnology interests say "no" to this last question, and do not agree that benefits from the use of that information must be shared. Developing countries with large amounts of biodiversity to be sequenced, or which have already been sequenced, say "yes." One current demand of developing countries raised in several intergovernmental forums, including the meeting of the contracting parties to the International Treaty on Plant Genetic Resources for Food and Agriculture, is to require disclosure of country of origin of the genetic resource for sequences stored in public databases.[181]

In contrast to the Kyoto Protocol and the Paris Agreement, the CBD contains no specific provisions to establish markets in biological diversity. In a way, there is no need—both the relationship between access and benefit-sharing fixed in the objectives of the treaty and the necessary reliance on sovereign or private control over genetic resources reflect the U.S. (neo)liberal approach of reliance on market mechanisms to bring about environmental protection. At the time the treaty was negotiated, countries of the Global South had some leverage in the negotiations, as developed countries wanted both an agreement to protect biological diversity and access to that diversity. The Global South was keen for sustainable development finance, while developed countries sought to limit such financial commitments. The promise of benefits from access to biodiversity was a valuable chip on the negotiating table—if the market could provide financial resources, developed countries and their financial institutions and agencies would not have to. However, the intervening years since the treaty was adopted have illustrated the unequal nature of the eventual bargain. Resources promised in Rio in 1992 were never delivered, while the trade and IPR regime in the World Trade Organization (WTO) was

---

*For a very detailed play-by-play of the negotiations, see Gurdial Singh Nijar, *The Nagoya Protocol on Access and Benefit Sharing of Genetic Resources: An Analysis*, Kuala Lumpur: CEBLAW, 2011.

Photo 5.1   The Captain Hook Awards

Captain Hook Awards at COP 13 organized by the Coalition Against Biopiracy, UN Convention on Biological Diversity COP13 in Cancún, México, December 2016.
Source: Photo by IISD/Francis Dejon (enb.iisd.org/biodiv/cop13/enb/9dec.html).

strengthened (see Box 5.4)—to the benefit of developed countries. It should not surprise anyone that the issue of access and sharing of benefits continues to dominate politics of the CBD.

HOW INTERNATIONAL ARE INTERNATIONAL ENVIRONMENTAL REGIMES?

In this chapter, we have seen that the environmental practices of particular states have, historically, been observed and copied by other states, albeit with various national modifications. Just as the political systems of some countries—mostly former colonies, but in more recent instances of "independence," as across the former Yugoslavia and Soviet Union—are modeled on the British parliamentary system, others have adopted forms and practices similar to the U.S. executive-legislative arrangement, and a very few retain the centrally directed party governments that originated with the Soviet Union. There are always national and cultural differences in the ways these institutional patterns are implemented, to be sure, but the basic features of state organization remain visible and similar.[182] "Social learning" among groups is the norm for human beings and has probably been the case since quite early in the history of the species.

BOX 5.4	The World Trade Organization and intellectual property protection for plants

Another intergovernmental institution with rules governing intellectual property protection for living organisms and their genes is the World Trade Organization (WTO) and its associated Agreement on Trade-Related Aspects of Intellectual Property Rights (otherwise known as the TRIPS agreement).* Article 27 of the TRIPS agreement defines patentable subject matter and allows for "plants and animals other than microorganisms" to be excluded from industrial patent protection rules. However, countries are required to provide for intellectual property protection of plant varieties "either by patents or by an effective *sui generis* system or by any combination thereof." To be clear, the "plant varieties" covered by this regime are breeder-developed varieties and genetically engineered plant varieties, not historically and farmer-developed varieties. In the negotiations on the TRIPS agreement, multinational seed companies were active lobbyists, with the U.S. government supporting the extension of full patent protection to *all* life-forms as is the case under U.S. law. *Sui generis* systems, such as those established under the International Union for the Protection of New Varieties of Plants, or UPOV after its French acronym, allow for possible exemptions to full patent protection, including the possibility of allowing farmers to save seed (otherwise planting saved seed would be a violation of a patent, as it would be illegally reproducing someone else's invention) or breeders to use a seed variety to modify and improve. These are, of course, traditional practices of farmers and breeders that are seen as appropriate and necessary by many, but not by biotechnology firms. Monsanto and allies sought to remove the possibility for these exemptions in the global rules on seed protection. The recognition that seeds are the basis of lives and livelihoods has allowed for some small cracks to open in the intellectual property fortress, although countries wishing to join the WTO still must engage in very hard negotiations on the degree of flexibility they have with UPOV exemptions. The story of Nepal's accession to the WTO and hard-won fight to preserve farmers' rights to seed is very instructive in this regard.

Sources: "UPOV," accessed August 16, 2017, http://www.upov.int/portal/index.html.en; Posh Raj Pandey, Ratnakar Adhikari, and Swarnim Waglé, "Nepal's Accession to the World Trade Organization: Case Study of Issues Relevant to Least Developed Countries," UN Department of Economic and Social Affairs, CDP Background Paper No. 23 ST/ESA/2014/CDP/23 (November 2014).

---

*There is yet another international organization dedicated to intellectual property rights: the World Intellectual Property Organization. As a rule, the TRIPS agreement tends to preempt the World Intellectual Property Organization (WIPO), with the latter more focused on the mechanics of intellectual property patents. See http://www.wipo.int/portal/en/index.html (accessed August 27, 2017).

Sometimes, however, individual countries have the power and wealth to impose their ways of doing things on much of the world. The ideological emphasis on capitalism and markets is not unique to the United States, but the United States, as the world's dominant power since the end of World War II, has been a leader in shifting the policies and practices of environmental internationalism toward growing reliance on both markets and economic growth as solutions to environmental damage. This observation raises questions about both the sources and effectiveness of IERs. As we shall see later in this chapter, certain assumptions are made in the literature about how such regimes are created and how they function. Although it is widely recognized that power (and wealth) do matter when international issues are at stake, many students of international relations nevertheless assume that negotiation and bargaining can redress many, if not all, of these inequalities. After all, in markets, we often find individuals with enormously disparate financial endowments engaging in peaceful and cooperative bargaining and exchange. But no one would argue that the Jaguar salesperson is, somehow, the equal of the multimillionaire interested in purchasing a car. The former does have the power to deny the latter her purchase, but at what cost? A similar logic applies to bargaining among states.

What has become increasingly common during the past two centuries is that specifically national practices have become the basis for international institutions, such as the CBD and the WTO. When the British Empire ruled the seas, national and colonial economies were run along the lines favored by London. When countries tried to defy the British Empire (for example, by repudiating their debts), gunboat diplomacy was born, ensuring that the rules would be followed.[183] Since 1945 the United States has played this central role in shaping and directing international politics and economics (the Soviet counterhegemony notwithstanding), one that has only become more prominent during the years following 1991.[184] To be sure, other countries and entities, such as China and the European Union, are major players in shaping the global political economy and IERs and may well become more central in the future. It is difficult, nonetheless, to construct and maintain multilateral and international initiatives without the participation of the United States (the Trump administration's proclivity for defecting or withdrawing from international treaties will put this hypothesis to the test over the coming years—although not playing the game can be as influential as joining it). The United States has made it clear that, if multilateral and international initiatives are not organized and pursued in a way that is acceptable to Washington, the United States will not participate.

The U.S. Mark on the Global Climate Regime

The fate of the Kyoto Protocol to the UNFCCC sheds some light on this general argument. The Kyoto Protocol set out binding emission reduction obligations only for developed countries, in keeping with the legal frame of its parent treaty, the UNFCCC, which directed that developed countries should take the lead in addressing climate change. The United States stood fast on its refusal to sign the protocol and the U.S. Senate made it clear that ratification would not happen without commitments by developing countries—as a result, the Clinton administration never submitted the protocol to the Senate for approval, and the George W. Bush administration actually rescinded the United States' signature. This was an untenable situation for the rest of the industrialized countries and set the stage for a strategy to negotiate a new agreement while letting the Kyoto Protocol die a slow death. A more detailed look at the path from the Kyoto Protocol to the Paris Agreement, starting with the historic failure of negotiations in Copenhagen in 2009, illustrates the key role of the United States in the design of the current climate change regime, notwithstanding its periodic refusal to commit to it.

In 2009, all eyes were on Copenhagen as climate negotiators were expected to agree on new rules for action under the UNFCCC. The meeting ended in spectacular failure, ascribed in part to insufficient diplomatic skills of the Danish prime minister who was guiding the meeting and framing the "Danish text."[185] While the prime minister was perhaps challenged by the complexity of multi-party negotiations, the fact is that countries were still far apart in their positions on what the next phase of the climate regime should look like when they arrived at the fifteenth meeting of the UNFCCC Conference of the Parties.

There were a number of axes of disagreement in Copenhagen but an enduring division was over the principles of equity and "historical responsibility." Most developing countries think that industrialized country fossil fuel emissions that enabled their development from the middle of the nineteenth century, and that are still in the atmosphere—their historical responsibility—ought to be counted in some way in establishing the emission reduction obligations of the rich countries. As well, poor countries argued that a new agreement must be based on distributional equity in deciding how to share the global burden of reducing greenhouse gas emissions.

Recall the discussion in chapter 3 and above about the carbon budget. The IPCC estimates that, collectively, humans have already put nineteen hundred gigatons of carbon into the atmosphere and, depending on how you calculate the remainder, there is roughly another four hundred to one thousand gigatons that can be emitted without exceeding the two-degrees-Celsius benchmark. The question of historical responsibility can thus be articulated as "in calculating responsibility moving forward, does it matter who put the nineteen hundred gigatons into the atmosphere?" Or do we only negotiate how to distribute the remaining one thousand gigatons without paying attention to historical emissions, the vast majority of which came from industrialized countries? In the coming decades, populous developing countries such as India and China will take up a large share of the remaining atmospheric space as their populations grow and standards of living, and consumption, increase. Most developing countries argue there is an inherent right to that development which, with existing technologies, still requires burning fossil fuels, and hence the right to a certain amount of growth in development-related emissions over and above current emissions.

Another important axis of disagreement in Copenhagen was about the possible architecture of the agreement: would it be "top-down" or "bottom-up"?[186] A top-down agreement considers how much atmospheric space is left, say, to keep warming below two degrees Celsius above pre-industrial levels.* That remaining space would then (ideally) be divided up equitably, with each country given an emissions reduction target, and establishing a collective timetable for action. Before calculating individual targets in this instance, however, the parties to the UNFCCC would need to decide whether to factor in historical emissions, from say 1850, or whether to start the clock at a more recent date, such as 1990. A bottom-up agreement, by contrast, would give countries the flexibility to decide what actions they would be able to take and to set their own targets, based on their own national circumstances.[†]

Flexibility in setting their own targets would give space to developed countries to effectively grandfather in their large historical and current per capita emissions—which would greatly disadvantage the Global South. Developed countries could pledge to take on large cuts toward 2050 and along the way still take up far more than a fair share of the remaining atmospheric space (see Figure 5.3).[‡] For example, in the lead-up to the Copenhagen meeting the European Union had proposed a target of an 80 to 95 percent reduction in developed country emissions by 2050, with an overall global goal of a 50 percent reduction.[§] Yet, to stay within a budget to remain below two degrees Celsius of warming, that scenario would result in a far less than fair apportionment of the remaining one thousand gigatons, dramatically cutting into developing

---

*Any temperature target would do here. The Paris Agreement sets as a target "holding the increase in the global average temperature to well below 2°C above pre-industrial levels and to pursue efforts to limit the temperature increase to 1.5°C above pre-industrial levels."

†Commentators often call the Kyoto Protocol a top-down agreement, which is not really accurate. While it is true that the Kyoto Protocol establishes legally binding targets for developed countries, those targets were the result of lengthy and difficult negotiations, not based on equitable shares of a global cap but on what the countries in the end would politically agree to, country by country. The legally binding emission reduction targets just for developed countries was the main driver of their (developed countries') effort to spend the next decade after the Kyoto Protocol came into force (in 2005) to push for negotiations on the Paris Agreement.

‡Two resources for understanding and calculating what might be a fair share of atmospheric space are the website Climate Fairshares, http://climatefairshares.org, and the Climate Equity Reference Calculator, https://calculator.climateequityreference.org/.

§Letter from the Council of Europe and the European Commission to Yvo de Boer, executive secretary of the UNFCCC, "Expression of Willingness to be Associated with the Copenhagen Accord and Submission of the Quantified Economy-wide Emissions Reduction Targets for 2020," accessed August 16, 2017, https://unfccc.int/files/meetings/cop_15/copenhagen_accord/application/pdf/europeanunioncphaccord_app1.pdf.

country prospects for any kind of sustainable development.[187] The bottom-up approach would allow industrialized countries to take up a huge, disproportionate share of atmospheric space into the future, not merely historically. This was hardly acceptable to the Global South.

A third critical axis of disagreement in Copenhagen, and since, has been over the financial commitments of developed countries to support emissions reductions and adaptation to climate impacts in the Global South. Staggering amounts of finance are needed to both address the impacts of climate change and to fund a transformation in energy systems away from reliance on fossil fuels.[188] The final draft text from Copenhagen promised thirty billion U.S. dollars in fast-start finance over the next three years, as well as a commitment to mobilize up to one hundred billion U.S. dollars a year by 2020 from private and public sources. Those amounts are far short of the trillions that the World Bank says are necessary.* In the dramatic final plenary in Copenhagen, the representative of Tuvalu likened the proposed agreement to a famous betrayal: "It looks like we are being offered 30 pieces of silver to betray our people and our future. Our future is not for sale. I regret to inform you that Tuvalu cannot accept this document."[189]

In the end, the Copenhagen meeting collapsed with no substantive outcome agreed to by consensus, although a document did emerge at the end of the two-week meeting, negotiated in back rooms by a handful of countries (twenty-six of the 190+ at the meeting), including the United States, China, India, and Brazil.† That document, the Copenhagen Accord, set out a framework for a bottom-up approach and included the finance numbers mentioned above. It did not receive official recognition by the meeting, though it set the frame for ongoing negotiations. Six years later, the Paris Agreement would codify the voluntary, bottom-up approach first set out in the Copenhagen Accord.

---

* "McKinsey estimates the total global cost of meeting the Paris targets to be €200-350 billion ($US 300-420 billion) per year by 2030. This is less than one percent of the forecasted global GDP in 2030." Hannah Ritchie, "How Much Will It Cost to Mitigate Climate Change?" *Our World in Data*, March 27, 2017, accessed August 27, 2017, https://ourworldindata.org/how-much-will-it-cost-to-mitigate-climate-change/. By contrast, official development assistance in 2015 totaled $131.6 billion U.S. dollars (not all of it to the Global South) while foreign direct investment in developing countries in 2016 was $646 billion, with more than 75 percent going to Asia. Organisation for Economic Co-operation and Development, "Development Aid Rises Again in 2015, Spending on Refugees Doubles," n.d., accessed August 27, 2017, http://www.oecd.org/dac/development-aid-rises-again-in-2015-spending-on-refugees-doubles.htm; UN Conference on Trade and Development, "World Investment Report 2017," Geneva, 2017, accessed August 27, 2017, http://unctad.org/en/PublicationsLibrary/wir2017_en.pdf.

†Former U.S. Secretary of State Hillary Clinton tells the story of how she and U.S. President Obama burst into the room where the BASIC countries (Brazil, China, India, and South Africa) were meeting and convinced them to make the Copenhagen deal that paved the way for the Paris Agreement: https://www.youtube.com/watch?v=BS3t BQVlR3Q.

Figure 5.3 Emissions Debt, Past and Future

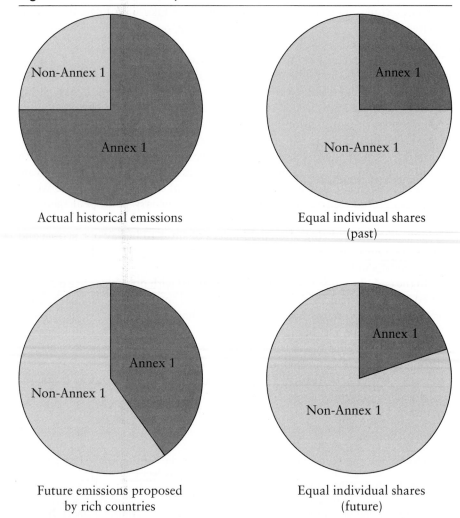

Actual historical emissions

Equal individual shares
(past)

Future emissions proposed
by rich countries

Equal individual shares
(future)

The extent of developed countries' emissions debt reflects their excessive past, present, and proposed use of shared atmospheric space. With less than 20 percent of the population, developed countries have produced more than 70 percent of historical emissions since 1850 (upper left), far more than their fair share based on equal per-person emissions (upper right). After diminishing the Earth's atmospheric space—denying it to poor countries and communities—the same rich countries now propose consuming a disproportionate share of the remaining space until 2050 (lower left) as compared to an equal per capita share (lower right).

Source: Matthew Stilwell, "Climate debt—a primer," in Niclas Hallström, ed., *What Next Volume III: Climate, Development and Equity*, accessed October 1, 2017, http://whatnext.org /resources/Publications/Volume-III/Single-articles/wnv3_stillwell_144.pdf.

There were a number of stops along the road from Copenhagen to Paris: Cancún, Durban, Doha, Warsaw, and Lima, the locations of each of the subsequent annual Conferences of the Parties to the convention. It was in 2011, at COP-17 in Durban, South Africa, where the issue of equity would make the headlines. For developing countries, a fair division of the remaining atmospheric space would require taking into consideration historical responsibility, their right to development and overriding priority of poverty eradication, and the current economic capabilities of developing countries. In shorthand in the negotiations, these ideas are captured in the word "equity." In Durban, then-U.S. Special Envoy for Climate Change Todd Stern was quite blunt about the U.S. position: "If equity's in, we're out."[190] The decision from Durban made no mention of equity and launched the final phase of negotiations toward the Paris Agreement.

In 2014, legal advisors to the U.S. negotiating team, in the introduction to a think tank document on "alternative models for the 2015 climate change agreement," reflected presciently on the possible design of the new regime:

> In many if not most countries, the climate change issue is driven more by national than by international politics, so the agreement needs to allow states to determine the content of their own commitments. This approach represents a concession to political and diplomatic realities, as well as to the limits of international agreements in influencing countries' behavior in an area so vital to their interests.
>
> If the overriding goal of the UN climate change regime is to achieve climate effectiveness, then the Durban Platform outcome must reflect climate justice in a manner that is acceptable to the major emitters, both developed and developing, so as not [to] discourage participation and compliance.[191]

In the end, U.S. fingerprints were all over the 2015 Paris Agreement.[192] Over the several years of negotiations between the Durban and Paris Conferences of the Parties, U.S. climate negotiators had made it clear that they were *only* ready to sign an agreement that could meet the approval of the Republican-led U.S. Congress. That would mean, among other things, an agreement whereby developed and developing countries had similar commitments, leaving no room for considerations of equity and historical responsibility. Early in 2015, legal scholars had begun to float alternative architecture options that could accommodate the requirements of the U.S. political situation, including a new possibility—that the final text would be something that the U.S. president could put in place by his (or her) own executive authority, without passing through the Senate or House of Representatives.[193] The architecture of the Paris Agreement is, consequently, the end result both of many years of negotiations and the demands of the U.S. negotiators for no new legally binding emissions limits or new legally binding financial commitments.

One key driver of the U.S. approach in the Paris Agreement negotiations was the deep hostility among many Republican legislators to any, much less a new, international legal agreement. A long-standing campaign, funded by conservative donors and fossil fuel interests, such as the Koch brothers, has fueled climate denialism in the United States and convinced legislators to obstruct and curtail U.S. participation in global agreements on climate change.[194] It was the Republican-controlled Senate that prevented the United States from ratifying the Kyoto Protocol after the Clinton administration signed onto the agreement in 1997. The Obama administration subsequently made a calculation that passing a new treaty or protocol through Congress would be impossible or cost too much in political terms. Legal experts seeking a way around the requirement for Congressional approval advised to push for an agreement that would not require changing any U.S. law, and therefore would not require ratification by the Senate or simple majority legislative approval as a congressional-executive agreement.[195]

Despite strong support among many other countries for a top-down, legally binding approach, the United States made clear all along the road from Copenhagen to Paris that such an approach was not an option. In the end, the agreement conformed rather closely to parameters that U.S. legal experts had outlined—to avoid legally binding emissions limits and new legally binding financial commitments, with procedural rather than substantive obligations—which would allow President Obama to accept the agreement on the basis of his existing authority, without submission to the U.S. legislative branch for approval. Of course, even this limited set of obligations was too much for U.S. conservatives and, shortly after taking office, President Trump announced the United States would withdraw from the agreement.

At the core of the Paris Agreement—which overall is a legally binding agreement—is a requirement for countries to develop and submit *voluntary* "Nationally Determined Contributions," which are national plans for emission reductions and other climate action. In other words, not only are emission reductions entirely voluntary, their scope and timing are too. What is legally binding is the requirement to submit a plan and report on its implementation (with no obligations to actually implement or achieve that plan, which also would have required a change to U.S. law).[196] Of course, there were likely many other countries beyond the United States—both developed and developing—that welcomed the flexible nature of the Paris obligations.

The agreement to determine emission reductions in a bottom-up, voluntary manner means then that there is no global cap, now or in the foreseeable future. As we saw in chapter 3, a cap is a necessary condition that creates the scarcity required to establish a market in emission reductions and value in traded units or emission allowances. The Paris Agreement contains very vague provisions for trading, with rules yet to be developed. But developed countries in particular would not give up on their demand to include market provisions in the agreement, even though it seemed to make little sense without a global cap. They settled on inclusion of a framework that would allow for regional, national, and sub-national markets to trade among each other. Caps would be set by individual political entities within their own markets—whether the European Union, New Zealand, or California. Basic rules for trading between those markets, such as accounting guidance to avoid double counting, prevent fraud, and ensure the standardization of carbon units traded (or as they are known in the Paris Agreement, internationally transferred mitigation outcomes) would be established at the global level.[197]

While many climate idealists remain supportive of binding cap and carbon allocations, the Paris Agreement has for now stifled that hope. Still, the challenge of two degrees Celsius remains. How to get there from here? There is a recognition among scholars and activists that the possibilities for international climate treaties and their yearly Conferences of the Parties to catalyze significant collective action at this moment are not great.[198] While the need for cooperation and collaboration among states remains critical, the failure to make significant progress, after almost thirty years of such conferences, has motivated cities, states, provinces, and other agencies and entities to embark on their own local initiatives.[199] The ostensible hope that these might add up to more than the sum of the parts rests on one of two logics: (1) that economic and social drivers will motivate many more of these initiatives, eventually leading to social practices that are broadly, if not universally shared; and/or (2) that governments can be shamed by citizens' actions and demands, via a discursive shift, into agreeing to hard commitments to cap and reduce emissions. At this point in time, it is difficult to say whether either of those logics will work. Meanwhile, the annual Conferences of the Parties will continue to meet, raising expectations and probably lowering them, as well.

Hegemons and Regime Autonomy

Let us return, now, to the argument made at the beginning of this chapter: that so-called IERs are more reflections of national practices than truly "international" in any meaningful sense. The case of U.S. behavior in the climate change regime since 1992 suggests some important insights into both the anatomy and autonomy of IERs. Most are of relatively recent provenance, having been established sometime during the past four decades (although a few date back to the beginning of the twentieth century).[200] Prior to 1970 there were a few of what we would now call international environmental agreements—such as, for example, the International Whaling Commission—but these were not then recognized as or called IERs.[201] In any event, such agreements were intended primarily to manage the exploitation of resources, such as whales, seals, and birds, through quotas, rather than to protect or preserve them. Many of the "next generation" of IERs (if we can use that term) were heavily influenced by the presence and participation of the United States. Another excellent example of this influence—followed by an earlier case of American defection—took place in the negotiations over and ratification of the Third UN Convention on the Law of the Sea (UNCLOS), a process that began in the early 1970s and ended in 1981.[202]

UNCLOS was designed to replace earlier conventions and to address the many new issues that had arisen since their negotiation. The United States was deeply involved in these discussions, and the final negotiated text addressed the major American objectives. These included concerns about several countries' unilateral extension of territorial jurisdiction over offshore areas beyond the traditional three-mile limit, military access to certain national waterways, and an institutional mechanism for exploitation of mineral nodules, containing nickel, manganese, and chromium, found on the deep ocean floor.[203] But in 1982, just when American ratification was expected, the Reagan administration backed away from UNCLOS. The United States was opposed, in particular, to the provisions governing nodule mining, seeing them as tantamount to "international socialism" and therefore anathema to the free market Reagan administration. The White House announced that it would neither sign nor ratify UNCLOS, and over the intervening years the United States has not done so, even though the treaty finally became international law in 1994. Between 1982 and 1994, UNCLOS came to be regarded as customary international law, and even the United States observed those provisions deemed relevant to its maritime interests. After a long period of up-and-down interest, a function of the international price of the minerals, there has recently been renewed interest in seabed mining as well as growing concern about its ecological impacts. But no one has, as yet, begun to mine.[204]

The point here is that the U.S. imprint on this IER looms quite large, *even though the United States has not ratified the convention* (162 countries are members). No IER can afford to forego the interests and desires of the United States, whether it ratifies or not, and such American influence extends to virtually all other IERs. This forces us to ask how autonomous IERs are of the states that comprise their membership. To what degree do they reflect some sort of collective international, multilateral, or global interest? Can international organizations, such as the UN Environment Programme, act independently of the wishes and preferences of the major powers involved in their establishment and operation? Can the members of an IER decide to act in opposition to the interests of a hegemonic state, whether member or not? Are IERs (and, by extension, other international regimes) simply superstructure to material bases of power and wealth?

Most realists, such as Stephen Krasner, and a few political economists, such as the late Susan Strange, have argued something close to the last point.[205] Regimes, they say, are mere reflections of the distribution of power in the international system. IERs are able to function, they argue, only with the forbearance of its dominant member(s), and they become dysfunctional if those states refuse to participate.* Liberal analysts are more likely to see the possibilities of autonomous action by IERs and associated treaty secretariats and

---

*The coming years are likely to provide a test of this proposition, as the United States seeks to withdraw from or renegotiate a number of international conventions and treaties.

international organizations. After all, the latter are ordinary bureaucracies, composed of individuals with different degrees of power and authority as well as access to resources. Secretariats and international organizations can play central roles in agenda setting, in managing the negotiating process, and in overseeing implementation.[206]

But the very discourse of regimes, which presents them as forms of liberal international cooperation, tends to obscure relations of power and domination internal to those institutions. What appears to be cooperation is normally a constrained expression of such power and domination. As we saw in chapter 2, power is not always displayed in visible form; sometimes it is embedded in institutions, language, and practice in ways that are rather subtle. Many of the institutions with which we are familiar (although not all of them) serve our ends and interests, even if power is inherent in their structure and functioning. Such institutions have historical and customary legitimacy as well. Most IERs, however, are of relatively recent provenance and have been created within a relatively underdeveloped social context. Neither history nor custom serves to buffer their hard edges or the fact that they exist primarily to serve the interests of the dominant members. What this means is that the autonomy of IERs is limited not only by the overt exercise of power by these states but also by their very structure. At best, relatively minor matters are left to IERs, such as conference oversight, information collection, and, perhaps, harmless and sometimes ineffectual inspections. Major issues that directly affect the interests of the dominant members are kept off the organization's agenda or shifted into the purview of another regime in which power can be exercised with less trouble.*

An example of this can be seen, once again, in the trials and tribulations of the UNFCCC and its subsidiary treaties, the Kyoto Protocol and the Paris Agreement. All three agreements are intended to address anthropogenic emission of all greenhouse gases. Carbon dioxide constitutes a major fraction of these emissions; most carbon dioxide emissions come from the burning of fossil fuels, and a substantial portion of those fossil fuels are combusted in vehicle engines. We might expect, therefore, that an effective agreement would address directly the energy problem and seek to moderate consumption of fossil fuels in the transportation sector, including emissions from air travel and shipping. Yet, none of the three include provisions along these lines, such as a direct carbon tax on fossil fuels. Indeed, such a tax has been adamantly opposed by that group of industrialized countries ("JUSCANZ"—Japan, the United States, Canada, Australia, New Zealand) and oil-producing ones (Organization of the Petroleum Exporting Countries) that would be best able to absorb it. Instead, under the Kyoto Protocol, tradable emission permits were being touted as the solution

---

*This is often called "forum shopping," as states and interested parties seek to bring issues to that regime likely to generate the most favorable response to their case. See, for example, Hannah Murphy and Aynsley Kellow, "Forum Shopping in Global Governance: Understanding States, Business and NGOs in Multiple Arenas," *Global Policy* 4, #2 (May 2013): 139–49.

(and as we note above, the nature and scope of market mechanisms under the Paris Agreement are still under negotiation). As we have seen throughout this book, however, trade in emission permits allows rich countries to purchase rights to pollute from poor ones and to continue their energy-consumptive lifestyles; turning forests into reserve land for mitigation effectively does the same. Reductions in energy use, according to this logic, will come about if the implicit cost of carbon emissions rises and if such cost increases are incorporated into the price of fossil fuels and carbon-intensive products. But in this case, the embedded "tax" accrues not to public coffers, where it might be used to buffer the impacts of higher fuel prices on the poor. Rather, it is the private possessors of the permits who will reap windfall rewards. This is "efficiency" at work.

Are all international regimes subject to the dictates and desires of the hegemon? The WTO might be seen as having a significant degree of autonomy, with a strong, mutually agreed set of rules which bind small and large countries alike.[207] The TRIPS agreement, mentioned earlier, is one of the agreements constituting the rules of the WTO, and incorporates monitoring, sanctioning, and enforcement powers not yet found in other international regimes.[208] These powers can be used even against the United States and its corporations, providing what at least on the face of it looks like a level playing field, where small countries may bring claims and regularly win cases against powerful global hegemons. However, corporations can also use those rules, through their national governments, to further control even domestic policy setting.[209]

IPRs were originally addressed under the provisions of the World International Property Organization (WIPO), successor to several earlier international agreements concluded toward the end of the nineteenth century. WIPO dealt only with the mutual recognition of national copyrights and patents by its member states. Signatories agreed to accept copyrights and patents granted in other member countries as binding. Any country that was not a member of WIPO could, with relative impunity, violate IPRs granted in others.[210] Because there was no way to enforce such rules except within the national courts of member states, violations and piracy were not uncommon. Not only could this result in major economic losses to the IPR holder, the pirated version of the IPR could leak into international markets, leading to additional losses (a fairly common problem these days with trademarked products, recordings, and software).

In an effort to remedy this problem, and under pressure from pharmaceutical and information technology corporations, among others, the United States led a move in the mid-1980s to "forum shift" IPRs from WIPO to a new regime, as one element of the agreements under negotiations leading to the then-new WTO.[211] The WTO regime incorporates many enforcement and sanctioning mechanisms not available to most other IERs. These include the right to bring trade disputes for adjudication before special panels as well as the imposition of damages on a country whose laws or behavior are found to be in violation of WTO rules and regulations.[212] The TRIPS agreement incorporates even more stringent remedies. For example, a TRIPS dispute resolution panel can order national judicial authorities to search for and seize evidence of IPR violations. The panel can then use such discoveries to make judgments about whether violations have occurred and what kinds of sanctions should be imposed. Such authority largely benefits corporations in rich countries—especially the United States and Europe—that own the vast majority of patents already granted and which want to halt the pirating of IPRs in poor ones. These mechanisms can also be used by rich countries against each other, should the occasion arise.[213] But because transnational corporations operate throughout the industrialized world and have an interest in mutual protection of IPRs, they are less likely egregiously to violate each other's patents.

An explicit example of the ways in which TRIPS is biased toward transnational corporations and rich countries was the controversy over the sale of drugs for AIDS in poor countries (especially Africa). TRIPS permits developing countries to produce generic versions of patented pharmaceuticals in situations of dire need for domestic use only (known as compulsory licensing). The export of such generics is expressly forbidden, on the principle that this would undermine international law and deprive companies of legitimate profits.* In the case of the AIDS drugs, transnational pharmaceutical companies not only opposed domestic production in South Africa, they also charged high prices for the drugs they did sell there. As a result, access to the drugs was quite limited. In response, the South African government announced its intention to begin production of generic versions of the drugs (even though it did not possess the capability to do so) and was hauled into court by a group of companies intent on protecting their patents. These companies were less concerned about generics being produced in South Africa than the chance they might be imported from India, where they were being manufactured. In that case, the companies would have demanded their governments to request the convening of a dispute resolution panel, whose decision would likely have favored them, but not without a great deal of bad press.† Eventually, public and other pressures led to a

---

*Countries may import medicines from third parties where patented, brand name drugs are sold at lower prices, a practice known as "parallel imports." WTO, "Obligations and Exceptions," Sept. 2006, accessed September 29, 2017, https://www.wto.org/english/tratop_e/trips_e/factsheet_pharm02_e.htm; Ellen M. F. 't Hoen, "TRIPS, Pharmaceutical Patents and Access to Essential Medicines: Seattle, Doha and Beyond," *Chicago Journal of International Law* 3, #1 (2002): Article 6, accessed August 28, 2017, http://chicagounbound.uchicago.edu/cgi/viewcontent.cgi?article=1171&context=cjil.

†The TRIPS agreement and the Dispute Settlement Understanding under the WTO are both interstate agreements, and only states have standing to bring a case, which they would do on behalf of companies operating under their jurisdiction. Many newer bilateral and multilateral investment treaties have expanded the rights of corporations to give them direct standing to sue countries, under what are know as investor-state dispute settlement (ISDS) provisions, such as chapter 11 of the North American Free Trade Agreement. The case brought by Bechtel against the government of Bolivia is one example of the working of an ISDS provision in the bilateral investment treaty between the Netherlands and Bolivia. See Nichols, "Transnational Capital and the Transformation of the State."

settlement of the suit, and an agreement to lower the cost of the drugs (which still remains too high for most people in South Africa). The companies won their battle, but garnered a great deal of unfavorable publicity in the process.[214]

Developing countries continued to press the public case, pushing for a declaration on TRIPS and public health at the WTO ministerial held several months later in Doha. The declaration affirmed the rights of all countries to use TRIPS' flexibilities with respect to compulsory licensing and parallel imports in ways that protect public health. The text also created a path to resolve an even stickier issue—whether countries without the pharmaceutical manufacturing capacity to produce their own generics could import generic versions of drugs produced under compulsory license. That resolution came in August 2003, with a favorable outcome for developing countries. While these types of wins are few and far between in institutions where global hegemons wield power, the rules-based regime and effective use of media by prominent NGOs, such as Médecins sans Frontières, publicized an issue that poked at the conscience of the world, and a united caucus of developing countries combined to create conditions for a successful outcome, both with the declaration and the resolution of the topic of parallel imports of generic drugs.[215]

The particular structure of both TRIPS and the WTO is nevertheless strongly biased toward the American position on trade matters. As we have seen, this reflects a higher value put on trade and economics than on environmental protection or ethics, as well as the abiding belief that free trade enriches countries and leads them to be more concerned about nature. But the structural bias built into the WTO does not mean that it will always respond to specific American preferences, if those deviate from the norm, as we see with the Doha declaration on TRIPS and public health. In another example, at the Seattle ministerial WTO meetings in 1999, President Clinton proposed that the organization's members consider the inclusion of labor and environmental provisions in its next negotiating round. He was roundly opposed by leaders of developing countries, who saw this as a ploy to establish new trade barriers against the entry of certain of their goods into rich country markets. As a result, the meeting collapsed. The George W. Bush administration showed almost no interest in making free trade more environmentally sensitive, while the Obama administration followed more or less the Clinton path, although it was opposed by other countries in attempting to introduce environmental dispute resolution into the Trans-Pacific Partnership Agreement (from which the Trump administration has withdrawn).[216]

The obstacles to environmental internationalism are further increased by the contradictory effects of globalization and global neoliberalism. As we saw in chapters 2 and 3, even as developing countries are being urged through IERs to protect and restore their environments, they are also being pressured to reduce their domestic expenditures on social and environmental problems in order to provide more attractive settings for foreign investment. Unless new, ecologically modernized industries or enterprises replace older, polluting ones, however, levels of environmental damage will increase.[217] Any additional revenues that accrue from foreign investment are more likely to be directed toward payment of interest on debts than environmental improvement. And the movement of people into the cities, drawn by the promises of jobs, will increase the stress on urban infrastructures and environments. The contradictions will continue. Both people and the environment will suffer.

A SUCCESSFUL CASE OF ENVIRONMENTAL INTERNATIONALISM

The European Union may offer a case of successful environmental internationalism. Its policies and practices are, in part, a product of those of the twenty-eight member states (soon to be twenty-seven after the United Kingdom's exit from the European Union) and the several hundreds of millions of members of their civil societies. These policies and practices are also the creations of the one legislative and two executive branches of the European Union.* These are the European Parliament, whose members are elected from individual states but who tend to organize into party-aligned groups and caucuses; the European Commission, composed of expert directorates (bureaucracies); and the European Council, which is composed of member states with weighted voting.[218] The commission is where regulations are formulated; the council is where they are ratified or rejected. The parliament does not have much say in all of this. The European Union's environmental agenda is generated with a heavier dose of technical expertise than is the norm for purely national politics. This is possible because the commission is a technocratic system of departments, insulated to some degree from both national and EU politics and holding considerable authority to override the wishes of the other EU institutions and member states.[219] At the same time, however, weighted voting in the council, and the political structures and preferences of individual member states and their jurisdictional subunits, affect commission directives and the extent to which they are implemented. All this is further complicated by the fact that NGOs are able to influence legislation through both the commission and the parliament, sometimes in opposition to member states and the council.[220] The European Union's environmental legislation also influences

---

*The European Union has evolved (and is still evolving as evidenced by the impending exit of one of its largest member states) from an economic union—the European Economic Community, with six founding member states—to a political union. However, that political union is still limited—member states accord responsibility to the EU institutions to develop common policy in a number of circumscribed core areas, including the environment and foreign affairs and security policy, but states still retain sovereignty and control over most of what we might consider the work of a national government. Simon Usherwood and John Pinder, *The European Union: A Very Short Introduction*, fourth edition, Oxford: Oxford University Press, 2018; John McCormick, *Understanding the European Union: A Concise Introduction*, seventh edition, London: Palgrave, 2017.

environmental lawmaking in its prospective members (for example, Montenegro, Serbia, and Turkey); a prerequisite for admission is the integration of EU directives into national law. If twenty-eight countries, comprising the world's largest economic bloc, all adopt comparable environmental laws, they could well become de facto international standards. How might this work, given disparate power relations? A good example can be seen in the recycling of nontoxic wastes and further steps being taken toward a "circular economy."

Both individual member states and the European Union have had waste management and recycling laws on the books since the 1970s. As more laws were passed, and they became more rigorous and extensive, they sometimes acted as barriers to trade between member states. For those producers operating in smaller national markets, recycling requirements might provide a competitive advantage, since it is expensive for bottlers in other states, seeking to sell goods throughout the European Union, to retrieve their containers. In 1981, for example, Denmark passed a law requiring that beer and soda be bottled in recyclable containers and mandating that Danish retailers take back all the containers. In order to facilitate recycling and limit costs to retailers, the government restricted the number of acceptable containers to about thirty types. Many of the containers produced in other member states did not meet these requirements, and the government prohibited the sale of such "foreign containers" within Denmark.[221] Bottlers in other member states then complained to the European Commission that the Danish law made it too costly for them to sell their beverages in Denmark.

The commission agreed that the Danish standards constituted a restriction on trade and were, therefore, in violation of European Community law. The Danish government amended the rules, but these changes were deemed unacceptable, too. At that point, the commission filed a complaint with the European Court of Justice, charging, in essence, that the Danish regulations constituted an illegal form of "environmental protectionism." The details of the case are rather complicated and need not concern us here.[222] The court, however, found the Danish law legal, with one exception: there could be no restrictions on container types, so long as *all* containers were recyclable. This meant that non-Danish bottlers had to find ways to collect their bottles and bring them back home. Nevertheless, the law stuck and set the stage for EU law and a growing recognition that waste disposal was a problem throughout the European Union.

In 1991 the German government began to implement a much more rigorous recycling law. This one was designed to reduce wastes destined for both landfills, whose capacity was limited, and incineration, which contributed to air pollution. The German law required producers to take back and recycle all packaging, eventually including even candy and butter wrappers. In order to comply with the regulations, three thousand companies created an enterprise to handle recovery and recycling, the Duales System Deutschland. In return for participating in this system, producers put a "Green Dot" on their products, certifying that the packaging is recyclable.[223] Needless to say, a torrent of complaints followed from companies based outside of Germany. They charged that this law, too, represented an illegal barrier to trade and was too costly to implement. Moreover, the plan's very success created another problem: more wastes were being collected than could be recycled within Germany. In some instances, recycled materials were sorted by consumers, collected by municipalities, and then dumped in landfills.[224] The surplus was exported to other EU countries, where prices for recycled materials and domestic recycling programs both collapsed. That surplus is now also being taken up by Energy-from-Waste incineration plants, serving instead as euphemistically termed "refuse-derived fuel."[225] These are hardly viable solutions to the municipal waste problem.

Recalling the Danish container case, and not wishing to appear anti-environmental, the European Commission was reluctant to challenge the German law. Consequently, in 1994, the commission approved a packaging directive for the entire European Union. It was much less stringent than the German law but, on average, more so than existed in most member states. Several countries, including Germany, opposed the EU directive, claiming it to be too weak, but to no avail.[226] The directive stood. In 1994, Germany trumped the commission and passed the "Closed Substance Cycle Waste Management Act."[227] This law encouraged producers to use fewer materials, create less waste, and avoid hazardous materials in the manufacture of goods. (In the chemical industry, for example, petroleum in plastics has been replaced in many products by sugar and starch, so that the resulting products are biodegradable.) Although the German law and the general trend toward closed-loop recycling was strongly opposed by manufacturers, European institutions sought to expand the process further. In 2000, for example, the European Union passed a directive placing complete responsibility for the environmental impacts of junked automobiles onto automakers, who are expected to ensure that virtually all auto parts are recyclable and appropriately marked as to content. Consumers will pay a portion of the recycling costs, but the manufacturers must put in place a system for accepting old cars for recycling.[228] Additional EU directives have been issued on waste (establishing a waste framework), batteries, and electrical and electronic waste.[229]

In 2015, the European Commission published an action plan on a transition to a circular economy, entitled "Closing the Loop," to update and unify the various directives on waste, recycling, and reuse.* The action plan sets 2030 targets for waste reduction, a common EU target for recycling of municipal (65 percent) and packaging (75 percent) waste, and a binding target to

---

* "The transition to a more circular economy, where the value of products, materials and resources is maintained in the economy for as long as possible, and the generation of waste minimised, is an essential contribution to the EU's efforts to develop a sustainable, low carbon, resource efficient and competitive economy. Such transition is the opportunity to transform our economy and generate new and sustainable competitive advantages for Europe." European Commission, *Communication from the Commission to the European Parliament, the Council, the European Economic and Social Committee and the Committee of the Regions, Closing the Loop—An EU Action Plan for the Circular Economy*, COM(2015) 614 final, Brussels, 2015, accessed September 28, 2017, http://eur-lex.europa.eu/legal-content/EN/TXT/PDF/?uri=CELEX:52015DC0614&from=en.

limit the amount of waste that ends up in landfills to 10 percent of the municipal waste stream.[230] However, in negotiations to finalize the action plan, many
states have pushed back against these targets.

What this example of recycling and other waste management steps toward
a circular economy illustrates is a tension between the environmental policies
and practices of member states and the "international" European Union, as
well as the ways in which there is, nonetheless, a process of exchange between
the two levels. Individual states have their own preferences, often determined
through a history of conflict and compromise among environmentalists, state,
and capital (a phenomenon common in the United States, as well). But different
standards can act as barriers to trade between members, and one of the charges
to the European Union is to facilitate such trade. Eliminating these barriers
requires some degree of regulatory harmonization among member states. Generally speaking, one might expect a "race to the bottom," as the weakest set
of rules becomes the group standard. This has not happened in the European
Union, for several reasons. First, the environment is one of the issue areas in
which the European Commission has been granted a considerable degree of
authority and autonomy by member states, and its Environmental Directorate
is, therefore, able to formulate directives that may have more basis in "science"
than politics, thereby overriding national standards and preferences.

Second, there is considerable pro-environment sentiment among the citizens of many of the larger EU members. NGOs and movements lobby at the
EU level in Brussels to put pressure on member governments, as is the case with
negotiations on the circular economy action plan.[231] The political organization
of the European Union opens up numerous opportunities for lobbying and
logrolling, for coalition building and blocking, and these are used frequently.
Finally, within the European Union, market access is of central importance to
producers, and it is in this context that garbage becomes something like a commodity. It is not, however, the garbage itself that is at issue but the packaging
and goods that must, eventually, be recycled. The cost of disposal is internalized into the cost of the product. This sets up a competitive dynamic in which
manufacturers strive to minimize that additional cost in order to make their
products more attractive to consumers. Here, the market is being used to environmental ends, but it is the interplay of politics within and among member
states that creates and structures that market.

But are EU environmental "regimes" really an example of environmental internationalism? Or are they more akin to an emergent environmental
"nationalism," along the lines we find in the United States and other large
countries? If the former is the case, what we are seeing is a type of regulatory
harmonization, in which individual regime members are being pressured to
adhere to higher standards than they might prefer, in the interest of inter-Union
integration. Because the European Union's institutions have been granted certain forms of conditional power over the individual members, such pressure
can protect the environment. In the longer run, this form of international

authority could work to create truly autonomous IERs and provide a model for broader environmental internationalism. If the latter is the case, however, the European Union will, gradually, find that it has certain Union-wide interests that conflict with other large and powerful states. The stories presented in this chapter are not conclusive. The jury remains out.

ARE THERE INTERNATIONAL ENVIRONMENTAL POLITICS?

In this chapter, as throughout this book, we have seen that "global environmental politics" includes much more than what is offered by the standard formulas of international relations. Moreover, the social relations of environmentalism and environmental politics do not divide themselves neatly into standardized categories or levels of analysis. In the long run, and depending on the degree and severity of environmental phenomena themselves, states could cede some greater portion of their sovereignty and endow IERs with a good deal more power and authority than they possess today, and environmental internationalism might flourish. Or many might rely heavily on local and regional efforts to address environmental problems, in the hope that this could substitute for national efforts.

Still, it would probably take a crisis of global proportions to generate the requisite international change. A sudden rise in sea levels from the collapse of the Ross Ice Shelf in Antarctica, which would quickly raise sea levels by fifteen feet, might do it.* Alternatively, a massive die-off of people and other living things as a result of rising temperatures and climatic changing conditions could compel the states of the world to agree to what would, in effect, be something akin to a world government. Inasmuch as the timing of such catastrophes is difficult to predict in advance, even if we are certain they will happen at some point in the future, it is not so clear how states might be pressured to become more environmentally sensitive and committed to true environmental internationalism. What, then, are we to do? This is the question we take up in the final chapter of this book.

FOR FURTHER READING

Bartley, Tim, et al. *Looking Behind the Label: Global Industries and the Conscientious Consumer*. Bloomington: Indiana University Press, 2015.

---

*In July 2017, a twenty-two-hundred-square-mile section (12 percent) of the Larsen C Ice Shelf calved off into the sea, following the collapse of Larsen A in 1995 and Larsen B in 2002, suggesting that perhaps such cataclysms are not out of the realm of possibility. See "Miles of Ice Collapsing into the Sea," *New York Times*, May 18, 2017, accessed August 17, 2017, https://www.nytimes.com/interactive/2017/05/18/climate/antarctica-ice-melt-climate-change.html?_r=0; and CNN, "Massive Iceberg Breaks Away from Antarctica," cnn.com, accessed August 17, 2017, http://www.cnn.com/2017/07/12/world/larsen-c-antarctica/index.html.

Chasek, Pamela S., David L. Downie, and Janet Welsh Brown. *Global Environmental Politics*, seventh edition. Boulder: Westview Press, 2016.

Ciplet, David, J. Timmons Roberts, and Mizan R. Khan. *Power in a Warming World: The New Global Politics of Climate Change and the Remaking of Environmental Inequality*. Cambridge: MIT Press, 2015.

Clapp, Jennifer. *Toxic Exports: The Transfer of Hazardous Wastes from Rich to Poor Countries*. Ithaca: Cornel University Press, 2001.

Dauvergne, Peter, and Jane Lister. *Timber*. Cambridge, UK: Polity, 2011.

Fowler, Cary, and Pat Mooney. *Shattering: Food, Politics, and the Loss of Genetic Diversity*. Tucson: University of Arizona Press, 1990.

Gaarde, Ingeborg. *Peasants Negotiating a Global Policy Space: La Vía Campesina in the Committee on World Food Security*. Oxon, UK: Routledge, 2017.

Guha, Ramachandra. *The Unquiet Woods: Ecological Change and Peasant Resistance in the Himalaya*. Berkeley: University of California Press, 2000.

Green, Jessica F. *Rethinking Private Authority*. Princeton: Princeton University Press, 2014.

Hoffman, Matthew J. *Climate Governance at the Crossroads: Experimenting with a Global Response after Kyoto*. Oxford: Oxford University Press, 2011.

Kjellén, Bo. *A New Diplomacy for Sustainable Development: The Challenge of Global Change*. Oxon, UK: Routledge, 2008.

Moreno, Camila, Daniel Speich Chassé, and Lili Fuhr. *Carbon Metrics: Global Abstractions and Ecological Epistemicide*. Berlin: Heinrich Böll Stiftung, Ecology Volume 42, 2015.

Nicholson, Simon, and Sikina Jinnah, eds. *New Earth Politics: Essays from the Anthropocene*. Cambridge: MIT Press, 2016.

Petras, James, and Henry Veltmeyer. *Extractive Imperialism in the Americas: Capitalism's New Frontier*. Leiden: Brill, 2014.

Rich, Bruce. *Mortgaging the Earth: The World Bank, Environmental Impoverishment, and the Crisis of Development*. Boston: Beacon Press, 1994.

Seto, Karen, and Anette Reenberg (eds.). *Rethinking Global Land Use in an Urban Era*. Cambridge: MIT Press, 2014.

6

Global Environmental Politics, Society, and You

★ ★ ★

INTO THE STREETS!

On March 26, 2000, a rather peculiar demonstration took place in Boston, Massachusetts. As seven thousand scientists and others attended "Bio2000," a five-day conference to discuss the latest advances in biotechnology, fifteen hundred people rallied outside the meeting venue in Copley Square to assert their opposition to the production of genetically modified organisms, especially in foods. Police and authorities feared another "Battle of Seattle,"[1] but they didn't get it. The protest remained nonviolent. A reporter for the *New York Times* wrote:

> [T]he demonstration was not wild but wildly creative. Street theater abounded; several protesters ran around dressed as white-coated fanatical scientists wielding giant syringes and several others fluttered about as butterflies to symbolize the monarchs that a study has shown were harmed by genetically altered corn. A man in a Frankenstein costume pushed a shopping cart bearing genetically engineered "Frankenflakes," and another wore a papier-mâché killer tomato on his head.[2]

In anticipation of the protest, some two thousand scientists signed a "Declaration in Support of Agricultural Biotechnology." According to the organizer of the declaration, Dr. C. S. Prakash of Tuskegee University in Alabama, "biotech crops allow farmers to grow more food on less land with less synthetic pesticides and herbicides."[3] Food fights, no less! In the years since, controversy and conflict over genetically modified organisms (GMOs) has waxed and waned, but it has not gone away.[4] What are the stakes?

For biotech proponents, it is feeding the hungry and malnourished; for biotech opponents, it is nothing less than the future of life on Earth. According to its advocates, agricultural biotechnology promises the twenty-first-century

255

equivalent of the Green Revolution, which will provide bounteous food supplies to the world and prevent global famine; its detractors worry that engineered genes, let loose into the food supply and environment, might wreak havoc on the health of both people and nature and contaminate both beyond repair. Meanwhile, many Americans seem largely indifferent to the presence of GMOs in their food, whereas in Europe, distaste for genetically engineered goods has become so great that cultivation and sale of such foods has, for decades, been stopped almost completely.[5] Observers were puzzled. What, exactly, did these demonstrators want, asked media pundits. Didn't they know that there are poor and hungry people out there in the world? How could these protestors say they were for justice, especially when so many were white and middle class and not hungry? Better to go back to their studies and jobs and let the experts and economists take care of business. Anyway, these problems were "global" and subject to the operation of markets, and not amenable to action on the ground.

In April 2017, no less than two climate-related protests took place over two consecutive weeks in opposition to the new Trump administration's and the Republican-dominated Congress's continuing denial and downplaying climate change, including one named a "March for Science"![6] While the atmosphere (no pun intended) at both Bio2000 and the latter was much the same, with the colorful diversity seen at the Boston march, the Marches for Science were much more serious and were replicated all over the world. Paradoxically, and notwithstanding the myriad of scientific, political, and regulatory initiatives launched in the United States (and around the world) since the early 1990s, the so-called climate deniers continued to maintain their steadfast position that climate change had not occurred, was not presently happening, and would not be a problem in the future. But why, in the face of seemingly incontrovertible evidence (and experience), was it necessary to demonstrate in support of science and politics? If science is not "truth," then what is it? And what is the role of such scientific truth in global environmental politics? Indeed, does science have a role?[7]

The paradox, or contradiction, arises because, while politics is a realm of "truth claims"—whose veracity is often open to questioning and challenge—these are no longer the *transcendent* claims associated with religious or totalitarian belief systems that seem to impose an absolute constraint on or limit to action. To a growing degree, science has acquired the patina of such transcendent truth claims in order to suppress debate and politics. This is not to argue that scientific claims are, somehow, false or fake,* but, rather, to emphasize that a politics which relies on science to determine social practice is no politics at all.[8]

---

*The work of Bruno Latour and scholars of science studies take on questions about the social construction of scientific facts. See Bruno Latour, *Pandora's Hope*; Bruno Latour, *Science in Action—How to Follow Scientists and Engineers through Society*, Cambridge, MA: Harvard University Press, 1987.

The demonstrations described above have been, and remain, common tactics of a broad range of environmental and other movements around the world. With the rise of the global justice (or anti-globalization) movement we discussed in chapter 4, such demonstrations became a staple of front-page news, activist hopes, internet strategizing, and editorial page dissing.[9] By the same token, so many similar demonstrations, actions, and occupations have taken place since 2000, that they are no longer exceptional (indeed, the mobilization by Donald Trump of large conservative crowds seemed to impress the media much more). Many observers gave these progressive demonstrations little credence, for their participants did not seem to have either a "strategy" or a "program" (a valid, if vacuous, critique).

At the same time that massive demonstrations attract attention, few, if any, media report on the much more numerous local movements and groups, such as those opposing the construction of big dams, working to restore local watersheds, protecting community gardens, opposing local fracking projects, or fighting to preserve critical habitats. Yet it is these latter actions and activists, rarely remunerated for their efforts, who are the bedrock of global environmentalism and who do have both strategies and programs. There is nothing particularly earth-shaking about projects to clean and protect creeks, streams, and rivers, or take on the issues and problems listed above, but it is such efforts that, in the long run, foretell growing political awareness about what we must do to protect the environment.

In this final chapter, therefore, we turn to two questions. First, "Where do you, dear reader, fit into all of this?" As consumers of goods and producers of pollution and waste, those of us in the Global North, in particular, are directly implicated in the damage being done to the earth's environment. We can make careful choices about what to buy and how much of it to use, with the hope that we might influence producers, encourage ecological modernization and corporate social responsibility, and even reduce the cradle-to-grave impacts of production and consumption. But even if others make similar selections and decisions, the choices you or I make, on our own, can never add up to more than the aggregated preferences of lone individuals (there's that collective action problem again!). Second, as asked in chapter 5, "What are *we* to do?"[10]

Antonio Gramsci, the Italian Communist who died in prison under Mussolini, argued,

> It will be said that what each individual can change is very little, considering his strength. This is true up to a point. But when the individual can associate himself with all the other individuals who want the same changes, and if the changes wanted are rational, the individual can be multiplied an impressive number of times and can obtain a change which is far more radical than at first sight ever seemed possible.[11]

How can we act, mobilize and motivate action, on the myriad of issues confronting us, to create "impressive numbers"? Can we find ways to make governments and states legislate effective law both domestically and internationally?

Or, if the burden is on us, how can we organize effectively to protect local and global environments?

The thrust of this book is that environmental problems are, first and foremost, political ones and are, therefore, about power. They have been caused by the often unjust exercise of various forms of power and, if they are to be dealt with, we will have to do so through the exercise of other forms of power. To exercise such power, we must act collectively, in concert with others. Moreover, *our actions must be political and have a political purpose.* We need to understand that politics is substantially absent from our systems of government, and we must rejuvenate or restore those missing elements through a new ethics and praxis.[12] Finally, we must undertake these actions with all kinds of people and through all kinds of institutions, ranging from the household to the global. The environment will not survive unless we do so.

In this chapter, we examine the form and content of such an ethic and what it implies in terms of politics and praxis. We begin with a discussion of the impoverishment of politics in a world dominated by economistic approaches to social problems such as environmental quality and justice and human health and hunger. The "capture" of environmental discourse and practices by neoclassical economics—in part, through some versions of environmental economics—has transformed politics from a matter of ethics to a problem in economics. Consequently, we no longer inquire into the deployment of power for environmental ends or the politics behind that power; rather, we ask "How much will it cost?" If the ratio of costs to benefits looks good, maybe a project will be done. If we find the "correct" price for a good, maybe we can achieve our desires without making any greater efforts than shopping as usual.

But what, exactly, is this "good" we accomplish through such practices? Consider the following thought experiment: what is the value of the last redwood on Earth? Its notional market value as a collector's item would be immense, although, paradoxically, it probably would have little, if any, ecological value.* Still, some people might be willing to pay exorbitant sums for a piece of the last redwood on Earth. For many, it would not be a "good" action to cut down the tree, but a few might decide it would be better to have a relic than to let the tree die a natural death and decompose into its constituent elements. According to the principles of the market, what people would be willing to pay would be the "correct" price for the last redwood, and it might well be of economic benefit to log it, especially if sold as one-inch mounted squares.[†] Surely, this would not be either ethical or our intention.

That is why an environmental (or ecological) ethic is so critical to global environmental politics, to which we turn in the second section of the chapter.

---

* "Probably" because redwood seedlings often sprout from the roots of old trees. The progeny of the last redwood would be genetically identical to each other.

† In practice, of course, the last redwood on Earth would probably be protected by legislation with a strong fence and armed guards, or perhaps reside in a corporate-sponsored theme park with the tree named after a generous donor.

An ethic is generally understood as a set of principles of right conduct. Defining "right conduct" is no easy task (and science is of very little help here). We could define it in terms of survival—certainly a critical issue—or utilitarianism—whatever makes the greatest number of people happy or secure. We could decide that what is most important to right conduct is recognition, respect, and empathy for people and for nature.[13] Or we could pursue what Tracy Osborne calls "public political ecology" organized around some notion of stewardship.[14] Whatever our conception, our ethic will be strongly contested and a focus of considerable conflict. It will be realizable only through politics and praxis.

So "what *are* we to do?" Demonstrations, such as those that took place in Boston, or the even larger demonstrations of Occupy, the Women's March, the Climate Marches, and the encampment by the Native American opponents of the Dakota Access Pipeline, are only a small part of a necessary strategy. They are essential to wider awareness of a broad range of environmental and social problems and deserve to be treated as a legitimate form of political action. But the question of strategy and praxis is one that needs to be addressed, albeit not quite in the way it is generally understood. The problem here is that we generally focus on somehow affecting or changing environmental policy, on the assumption that public opinion as well as scientific and other forms of knowledge are tools of persuasion and change. But, as Deborah Stone has pointed out, "public policy" is not about politics, per se, but about accomplishing particular ends in a rational (read "market-oriented") manner. That is, for policymakers, social goals are already known, and all that matters is how to achieve them most efficiently and with least resistance.[15] That is not politics. Stone looks to Plutarch to make the distinction: "They are wrong who think that politics is like an ocean voyage or a military campaign, something to be done with some end in view, or something which levels off as soon as that end is reached. It is not a public chore, to be got over with; it is a way of life."[16] *Politics*, as we shall see, is about engagement with the principles and conditions of life and about the explicit processes whereby people make decisions collectively and act on them. Strategy and praxis must, therefore, be related to politics in this elemental sense.

The propositions offered in the final section of this chapter as alternatives to politics as usual are, as much out of prudence as a recognition of the difficulties in strategizing and praxis, more in the realm of possible actions than certain outcomes. Not all strategies are "global" ones with grand objectives; indeed, grand strategies as often as not produce unintended consequences and failure.* It is less important to think big than to act, to decide with others of

---

*Michel Foucault's concept of "genealogy" may be useful here: "The forces operating in history do not obey destiny or regulative mechanisms, but the destiny of the battle. They do not manifest the successive forms of a primordial intention and their attention is not that of a conclusion, for they always appear through the singular randomness of events. . . . The world such as we are acquainted with it is not this ultimately simple

like mind (or those who can be persuaded) on praxis, to become political as a way of life rather than to treat politics as a chore and bore. Such praxis can be as simple as working toward a greener community or as complex as coordinating watershed groups around the world. The important step is to recognize that power can be a productive force and to act on that insight.

MARKETS ARE NOT POLITICS

Throughout this book, especially in chapters 3 and 4, we have noted the trend toward growing reliance on markets to address political problems and injustices. While various emergent nationalist and right-wing movements around the world are mobilizing, in part, against the domestic impacts of globalization, international institutions and policymakers continue to rely on market-based approaches to social and environmental problems. This reliance on economic principles and practices to "protect" the global environment raises troubling questions, as we have seen throughout this book. Most problematic, perhaps, is the eclipse of politics.

Recall Harold Lasswell's classic definition of politics, cited in chapter 5: "who gets what, when, and how."[17] His conception of politics has been broadly accepted in the discipline of political science, but inasmuch as it addresses only the distribution of resources and its justification, it would seem that the market would be the preferred mechanism for implementing Lasswell's formula. Because environmental protection is seen largely in utilitarian terms, why not apply that same formula to ostensibly noneconomic matters, such as environmental protection, labor rights, and social welfare? After all, is not global climate change simply an issue of how to distribute the costs and benefits of a limited supply of beneficial climate? What better means, then, to reduce damage to the environment than through the market? Obviously, we are strong critics of market-based approaches to political issues.

Is Everything for Sale?

As we saw in chapters 3 and 5, the rush to commodify nature in order to deal with human insults to the environment has two important implications: first, it reinforces the very status quo that is causing the problem, and second, it ranks efficiency and returns on investment above fairness, justice, and the well-being of both humans and nature. Reliance on existing institutions and

---

configuration where events are reduced to accentuate their essential traits, their final meaning, or their initial and final value. On the contrary, it is a profusion of entangled events," p. 361. For further discussion of the relationship between genealogy and strategy, see: Ronnie D. Lipschutz, "The Assemblage of American Imperium: Hybrid Power, World War and World Government(ality) in the Twenty-first Century," in *Legacies of Empire—Imperial Roots of the Contemporary Global Order*, edited by Sandra Halperin and Ronen Palan, pp. 221–42, cite on pp. 223–25, Cambridge: Cambridge University Press, 2015.

practices to solve pressing social problems tends to reify and reinforce the practices of those institutions as being able to provide "the answers," even if those approaches are at the core of the problems being addressed and serve only to exacerbate those problems. New institutions are often founded by individuals and groups who have left older ones but, for reasons of money and influence, are rarely willing to offer radical innovations on the principles and practices of old ones; at times, these initiatives even close down spaces of opportunity offered by those that already exist.*

"More efficient use" of financial and other resources is premised on the goal of "more bang for the buck," that is, saving money by maximizing reductions in environmental damage rather than protecting those who experience the greatest exposure and suffer the most. This might make sense in a world of finite resources, but all it does is postpone expenditures to an imagined, and often utopian, future. Efficiency has no effect at all on the power relations or institutions responsible for the insults in the first place. To put this another way, the substitution of market-based mechanisms for politics serves a larger goal: maintenance of status quo and systemic stability, even in the face of potential catastrophe (this can be seen in the response to the "Great Recession" in 2008 and the subsequent chipping away at regulations intended to prevent the activities that contributed to that event). Although "free" markets are inherently risky for investors, producers, and even consumers, reliance on them for adaptation to challenging conditions serves to reinforce the very structural framework that is the source of the problem (or externality) in the first place. Markets alone cannot repair the flaws generated by markets.

Moreover, in applying market mechanisms to environmental problems and preservation of nature, we come face-to-face with what can only be described as sleight of hand. Under more traditional "command-and-control" regulatory schemes, offenders were fined, taxed, or required to retrofit their industry, the costs of which were, perhaps, passed on to their customers. In return, a benefit accrued to the general public and the commons in the area affected by the pollutant.† To be sure, the improvement might have been small and diffuse, but the benefit was understood to be greater for those who were unduly burdened because of their proximity to the source (that is, those living near polluting industries experienced greater relative improvement than those living

---

*This is not a call for the "disruption" championed by tech gurus in Silicon Valley and elsewhere, which draws on Joseph Schumpeter's work on "creative destruction" and the periodic renewal of capitalism. That renewal does not, and cannot, address equity or justice. See Jill Lapore, "The Disruption Machine—What the Gospel of Innovation Gets Wrong," *The New Yorker*, June 23, 2014, accessed September 6, 2017, http://www.newyorker.com/magazine/2014/06/23/the-disruption-machine.

†But not always: see Tony Barboza and Jon Schleuss, "L.A. Keeps Building Near Freeways, Even Though Living There Makes People Sick," *Los Angeles Times*, March 2, 2017, accessed September 6, 2017, http://www.latimes.com/projects/la-me-freeway-pollution/.

far away). That seemed only fair. The move to market-based mechanisms, however, dispenses with this logic. It turns offending acts into commodities from which polluters may profit (or, at least, though which they can reduce their losses). To be sure, in the case of markets for air pollution, for example, polluters can choose whether to pay for equipment to reduce the polluting activities or for permits to continue them. But a permit-based arrangement may well diffuse or privatize the benefits in a way that those most affected see no local improvement in air quality at all (the same applies to conservation off-sets, discussed in chapter 2). This short-circuiting of a vaguely political process not only turns citizens into consumers of environmental quality, it also cuts most people entirely out of the decision-making loop. What little power in the political arena they may have possessed is stripped from them as decisions are moved to the market where the rich get richer and the poor get it in the neck.

This outcome is seen most clearly in regard to environmental justice in urban areas, particularly under the so-called bubble system of air quality. The bubble system treats an entire air basin, or "airshed," as a single physical unit (just as, in effect, the European Emissions Trading System treats the entire atmosphere above the twenty-eight member states as a "bubble"). The logic behind this type of chemistry is that, although much air pollution originates from point sources, such as power plants and factories, and may dispropor-tionately affect those near to or downwind from the emissions source, all pol-lutants tend to mix and mingle across airsheds. (There is also considerable pollution from diffuse sources, such as cars and furnaces, but these should be controlled at the factory, so to speak, through development and deployment of less-polluting devices.) Atmospheric mixing is especially evident in places such as the Los Angeles basin and Mexico City, where circulation is restricted by the surrounding mountain ranges and temperate weather.

By treating the airshed as a single unit, regulators assume that the pre-cise location of emission reductions, or the benefits of reduced emissions near the point source, do not matter. Instead, it is the aggregate reduction across the airshed that counts.* Hence, so long as the total volume of emissions is reduced, it does not matter which facility does the reduction or where a facility is located. Those plants able to reduce emissions to less than some stipulated level at an acceptable cost can sell their "unused" emission rights to those for whom the cost of reductions would be greater than the benefits. The end result, it is argued, is that everyone in the airshed enjoys lower pollution levels than would be the case under a "command-and-control" regulatory system.

A little reflection on this "solution," however, highlights serious ethical difficulties and consequent injustices. Residents of Beverly Hills or the Santa Monica Mountains, both of which are within the Los Angeles airshed, already

---

*This process replicates the distributive consequences of economic growth in liberal economies: it is aggregate growth that counts, and not the resulting distribution of increased income.

enjoy cleaner air, especially since neither is an industrial zone and both are relatively distant from most highways. Living far from and high above the offending emission sources, they can avoid some of the worst effects (except when they have to drive to Hollywood or the University of California at Los Angeles). What they receive as a result of the bubble system is cleaner air, better health, and a more attractive view. Residents of South Central, Southgate, or the City of Industry, having both lower incomes and elevations, and more industry, live in closer proximity to emission sources and freeways and are, hence, exposed to higher concentrations of pollutants, poorer health, with no view to compensate. Furthermore, those living in these areas have very little political voice with respect to any regulatory processes, limited time (if any) to participate in whatever public hearings and comment periods might be offered and no wealth that can be used to move away from the pollution. Even under the bubble system, emission levels will remain highest near the pollution sources, and those nearby will suffer. But they might be able to see the mountains on a clear day.*

Similarly, a climate offset system creates its own sort of global bubble. Consider the International Civil Aviation Organization's decision, described in chapter 3, to "cap" emissions from airplanes, assisted by a program of significant carbon offsets for steadily growing emissions. As the cost of these offsets rise and are added to ticket prices, individuals wealthy enough to afford air travel will continue to fly while, at some point, the cost will become too great for others (although we are a long way from that, so far). Somewhere else, very probably in a developing country, someone will be encouraged to vacate their forest home, or give up access to land, in order to preserve the carbon it contains, or they will be moved off their "marginal" plots to make way for tree plantations that are claimed to sequester offset carbon.[18] In the bubble of the global atmosphere, geoengineering technologies, whether solar radiation management or carbon dioxide removal, will offload risks onto the most vulnerable while enabling wealthy consumers (that is, those of us in the Global North) to continue to live high-carbon lifestyles.

There is a rather tricky philosophical aspect to the bubble system that is also worth recognizing. If, as suggested above, we regard the local atmosphere around a pollution source as an open-access commons, similar to that written about by Garrett Hardin (even if he was incorrect), a "tragedy" will arise not because of crowding by those living near the facility—their contribution is minimal—but because the plant's owners (or airlines and their customers, or

---

*Indeed, this is not to argue that the bubble system does not work: Los Angeles air and visibility are generally much improved over three or four decades ago; see Tony Barboza, "Will Trump Erect a Roadblock to Southern California's Decades-long Fight Against Smog?" *Los Angeles Times*, March 17, 2017, accessed September 6, 2017, http://www.latimes.com/local/lanow/la-me-ln-vehicles-smog-20170316-story.html; *Los Angeles Times*, "Photos: Los Angeles under Cover—Smog through the Years," April 28, 2017, accessed September 6, 2017, http://www.latimes.com/local/la-me-air-pollution -0428-pictures-photogallery.html.

other enterprises purchasing permits) have chosen to treat the commons as their own private property (this is rather like one of Hardin's villagers owning a hundred cows but blaming those who each own only one for ruining the commons).[19] In effect, bubbles and emission permits formalize such "resource grabs" by making legally private that which was previously treated as informally private. Offsets similarly facilitate the private use of the atmosphere but in a less direct albeit legal way.

So what? we might ask. After all, is not emissions reduction a form of investment in the planet, and are not corporations entitled to receive returns on their investments? If they invest in the most efficient means of pollution reduction, they should see the greatest returns, and that only seems fair. It is not. For centuries, polluters have been pouring toxic and unhealthy wastes into what are, after all, public commons, resources owned by no one but vital to everyone (a true "tragedy of the commons"). Such pollution represents the private appropriation of a resource—enclosure of the commons, if you will—without compensation to people (or other species) who have relied on the atmosphere and biosphere since life began.* This ethical lapse is hardly something that can be addressed through markets.†

Indeed, the question of whether it is right or ethical to pollute the environment at all is usually dismissed out of hand.[20] Some degree of damage must be accepted, we are told, if polluters are not to go out of business and suffer economic harm that will also impose costs on others and lead to unemployment. Because all human activities affect the environment, it is assumed that acceptance of such environmental impacts is merely a question of utility that, perhaps, can be compensated through what economists call "side-payments." But this is at heart an ethical matter and, therefore, political, for allowing pollution denies recognition to those who suffer from the impacts. More to the point, selling rights to pollute also treats the environment as a means to human happiness and well-being for some, rather than an end deserving consideration in itself. We shall return to this point later in this chapter.

The Deal We Can't Refuse?

It is in this context that the capture of environmental discourse by economics becomes so troubling: when everything must be given a monetary value in order

---

*In a local context—the Los Angeles Basin—it is, perhaps, true that the public realizes benefits from the products that come along with the pollution. Whether this is adequate compensation for the resulting health and environmental impacts is not at all clear. In the global bubble, whether the pollution exported is carbon or the toxic wastes generated by electronics production or disposal, this argument does not hold. See the discussion in the next section of this chapter.

†We might notice the similarity here to Marx's arguments about capital and labor: the former appropriates the value of the latter and becomes the owner of work by an individual's body through payment of a wage; in our example, capital appropriates the health and well-being of exposed individuals and transfers those costs to the public while benefiting from privatization of living space.

to be considered worthy of recognition, even those things that are "priceless" have no value at all. This will to monetize is a particular commodity fetish of liberalism.[21] It contributes, moreover, to an obsession with "value," and pays little, if any, attention to ethics or politics. Throughout much of the world, and especially in the United States, there is incessant chatter about all kinds of values: family values, national values, democratic values, moral values, environmental values, good value, comparative values, relative values, efficiency as a value, and so on. This discourse of values is hardly an accident or coincidence, although it would be a functionalist error to argue that it is wholly a result of intentional decision making with respect to marketization.

No, this economization of everyday life arises from certain foundational principles embedded in liberalism as a philosophy and capitalism as a practice, as they have developed over the centuries, and diffused throughout the world over the five hundred years of globalization. Without romanticizing them, in so-called traditional societies, people undoubtedly have what we would consider both values and preferences, but these are strongly moderated or even suppressed by hierarchies of power, bonds of obligation, and relatively fixed social roles.[22] If we were medieval peasants, we might think it important to receive food, family, and shelter from the local lord, but those hardly constitute "values" as we understand the term today. In any event, as peasants, we would not have much discretion over what we most "valued."

Liberalism, by contrast, relies on the myth, if not the fact, of social mobility, voter choice and consumer freedom. We need not be bound by the circumstances of our birth, our status, or our income; opportunities to improve our lives abound, and making the right choice can change our situations radically. And this is key: *liberalism is about the freedom to choose*, as Milton Friedman argues, but it is *only* that freedom and related ones that count.[23] That kind of freedom is not about the freedom or equality or justice or equity of consequences and outcomes, the circumstances under which choices are offered and choices are made or emancipation.[24] In order to make a choice within a liberal-capitalist context, however, we must have some way to compare alternatives, and that is where value comes in. If we are socialized into a particularly rigid set of moral perspectives, such as damnation following from a failure to be devout, the freedom to choose hardly matters. And if we are socialized into choosing the less expensive of two items—the one that gives "better value"— the freedom to choose is only notional. But what happens if we must choose between two quite different possibilities, say, between going to medical school and running off to the circus? Or protecting a stand of old-growth redwoods as opposed to turning them into decking for a house in the suburbs? Which provides greater value? Which do we value more? Which should society choose?

As we saw earlier in this book, the "father" of classical liberal political economy, Adam Smith, believed that individual appetite would be constrained by "moral sentiments," that is, religious beliefs and strictures.[25] He hardly reckoned with a society in which almost every choice made by an individual would be measured by markets and prices alone. He thought that there were choices that could not be measured in the market, and that a common morality, based in Christianity, would see to such noneconomic choices while restraining economic appetites. For a host of reasons, both cognitive and materialist, idealist and realist, capitalist industrial societies are no longer constrained by such moralities, notwithstanding the rise of conservative religions and nationalist ideologies around the world. And, so, choices generally come down to the bottom line: "Which deal can I not refuse?" This monetization of everything has been extended to consideration of the environment and nature no less than other things. Indeed, the application of market-based values includes even those matters that, in some instances, involve life or death (what is a human life worth?).[26]

Thus, according to this logic, even if global warming poses an unprecedented threat to human civilization and the environment, there is no monetary or financial value to be realized in concerted action at this time, and the problem can be addressed only by commodifying the very behaviors and practices that are the source of the threat. Rather than restricting or even banning offending practices, we look for ways to make a buck on the deal. In a somewhat Swiftian fashion,* this is rather as if criminals had to bid for scarce death contract permits in order to commit first-degree murder, under the theory that very high prices would, through high elasticity of demand, reduce the incidence of homicide. Certainly, it could be argued that such an approach would be much more efficient and financially sound than present arrangements: death row could become self-financing, the state could generate funds to pay for more policing, there would be fewer appeals and fewer convicts awaiting execution. Much would be saved in insurance and other payouts on the decedents, too. But it is hard to imagine that this form of legalized murder would generate much public support outside, perhaps, of a few libertarian think tanks.[†]

The point here is that market-based solutions to market-generated problems do not eliminate those problems, they only shift the costs elsewhere, usually onto those who have less power and wealth. The answer to this dilemma is not necessarily to get rid of markets, as such; rather, it is to create the conditions for structural change by reasserting the primacy of politics over markets. And this must happen in two respects. First, we must recognize the importance of an ethical stance, rather than an economistic one, in respect to nature, the environment, and other social problems. Second, we must revive the practice of politics in ways that transcend voting, lobbying, and the occasional letter to

---

*A rather quick read if you are not familiar with Swift's biting social commentary: Jonathan Swift, "A Modest Proposal: For Preventing the Children of Poor People in Ireland from Being a Burden to their Parents or Country, and for Making Them Beneficial to the Public," 1729, accessed September 2, 2017, http://art-bin.com/art/omodest.html.

†Of course, since contract killers act as agents for principals, who pay for the deed to be done, the costs of permits would probably be internalized in the contract payment. When death contract permits are banned, only the wealthy will be able to get death contract permits.

representatives. It is to these arguments that we turn in the final two parts of this chapter.

TOWARD AN ENVIRONMENTAL ETHIC

The various philosophies and worldviews described in chapter 2 all reflect particular ethical perspectives on humans and the environment, especially on what is often called the relationship between nature and culture as well as the ideal and the real.[27] In a broad sense, these approaches can be separated into those that are, on the one hand, anthropocentric, or human-centered, and those that are biocentric, or nature-centered, and, on the other, those that are liberal-ideational and those that are critical-materialist (Table 6.1).

Those in the top row of the table tend to place a higher value on human beings and social systems, in the second row, on the equality of species (including humans) and ecosystems. But those in the first column believe that imaginaries can change the real world, while those in the second column argue that reality, both historical and contemporary, must shape ideas and what is possible. As we have suggested above, and as John Meyer and others have noted, such distinctions reflect a relentless search for some authority—*any* authority—that could replace the religion so important to Adam Smith's political economy and so lacking in ours today.[28] Some search for transcendent "laws of nature" arising from science or a rationalist logic that can tell us what we must do to save the world.* Discovery and affirmation of physical and biological rules and limits on human behavior—basically, in the anthropocentric-idealist box—would, some seem to hope and believe, bring rationality to the human world and restore Order and Harmony to Nature and our relationships with it. If the public and its leaders can, through the relentless logic of facts and their implications, be made to see the consequences of our excesses—injury to and death of both nature and humans—surely they will find it in their interest to regulate or eliminate offending activities, either voluntarily or through coercion. (This, it should be noted, is precisely Garrett Hardin's conclusion about the "tragedy" of unlimited breeding as well as Thomas Hobbes's regarding the need for a Leviathan to overcome the State of Nature.[29])

By contrast, the biocentric-idealist search, at its extreme, imagines an immanent Order in Nature that can be reached or restored through a return

Table 6.1. Environmental and ecological worldviews

	Liberal-ideational	Critical-materialist
Anthropocentric	Realism/liberalism; science-based	Critical political economy
Biocentric	Social naturalism	Critical transhumanism

---

*The term "laws of nature" does not mean the same thing as "natural law," which, as one of the bases for human and other rights, is a social construction.

to eternal verities, which must take place through individual and collective recognition of our place in nature and adjustment to that role. This is a rather utopian vision, based as it is on the idea that, if everyone changes what they believe via self-enlightenment, they will also alter their behaviors and practices. We have argued against such idealism for two reasons: first, it is easy to defect and backslide and there is nothing to compel the required behavioral changes; and second, the material world, as it exists, acts as a serious (historical) constraint against such changes.

A somewhat more reality-based, social constructivist perspective eschews the existence of an immanent universe with immutable laws and argues that social order is whatever human beings decide ought to be.[30] That is, it is up to us to decide on limits, even if nature's laws alert us to the possibility of disaster. We should, of course, be prudent, but there is no one and nothing to force us to behave in a particular way. Again, if we are sensible and rational, we will see that it is in our interest to impose limits on what we do—*but we can make a choice about whether we want to observe limits.* The very idea that such unauthorized choices might be possible is extremely disturbing to some biocentric-idealists, who believe that such freedom would permit license to do anything and everything to nature (and each other). As a result, they find themselves positioned in a rather curious stance, not so far from the conservative one that would prefer a transcendent source of authority and law.[31]

Worldviews rooted in political economy tend to put humans before nature and are particularly concerned about distributions of economic and political power that allow one class or fraction to dominate another. Those who, unknowingly or not, are actively engaged in such domination are likely to also treat nature in the same way. From this, critical political economists conclude that only the overthrow of systems of domination can eliminate oppression and restore social relations to a more equitable form. A higher regard for nature will (should) follow.

Many political ecologists are not biocentric but are more apt to see the domination of humans and nature as of a piece, rather than as a consequence of ignorance. Given the political economy in which we live, however, this process is largely obscured (we could call this mystification or false consciousness or naturalization or even, in Gramsci's terms, "common sense"). Our material world has been organized and constructed over long historical periods, and the oppressive practices of the past have shaped those of today, in ways that have become internalized and normal. But change is not only a matter of raised consciousness and new technologies; it also takes, as we have argued throughout this book, concerted and strategic *political* action, organized around an environmental ethics that denormalizes oppressive structures and replaces them with new, emancipatory ones.

In a world of more than 7.5 billion people (rising to perhaps ten billion or more by the end of the twenty-first century), thousands of distinctive cultures, and 190+ countries, we might doubt that any kind of environmental ethic, whether transcendent or immanent, natural or constructed, will find universal appeal on the basis of either logic or passion or, for that matter, fear alone. There continues to be, as well, little interest in, let alone possibility of, imposing such an ethic through some form of direct coercion.* For the most part, the debates among conflicting worldviews thus become rather empty ones: they are *oughts*, confronted by a deeply embedded *is*.

Some have attempted to finesse this problem by adopting familiar discourses and anthropomorphizing the environment, as if to say, "If we cannot become a part of nature, then let nature become a part of us." The results are, again, rather odd and contrary. Christopher Stone asks, "Should trees have standing?" and argues that nature should have legal rights or, at least, an ombudsperson.[32] Ulrich Beck proposes that nature has, in effect, been absorbed by a technical-regulatory "subpolitics" whose object is management in the interest of something like a vast world-spanning machine.[33] Donna Haraway suggests that we have so transformed nature through technology that we and our civilization have become "cyborgs" against which we should not struggle.[34] Others have made similar arguments and such critique is central to, for example, the work of Naomi Klein.[35] Our central argument in this book is that the environmental problematic is not, strictly speaking, about *either* humans or Nature. Rather, it is very much an effect of the contemporary relationships between politics and economics, between the public and the private, between power and public policy, particularly in industrialized market democracies. The same processes that generate environmental damage also generate a host of other externalities, including poverty, hunger, refugees, human rights violations, and even violence.[36] That these are, for the most part, treated separately from damage to the environment has mostly to do with the reductionism characteristic of a liberal ontology, which treats causes as independent phenomena leading to independent consequences.†

But what does this mean, exactly? In *The Great Transformation*, Karl Polanyi argued that the two world wars were a direct result of the "disembedding" of economics from social life—that is, making economy an independent field of knowledge and practice accountable *only* to individuals. Attempts to make these two spheres completely distinct through unregulated markets could only result in what he called a "stark utopia," which "would have physically destroyed man and transformed his surroundings into a wilderness."[37]

---

*One possibility, not addressed here, is that an emerging system of digital governmentality will be applied to manage and limit individuals' consumption and environmentally destructive activities (such as flying). See Lipschutz, "Property and Privatization."

†Whether such a problematic would arise in a similar form or in such magnitude under other socioeconomic systems is largely an academic matter (note that the centrally planned economies of really existing socialism, present and past, do not constitute a "different" system in the sense meant here). On causes and consequences, see, for example, Paul Feyerabend, *Against Method: Outline of an Anarchistic Theory of Knowledge*, London: Verso, 1993.

Marxists point out that capitalism is a social system unlike any other in human history, in that it requires the analytical separation of politics and economics, without ever actually freeing the latter from the former.[38]

It might help to unpack what this means. The individualism that lies at the center of this effort to separate politics and economics also sets boundaries between "mine" and "yours." As we pointed out in chapter 3, liberalism has a difficult time dealing with "ours," since common pool resources are hedged about with all kinds of social rules and social relationships meant to govern access to goods. Indeed, historically and even today, there is no such thing as an individual; we are constituted by our relationships to actors and actants in the material world, as articulated by the "new materialism" and actor-network theory. The reification of the individual is thus both a myth and a trick, whose effect is to divide society and the world into parts and make them susceptible to the analytical tools of neoclassical economics. But these tools are blunt instruments and are hardly up to the task of the micro-surgery that would be required to deal with each distinct person and her constitution through relationships with people and the world. Such reductionism leads to Polanyi's "stark utopia" which, paradoxically, appears in both neoclassical and eco-Marxist worldviews.

States, as the grantors and guarantors of private property rights, in particular, are in a position to define legal and economic relationships between individual and corporate actors and the material world, giving the owners of property what is, in effect, a private grant of political authority within delimited domains of ownership. In theory, owners of such property are empowered to do whatever they wish with it, including eliminating it, even if such destruction destroys things vital to life and well-being in the process. The state is enjoined from intervening in those privatized domains, having given up its prerogative to impose rules there.* Of course, states find it necessary or desirable to constrain individuals in what they can do, but this is frequently decried as unacceptable "political intervention in markets." Hence, benefits to private parties are counterposed to costs to others and, because the former are concentrated and the latter usually diffuse, private interests often, if not always, trump the public good.

Eco-Marxists argue that only the abolition of capitalism and its replacement by socialism can address this contradiction; environmental economists believe that correct prices will do the trick. Neither solution appears terribly plausible. In the near term, at least, socialism's prospects are dim, and the prices required to fully internalize social externalities could never stand the test of public acceptability (consider that between August 2017 and March 2018, the average price of a gallon of gasoline in California moved back

---

*This is not, in actuality, what is practiced, but neoliberal states make every effort to maintain the illusion, even restructuring the rule systems that make private property possible. See, for example, Lipschutz, "Property and Privatization"; Hurt and Lipschutz, *Hybrid Rule & State Formation*.

and forth between $2.50 and $4.00, a level considered confiscatory by many drivers, even though it was half or less of what European drivers pay, and is probably as low a price as it has *ever* been historically).[39] Both ecosocialism and internalization view environmental problems in distributive terms, which can be addressed either through the appropriate economic tools or the elimination of politics. Neither is particularly interested in the extent to which politics has been displaced or indeed, how this state of affairs has come about.

The argument we present in this book is about *politics*. What we regard as politics, contrary to the definition found in most texts on the global environment and other social matters, is *not* about distribution, the making of rules and deals through bargaining and negotiation or who gets what share of the pie. Politics is, instead, about *constitution* of the world, about how, and to what ends, power is to be used through the relationships discussed above. As we saw in chapter 1, power has multiple facets; not only can it be used to persuade or coerce (in an especially impoverished, economistic fashion), it can also be used to construct and produce society and its members, albeit not often in fair and just fashion. Nevertheless, in practicing politics, we seek to do more than bribe or reward others for their support for our project; we also try to persuade them that our project is worthy of their participation and will include them in all its stages, that it will recognize their needs, identities, and status, and that it will generate socially productive outcomes that are just and beneficial for both people and nature.

At the end of the day, we believe, most people are concerned about fairness and justice—to people, to nature, to Earth. Most people, knowingly or not, believe that, if politics as we now practice it does produce fair and just outcomes, there is no reason to fix the system.* But liberalism and capitalism address opportunities, rather than means or ends, arguing that the entrepreneurial individual can always find ways to profit from circumstances and conditions. Accordingly, we believe that the environmental problematic, and all the other externalities listed above, can be addressed only through political ethics that considers means as well as ends—one that pays attention to the justice and fairness of both means and ends. To put this another way, there is no philosophy, worldview, or ethic that is specific to the environment and nature and that can provide a program for their rescue. If we can construct a fair and just society, it is a good bet, we think, that we will also protect the environment. An ethic of justice and fairness will, of necessity, consider relations among people as well as relations between people and nature—and it will, of necessity, gore many oxen, especially relations of private property. But, rather than springing, like mythical mushrooms on a rainy day, from the soil

---

*The U.S. presidential election of 2016 (and elections in other countries around the world) seem to indicate that a substantial fraction of both voters and non-voters do not believe this—although it is the aspirational goal they seek: to restore "fairness" to the system for them, as they believe it once was.

of a suddenly dysfunctional worldview, such an ethic will emerge slowly and painfully out of what is, for the moment, rather unfertile ground. To push the metaphor just a bit further, what an environmental ethic requires is appropriate compost, and compost is created through action, through praxis, and not through beliefs, ideas, and imaginaries.

TOWARD AN ENVIRONMENTAL PRAXIS

To repeat the question: what are we to do? Tracy Osborne, drawing on the work of many political ecologists and the experiential education projects she has established at the University of Arizona, argues for what she calls a "public political ecology [that] . . . serves as a *community of praxis* . . . [and] joins public debates on environmental change and strives to shape the consciousness of civil society towards an integrated view of 'earth stewardship.'"[40] Osborne is concerned about altering the relatively passive stance of most scholars, who may study, analyze, and recommend but who do not engage with the public and marginalized communities who are experiencing all of the externalities arising from the contemporary social system. We believe that such a "public political ecology" must extend beyond the academy and that scholars and students must become engaged with and active participants in the actions and struggles required for real, material change.

What is this "praxis?" How might we enrich the metaphorical soil and coax from it an environmental ethic? Praxis is a type of politics, in a fundamental sense. It involves learning about power and how to use it productively. According to Paulo Freire, "Human activity consists of action and reflection; it is *praxis*; it is a transformation of the world. And as praxis, it requires theory to illuminate it. Human action is theory and practice; it is reflection and action. It cannot be reduced to either verbalism or activism."[41]

Michel Foucault argues that power is constitutive of contemporary social relations. The (post)modern subject—each of us—is a product of relations of power, power that is diffused in "capillary" fashion throughout human civilizations and societies. In a sense, power is about management, but about people managing and disciplining themselves in concert with institutions of governance, or what Foucault called "governmentality" and "biopolitics" (see chapters 1 and 2). Rules, laws, norms, practices, and customs are the articulation of this scheme of self-management from which there appears to be no escape. We participate actively in weaving the web in which we are ensnared. But that does not mean that we are flies rather than spiders.

Contemporary "advanced" liberalism (that is, "neoliberalism") is a system of discourse and practice that constitutes a particular type of governmentality, both ideologically and materially.[42] This discourse proposes to people how they ought to behave and disciplines and punishes them if they do otherwise. "Discourse" has several different definitions. Most commonly it is understood as the utterance of language: a speech. Often, it is used to describe a form of authority rooted in language: the right to name things, actions, and processes. A third understanding of discourse is as a structure of language, actions, and things that constitute specific, power-based ways of living in the world: a way of life, in other words. The power of discourse, framed as the "right" way of life, helps to make us who we are: worker, student, teacher. We can resist or oppose that power, but for the most part the result is either reincorporation or marginalization. We can change our identities, our jobs, our goals as individuals, but if we carry such change to an extreme, we might well find ourselves treated as an "odd case" or even marginalized entirely. Acting against power in this fashion is much like tilting against windmills: it looks brave, but it hardly makes a difference.

Fortunately, Foucault also points out that power is productive, and not only a mode of oppression or a tool of consensus. As he famously wrote,

> If power were never anything but repressive, if it never did anything but say no, do you really think one would be brought to obey it? What makes power hold good, what makes it accepted, is simply the fact that it doesn't only weigh on us as a force that says no, but that it traverses and produces things, it induces pleasure, forms knowledge, produces discourse. It needs to be considered as a productive network that runs through the whole social body, much more than as a negative instance whose function is repression.[43]

Although Foucault was nowhere very explicit about how power, in his understanding, could reshape this "productive network"—indeed, some read him as arguing that "resistance is futile" and condemn him for it—we might recognize that power can "traverse and produce things" in more than one fashion, through praxis, even if we are bonded to the webs of neoliberal governmentality.[44]

But working with this type of power does not mean simply rearranging that web so as to create a different organization of governmentality or destroying it so as to create a chaos out of which a new system might be spun. Rather, it is about generating, through politics, new or more creative productive networks. To put this another way, on the one hand, power "produces" the subject through the relations it fosters, in the sense described above; on the other hand, power can also produce discursive ruptures in the web of governmentality, small discontinuities that are hardly noticeable at first. A spider's web can tolerate many such small ruptures, and these are of no real concern to the spider until the web falls apart. Then, the spider spins a new one. What happens as such ruptures accumulate within the system of governmentality is much less evident.* This metaphor of a web is a very crude one, of course, but it begins to suggest something about politics, power, and praxis. Power must be applied, and praxis exercised, within the micropolitical spaces of contemporary life.

---

*These ruptures sound a lot like the dialectics of contradictions, via which thesis and antithesis lead to synthesis, but we think they are not.

Here, we turn to Hannah Arendt and her view that politics "rises directly out of acting together."[45] Although Arendt is hardly a logical complement to Foucault in this discussion, she had much to say about politics, power, and action that is germane here. Writing about "action" in *The Human Condition*, Arendt noted that "What first undermines and then kills political communities is loss of power and final impotence; and power cannot be stored up and kept in reserve for emergencies, like the instruments of violence, but exists only in its actualization." And, she continued, "Power is actualized only where word and deed are not parted company, where words are not empty and deeds not brutal, where words are not used to veil intentions but to disclose realities, and deeds are not used to violate and destroy but to *establish relations and create new realities*."[46] Where can such politics take place? In the "space of appearance," according to Arendt, which "comes into being wherever men are together in the manner of speech and action, and therefore [it] predates and precedes all formal constitution of the public realm and the various forms of government, that is, the various forms in which the public realm can be organized."[47]

For Arendt, politics could take place only through the *polis*, as it had been constituted in ancient Greece and, especially, Athens. Here, however, she points out that "[t]he *polis*, properly speaking, is not the city-state in its physical location; *it is the organization of the people as it arises out of acting and speaking together*, and its true space lies between people living together for this purpose, no matter where they happen to be. . . . [A]ction and speech create a space between the participants which can find its proper location almost any time and anywhere."[48] This particular conception of the *polis* is an interesting one, for it suggests that place is not an essential concomitant to politics, action, and praxis. We shall return to this point later, but first we have to consider how politics, in the space of appearance, can create an alternative to governmentality that does not serve to reproduce the relations of coercive power that constitute both it and the political subject.

Within such a "space," democracy becomes possible in a form that is radically different from its diluted representational form. And engaging in such democratic action is, perhaps, the most important step in challenging governmentality and recognizing just how limited our representative democracies really are. After all, how can we know what democratic politics are if we have never participated in them? How can we comprehend what is missing from our "democratic" systems if we have not experienced democratic politics? And how can we challenge the marketization of politics if our only concern is about the monetary cost of decision making rather than the disposition of power in politics?

This is why, in our estimation, environmental ethics and praxis are not to be found in an ever-greater proliferation of international nongovernmental organizations (INGOs), bureaucracies, and laws.[49] Using the American model of institutional politics, increasingly based on the deployment of money and lobbyists, many NGOs have adopted a neocorporate approach to organization and practice. When the media reports on national or international environmental matters, it is these NGOs that are usually consulted. When governments feel the need to include environmentalists in delegations and on commissions, it is from these NGOs that individuals are seconded. When corporations want to green their image, they donate large sums to these organizations or form "cause marketing partnerships."[50] And when legislatures formulate and debate laws intended to address environmental problems, it is these NGOs who provide the legal expertise and testimony in support of or in opposition to the legislation under debate. Their participation in "global environmental politics" has become routine and bureaucratic. They seek to participate in policymaking as full-fledged "stakeholders." They search for operating revenues through various types of projects supported by a broad range of funders, including both government agencies and private corporations. They differ from those they criticize only in regard to "what is to be done," and rarely in regard to "what must not be done any longer." These NGOs have become part of the very structure of neoliberal governmentality that is the source of the problem.

To where, then, can we look for answers?

GLOBAL ENVIRONMENTAL POLITICS
FROM THE GROUND UP

In her 1990 book, *Justice and the Politics of Difference*, Iris Marion Young argues

> One important purpose of critical normative theory [and speculation] is to offer an alternative vision of social relations which, in the words of [Herbert] Marcuse, "conceptualizes the stuff of which the experienced world consists . . . with a view to its possibilities, in the light of their actual limitation, suppression, and denial." Such a positive normative vision can inspire hope and imagination that motivate action for social change. It also provides some of the reflective distance necessary for the criticism of existing social circumstances.[51]

Young is focused here on domestic politics, striving for a realistic vision of what is possible from what already is: "A model of a transformed society must begin from the material structures that are given to us at this time in history."[52] We cannot create new societies or even practices out of ideas alone; we must work with what material conditions we have. When we begin to look around, we discover that there is, in fact, much to work with.

One form of social power and politics is exemplified by the watershed organization.[53] Over the past couple of decades, watershed protection groups have become ubiquitous throughout the world. Focused on a single stream, river, or other body of water, they nonetheless share an epistemic vision of the place of watersheds in both the local and global environment. These organizations look much like standard NGOs, seeking to solve environmental problems through standard techniques and practices. Yet there is some reason to think that very few of the practicing watershed groups are actually standard in the NGO fashion. Individual groups all share a view that the creek, the stream, the river are central to where they live and merit more attention and care than is being given to them. At the same time, these groups are different, in their political culture, their economies, their watersheds, their projects.

Watershed organizations—the *U.S. Non-Profit Water Directory* lists over seven hundred in the United States alone—exemplify social power and politics in action, at a local scale, but with global implications.[54] Thus, for example, although there are rivers and creeks all over the world, and groups dedicated to their protection and restoration, we do not find activists dedicating themselves to some universal conception of river or creek. Instead, it is the specific river or creek that runs through their community and the meaning of that waterway to place that are important. That meaning may have economic content, but it is also as likely to involve some notion of community identity. In other words, it is rooted in the history and political economy of that place.[55] At the same time, there are rivers and creeks in other places, in other countries, and groups in those locations seek to protect and restore their waterways, which have specific meanings for their communities. All watersheds are similar in some respects, and they all channel water, which is essential to life. So there are also meanings, as well as actions, that are shared globally.[56]

Governments have not been insensitive to local concerns about watersheds, especially insofar as they are required by law to clean them up and keep them clean. Thus, along with local activism, there have emerged government projects as well as transnational coordinating organizations focused on watershed stewardship and protection.[57] These projects seek to improve water quality, prevent stream bank erosion, protect fish-spawning areas, limit damaging activities such as logging and cattle grazing, and encourage streamside property owners to treat the watershed as something of a public good. Nor have responsible agencies been blind to the role such groups can play in furthering governmental goals. Consequently, in many places "official" state-sanctioned watershed projects have been launched, and in others, independent groups have been given a role to play in official programs.[58] But those state agencies tasked with water-related responsibilities are not entirely comfortable with these independent groups. Watershed protection is not a political act in the conventional understanding of the term, but it becomes political when it challenges established economic and social institutions and artifacts. Watershed groups often tend to be rather more radical, less manageable, more impulsive, and less systematic than bureaucrats and technocrats would like. They ignore or even trample on private property rights. They have no respect for the legal niceties and procedures of the regulatory process. They do not pay adequate attention to scientific principles and evidence.

They are too political.

"Too political" is code for the creation of a space of appearance in which people can engage in environmental politics and praxis.* In such spaces, people experience what is possible and how action is a form of productive power. Politics in the space of appearance, whether focused on the watershed, the urban community garden or park, toxic wastes, climate change, or environmental justice or displacement, is not only about the pursuit of shared interests, as collective action theorists generally describe it, or the mobilization of resources, as social movement theorists would have it.[59] It is also about the *productive* application of power in spaces of appearance. People choose. People decide. This is an experience that institutionalized political processes— voting, lobbying, emailing representatives—never offer. It is an experience that illuminates the possibilities of politics in all of its raw, elemental form.[60] It is disruptive and aggravating, but in terms of praxis, it involves essential action. It is not a "solution" to a problem; rather, it is a means of engaging with those things that ought not to be, but are.

Being "too political" ruptures the web of governmentality. These are small ruptures, and they are not very conspicuous. No one in San José or São Paulo, in Delhi or Davos, cares very much about watershed groups causing small ruptures nearby. They have their own problems to worry about, thank you. No one fears that environmental praxis poses a challenge to the stability of the republic or kingdom or union. At most, they might be a nuisance for municipal and civic sensibilities (and who, in the capitals of the world's great nation-states, cares about that?). They are hardly a threat to Western or even world civilization.

But perhaps they are.

After all, if governmentality is about management, environmental politics and praxis, as we envision them here, are not. They are, inevitably, contentious and conflictual, for they challenge the very constitutive basis of neoliberal governmentality.[61] They challenge exactly those principles, practices, and policies that seek to "manage" the environment even as they pollute and destroy it in the first place. They offer a different way of thinking about and acting for people and nature. If neoliberal governmentality is about order, environmental politics and praxis are not. Order is the enemy of change and, without change, we and nature will not survive. Then, there will be order, but it will be the Order of Death.

WHAT IS TO BE DONE AND WHO IS TO DO IT?

At the beginning of this book, we confronted the argument that the "structures that are given to us at this time in history" are both social and natural and,

*Being "too political" is a charge also levied against the high school students who started the #NeverAgain movement after the shootings at Marjory Stoneman Douglas High School in Parkland, Florida. https://twitter.com/cameron_kasky/status/9717814 24717156353.

BOX 6.1 | **A concluding manifesto**

The authors of this book are both in the second half-century of their lives. While so many decades of living on the planet affords some amount of wisdom, we thought that our concluding manifesto should be written by someone with a different vantage point—a young person looking toward the future. Nathan Thanki is a former student who read the first edition of this book in a class not very many years ago.

A MANIFESTO FOR ACTION

"To love. To be loved. To never forget your own insignificance. To never get used to the unspeakable violence and the vulgar disparity of life around you. To seek joy in the saddest places. To pursue beauty to its lair. To never simplify what is complicated or complicate what is simple. To respect strength, never power. Above all, to watch. To try and understand. To never look away. And never, never to forget." —Arundhati Roy.*

★ ★ ★

We face complex and terrifying social and ecological problems. They have no simple or halfhearted solutions; **there are no silver-bullet political moves or technological answers, and the market will not save us.** We must cut our way out of this thicket. Luckily, there is much in human history and experience from which we can learn invaluable lessons about what we can do and what we should not do. *"The struggle of man against power is the struggle of memory against forgetting."*† Memory can prevent us from succumbing to despair and recognizing that the challenge of apocalypse is nothing new. But we should not obsess over the past. We must also look to the future itself and the new world that is possible and that we are fighting to build. But most of all, we must engage in struggle NOW. TODAY.

★ ★ ★

Victories do not happen by themselves. While there may not be any final victories,‡ nor are there any final defeats. **To win, even if only once, even if only for a while, we must fight and struggle.** And we must hope. The allure of despair is that it absolves us of our responsibility to imagine a better world. Much like Walter Benjamin's "Angel of History," we have our backs to the future—meaning that the future could be anything. It is up to us to imagine what that could be. What are we struggling and fighting for? What we must do?

---

*Arundhati Roy, "The End of Imagination," in *New Nukes: India, Pakistan and Global Nuclear Disarmament*, edited by Praful Bidwai and Achin Vanaik, p. xxiii, Oxford: Signal Books, 2000.

†Milan Kundera, *The Book of Laughter and Forgetting*, p. 4, London: Faber and Faber Limited, 1996.

‡Tony Benn, *Letters to My Grandchildren*, p. 120, London: Arrow, 2010.

(Continued)

★ ★ ★

We are fighting for life itself and for the diversity of life. *"The world we want is one where many worlds fit."** We cannot and must not fight for a singular vision to be imposed everywhere; we must fight and struggle for equality, freedom, community, and a world worth living in. **These rights are non-negotiable.**

We know how to fight for these goals as well: we know that our struggles must make room for poetry and art and the chance to enjoy life while also bearing witness to its "unspeakable violence and vulgar disparity."† We know that our struggles must lead by example rather than replicate the injustices of the system we are trying to change.

★ ★ ★

Everything must change.‡ On a practical and policy level, we have many partial solutions to the ecological crises: redirecting funds away from fossil fuel corporations and into renewables; insisting that decision makers and corporations leave fossil fuels in the ground; starting energy and food cooperatives; ameliorating and eliminating concrete social and environmental injustices; steering technological shifts toward collective liberation rather than oppression. We know that tackling the crisis involves dismantling the prison and military-industrial complex, stopping deportations, and securing a livable wage for workers, land rights for indigenous peoples, and access to food for those without. We cannot list everything that can and must be done, but we must commit to seeing that these things and many others can be and indeed are done.

★ ★ ★

Together we are more than the sum of our individual parts. You may think nothing you do matters, but that would be to think only as an individual rather than as part of the collective. Social change is difficult to explain, measure, and predict, but one thing is certain: no one ever credited with enacting change did so alone. *"Resistance to destructive extraction has to be built one block at a time. When all blocks link together, a wall forms—sometimes protecting a whole nation—to block the tide of rapacious exploitation."§* If that sounds idealistic, it

---

*Ejército Zapatista de Liberación Nacional, "Fourth Declaration of the Lacondan Jungle," Lacondan Jungle, 1996, accessed October 1, 2017, http://schoolsforchiapas.org/wp-content/uploads/2014/03/Fourth-Declaration-of-the-Lacandona-Jungle-.pdf.

†Roy, "The End of Imagination," p. xxiii.

‡Naomi Klein, *This Changes Everything: Capitalism Vs the Climate*, New York: Simon & Schuster, 2014.

§Nnimmo Bassey, "Leaving the Oil in the Soil—Communities Connecting to Resist Oil Extraction and Climate Change," in *What Next Vol III: Climate, Development and Equity*, edited by Niclas Hällström, p. 332, Uppsala: What Next Forum, 2012.

(Continued)

is. But it is an idealism steeped in pragmatism, realism, and possibility. Social movements take specific, planned concrete steps toward their ambitions. Utopia is not an actual destination; it is a reason to continue walking.*

★ ★ ★

Your home is in *this* world. Environmental movements, past and present, are often painted as exceedingly either/or: either too pragmatic and complicit in sustaining the status quo for minor reforms, or too idealistic with no impacts or any actual material improvements in people's lives. Either too "not in my backyard" or locally irrelevant. Action and struggle must not only imagine a new world but also take care of seemingly minute problems. They must be focused on both social justice and preserving nature. After all, *"there is no thing as a single-issue struggle because we do not live single-issue lives."*[†] Environmental and social issues are global, local, and everything in between at the same time.

Environmental activists must fight everywhere, all at once, in coalitions that may bristle with contradictions even as common ground is found. There are many fronts in the defense of life, and they are all connected. Each requires different strategies and tactics; each must be recognized as linked to others. The environmental movements we need must identify opportunities on each front for gaining ground, opportunities that open up the possibility of other opportunities, even if victory is not the result. These movements must recognize, too, and accept that actions today may only yield results when they no longer exist. These movements must be committed to collaboration rather than competition, negotiate through their differences, make common plans and strategies, and build collective social power.

★ ★ ★

The greatest source of social power is hope. The successes of hegemony render its subjects unable to see alternatives, to question the order of things. Reclaiming power requires breaking hegemony's control over our imaginations and hope and creating alternatives that can be.

★ ★ ★

Change always seems impossible until it has been achieved.[‡] The crises we face are so severe that they often overwhelm us. Amid endless

---

*Eduardo Galeano, *Las Palabras Andantes*, p. 230, Buenos Aires: Catálogos S.R.L., 1993.

†Audre Lorde, "Learning from the 60s," presentation, Malcolm X weekend at Harvard University, Boston, MA, February, 1982, accessed October 1, 2017, http://www.blackpast.org /1982-audre-lorde-learning-60s.

‡This is a paraphrase of a famous quote, sometimes attributed to Nelson Mandela, Pliny the Elder, and others. Lots of people in current day movements use it and attribute it to Mandela.

(Continued)

talk of the "Anthropocene" and civilizational collapse, it becomes easier to imagine the end of the world than the end of capitalism. But we must not allow such fatalism to triumph. *"Again and again, far stranger things happen than the end of the world."** Let us act to make them happen.

★ ★ ★

Nathan Thanki is a graduate of College of the Atlantic, trained as a human ecologist. He helps to coordinate the Global Campaign to Demand Climate Justice, among other initiatives. He has lived in Sudan, Canada, Peru, the Solomon Islands, and the United States, and is currently based in Belfast, Ireland. A closet optimist, he thinks we should be realistic and demand the impossible.

---

*Rebecca Solnit, *Hope in the Dark: Untold Histories, Wild Possibilities*, p. 1, Edinburgh: Canongate Books, 2016.

together, constitute a social and political ecology whose normal functioning has transformed and continues to transform nature and the environment. That social ecology is both local and global. It is found in the places in which people live, work, play, and love. It is world girdling, too.

We also dealt with the proposition that, in order to understand global environmental politics, and to act, we must take greater cognizance of power and how it is constituted through ideological, historical, and material processes. In particular, power as exercised through the structures and processes associated with global capitalism constitutes an obstacle to protection of nature. At the same time, however, capitalism stands as the operating system of the material structures we are given at this particular time in history.

Finally, throughout this book, we have argued that markets and capitalism, as they are presently constituted, cannot solve the global environmental problematic, for several reasons. First, the imperatives of uncontrolled growth, which are so important to contemporary capitalism, contradict the physical realities of nature and the social realities of the world's poor. We cannot grow or consume or technologically find our way out of the crisis.

Second, a reliance on markets and capitalism as the "solution" results in the obscuring of both power and its exercise, and the disappearance of politics from the scene. Problems are transmogrified from the social to the technical realm, and left to the "experts." But however wise these experts may be, their solutions are more likely to be bent in ways that reinforce the status quo, rather than change it.

Third, managing people from above is rather like herding cats. It cannot be done. We are each part of a social body, but we are also individuals. The virtue of the market is that it permits us to be individuals; the cost of the market is that, as individuals, we have no concept of the public good and no reason to find one. That is the job of politics.

And that is what this book is, finally, about: environmental politics for a changing world, in a world of *global* environmental politics. A truly democratic, effective, and meaningful global environmental politics can begin only where politics is possible: habitat as home. This is not an admonition to "think globally, act locally." It means that changing the status quo, reasserting control over markets and capitalism, and protecting nature and the environment all require direct engagement with politics *where it begins*: in the spaces of appearance, in the places where we live.

Shall we begin?

FOR FURTHER READING

Abbey, Edward. *The Monkey Wrench Gang*. New York: Perennial, 2000.

Callenbach, Ernest. *Ecotopia*. New York: Bantam, 1975.

Dauvergne, Peter. *Environmentalism of the Rich*. Cambridge: MIT Press, 2016.

Ghosh, Amitav. *The Great Derangement: Climate Change and the Unthinkable*. Chicago: University of Chicago Press, 2016.

Li, Tania Murray. *The Will to Improve: Governmentality, Development, and the Practice of Politics*. Durham: Duke University Press, 2007.

Magnusson, Warren. *Politics of Urbanism*. Oxon: Routledge, 2011.

Robinson, Kim Stanley. *Pacific Edge*. New York: Tor, 1990.

Solnit, Rebecca. *Hope in the Dark: Untold Histories, Wild Possibilities*, updated edition. Chicago: Haymarket Books, 2016.

Stone, Deborah. *Policy Paradox: The Art of Political Decision Making*, third edition. New York: Norton, 2011.

Notes

CHAPTER 1: WHAT ARE "GLOBAL ENVIRONMENTAL POLITICS"?

1. Monica Anderson, "Technology Device Ownership: 2015," Pew Research Center, Oct. 29, 2015, p. 3, accessed January 8, 2017, http://www.pewinternet.org/files/2015/10/PI_2015-10-29 _device-ownership_FINAL.pdf.

2. Statista.com, "Number of smartphones sold to end users worldwide from 2007 to 2015," Statista 2017, accessed Feb. 10, 2017, https://www.statista.com/statistics/263437/global -smartphone-sales-to-end-users-since-2007/.

3. *The Economist*, "Planet of the Smartphones," Feb. 26, 2015, accessed Feb. 10, 2017, http:// www.economist.com/news/leaders/21645180-smartphone-ubiquitous-addictive-and-transformative -planet-phones.

4. Brian Rohrig, "Smartphones Smart Chemistry," *ChemMatters* (April/May 2015), p. 10, accessed Feb. 10, 2017, https://www.acs.org/content/dam/acsorg/education/resources/highschool /chemmatters/archive/chemmatters-april2015-smartphones.pdf.

5. Cécile Bontron, "Rare-earth mining in China comes at a heavy cost for local villages," *The Guardian*, Aug. 7, 2012. accessed Feb. 10, 2017, https://www.theguardian.com/environment /2012/aug/07/china-rare-earth-village-pollution; Max Plenke, "14 Shocking Photos Show What Your Smartphone is Really Doing to the World," *Mic.com*, May 19, 2015, accessed Feb. 10, 2017, https://mic.com/articles/118620/14-powerful-photos-show-what-your-smartphone-is-really -doing-to-the-world#.AyCVpJzjW.

6. Mina Ercan, et al., "Life Cycle Assessment of a Smartphone," Fourth International Conference on ICT for Sustainability, Aug. 29–Sept. 1, 2016, Amsterdam, accessed Feb. 10, 2017, http:// www.atlantis-press.com/php/download_paper.php?id=25860375. This article offers a number of other smartphone life cycle assessment indices of hazard and pollution.

7. Farhad Manjoo, "A Wild Idea: Making our Smartphones Last Longer," *New York Times*, March 12, 2014, accessed Feb. 10, 2017, https://www.nytimes.com/2014/03/13/technology /personaltech/the-radical-concept-of-longevity-in-a-smartphone.html?_r=0; Leyla Acaroglu, "Where Do Old Cellphones Go to Die?" *New York Times*, May 4, 2013, accessed Feb. 10, 2017, http:// www.nytimes.com/2013/05/05/opinion/sunday/where-do-old-cellphones-go-to-die.html.

8. Jay Greene, "The environmental pitfalls at the end of an iPhone's life," *CNET.com*, Sept. 26, 2012, accessed August 15, 2017, https://www.cnet.com/news/the-environmental-pitfalls-at-the -end-of-an-iphones-life/. Greene goes on to note that many of the discarded electronic devices and components that end up in Guiyu were manufactured a few hundreds of miles away.

9. Justia, "Mother Natures' Perfect Food: Premium Bananas from Ecuador—Trademark Details," accessed June 21, 2017, https://trademarks.justia.com/863/62/mother-nature-s-perfect -food-premium-bananas-from-86362788.html. But see also: Robert J. Szczerba, "Ingredients of

'Nature's Perfect Foods' Will Terrify You," *Forbes*, July 24, 2014, accessed June 21, 2017, https://www.forbes.com/sites/robertszczerba/2014/07/24/ingredients-of-natures-perfect-foods -will-terrify-you/#6fab7e59551d.

10. CGIAR, "Banana and plantain," n.d., accessed Jan. 15, 2017, http://www.cgiar.org/our -strategy/crop-factsheets/bananas/; "Banana Market," in *The State of Sustainability Initiatives Review 2014*, edited by Jason Potts, et al., pp. 97–114, Winnipeg: International Institute for Sustainable Development, 2014, accessed January 10, 2017, https://www.iisd.org/pdf/2014/ssi_2014 _chapter_5.pdf.

11. Alanna Petroff, "There's a global banana crisis," CNN.com, April 20, 2016, accessed Jan. 10, 2017, http://money.cnn.com/2016/04/19/news/bananas-banana-crisis-disease/; BananaLink .org, "Where Bananas are Grown," n.d., accessed Jan. 10, 2017, http://www.bananalink.org.uk /content/where-bananas-are-grown.

12. BananaLink.org, "Why Bananas Matter," n.d., accessed Jan. 10, 2017, http://www.banana link.org.uk/why-bananas-matter; Lorna Piatti-Farnell, *Banana: A Global History*, London: Reaktion Books, 2016; Tim Denham, Edmond De Langh, and Luc Vrydaghs, "Special Issue: History of Banana Domestication," *Ethnobotany Research and Applications* 7 (2009): 163–380.

13. Intergovernmental Group on Bananas and Tropical Fruits, "Banana Market Review and Banana Statistics 2012–2013," FAO, Rome, 2014, p. 3, accessed Jan. 10, 2017, http://www.fao .org/docrep/019/i3627e/i3627e.pdf.

14. Pedro Arias, et al., "Banana Exporting Countries," in *The World Banana Economy, 1985– 2002*, chapter 2, Rome: FAO Economic and Social Development Department, 2003, accessed June 22, 2017, http://www.fao.org/docrep/007/y5102e/y5102e05.htm#bm05.

15. Intergovernmental Group on Bananas and Tropical Fruits, "The Changing Role of Multinational Companies in the Global Banana Trade," Rome: FAO, 2014, accessed Jan. 10, 2017, http:// www.fao.org/docrep/019/i3746e/i3746e.pdf.

16. Berna van Wendel de Joode, et al., "Indigenous children living nearby plantations with chlorpyrifos-treated bags have elevated 3,5,6-trichloro-2-pyridinol (TCPy) urinary concentrations," *Environmental Research* 117 (2012): 17; Catharina Wesseling, et al., "Hazardous Pesticides in Central America," *International Journal of Occupational and Environmental Health* 7, #4 (2001): 287–94.

17. See, for example, Alison Hope Alkon and Julian Agyeman, eds. *Cultivating Food Justice— Race, Class, and Sustainability*, Cambridge, MA: MIT Press, 2011.

18. Sometimes a distinction is made between "first" nature, untouched by humanity, and "second" nature, which is nature as transformed by human action. A few postmodernists also write about "third" nature, which results from the fusion of humans, machines, and nature. See: James O'Connor, "Three Ways to Look at the Ecological History and Cultural Landscapes of Monterey Bay," in *Natural Causes: Essays in Ecological Marxism*, edited by James O'Connor, pp. 71–93, New York: Guilford Press, 1998; Donna Haraway, "Manifesto for Cyborgs: Science, Technology, and Socialist Feminism in the 1980s," *Socialist Review* 80 (1985): 65–108.

19. Joanna Klein, "Long Before Making Enigmatic Earthworks, People Reshaped Brazil's Rain Forest," *New York Times*, Feb. 10, 2017, accessed Feb. 11, 2017, https://www.nytimes .com/2017/02/10/science/amazon-earthworks-geoglyphs-brazil.html; see also Kathleen D. Morrison, "The Human Face of the Land: Why the Past Matters for India's Future," Nehru Memorial Museum and Library, History and Society Occasional Paper, New Series 27, 2013, accessed Feb. 13, 2017, http://125.22.40.134:8082/jspui/bitstream/123456789/819/1/27_Kathleen_D_Morrison .pdf.

20. Hence the term "anthropocene." See Working Group on the "Anthropocene," "What is the 'Anthropocene'—current definition and status? Subcommission on Quaternary Stratigraphy," accessed Feb. 13, 2017, https://quaternary.stratigraphy.org/workinggroups/anthropocene/.

21. David M. Konisky, Llewelyn Hughes, and Charles H. Kaylor, "Extreme weather events and climate change concern," *Climatic Change* 134 (2016): 533–47.

22. Anthony Giddens has called this the "double hermeneutic." See, for example, Hans Harbers and Gerard de Vries. "Empirical consequences of the 'double hermeneutic,'" *Social Epistemology* 7, #2 (1993): 183–92.

23. Will M. Bennis, Douglas L. Medin, and Daniel M. Bartels, "The Costs and Benefits of Calculation and Moral Rules," *Perspectives on Psychological Science* 5, #2 (2010): 187–202; Simon Dietz, Ben Groom, and William A. Pizer, "Weighing the Costs and Benefits of Climate Change to Our Children," *The Future of Children*, 26, #1 (Spring 2016): 133–55.

24. Manjana Milkoreit, "Hot deontology and cold consequentialism—an empirical exploration of ethical reasoning among climate change negotiators," *Climatic Change* 130 (2015): 397–409.

25. Kai M.A. Chan, et al., "Why protect nature? Rethinking values and the environment," *PNAS* 113, #6 (Feb. 9, 2016): 1462–65; Mark D. Davidson, "On the relation between ecosystem services, intrinsic value, existence value and economic valuation," *Ecological Economics 95* (2013): 171–77.

26. We do not, in this book, pursue these two concepts in much further detail. For more information about them, one place to start is: Briony MacPhee Rowe. "Anthropocentrism vs. Biocentrism," in *Green Issues and Debates: An A-to-Z Guide*, edited by Howard S. Schiffman, 30–35, Los Angeles: Sage Publishing, 2011.

27. Karl Polanyi, *The Great Transformation*, Boston: Beacon Press, 2001, second edition.

28. Joseph Goldstein, "The New Age Caveman and the City," *New York Times*, Jan. 8, 2010, accessed Feb. 14, 2017, http://www.nytimes.com/2010/01/10/fashion/10caveman.html; Philip Bethge, "Modern Day Flintstones: A Stone Age Subculture Takes Shape in the US," *Speigel Online*, Feb. 11, 2010, accessed Feb. 14, 2017, http://www.spiegel.de/international/zeitgeist/a-677121.html.

29. Patrick Baard and Karin Edvardsson Björnberg, "Cautious Utopias: Environmental Goalsetting with Long Time Frames," *Ethics, Policy and Environment* 18, #2 (2015): 187–201; David Harvey, *Spaces of Hope*, Berkeley, CA: UC Press, 2000.

30. Steve Rubenstein, "Sierra Sequoia's Toppling: One Less Tree to Walk Through," *San Francisco Chronicle*, Jan. 9, 2017, accessed Jan. 9, 2017, http://www.sfchronicle.com/science/article /Sierra-sequoia-s-toppling-One-less-tree-to-10846026.php.

31. But see Bruno Latour and Steve Woolgar, *Laboratory Life: The Construction of Scientific Facts*, Princeton: Princeton University Press, 1979.

32. As proposed by Stephen J. Gould, *Wonderful Life: The Burgess Shale and the Nature of History*, New York: Norton, 1990.

33. One of the most well-known quotes of Hobbes is on the state of nature, where "the life of man is solitary, poor, nasty, brutish, and short." Thomas Hobbes, *Leviathan*, New York: Macmillan, 1962, Oakeshott ed. chapter 14.

34. John Locke, *Two Treatises of Government*, Cambridge: Cambridge University Press, 1689–1690/1960.

35. Robert Dahl, "The concept of power," *Systems Research and Behavioral Science* 2, #3 (1957): 210–15.

36. John Gaventa, *Power and Powerlessness: Rebellion and Quiescence in an Appalachian Valley*, Urbana-Champaign: University of Illinois Press, 1980.

37. Karen Litfin, *Ozone Discourses: Science and Politics in Global Environmental Cooperation*, New York: Columbia University Press, 1994, p. 13. An excellent discussion of manipulation of language can be found in the appendix on "Newspeak" in George Orwell's novel *Nineteen Eighty-Four*.

38. Brandon Howard, "Timeline: History of the Dakota Access Pipeline," *Chicago Tribune*, Jan. 24, 2017, accessed Feb. 16, 2017, http://www.chicagotribune.com/news/nationworld/ct-dakota -access-pipeline-timeline-dapl-20161219-htmlstory.html. A new documentary from Bullfrog Films, "AWAKE: A Dream of Standing Rock," provides history and context for the encampment and struggle; see: http://www.bullfrogfilms.com/catalog/awake.html (accessed September 11, 2017).

39. Until completion of the pipeline, highly flammable Bakken shale oil was being moved around the country in railway tank cars that, in a few instances, had the unfortunate tendency to catch fire and blow up. See Ian Austen, "Train Blast Kills at Least One and Forces Evacuations in Canada," *New York Times*, July 6, 2013, accessed Feb. 16, 2017, http://www.nytimes.com/2013/07 /07/world/americas/people-missing-and-buildings-destroyed-in-canada-rail-blast.html. Forty-seven people died.

40. Donald J. Trump, "Presidential Memorandum Regarding Construction of the Dakota Access Pipeline," Office of the Press Secretary, The White House, Jan. 24, 2017, accessed Feb. 16, 2017, https://www.whitehouse.gov/the-press-office/2017/01/24/presidential-memorandum -regarding-construction-dakota-access-pipeline. Litigation in the case is ongoing; see Earth Justice, "Standing Rock Litigation," accessed August 21, 2017, http://earthjustice.org/features/faq -standing-rock-litigation.

41. Associated Press, "Pipeline Exec Compares Dakota Protestors to Terrorists," *New York Times*, Feb. 15, 2017, accessed Feb. 16, 2017, https://www.nytimes.com/aponline/2017/02/15/us /politics/ap-us-congress-oil-pipeline.html.

42. Michel Foucault, "Truth and Power," in *Power/Knowledge*, New York: Pantheon, 1980, trans. Colin Gordon, pp. 109–33.

43. William H. Sewell, Jr., "A Theory of Structure: Duality, Agency and Transformation," *American Journal of Sociology* 98, #1 (July 1992): 1–29; Warren Magnusson and Karena Shaw, eds., *A Political Space: Reading the Global through Clayoquot Sound*, Minneapolis: University of Minnesota Press, 2003.

44. Sewell, "A Theory of Structure."

45. Paul F. Steinberg, *Who Rules the Earth? How Social Rules Shape Our Planet and Our Lives*, Oxford: Oxford University Press, 2015.

46. Pierre Bourdieu, *The Logic of Practice*, Stanford, CA: Stanford University Press, 1990, trans. Richard Nice.

47. "The Eighteenth Brumaire of Louis Napoleon," in *The Karl Marx Library*, edited by Saul K. Padover, New York: McGraw Hill: 1972, vol. 1, p. 245.

48. An excellent place to begin an exploration of political economy is Polanyi, *The Great Transformation*.

49. Richard Peet, Paul Robbins, and Michael J. Watts (eds.), *Global Political Ecology*, London: Routledge, 2011.

50. Until January 2018, Rex Tillerson, chairman and chief executive officer of ExxonMobil from 2006 to 2016, was the U.S. Secretary of State in the Trump administration.

51. Lynn White Jr., "The Historical Roots of Our Ecologic Crisis," *Science* 155 (1967): 1203–07.

52. International law, reflected in the *Trail Smelter* case of 1941 and the Stockholm Declaration on the Human Environment of 1972, does forbid such cross-boundary pollution, although it is dependent on the offending state to obey that law. See, for example, Ved. P. Nanda, *International Environmental Law and Policy*, Irving-on-Hudson, NY: Transnational, 1995, pp. 76–78, 83–86.

53. World Commission on Environment and Development, *Our Common Future*, Oxford: Oxford University Press, 1987, p. 27.

54. Anil Agarwal and Sunita Narain, *Global Warming in an Unequal World: A Case of Environmental Colonialism*, New Delhi: Centre for Science and Environment, 1991; Christian Holz, Sivan Kartha, and Tom Athanasiou, "Fairly Sharing 1.5: National Fair Shares of a 1.5 °C-compliant Global Mitigation Effort," *International Environmental Agreements* (2017), accessed October 1, 2017, https://link.springer.com/article/10.1007/s10784-017-9371-z.

55. Andrew Goudie, *The Human Impact on the Natural Environment*, Cambridge: MIT Press, 2000.

56. Polanyi, *The Great Transformation*.

57. See Ronnie D. Lipschutz and Ken Conca, eds., *The State and Social Power in Global Environmental Politics*, New York: Columbia University Press, 1993; Ronnie D. Lipschutz, with Judith Mayer, *Global Civil Society and Global Environmental Governance*, Albany: SUNY Press, 1996; Gabriella Kütting and Ronnie D. Lipschutz, eds., *Global Environmental Governance—Power and Knowledge in a Local-Global World*, London: Routledge, 2009.

58. Sheldon Wolin, "Fugitive Democracy," in *Democracy and Difference*, edited by Seyla Benhabib, Princeton, NJ: Princeton University Press, 1996, pp. 31–45.

59. Penina M. Glazer and Myron P. Glazer, *The Environmental Crusaders: Confronting Disaster and Mobilizing Community*, University Park: Pennsylvania State University Press, 1998; Christopher Rootes, ed., *Environmental Movements: Local, National and Global*, London: Frank Cass, 1999; Patrick Novotny, *Where We Live, Work, and Play: The Environmental Justice Movement and the Struggle for a New Environmentalism*, Westport, CT: Praeger, 2000; David N. Pellow, *Garbage Wars: The Struggle for Environmental Justice in Chicago*, Cambridge: MIT Press, 2002.

60. Timothy Luke, *Ecocritique: Contesting the Politics of Nature, Economy, and Culture*, Minneapolis: University of Minnesota Press, 1997.

61. Peter Gourevitch, *Politics in Hard Times*, Ithaca, NY: Cornell University Press, 1986.

62. Jonathan Schell, *The Fate of the Earth*, New York: Knopf, 1982.

CHAPTER 2: DECONSTRUCTING THE GLOBAL ENVIRONMENT AND "GLOBAL ENVIRONMENTAL POLITICS"

1. White, "Historical Roots of Our Ecologic Crisis."

2. Max Weber, *The Protestant Ethic and the Spirit of Capitalism*, New York: Scribner; translated by Talcott Parsons, 1958.

3. Bagehot, "There is No Alternative," *The Economist*, May 7, 2009, accessed June 29, 2017, http://www.economist.com/node/13610705.

4. White, "Historical Roots of Our Ecologic Crisis."

5. Konisky, et al., "Extreme Weather Events."

6. Kenneth N. Waltz, *Man, the State and War*, New York: Columbia University Press, 1958.

7. Ronnie D. Lipschutz, *When Nations Clash: Raw Materials, Ideology and Foreign Policy*, New York: Harper and Row, 1989.

8. Harold Sprout and Margaret Sprout, *The Ecological Perspective on Human Affairs*, Princeton, NJ: Princeton University Press, 1965; Michael T. Klare, *Resource Wars: The New Landscape of Global Conflict*, New York: Metropolitan Books, 2001.

9. The Climate and Security Advisory Group, "Briefing Book for a New Administration," Center for Climate and Security, Washington, DC, Sept. 14, 2016, p. 6, accessed Feb. 24, 2017, https://climateandsecurity.files.wordpress.com/2016/09/climate-and-security-advisory-group_briefing-book-for-a-new-administration_2016_11.pdf.

10. Peter Gleick, "Water and U.S. National Security," *War Room*, June 15, 2017, accessed June 29, 2017, https://warroom.armywarcollege.edu/articles/water-u-s-national-security/.

11. Intelligence Community Assessment, "Global Water Security," National Intelligence Council, Washington, DC, Feb. 2, 2012, ICA 2012-08, accessed June 29, 2017, https://fas.org/irp/nic/water.pdf.

12. Joshua Hammer, "Is a Lack of Water to Blame for the Conflict in Syria?" *Smithsonian Magazine*, June 2013, accessed June 29, 2017, http://www.smithsonianmag.com/innovation/is-a-lack-of-water-to-blame-for-the-conflict-in-syria-72513729/. But see also: Jan Selby, et al., "Climate Change and the Syrian Civil War Revisited," *Political Geography* 60 (2017): 232–44.

13. Daniel C. Esty, et al., "State Failure Task Force Report: Phase II Findings," *Environmental Change and Security Project Report* 5 (Summer 1999): 64, accessed June 29, 2017, https://www.wilsoncenter.org/sites/default/files/Phase2.pdf.

14. Thomas Malthus, *An Essay on the Principle of Population*, 1778; Amherst, NY: Prometheus Books, 1998.

15. Garrett Hardin, "Lifeboat Ethics." Since the 1960s, Paul and Anne Ehrlich have repeatedly warned of the perils of population growth. See Clyde Haberman, "The Unrealized Horrors of Population Explosion," *New York Times*, May 31, 2015, accessed June 29, 2017, https://www.nytimes.com/2015/06/01/us/the-unrealized-horrors-of-population-explosion.html.

16. Eric B. Ross, *The Malthus Factor*, London: Zed Books, 1998.

17. Amartya Sen, *Poverty and Famines: An Essay on Entitlement and Deprivation*, Oxford: Oxford University Press, 1981; Amartya Sen and Jean Dreze, eds., *The Amartya Sen and Jean Dreze Omnibus*, New Delhi: Oxford University Press, 1999.

18. Food and Agriculture Organization, *The Future of Food and Agriculture—Trends and Challenges*, Rome, 2017, pp. 47–48, accessed July 3, 2017, http://reliefweb.int/sites/reliefweb.int/files/resources/a-i6583e.pdf.

19. C. B. Macpherson, *The Political Theory of Possessive Individualism: Hobbes to Locke*. Oxford: Oxford University Press, 1962.

20. Robert Gilpin, *War and Change in World Politics*, Cambridge: Cambridge University Press, 1981.

21. Kenneth W. Waltz, *Theory of International Politics*, Boston: Addison-Wesley, 1979.

22. Adam Smith, *The Wealth of Nations*, 1776; New York: Knopf, 1991; Emma Rothschild, *Economic Sentiments: Adam Smith, Condorcet, and the Enlightenment*, Cambridge: Harvard University Press, 2001.

23. Adam Smith, *The Theory of Moral Sentiments*, 1761; New York: A.M. Kelley, 1966; Fred Hirsch, *Social Limits to Growth*, Cambridge: Harvard University Press, 1976.

24. Strategic-metal.com, "Scandium Price," Strategic Metals Investments Ltd., 2017, accessed September 11, 2017, http://strategic-metal.com/products/scandium/scandium-price/.

25. Ronald Coase, "The Problem of Social Cost," *Journal of Law and Economics* 3 (1960): 1–44.

26. Terry L. Anderson and Donald R. Leal, *Free Market Environmentalism for the Next Generation*, New York, Palgrave Macmillan, 2015.

27. Amir Shani, "From Threat to Opportunity: The Free Market Approach for Managing National Parks and Protected Areas," *Tourism Recreation Research* 38, #3 (2013): 371–74.

28. Mancur Olson, *The Logic of Collective Action*, Cambridge: Harvard University Press, 1965; Russell Hardin, *Collective Action*, Baltimore: Johns Hopkins University Press, 1982.

29. Ronald B. Mitchell, et al., "International Environmental Agreements Database Project, 1850–2016," Eugene, OR: University of Oregon, 2017, accessed July 3, 2017, https://iea.uoregon.edu/.

30. Sabin Roman, Seth Bullock, and Markus Brede, "Coupled Societies are More Robust Against Collapse: A Hypothetical Look at Easter Island," *Ecological Economics* 132 (2017): 264–78.

31. This is often explained by reference to something called the "Kuznets Curve." See, for example, David I. Stern, "The Environmental Kuznets Curve after 25 Years," *Journal of Bioeconomics* 19 (2017): 7–28.

32. E. A. Rehfuess, et al., "Enablers and Barriers to Large-Scale Uptake of Improved Solid Fuel Stoves: A Systematic Review," *Environmental Health Perspectives* 122 (2014): 120–30.

33. Stern, "Environmental Kuznets Curve."

34. See, for example, Ronnie D. Lipschutz, "Can Climate Change Us?" *Development and Change* 48, #3(2017): 623–35, esp., 627–29.

35. International Union for the Conservation of Nature and Natural Resources, United Nations Environment Programme, and World Wildlife Fund, *World Conservation Strategy: Living Resource Conservation for Sustainable Development*, Moreges, Switzerland: International Union for the Conservation of Nature and Natural Resources/World Wildlife Fund; Nairobi, Kenya: United Nations Environment Programme, 1980. Two views of sustainable development can be found in Ronnie D. Lipschutz, "Wasn't the Future Wonderful? Resources, Environment, and the Emerging Myth of Global Sustainable Development," *Colorado Journal of International Environmental Law and Policy* 2 (1991): 35–54; and Ronnie D. Lipschutz, "The Sustainability Debate: Déjà Vu All Over Again?" *Global Environmental Politics* 9, #4 (Nov. 2009): 136–41.

36. WCED, *Our Common Future*, p. 8.

37. See references cited in Lipschutz, "Sustainability Debate."

38. Winnie Hu, "Your Uber Car Creates Congestion. Should You Pay a Fee to Ride?" *New York Times*, December 26, 2017, accessed March 21, 2018, https://www.nytimes.com/2017/12/26/nyregion/uber-car-congestion-pricing-nyc.html.

39. Weeberb J. Requia, et al., "Carbon Dioxide Emissions of Plug-in Hybrid Electric Vehicles: A Life-cycle Analysis in Eight Canadian Cities," *Renewable and Sustainable Energy Reviews* 78 (2017): 1390–96.

40. Rachael L. Shwom, "A Middle Range Theorization of Energy Politics: The Struggle for Energy Efficient Appliances," *Environmental Politics* 20:5 (2011): 705–26.

41. Hiroko Tabuchi, "Behind the Quiet State-by-State Fight Over Electric Vehicles," *New York Times*, March 11, 2017, accessed March 16, 2017, https://www.nytimes.com/2017/03/11/business/energy-environment/electric-cars-hybrid-tax-credits.html.

42. See, for example, Ronnie D. Lipschutz with James K. Rowe, *Globalization, Governmentality, and Global Politics: Regulation for the Rest of Us?* London: Routledge, 2005.

43. Polanyi, *The Great Transformation*.

44. Peter Dauvergne, *The Shadows of Consumption: Consequences for the Global Environment*, Cambridge: MIT Press, 2008.

45. Robert Gilpin, *The Political Economy of International Relations*, Princeton, NJ: Princeton University Press, 1987.

46. See, for example, Ivan Moscati, "Were Jevons, Menger, and Walras Really Cardinalists? On the Notion of Measurement in Utility Theory, Psychology, Mathematics, and Other Disciplines, 1870–1910," *History of Political Economy* 45, #3 (2013): 373–414.

47. For one view of the relationship between capital and power, see Jonathan Nitzan and Shimshon Bichler, *Capital as Power: A Study of Order and Creorder*, London: Routledge, 2009.

48. An extended discussion of this can be found in Lipschutz, with Rowe, *Globalization, Governmentality and Global Politics*, in chapter 2: "Creating a Stark Utopia? Self-Regulating Markets and the Disappearance of Politics."

49. Credit Suisse Research Institute, "Global Wealth Report 2016," Zurich, 2016, accessed June 13, 2017, http://publications.credit-suisse.com/tasks/render/file/index.cfm?fileid=AD783798-ED07-E8C2-4405996B5B02A32E.

50. Peter Dauvergne, *Environmentalism of the Rich*, Cambridge: MIT Press, 2016.

51. Intergovernmental Panel on Climate Change, *Climate Change 2014: Synthesis Report*, Geneva, Switzerland: Intergovernmental Panel on Climate Change, 2015, p. 64. We provide more detail on these numbers and the scale of this challenge in chapters 3 and 5.

52. Lipschutz, "Can Climate Change Us?" p. 628.

53. Nancy Birdsall, "Oops: Economists in Confused Search for the Middle Class in the Developing World," Center for Global Development, May 31, 2012, accessed June 22, 2016, http://www.cgdev.org/blog/oops-economists-confused-search-middle-class-developing-world.

54. CNNMoney, "Median Income Jumps 5.2% in 2015," Sept. 13, 2016, accessed July 12, 2017, money.cnn.com/2016/09/13/news/economy/median-income-census/.

55. CNNMoney, "America's Upper Middle Class is Thriving," June 21, 2016, accessed July 12, 2017, money.cnn.com/2016/06/21/news/economy/upper-middle-class/.

56. Kristie Dotson, "Tracking Epistemic Violence, Tracking Practices of Silencing," *Hypatia* 26, #2 (Spring 2011): 236–57.

57. As a proxy for the current global state of environmental injustice, see the recent assessment of Agenda 2030 and its Sustainable Development Goals: J. Sachs, et al., "SDG Index and Dashboards Report 2017," New York: Bertelsmann Stiftung and Sustainable Development Solutions Network, 2017, accessed July 11, 2017, http://www.sdgindex.org/assets/files/2017/2017-SDG-Index-and-Dashboards-Report--full.pdf.

58. Ronnie D. Lipschutz, "Getting Out of the CAR: DeCARbonization, Climate Change and Sustainable Society," *International Journal of Sustainable Society* 4, #4 (2012): 336–56.

59. Herman Daly and John Cobb, Jr., *For the Common Good—Redirecting the Economy Toward Community, the Environment, and a Sustainable Future*, Boston: Beacon Press, 1989.

60. Daly and Cobb, *For the Common Good*, p. 76, chap. 11.

61. This argument is presented in greater detail in Ronnie D. Lipschutz, "Property and Privatization," pp. 537–46, in *Research Handbook on Climate Governance*, edited by Karin Bäckstrand and Eva Lövbrand, Cheltenham, UK: Edward Elgar, 2016.

62. Anderson and Leal, *Free Market Environmentalism*.

63. Hardin, "Tragedy of the Commons."

64. See, for example, Elinor Ostrom, *Governing the Commons—The Evolution of Institutions for Collective Action*, Cambridge: Cambridge University Press, 1990, 2015; Elinor Ostrom, "Collective Action and the Evolution of Social Norms," *Journal of Natural Resources Policy Research* 6, #4 (2014): 235–52.

65. Doreen Stabinsky, "Rich and Poor Countries Face Off Over 'Loss and Damage' Caused by Climate Change," *The Conversation*, December 7, 2015, accessed August 10, 2017, https://theconversation.com/rich-and-poor-countries-face-off-over-loss-and-damage-caused-by-climate-change-51841.

66. Lipschutz, "Property and Privatization," from which these two paragraphs are taken.

67. For one argument in this vein, see Gary Flomenhoft, "Earth Inc., Shareholders' Report," Gund Institute for Ecological Economics, University of Vermont, 2003, accessed September 12, 2017, http://www.earthinc.org/earthinc.php?page=shareholder_report.

68. Christopher Stone, "Should Trees Have Standing? Toward Legal Rights for Natural Objects," *Southern California Law Review* 45 (1972): 450–77; Julie Turkewitz, "Corporations Have Rights. Why Shouldn't Rivers?" *New York Times*, Sept. 26, 2017, accessed September 28, 2017, https://nyti.ms/2jZ2Jx1.

69. Marx, "The Eighteenth Brumaire of Louis Napoleon."

70. Michael Watts, "Political Ecology," pp. 257–74, in *A Companion to Economic Geography*, edited by Eric Sheppard and Trevor Barnes, Oxford: Blackwell, 2000, p. 257

71. Tracy Osborne, "Public Political Ecology: A Community of Praxis for Earth Stewardship," *Journal of Political Ecology* 24 (2017): 843–60.

72. Dianne E. Rocheleau, "Political Ecology in the Key of Policy: From Chains of Explanation to Webs of Relation," *Geoforum* 39 (2008): 716–27, esp. p. 722.

73. See the discussion on the Frankfurt School in the entry on "Critical Theory," Stanford Encyclopedia of Philosophy, accessed August 24, 2017, https://plato.stanford.edu/entries/critical-theory/.

74. Michel Foucault, "Truth and Power," in *Power/Knowledge*, translated by Colin Gordon, pp. 109–33, New York: Pantheon, 1980; Hannah Arendt, *The Human Condition*, Chicago: University of Chicago Press, 1958, second edition.

75. Michel Foucault, "Governmentality," in *The Foucault Effect: Studies in Governmentality*, edited by Graham Burchell, Colin Gordon and Peter Miller, pp. 87–104, Chicago: University of Chicago Press, 1991.

76. Foucault, "Governmentality," p. 93.

77. Mitchell Dean, *Governmentality: Power and Rule in Modern Society*, London: Sage, 1999, p. 99.

78. Dean, *Governmentality*, p. 99.

79. Foucault, "Truth and Power," p. 119.

80. MacPhee Rowe, "Anthropocentrism vs. Biocentrism."

81. Henry David Thoreau, "Walking," *The Atlantic Monthly* (June 1862), accessed July 14, 2017, https://www.theatlantic.com/magazine/archive/1862/06/walking/304674/.

82. Josef Keulartz, *The Struggle for Nature: A Critique of Radical Ecology*, translated by Rob Keuitenbrouwer, London: Routledge, 1998.

83. Murray Bookchin, *The Ecology of Freedom*, Montreal: Black Rose Books, 1992; revised edition.

84. Bookchin, *The Ecology of Freedom*, pp. xx–xxii.

85. Anna Bramwell, *Ecology in the 20th Century—A History*, New Haven: Yale University Press, 1989.

86. Peter Kropotkin, *Mutual Aid: A Factor of Evolution*, edited by Paul Avrich, chapter 2, London: Allen Lane, 1972.

87. John Clark, "A Social Ecology," in *Environmental Philosophy: From Animal Rights to Radical Ecology*, edited by Michael E. Zimmerman, et al., pp. 416–40, Upper Saddle River, NJ: Prentice Hall, 1998, second edition.

88. Clark, "A Social Ecology," p. 421.

89. Clark, "A Social Ecology," p. 421.

90. See, for example, Umeek of Ahousaht (E. Richard Atleo), "Commentary: Discourses in and about Clayoquot Sound: A First Nations Perspective," in *A Political Space: Reading the Global through Clayoquot Sound*, edited by Warren Magnusson and Karena Shaw, pp. 199–208, Minneapolis: University of Minnesota Press, 2003.

91. Clark, "A Social Ecology," p. 433.

92. Clark, "A Social Ecology," p. 435.

93. G. W. F. Hegel, *The Philosophy of History*, translated by J. Sirbee, 1837; New York: Dover, 1956.

94. James E. Lovelock and Lynn Margulis, "Atmospheric Homeostasis by and for the Biosphere: The Gaia Hypothesis," *Tellus*, Series A, 26 (1974): 2–10, Stockholm: International Meteorological Institute.

95. "A Globe, Clothing Itself with a Brain," *Wired*, June 1, 1995, accessed August 24, 2017, https://www.wired.com/1995/06/teilhard/.

96. Karen Litfin, *Ecovillages: Lessons for Sustainable Community*, Cambridge: Polity, 2013.

97. Global Ecovillage Network, "What is an Ecovillage?" accessed July 17, 2017, https://ecovillage.org/about/gen/; "four dimensions of sustainability," accessed July 17, 2017, https://ecovillage.org/projects/dimensions-of-sustainability/.

98. Rhydian Fôn James and Molly Scott Cato, "A Bioregional Economy: A Green and Post-Capitalist Alternative to an Economy of Accumulation," *Local Economy* 29, #3 (2014): 173–80, quote on pp. 177–78.

99. James and Cato, "A Bioregional Economy," p. 178.

100. Ronnie D. Lipschutz, "Bioregionalism, Civil Society, and Global Environmental Governance," in *Bioregionalism*, edited by Michael Vincent McGinnis, pp. 102–20, London: Routledge, 1999.

101. Michael Mann, *The Sources of Social Power: The Rise of Classes and Nation-states, 1760–1914*, chapter 2, Cambridge: Cambridge University Press, 1993, vol 2.

102. David Pepper, *Eco-socialism: From Deep Ecology to Social Justice*, London: Routledge, 1993, p. 67.

103. Karl Marx, *Critique of the Gotha Programme*, Peking: Foreign Languages Press, 1972, p. 17, accessed July 18, 2017, https://archive.org/stream/CritiqueOfTheGothaProgramme/cgp_djvu.txt.

104. Philip R. Pryde, *Environmental Management in the Soviet Union*, Cambridge: Cambridge University Press, 1991.

105. James O'Connor, "On the Two Contradictions of Capitalism," *Capitalism Nature Socialism* 2, #3 (1991): 107–09.

106. O'Connor, "On the Two Contradictions of Capitalism," p. 108.

107. Pepper, *Eco-socialism*, p. 234.

108. Pepper, *Eco-socialism*, pp. 235–36.

109. Daniel Miller, ed., *Acknowledging Consumption: A Review of New Studies*, London: Routledge, 1995.

110. Rocheleau, "Political Ecology in the Key of Policy," p. 724.

111. Graham Huggan and Helen Tiffin, "Green Postcolonialism," *Interventions* 9, #1 (2007): 1–11.

112. Huggan and Tiffin, "Green Postcolonialism," p. 2.

113. Huggan and Tiffin, "Green Postcolonialism," p. 2.

114. Pablo Mukherjee, "Surfing the Second Waves: Amitav Ghosh's Tide Country," *New Formations* 59 (Autumn 2006): 144–57, quote on p. 144.

115. See, for example, Jason W. Moore, "Ecology, Capital, and the Nature of Our Times: Accumulation and Crisis in the Capitalist World-Ecology," *American Sociological Association* 17, #1 (2011): 107–46.

116. Niamh Moore, "Eco/feminism and Rewriting the Ending of Feminism: From the Chipko Movement to Clayoquot Sound," *Feminist Theory* 12, #1 (2011): 3–21; Greta Gaard, "Ecofeminism Revisited: Rejecting Essentialism and Re-Placing Species in a Material Feminist Environmentalism," *Feminist Foundations* 23, #2 (Summer 2011): 26–53.

117. Ariel Salleh, *Ecofeminism as Politics: Nature, Marx, and the Postmodern*, London: Zed Books, 1997.

118. Salleh, *Ecofeminism as Politics*, p. 14.

119. Salleh, *Ecofeminism as Politics*, pp. 13, 17.

120. Salleh, *Ecofeminism as Politics*, pp. 12–14, 17, 53–54, 192.

121. Gaard, "Ecofeminism Revisited."

122. Stacy Alaimo and Susan Heckman, eds., *Material Feminisms*, Bloomington: Indiana University Press, 2007, p. 4, cited in Gaard, "Ecofeminism Revisited," p. 42.

123. Andrea J. Nightingale, "Environment and Gender," in *The International Encyclopedia of Geography: People, the Earth, Environment, and Technology*, edited by Douglas Richardson, et al., New York: John Wiley and Sons, Ltd., 2017.

124. Dianne Rocheleau, Barbara Thomas-Slayter, and Esther Wangari (eds.), *Feminist Political Ecology: Global Issues and Local Experiences*, London: Routledge, 1996, p. 4.

125. Rocheleau, Thomas-Slayter, and Wangari, *Feminist Political Ecology*, p. 27.

126. Catriona Mortimer-Sandilands, "Queer Ecology," in *Keywords for Environmental Studies*, edited by Joni Adamson, William A. Gleason, and David N. Pellow, New York: NYU Press, 2016, accessed July 24, 2017, http://keywords.nyupress.org/environmental-studies/essay/queer-ecology/.

127. Rachel Stein, "Introduction," in *New Perspectives on Environmental Justice: Gender, Sexuality, and Activism*, edited by Rachel Stein, et al., New Brunswick, NJ: Rutgers University Press, 2004, pp. 1–17, p. 7, cited in Mortimer-Sandilands, "Queer Ecology."

128. Catriona Mortimer-Sandilands, "Unnatural Passions? Notes Toward a Queer Ecology," *Invisible Culture* 9 (2005), accessed July 24, 2017, https://www.rochester.edu/in_visible_culture /Issue_9/issue9_sandilands.pdf.

129. Mortimer-Sandilands, "Unnatural Passions?"

130. Mortimer-Sandilands, "Unnatural Passions?"

131. Bruno Latour, *Politics of Nature: How to Bring the Sciences into Democracy*, Cambridge, MA: Harvard University Press, 2004; Bruno Latour, *Reassembling the Social: An Introduction to Actor-Network-Theory*, Oxford: Oxford University Press, 2007.

132. Deborah Stone, *Policy Paradox: The Art of Political Decision Making*, New York: Norton, 2011, third edition.

CHAPTER 3: CAPITALISM, GLOBALIZATION, AND THE ENVIRONMENT

1. Sven Beckert, *Empire of Cotton: A Global History*, New York: Alfred A. Knopf, 2014.

2. Fred Hirsch, *Social Limits to Growth*, 1977; Dean Curran, "The Treadmill of Production and the Positional Economy of Consumption," *Canadian Review of Sociology*, 54, #1 (Feb. 2017): 28–47.

3. Polanyi, *The Great Transformation*.

4. Polanyi, *The Great Transformation*.

5. Smith, *The Wealth of Nations*; Polanyi, *The Great Transformation*; Ha-Joon Chang, *23 Things They Don't Tell You about Capitalism*, New York: Bloomsbury Press, 2012.

6. Hirsch, *Social Limits to Growth*.

7. Ha-Joon Chang, *Economics: The User's Guide*, New York: Bloomsbury Press, 2014.

8. Smith, *Wealth of Nations*; Rothschild, *Economic Sentiments*.

9. Kate Murphy, "America is Running Out of Bomb-Sniffing Dogs," *New York Times*, August 4, 2017, accessed August 4, 2017, https://nyti.ms/2ur2ZEL.

10. Lawrence D. Meinert, Gilpin R. Robinson, Jr., and Nedal T. Nassar, "Mineral Resources: Reserves, Peak Production and the Future," *Resources* 5, #1 (2016), accessed August 1, 2017, http://www.mdpi.com/2079-9276/5/1/14/htm; H. E. Goeller and Alvin M. Weinberg, "The Age of Substitutability," *Science* 191, #4228 (Feb. 20, 1976): 683–89.

11. K. V. Ragnarsdóttir, H. U. Sverdrup, and D. Koca, "Assessing Long Term Sustainability of Global Supply of Natural Resources and Materials," pp. 83–116, in *Sustainable Development—Energy, Engineering and Technologies—Manufacturing and Environment*, edited by Chaouki Ghenai, Rijeka, Croatia: InTech, accessed August 1, 2017, https://www.intechopen.com/books /sustainable-development-energy-engineering-and-technologies-manufacturing-and-environment; Julie Butters, "This is Dysprosium—if We Run Out of It, Say Goodbye to Smartphones, MRI Scans and Hybrid Cars," *Phys.Org*, June 6, 2016, accessed August 1, 2017, https://phys.org /news/2016-06-dyprosiumif-goodbye-smartphones-mri-scans.html.

12. For a survey of contrasting and conflicting assessments of the future of natural resource supplies, see World Economic Forum, "The Future Availability of Natural Resources—A New Paradigm for Global Resource Availability," Geneva, Switzerland, Nov. 2014, accessed August 1, 2017, http://www3.weforum.org/docs/WEF_FutureAvailabilityNaturalResources_Report_2014.pdf.

13. Stefan Giljum and Friedrich Hinterberger, "The Limits of Resource Use and Their Economic and Policy Implications," pp. 3–16, in *Factor X: Policy, Strategies and Instruments for a Sustainable Resource Use*, edited by Michael Angrick, Andreas Burger and Harry Lehmann, Dordrecht, Netherlands: Springer Science, 2014; Ralf Seppelt, et al., "Synchronized Peak-rate Years of Global Resources Use," *Ecology and Society* 19, #4 (2014): 50; Johan Rockstrom, et al., "Planetary Boundaries: Exploring the Safe Operating Space for Humanity," *Ecology and Society* 14 (2009): 32, accessed August 30, 2017, http://www.ecologyandsociety.org/vol14/iss2/art32/.

14. James Petras and Henry Veltmeyer, eds., *The New Extractivism: A Post-Neoliberal Development Model or Imperialism of the Twenty-First Century?* London: Zed, 2014.

15. David W. Pearce and R. Kerry Turner, *Economics of Natural Resources and the Environment*, Baltimore: Johns Hopkins University Press, 1990, chapter 14; Jonathan Harris and Brian Roach, *Environmental and Natural Resource Economics: A Contemporary Approach*, London: Routledge, 2018, fourth edition.

16. Coase, "Problem of Social Cost."

17. Megan Martorana, "Pareto Efficiency," *Region Focus* (Federal Reserve Bank of Richmond), Winter 2007, p. 8, accessed August 2, 2017, https://www.richmondfed.org/~/media/richmond fedorg/publications/research/region_focus/2007/winter/pdf/jargon_alert.pdf.

18. Ramachandra Guha, *The Unquiet Woods*; Allan Greer, "Commons and Enclosure in the Colonization of North America," *American History Review* (2012): 365–86.

19. Thom Kuehls, "The Environment of Sovereignty," in *A Political Space: Reading the Global through Clayoquot Sound*, edited by Warren Magnusson and Karena Shaw, pp. 179–98, Minneapolis, MN: University of Minnesota Press, 2003.

20. Thomas Princen, "Consumption and Its Externalities: Where Economy Meets Ecology," *Global Environmental Politics* 1 (Aug. 2001): 11–30; Thomas Princen, Michael Maniates, and Ken Conca, *Confronting Consumption*, Cambridge: MIT Press, 2002; Peter Dauvergne, "The Problem of Consumption," *Global Environmental Politics* 10, #2 (May 2010): 1–10.

21. Leisa Reinecke Flynn, Ronald E. Goldsmith, and Wesley Pollitte, "Materialism, Status Consumption and Market Involved Consumers," *Psychology and Marketing* 33, #9 (Sept. 2016): 761–76.

22. Ruut Veenhoven and Floris Vergunst, "The Easterlin Illusion: Economic Growth Does Go with Greater Happiness," *International Journal of Happiness and Development* 1, #4 (2014): 311–43; Stefano Bartolini and Francesco Sarracino, "Happy for How Long? How Social Capital and Economic Growth Relate to Happiness Over Time," *Ecological Economics* 108 (2014): 242–56.

23. Mathis Wackernagel and William Rees, "Ecological Footprints for Beginners," in *The Ecological Design and Planning Reader*, edited by Foster O. Ndubisi, pp. 501–5, Washington, DC: Island Press, 2014; Kai Fang, Reinout Heijungs, and Geer R. de Snoo, "Theoretical Exploration

for the Combination of the Ecological, Energy, Carbon, and Water Footprints: Overview of a Footprint Family," *Ecological Indicators* 36 (2014): 508–18.

24. Allesandro Galli, et al., "Questioning the Ecological Footprint," *Ecological Indicators* 69 (2016): 224–32.

25. Stefanie Hellweg and Lorenç Milà I. Canals, "Emerging Approaches, Challenges and Opportunities in Life Cycle Assessment," *Science* 344, #6188 (June 6, 2014): 1109–13.

26. The literature on consumption has grown by leaps and bounds since the first edition of this book. See, for example, Tania Briceno and Sigrid Stagl, "The Role of Social Processes for Sustainable Consumption," *Journal of Cleaner Production* 14 (2006): 1541–51; Jesse Bricker, Rodney Ramcharan, and Jacob Krimmel, "Signaling Status: The Impact of Relative Income on Household Consumption and Financial Decisions," Federal Reserve Board, Washington, DC, 2014–76, accessed September 26, 2014, https://www.federalreserve.gov/econresdata/feds/2014/files/201476pap.pdf; Greger Henriksson and Miriam Börjesson Rivera, "Why do We Buy and Throw Away Electronics?" Paper presented to ISDRC 2014: Resilience—The New Research Frontier, Trondheim, 2014, accessed August 6, 2015, https://www.diva-portal.org/smash/get/diva2:737102/FULLTEXT01.pdf; David Clinginsmith and Roman Sheremta, "Status and the Demand for Visible Goods: Experimental Evidence on Conspicuous Consumption," Munich Personal RePEc Archive, December 4, 2015, accessed August 3, 2017, https://mpra.ub.uni-muenchen.de/68202/.

27. eMarketer, "Worldwide Ad Spending Growth Revised Downward," April 21, 2016, accessed September 14, 2017, https://www.emarketer.com/Article/Worldwide-Ad-Spending-Growth-Revised-Downward/1013858.

28. Malcolm Gladwell, "The Science of Shopping," *New Yorker*, Nov. 1996, pp. 66–75; David Lewis, *The Brain Sell: When Science Meets Shopping*, London: Brealey, 2013.

29. Nicole Burrows, "How Consumers Drive the Economy," *Nassau Guardian*, June 11, 2014, accessed August 4, 2017, http://www.thenassauguardian.com/opinion/op-ed/48000-how-consumers-drive-the-economy-.

30. Julio L. Rivera and Amrine Lallmahomed, "Environmental Implications of Planned Obsolescence and Product Lifetime: A Literature Review," *International Journal of Sustainable Engineering* 9, #2 (2016): 119–29.

31. Experts Exchange, "Processing Power Compared," n.d., accessed August 4, 2017, http://pages.experts-exchange.com/processing-power-compared/.

32. Underwriters Laboratory, "The Life Cycle of Materials in Mobile Phones," 2011, accessed August 4, 2017, http://services.ul.com/wp-content/uploads/sites/4/2014/05/ULE_CellPhone_White_Paper_V2.pdf; Manivannan Senthil Velmurugan, "Environmental and Health Aspects of Mobile Phone Production and Use: Suggestions for Innovation and Policy," *Environmental Innovation and Societal Transitions* 21 (2016): 69–79.

33. Elizabeth Jardim, "From Smart to Senseless: The Global Impact of 10 Years of Smartphones," Greenpeace, Washington, DC, 2017, accessed August 4, 2017, http://www.greenpeace.org/usa/wp-content/uploads/2017/03/FINAL-10YearsSmartphones-Report-Design-230217-Digital.pdf.

34. WCED, *Our Common Future*, p. 20.

35. Stern, "The Environmental Kuznets Curve."

36. Rachel S. Franklin and Matthias Ruth, "Growing Up and Cleaning Up: The Environmental Kuznets Curve Redux," *Applied Geography* 32 (2012): 29–39, p. 35.

37. Franklin and Ruth, "Growing Up and Cleaning Up," p. 35.

38. C. Kenny, "Why Ending Extreme Poverty Isn't Good Enough," *Bloomberg Businessweek*, April 29, 2013, accessed June 22, 2016, http://www.bloomberg.com/news/articles/2013-04-28/why-endingextreme-poverty-isnt-good-enough.

39. Jintai Lin, et al., "China's International Trade and Air Pollution in the United States," *PNAS* 111, #5 (Feb. 4, 2014): 1736–41; Li Yanmei, et al., "Sources and Flows of Embodied CO-2 Emissions in Import and Export Trade of China," *Chinese Geographical Science* 24, #2 (2014): 220–30.

40. Curran, "The Treadmill of Production."

41. Hirsch, *Limits to Growth*.

42. Stephanie Clifford, "That 'Made in U.S.A.' Premium," *New York Times*, November 30, 2013, accessed August 8, 2017, http://www.nytimes.com/2013/12/01/business/that-made-in-usa-premium.html.

43. Peter Dauvergne, "Problem of Consumption."

44. Chukwumerije Okereke, "Climate Justice and the International Regime," *WIREs Climate Change* 1 (May/June, 2010): 462–74; Lukas H. Meyer and Dominic Roser, "Climate Justice and Historical Emissions," *Critical Review of International Social and Political Philosophy* 13, #1 (March 2010): 229–53.

45. Kevin J. Boyle, "Contingent Valuation in Practice," in *A Primer on Nonmarket Valuation*, edited by Patricia A. Champ, Thomas C. Brown, and Kevin J. Boyle, pp. 83–132, Dordrecht, Netherlands: Springer, 2017, second edition.

46. See, for example, Alex Y. Lo and C.Y. Jim, "Protest Response and Willingness to Pay for Culturally Significant Urban Trees: Implications for the Contingent Valuation Method," *Ecological Economics* 114 (2015): 58–66.

47. Juliet Eilperin and Brady Dennis, "Obama Names Five New National Monuments, Including Southern Civil Rights Sites," *The Washington Post*, Jan. 12, 2017, accessed August 8, 2017, https://www.washingtonpost.com/national/health-science/obama-names-five-new-national-monuments-including-southern-civil-rights-sites/2017/01/12/7f5ce78c-d907-11e6-9a36-1d296534b31e_story.html. See also U.S. Environmental Protection Agency, "Waters of the United States (WOTYS) Rulemaking," n.d., accessed August 14, 2017, https://www.epa.gov/wotus-rule.

48. *Los Angeles Times*, "Here are the National Monuments Being Reviewed Under Trump's Order," May 5, 2017, accessed August 8, 2017, http://www.latimes.com/nation/la-na-national-monuments-pictures-20170426-htmlstory.html.

49. Land Trust Alliance, "2016 Annual Report," Washington, DC, accessed August 9, 2017, http://s3.amazonaws.com/landtrustalliance.org/2016AnnualReport.pdf; The Wilderness Society, "America's Public Lands," Washington, DC, n.d., accessed August 9, 2017, http://wilderness.org/sites/default/files/Fact%20Sheet%20America%27s%20Public%20Lands%20.pdf.

50. Marissa Bongiovanni and Erin Clover Kelly, "Ecosystem Service Commodification: Lessons from California," *Global Environmental Politics* 16, #4 (Nov. 2016): 90–110; Jessica Owley, "Cultural Heritage Conservation Easements: Heritage Protection with Property Law Tools," *Land Use Policy* 49 (2015): 177–82; Kelly Kay, "Breaking the Bundle of Rights: Conservation Easements and the Legal Geographies of Individuating Nature," *Environment and Planning A* 48, #3 (2015): 504–22.

51. Robert Costanza, et al., "The Value of the World's Ecosystem Services and Natural Capital," *Nature*, May 15, 1997, pp. 253–60.

52. Robert Costanza, et al., "Changes in the Global Value of Ecosystem Services," *Global Environmental Change* 26 (2014): 152–58.

53. World Bank DataBank, "Gross Domestic Product 2016," accessed August 14, 2017, http://databank.worldbank.org/data/download/GDP.pdf.

54. See, for example, Silvia Federici, "The Reproduction of Labour-power in the Global Economy, Marxist Theory and the Unfinished Feminist Revolution," reading for Jan. 27, 2009 University of California Santa Cruz seminar "The Crisis of Social Reproduction and Feminist Struggle," accessed August 14, 2017, https://caringlabor.wordpress.com/2010/10/25/silvia-federici-the-reproduction-of-labour-power-in-the-global-economy-marxist-theory-and-the-unfinished-feminist-revolution/.

55. Christopher M. Rea, "Commodifying Conservation," *Contexts* 14, #1 (Feb. 2013): 72–73; Clive L. Spash and Iulie Aklaken, "Re-establishing an Ecological Discourse in the Policy Debate Over How to Value Ecosystems and Biodiversity," *Journal of Environmental Management* 159 (2015): 245–53; Bram Büscher and Robert Fletcher, "Nature is Priceless, Which is Why Turning It into 'Natural Capital' is Wrong," *The Conversation*, September 21, 2016, accessed December 3, 2017, https://theconversation.com/nature-is-priceless-which-is-why-turning-it-into-natural-capital-is-wrong-65189.

56. Marianna Fenzi and Christophe Bonneuil, "From 'Genetic Resources' to 'Ecosystems Services': A Century of Science and Global Policies for Crop Diversity Conservation," *Culture, Agriculture, Food and Environment* 38, #2 (Dec. 2016): 72–83.

57. Carbon Trade Watch, "Carbon Trading, REDD+ and the Push for Pricing Forests," June 19, 2012, accessed September 29, 2017, http://www.carbontradewatch.org/articles/carbon-trading-redd-and-the-push-for-pricing-forests.html.

58. N. Carroll, et al., eds., *Conservation and Biodiversity Banking: A Guide to Setting Up and Running Biodiversity Credit Trading Systems*, London, UK: Earthscan, 2008.

59. Spash and Aklaken, "Re-establishing an Ecological Discourse."

60. Pam Belluck, "In Breakthrough, Scientists Edit a Dangerous Mutation from Genes in Human Embryos," *New York Times*, Aug. 2, 2017, accessed August 14, 2017, https://www.nytimes.com/2017/08/02/science/gene-editing-human-embryos.html.

61. Ker Than, "7 Takeaways From the Supreme Court's Gene Patent Decision," *National Geographic*, June 15, 2013, accessed August 14, 2017, http://news.nationalgeographic.com/news/2013/06/130614-supreme-court-gene-patent-ruling-human-genome-science/.

62. Locke, *Two Treatises*, chapter 5, "On Property," Sec. 32.

63. Wen Zhou, "The Patent Landscape of Genetically Modified Organisms," *Signal to Noise Special Edition: GMOs and Our Food*, Harvard Graduate School of Arts and Sciences," Aug. 10, 2015, accessed August 14, 2017, http://sitn.hms.harvard.edu/flash/2015/the-patent-landscape-of-genetically-modified-organisms/.

64. Phillips McDougall, "The Cost and Time Involved in the Discovery, Development and Authorisation of a New Plant Biotechnology Derived Trait," Consultancy Study for Crop Life International, Sept. 2011, accessed August 14, 2017, https://croplife.org/wp-content/uploads/2014/04/Getting-a-Biotech-Crop-to-Market-Phillips-McDougall-Study.pdf.

65. Matthew Herper, "The Cost of Creating a New Drug Now $5 Billion, Pushing Big Pharma to Change," *Forbes*, Aug. 11, 2013, accessed August 14, 2017, https://www.forbes.com/sites/matthewherper/2013/08/11/how-the-staggering-cost-of-inventing-new-drugs-is-shaping-the-future-of-medicine/#12c3d92213c3. But see also: Vinay Prasad and Sham Mailankody, "Research and Development Spending to Bring a Single Cancer Drug to Market and Revenues after Approval," *JAMA Internal Medicine*, Sept. 11, 2017, accessed September 13, 2017, http://jamanetwork.com/journals/jamainternalmedicine/data/journals/intemed/0/jamainternal_prasad_2017_oi_170062.pdf.

66. Lin Tao, et al., "Nature's Contribution to Today's Pharmacopeia," *Nature Biotechnology* 32, #10 (Oct. 2014): 979–80.

67. Janna Rose, "Biopiracy: When Indigenous Knowledge is Patented for Profit," *The Conversation* March 7, 2016, accessed August 15, 2017, http://theconversation.com/biopiracy-when-indigenous-knowledge-is-patented-for-profit-55589.

68. See, for example, Shane Greene, "Indigenous People Incorporated? Culture as Politics, Culture as Property in Pharmaceutical Bioprospecting," *Current Anthropology* 45, #2 (April 2004): 211–37; Ian Vincent McGonigle, "Patenting Nature or Protecting Culture? Ethnopharmacology and Indigenous Intellectual Property Rights," *Journal of Law and the Biosciences* (Feb. 2016): 217–26.

69. Convention on Biological Diversity, "About the Nagoya Protocol," n.d., accessed August 15, 2017, https://www.cbd.int/abs/about/.

70. Laura A. Foster, "Privatizing Hoodia—Patent Ownership, Benefit-sharing, and Indigenous Knowledge in South Africa," in *Privatization, Vulnerability and Social Responsibility: A Comparative Perspective*, edited by Martha Albertson Fineman, Titti Mattsson, and Ulrika Andersson, pp. 180–23, London: Routledge, 2017.

71. World Intellectual Property Organization Background Brief, "The WIPO Intergovernmental Committee on Intellectual Property and Genetic Resources, Traditional Knowledge and Folklore," Geneva, 2016, accessed August 15, 2017, http://www.wipo.int/edocs/pubdocs/en/wipo_pub_tk_2.pdf.

72. James Fairhead, Melissa Leach, and Ian Scoones, "Green Grabbing: A New Appropriation of Nature?" *The Journal of Peasant Studies* 39, #2 (2012): 237–61; Catherine Corson and Kenneth Iain MacDonald, "Enclosing the Global Commons: The Convention on Biological Diversity and Green Grabbing," *The Journal of Peasant Studies* 39, #2 (2012): 263–83.

73. Property and Environment Research Center, "Free Market Environmentalism," 2017, accessed August 17, 2017, https://www.perc.org/about-perc/free-market-environmentalism.

74. Kent H. Redford, William Adams, and Georgina M. Mace, "Synthetic Biology and Conservation of Nature: Wicked Problems and Wicked Solutions," *PLOS Biology* 11, #4 (April 2013), accessed August 15, 2017, http://journals.plos.org/plosbiology/article?id=10.1371/journal.pbio.1001530.

75. See, for example, DivSeek, "Working Groups," accessed August 15, 2017, http://www.divseek.org/; Benjamin D. Neimark, "Industrializing Nature, Knowledge, and Labour: The Political Economy of Bioprospecting in Madagascar," *Geoforum* 75 (2012): 274–85.

76. Alan L. Harvey, RuAngelie Edrada-Ebel, and Ronald J. Quinn, "The Re-emergence of Natural Products for Drug Discovery in the Genomics Era," *Nature Reviews* 14 (Feb. 2015): 110–20.

77. Nico Schrijver, "Managing the Global Commons: Common Good or Common Sink?" *Third World Quarterly* 37, #7 (2016): 1252–67.

78. Todd Sandler, "Collective Action: Fifty Years Later," *Public Choice* 164 (2015): 195–216.

79. For example, Paul Ehrlich, *The Population Bomb*, New York: Ballentine, 1968.

80. Hardin, "Lifeboat Ethics."

81. Derek Wall, *The Commons in History*, Cambridge, MA: MIT Press, 2014.

82. Ostrom, *Governing the Commons*; Elinor Ostrom, "Collective Action and the Evolution of Social Norms." A recent review of scholarship on the commons can be found in "The Commons and the Common," *Review of Radical Political Economics* 48, #1 (March 2016, special issue).

83. Andriana Vlachou and Georgios Pantelias, "The EU's Emissions Trading System, Part 1: Taking Stock," *Capitalism Nature Socialism* 28, #2 (2017): 84–102; Andriana Vlachou and Georgios Pantelias, "The EU's Emissions Trading System, Part 2: A Political Economy Critique," *Capitalism Nature Socialism* 28, #3 (2017): 108–27.

84. Daniel Altman, "Just How Far Can Trading of Emissions Be Extended?" *New York Times*, May 31, 2001, accessed August 15, 2017, http://www.nytimes.com/2002/05/31/business/just -how-far-can-trading-of-emissions-be-extended.html.

85. Richard Schmalensee and Robert N. Stavins, "The SO_2 Allowance Trading System: The Ironic History of a Grand Policy Experiment," *Journal of Economic Perspectives* 27, #1 (Winter 2013): 103–21.

86. For a series of case studies on carbon trading markets, see International Emissions Trading Association, "The World's Carbon Markets—A Case Study Guide to Emissions Trading," 2016, accessed August 15, 2017, http://www.ieta.org/The-Worlds-Carbon-Markets; also World Bank Group, "State and Trends of Carbon Pricing," Washington, DC, Oct. 2016, accessed August 15, 2017, https://openknowledge.worldbank.org/bitstream/handle/10986/25160/9781464810015.pdf ?sequence=7andisAllowed=y.

87. Lawrence H. Goulder and Andrew Scheim, "Carbon Taxes vs. Cap and Trade: A Critical Review," NBER Working Paper #19338, August 2013, accessed August 15, 2017, http://www .nber.org/papers/w19338.pdf.

88. Vlachou and Pantelias, "EU's Emission Trading System," parts 1 and 2.

89. U.S. Environmental Protection Agency, "The Social Cost of Carbon," n.d., accessed August 16, 2017, https://19january2017snapshot.epa.gov/climatechange/social-cost-carbon_.html.

90. Uma Outka, "Fairness in the Low-Carbon Shift: Learning from Environmental Justice," *Brooklyn Law Review* 82, #2 (2017): 729–824, accessed August 16, 2017, https://ssrn.com /abstract=2989256; Mehdi Benatiya Andaloussi and Elisabeth Thuestad Isaksen, "The Environmental and Distributional Consequences of Emissions Markets: Evidence from the Clean Air Interstate Rule," June 25, 2017, accessed August 16, 2017, https://ssrn.com/abstract=2992180.

91. Larry Lohmann, "Uncertainty Markets and Carbon Markets: Variations on Polanyian Themes," *New Political Economy* 15, #2 (2010): 225–54.

92. *The Economist*, "Sins of Emission," August 3, 2006, accessed August 16, 2017, http://www .economist.com/node/7252897.

93. Steffen Dalsgaard, "The Commensurability of Carbon: Making Value and Money of Climate Change," *HAU: Journal of Ethnographic Theory* 3, #1 (2013): 80–98; Cheat Neutral YouTube advertisement, accessed August 25, 2017, https://www.youtube.com/watch?v=f3_CYdYDDpk.

94. Craig A Hart, *Climate Change and the Private Sector: Scaling Up Private Sector Response to Climate Change*, New York: Routledge, 2013.

95. Martin Cames, et al., "How Additional is the Clean Development Mechanism?" Öko -Institut e.V., Berlin, March 2016, p. 11, accessed August 16, 2017, https://ec.europa.eu/clima /sites/clima/files/ets/docs/clean_dev_mechanism_en.pdf.

96. Alla A. Golub, et al., "Global climate policy impacts on livestock, land use, livelihoods, and food security," *PNAS* 110, #52 (Dec. 24, 2013): 20894–99; Thomas Sikor, et al., "Global Land Governance: From Territory to Flow?" *Current Opinion in Environmental Sustainability* 5 (2013): 522–27; Adeniyi P. Asiyanbi, "A Political Ecology of REDD+: Property Rights, Militarised Protectionism, and Carbonised Exclusion in Cross River," *Geoforum* 77 (2016): 146–56; Susan Chomba, Juliet Kariuki, Jens Friis Lund, and Fergus Sinclair, "Roots of Inequity: How the Implementation of REDD+ Reinforces Past Injustices," *Land Use Policy* 50 (2016): 202–13.

97. Nathalie Berta, Emmanuelle Gautherat, and Ozgur Gun, "Transactions in the European Carbon Market: A Bubble of Compliances in a Whirlpool of Speculation," *Cambridge Journal of Economics* 41 (2017): 575–93.

98. UN Framework Convention on Climate Change, "Adoption of the Paris Agreement," Article 2.1.a, p. 22, Dec. 12, 2017, Draft decision -/CP.21, FCCC/CP/2015/L9/Rev.1, https://unfccc.int/resource/docs/2015/cop21/eng/l09r01.pdf, accessed August 17, 2017.

99. Intergovernmental Panel on Climate Change, *Climate Change 2014: Synthesis Report. Contribution of Working Groups I, II, and III to the Fifth Assessment Report of the Intergovernmental Panel on Climate Change*, Geneva: Intergovernmental Panel on Climate Change, 2014, p. 64.

100. Marcia Rocha, et al., "Implications of the Paris Agreement for Coal Use in the Power Sector," *Climate Analytics*, 2016, accessed August 16, 2017, http://climateanalytics.org/files/climateanalytics-coalreport_nov2016_1.pdf.

101. Intergovernmental Panel on Climate Change, Working Group III: Mitigation, "Historic Trends and Driving Forces," Section 3.8.3, 2001, accessed August 16, 2017, http://www.ipcc.ch/ipccreports/tar/wg3/index.php?idp=125.); Vivian Scott, et al., "Fossil Fuels in a Trillion Tonne World," *Nature Climate Change* 5 (May 2015): 419–23.

102. Chris Lang, "A Plan to Burn the Planet: The Aviation Industry is in Talks with the World Bank about using REDD to Offset its Emissions," REDD-Monitor.org, May 17, 2017, accessed August 16, 2017, http://www.redd-monitor.org/2017/05/17/a-plan-to-burn-the-planet-the-aviation-industry-is-in-talks-with-the-world-bank-about-using-redd-to-offset-its-emissions/.

103. U.S. Fish and Wildlife Service, "For Landowners—Conservation Banking," accessed August 16, 2017, https://www.fws.gov/endangered/landowners/conservation-banking.html; and Speciesbanking.com, "Saving Biodiversity by Integrating it into All Spheres of Society," accessed August 16, 2017, http://us.speciesbanking.com/.

104. Nicolás Kosoy and Esteve Corbera, "Payments for Ecosystem Services as Commodity Fetishism," *Ecological Economics* 69 (2010): 1228–36; J. Pawliczek and S. Sullivan, "Conservation and Concealment in SpeciesBanking.com, USA: An Analysis of Neoliberal Performance in the Species Offsetting Industry," *Environmental Conservation* 38, #4 (2011): 435–44.

105. Pawliczek and Sullivan, "Conservation and Concealment," p. 435.

106. Pawliczek and Sullivan, "Conservation and Concealment," p. 435; Jonas Hein and Jean Carlo Rodriguez, "Carbon and Biodiversity Offsetting: A Way Towards Sustainable Development?" *The Current Column*, March 21, 2016, Bonn: German Development Institute.

107. Pavan Sukhdev, et al., "The Economics of Ecosystems and Biodiversity—An Interim Report," European Communities, 2008, accessed August 17, 2017, http://www.teebweb.org/media/2008/05/TEEB-Interim-Report_English.pdf.

108. James T. Mandel, C. Josh Dolan, and Jonathan Armstrong, "A Derivative Approach to Endangered Species Conservation," *Frontiers in Ecology and the Environment* 8, #1 (Feb. 2010): 44–49; John Chiang, "Winter is Coming, but 1 idea may Help save Planet Earth," *San Francisco Chronicle*, September 13, 2017, accessed September 13, 2017, http://www.sfchronicle.com/opinion/openforum/article/Winter-is-coming-but-1-idea-may-help-save-Planet-12195996.php.

109. Sian Sullivan, "Banking Nature? The Spectacular Financialisation of Environmental Conservation, *Antipode* 45, #1 (2013): 198–217; Jessica Dempsey, "The Financialization of Nature Conservation?" in *Money and Finance After the Crisis: Critical Thinking for Uncertain Times*, edited by Brett Christophers, Andrew Leyshon, and Geoff Mann, pp. 191–216, New York: Wiley Online Library, 2017.

110. Desiree Fields, "Unwilling Subjects of Financialization," *International Journal of Urban & Regional Research* 41, #4 (2017): 588–603.

111. Jeffrey Rothfeder, "The Great Unraveling of Globalization," *Washington Post*, April 24, 2015, accessed August 17, 2017, https://www.washingtonpost.com/business/reconsidering-the-value-of-globalization/2015/04/24/7b5425c2-e82e-11e4-aae1-d642717d8afa_story.html?utm_term=.4d4271023479; Thomas L. Friedman, *The Lexus and the Olive Tree: Understanding Globalization*. New York: Farrar Straus Giroux, 1999.

112. Naomi Klein, *The Shock Doctrine: The Rise of Disaster Capitalism*, Toronto: Random House Canada, 2007.

113. Lipschutz, with Rowe, *Globalization, Governmentality and Global Politics*, p. 25. References removed.

114. Leonid E. Grinin and Andrey V. Korotayev, "Origins of Globalization in the Framework of the AfroEurasian World-system History," *Journal of Globalization Studies* 5, #1 (May 2014): 32–64.

115. World Bank DataBank, "Gross Domestic Product 2016."

116. GlobalEconomy.com, "Global Exports," 2017, accessed August 17, 2017, http://www.theglobaleconomy.com/guide/article/65/.

117. MahiFX, "MahiFX Infographic: The $5.3 trillion Forest Market Explained," *Cision PR Newswire*, Oct. 24, 2013, accessed August 17, 2017, http://www.prnewswire.com/news -releases/mahifx-infographic-the-53-trillion-forex-market-explained-229134291.html. But see also Anirban Nag and Jamie McGeever, "Foreign Exchange, the World's Biggest Market, is Shrinking," *Reuters*, Feb. 11, 2016, acessed August 17, 2017, http://www.reuters.com/article/us -global-fx-peaktrading-idUSKCN0VK1UD.

118. Jacob Davidson, "Here's How Many Internet Users There Are," *Time.com Money*, May 26, 2015, accessed August 17, 2017, http://time.com/money/3896219/internet-users-worldwide/.

119. The Radicati Group, "Email Statistics Report, 2015-2019," March 2015, accessed August 17, 2017, http://www.radicati.com/wp/wp-content/uploads/2015/02/Email-Statistics -Report-2015-2019-Executive-Summary.pdf.

120. Domo, "Data Never Sleeps 5.0," *Domo.com*, accessed August 17, 2017, https://web-assets .domo.com/blog/wp-content/uploads/2017/07/17_domo_data-never-sleeps-5-01.png.

121. World Bank, "Air Transport, Passengers Carried," accessed August 17, 2017, http://data .worldbank.org/indicator/IS.AIR.PSGR; *Statista.com*, "Leading Airlines Worldwide in 2016, by Passenger Kilometers Flown (in Billions)," accessed August 17, 2017, https://www.statista.com /statistics/270986/airlines-by-passenger-kilometers-flown/; U.S. Energy Information Administration, "International Energy Outlook 2016: Chapter 8. Transportation sector energy consumption," Sept. 2017, accessed August 17, 2017, https://www.eia.gov/outlooks/ieo/transportation.php.

122. Fields, "Unwilling Subjects of Financialization."

123. John P. Tang, "Pollution Havens and the Trade in Toxic Chemicals: Evidence from U.S. Trade Flows," *Ecological Economics* 112 (2015): 150–60; Steven Poelhekke and Frederick van der Ploeg, "Green Havens and Pollution Havens," *The World Economy* 38, #7 (July 2015): 1159–78.

124. William J. Kelly, "U.S. Trade Deals from the 90's Set up China as a Pollution Haven," *Inside Climate News*, March 6, 2014, accessed August 17, 2017, https://insideclimatenews.org /news/20140306/us-trade-deals-90s-set-china-pollution-haven; Haidong Kan, "Globalisation and Environmental Health in China," *The Lancet* 384 (Aug. 30, 2014): 721–23.

125. Gary Gereffi and Olga Memedovic, "The Global Apparel Value Chain: What Prospects for Upgrading by Developing Countries?" Vienna: UNIDO, 2003, accessed August 17, 2017, https:// www.unido.org/uploads/tx_templavoila/Global_apparel_value_chain.pdf; Pietra Rivoli, *The Travels of a T-shirt in the Global Economy*, Hoboken, NJ: Wiley, 2015, second edition.

126. Josh Lepawsky, "The Changing Geography of Global Trade in Electronic Discards: Time to Rethink the E-waste Problem," *The Geographical Journal* 181, #2 (June 2015): 147–59.

127. Philip McMichael, "Global Development and The Corporate Food Regime," in *New Directions in the Sociology of Global Development*, edited by Frederick H. Buttel and Philip McMichael, pp. 265–99, Emerald Group Publishing Limited, 2005; Carmen G. Gonzalez, "Food Justice: An Environmental Justice Critique of the Global Food System," in *International Environmental Law and the Global South*, edited by Shawkat Alam, et al., pp. 401–34, Cambridge: Cambridge University Press, 2015.

128. Edward F. Fisher and Peter Benson, *Broccoli and Desire*, Stanford: Stanford University Press, 2006.

129. Piers Blaikie and Harold Brookfield, *Land Degradation and Society*, London: Methuen, 1987; Rod Burgess, Marisa Carmona, and Theo Kolstee, eds., *The Challenge of Sustainable Cities: Neoliberalism and Urban Strategies in Developing Countries*, London: Zed Books, 1997.

130. See, for example, Vaclav Smil, *Making the Modern World: Materials and Dematerialization*, Chichester, West Sussex: Wiley and Sons, 2014; Vlad C. Corama, Åsa Moberg, and Lorenz M. Hilty, "Dematerialization Through Electronic Media," in *ICT Innovations for Sustainability*, edited by L. M. Hilty and B. Aebischer, pp. 405–21, New York: Springer, 2015.

131. L. M. Hilty and B. Aebischer, eds., "Part II—The Energy Cost of Information Processing," *ICT Innovations for Sustainability*, pp. 171–205, New York: Springer, 2015.

132. Leo Hickman, "Wind Myths: Turbines Kill Birds and Bats," *The Guardian*, Feb. 27, 2012, accessed September 14, 2017, https://www.theguardian.com/environment/2012/feb/27 /wind-energy-myths-turbines-bats.

133. But see Winnie Hu, "Your Uber Car Creates Congestion."

134. United Nations, "The Millennium Development Goals Report 2015," accessed August 18, 2017, http://www.un.org/millenniumgoals/2015_MDG_Report/pdf/MDG%202015%20rev%20 (July%201).pdf.

135. Martin Sandbu, "Critics Question Success of UN's Millennium Development Goals," *Financial Times*, Sept. 14, 2015, accessed August 18, 2017, https://www.ft.com/content/51d1c0aa-5085-11e5-8642-453585f2cfcd; Maya Fehling, Brett D. Nelson, and Sridhar Venkatapuram, "Limitations of the Millennium Development Goals: A Literature Review," *Global Public Health* 8, #1 (2013): 1109–22.

136. United Nations, "Sustainable Development Goals Kick Off with Start of New Year," Dec. 30, 2015, accessed August 17, 2017, http://www.un.org/sustainabledevelopment/blog/2015/12/sustainable-development-goals-kick-off-with-start-of-new-year/.

137. Tomáš Hák, Svatava Janoušková, and Bedřich Moldan, "Sustainable Development Goals: A Need for Relevant Indicators," *Ecological Indicators* 60 (2016): 565–73.

138. Filka Sekulova, et al., "Special Issue: Degrowth: From Theory to Practice," *Journal of Cleaner Production* 38, (Jan. 2013): 1–98; Arturo Escobar, "Degrowth, Postdevelopment, and Transitions: A Preliminary Conversation," *Sustainability Science* 10 (2015): 451–62.

139. Anders Fremstad, "Does Craigslist Reduce Waste? Evidence from California and Florida," *Ecological Economics* 132 (2017): 135–42.

140. Silja Samerski, "Tools for Degrowth? Ivan Illich's Critique of Technology Revisited," *Journal of Cleaner Production* (in press), accessed August 18, 2017, http://www.sciencedirect.com/science/article/pii/S0959652616316377?via%3Dihub; Kersty Hobson and Nichlas Lynch, "Diversifying and De-growing the Circular Economy: Radical Social Transformation in a Resource-scarce World," *Futures* 82 (2016): 15–25.

141. Lipschutz, "Getting Out of the CAR."

CHAPTER 4: CIVIC POLITICS AND SOCIAL ACTIVISM

1. Justin Gillis, "The Real Unknown of Climate Change: Our Behavior," *New York Times*, September 18, 2017, accessed September 23, 2017, https://nyti.ms/2y99xuj.

2. Karen Akerlof, et al., "Do People 'Personally Experience' Global Warming, and if so How, and Does It Matter?" *Global Environmental Change* 23 (2013): 81–91; Joel Achenbach, "Why Do Many Reasonable People Doubt Science?" *National Geographic*, March 2015, accessed September 25, 2017, http://ngm.nationalgeographic.com/2015/03/science-doubters/achenbach-text.

3. Arendt, *The Human Condition*.

4. Lipschutz and Conca, *The State and Social Power*; Lipschutz with Rowe, *Globalization, Governmentality, and Global Politics*.

5. Ronnie D. Lipschutz, "Environmental History, Political Economy and Policy: Re-discovering Lost Frontiers in Environmental Research." *Global Environmental Politics* 1, no. 3 (August 2001): 72–91; Doreen Massey, "Places and Their Pasts," *History Workshop Journal* 39 (1995): 182–92; Denis Cosgrove, "Geography Is Everywhere: Culture and Symbolism in Human Landscapes," in *Horizons in Human Geography*, edited by Derek R. Gregory and Rex Walford, pp. 118–35, London: Macmillan, 1985.

6. Whose *locus classicus* is Rachel Carson's *Silent Spring*.

7. Arendt, *The Human Condition*, p. 199.

8. Alejandro Colas, *International Civil Society: Social Movements in World Politics*, Cambridge, MA: Blackwell, Polity Press, 2002; Michael Edwards, ed., *The Oxford Handbook of Civil Society*, Oxford: Oxford University Press, 2011; Michael Edwards, *Civil Society*, Cambridge: Polity Press, 2014.

9. Robert Putnam, *Bowling Alone: The Collapse and Revival of American Community*, New York: Simon and Schuster, 2000.

10. Lipschutz, with Mayer, *Global Civil Society*.

11. Lipschutz, "Environmental History."

12. Karl Marx, Capital, volume 1, *The Process of Production of Capital*, chapter 7, accessed August 18, 2017, https://www.marxists.org/archive/marx/works/1867-c1/ch07.htm.

13. H. H. Gerth and C. Wright Mills, eds., *From Max Weber: Essays in Sociology*, p. 200, New York: Oxford University Press, 1946.

14. Karl Marx, *Eighteenth Brumaire*, chapter 1.

15. Pierre Bourdieu, *Outline of a Theory of Practice*, Cambridge: Cambridge University Press, 1972, 1977.

16. William Cronon, *Nature's Metropolis: Chicago and the Great West*, New York: Norton, 1991.

17. Morrison, "The Human Face of the Land."

18. David E. Nye, ed., *Technologies of Landscapes: From Reaping to Recycling*, Amherst: University of Massachusetts Press, 1999; Morrison, "The Human Face of the Land."

19. Thomas Dunlap, *Nature and the English Diaspora*, Cambridge: Cambridge University Press, 1999.

20. O'Connor, "Three Ways to Look at the Ecological History."

21. Robert Elliot, *Faking Nature: The Ethics of Environmental Restoration*, London: Routledge, 1997.

22. Lipschutz, with Mayer, *Global Civil Society*, chapter 4; Sheila Jasanoff, "Future Imperfect: Science, Technology, and the Imaginations of Modernity," in *Dreamscapes of Modernity: Sociotechnical Imaginaries and the Fabrication of Power*, edited by S. Jasanoff and S.-H. Kim, pp. 1–33, Chicago: University of Chicago Press, 2015.

23. Jack M. Hollander, *The Real Environmental Crisis: Why Poverty, Not Affluence, Is the Environment's Number One Enemy*, Berkeley: University of California Press, 2003.

24. Piers Blaikie, *The Political Economy of Soil Erosion in Developing Countries*, New York: Wiley, 1985; Piers Blaikie and Harold Brookfield, *Land Degradation and Society*, London: Methuen, 1987.

25. Lipschutz, with Mayer, *Global Civil Society*, pp. 242–245; John Agnew, "Representing Space: Space, Scale, and Culture in Social Science," in *Place/Culture/Representation*, edited by James Duncan and David Ley, pp. 251–71, 262, London: Routledge, 1993.

26. Olson, *Logic of Collective Action*; Ostrom, *Governing the Commons*.

27. Jean-Jacques Rousseau, *The Social Contract and Discourses*, translated by G. D. H. Cole, p. 238, New York: Dutton, 1950.

28. Olson, *Logic of Collective Action*; Waltz, *Man, the State, and War*.

29. C. Cramer, "*Homo Economicus* Goes to War: Methodological Individualism, Rational Choice and the Political Economy of War," *World Development* 30, no. 11 (Nov. 2002): 1845–64.

30. Emma Rothschild, *Economic Sentiments*, chapter 5.

31. Smith, *Theory of Moral Sentiments*; Hirsch, *Social Limits to Growth*.

32. Ophuls and Boyan, *Ecology and the Politics of Scarcity Revisited*; Heilbroner, *An Inquiry into the Human Prospect*.

33. Susan J. B. Buck, "No Tragedy on the Commons," *Environmental Ethics* 7 (Spring 1985): 49–61; David Feeny, Fikret Berkes, Bonnie J. McCay, and James M. Acheson, "The Tragedy of the Commons: Twenty-Two Years Later," *Human Ecology* 18 (1990): 1–19; Ostrom, "Collective Action and the Evolution of Social Norms."

34. Ostrom, *Governing the Commons*; Bromley, *Making the Commons Work*, San Francisco: ICS Press, 2002; "The Commons and the Common."

35. Philippe Fontaine, "Making Use of the Past: Theorists and Historians on the Economics of Altruism," *European Journal of the History of Economic Thought* 7 (2000): 407–22; Shankar Vedantam, "Does Studying Economics Make You Selfish?" *NPR*, February 21, 2017, accessed August 29, 2017, http://www.npr.org/2017/02/21/516375434/does-studying-economics-make-you-selfish.

36. Michel Crozier, Samuel P. Huntington, and Joji Watanuki, *The Crisis of Democracy: Report on the Governability of Democracies to the Trilateral Commission*, New York: New York University Press, 1975; Samuel P. Huntington, *American Politics: The Promise of Disharmony*, Cambridge: Harvard University Press, Belknap Press, 1981.

37. For an early, conservative expression of this sentiment, see Lewis J. Powell, Jr., "Attack on American Free Enterprise System," U.S. Chamber of Commerce, August 23, 1971, accessed August 29, 2017, http://law2.wlu.edu/powellarchives/page.asp?pageid=1251.

38. Avner de-Shalit, *Why Posterity Matters: Environmental Policies and Future Generations*, London: Routledge, 1995; Edith Brown Weiss, *In Fairness to Future Generations: International Law, Common Patrimony, and Intergenerational Equity*, Dobbs Ferry, NY: Transnational, 1989.

39. Thomas Princen, "Consumption and Its Externalities," p. 19.

40. This point is similar to John Rawl's "Veil of Ignorance," described in *A Theory of Justice*, Cambridge, MA: Harvard University Press, 1971.

41. Carolyn Raffensperger and Joel A. Tickner, *Protecting Public Health and the Environment: Implementing the Precautionary Principle*, Washington, DC: Island Press, 1999; Tim O'Riordan,

James Cameron, and Andrew Jordan, eds., *Reinterpreting the Precautionary Principle*, London: Cameron May, 2001.

42. This, in essence, is the argument presented in John McMurtry, *Value Wars: The Global Market versus the Life Economy*, London: Zed Books, 2002.

43. James Coleman, *Foundations of Social Theory*, p. 242, Cambridge: Harvard University Press, Belknap Press, 1990; Steinberg, *Who Rules the Earth?*

44. Magnusson and Shaw, *Political Space*.

45. Stone, *Should Trees Have Standing?*; Turkewitz, "Corporations Have Rights"; Alessandro Pelizzon and Monica Gagliano, "The Sentience of Plants: Animal Rights and Rights of Nature Intersecting?" *Australian Animal Protection Law Journal* 11, no. 5 (2015): 5–13.

46. Karl Marx, *The German Ideology*, London: Lawrence and Wishart, 1938.

47. Rosenberg, *Empire of Civil Society*.

48. Rosenberg, *Empire of Civil Society*, chapter 5.

49. Carole Pateman, "Feminist Critiques of the Public/Private Dichotomy," in *Public and Private in Social Life*, edited by Stanley I. Benn and Gerald F. Gaus, pp. 281–303, London: Croom Helm. 1983.

50. Pateman, "Feminist Critiques of the Public/Private Dichotomy."

51. Bourgeois activity: Karl Marx, "On the Jewish Question," in *Karl Marx and Frederich Engels: Collected Works*, vol. 3, *Marx and Engels: 1843–1844*, London: Lawrence and Wishart, 1844, 1975. Central to public polities: Hegel, *The Philosophy of History*; Putnam, *Bowling Alone*. See also: Ronnie D. Lipschutz, "Power, Politics, and Global Civil Society," *Millennium* 33, no. 3 (2005): 747–69.

52. Colas, *International Civil Society*.

53. Ronnie D. Lipschutz, "Reconstructing World Politics: The Emergence of Global Civil Society," *Millennium* 21 (winter 1992/93): 389–420; Helmut Anheier, Marlies Glasius, and Mary Kaldor, eds., *Global Civil Society*, Oxford: Oxford University Press, 2001.

54. Colas, *International Civil Society*.

55. Crozier, Huntington, and Watanuki, *The Crisis of Democracy*; Huntington, *American Politics*.

56. Timothy W. Luke, "On the Political Economy of Clayoquot Sound: The Uneasy Transition from Extractive to Attractive Models of Development," in Magnusson and Shaw, *A Political Space*, pp. 91–112.

57. Ronnie D. Lipschutz, "The Environment and Global Governance," in *Global Governance in the Twenty-First Century*, edited by John N. Clarke and Geoffrey R. Edwards, Basingstoke, England: Palgrave Macmillan, 2004.

58. Samuel P. Hays, *Beauty, Health, and Permanence: Environmental Politics in the United States, 1955–1985*, Cambridge: Cambridge University Press, 1987; Linda Lear, *Rachel Carson: Witness for Nature*, New York: Henry Holt, 1997; William Souder, *On a Farther Shore: The Life and Legacy of Rachel Carson*, New York: Broadway Books, 2013.

59. Gerard J. DeGroot, ed., *Student Protest: The Sixties and After*, London: Longman, 1998.

60. Victor Cohn, *1999: Our Hopeful Future*, Indianapolis: Bobbs-Merrill, 1956.

61. See, for example, Harrison Brown, *The Challenge of Man's Future*, New York: Viking, 1954.

62. Lear, *Rachel Carson*; Souder, *Farther Shore*.

63. Sidney Tarrow, *Power in Movement: Social Movements and Contentious Politics*, Cambridge: Cambridge University Press, 2011, third edition, revised and updated.

64. Chris Harman, *The Fire Last Time: 1968 and After*, London: Bookmarks, 1998, second edition; Barbara Ehrenreich and John Ehrenreich, *Long March, Short Spring: The Student Uprising at Home and Abroad*, New York: Monthly Review Press, 1969.

65. Lipschutz with Rowe, *Globalization, Governmentality and Global Politics*.

66. Richard J. Lazarus, "A Different Kind of 'Republican Moment' in Environmental Law, *Minnesota Law Review* 87 (2003): 999–1035; Daniel A. Farber, "The Conservative as Environmentalist: From Goldwater and the Early Reagan to the 21st Century," University of California Berkeley Public Law Research Paper, July 14, 2017, accessed August 30, 2017, https://papers.ssrn.com/sol3/Delivery.cfm/SSRN_ID3000045_code102259.pdf?abstractid=2919633andmirid=1.

67. Lipschutz, with Mayer, *Global Civil Society*, chapter 5; Fred Rose, *Coalitions across the Class Divide: Lessons from the Labor, Peace, and Environmental Movements*, Ithaca, NY: Cornell University Press, 2000; Jane I. Dawson, *Eco-Nationalism: Anti-Nuclear Activism and National Identity in Russia, Lithuania, and Ukraine*, Durham, NC: Duke University Press, 1996.

68. Richard Howitt, *Rethinking Resource Management: Justice, Sustainability, and Indigenous Peoples*, London: Routledge, 2001; Rob Nixon, "Slow Violence, Gender, and the Environmentalism of the Poor," *Journal of Commonwealth and Postcolonial Studies* 13, no. 2/14, no. 1 (2006–2007): 14–37.

69. Ronald Inglehart, *Culture Shift in Advanced Industrial Society*, Princeton, NJ: Princeton University Press, 1990; Francesca Ferrando, "The Party of the Anthropocene: Post-humanism, Environmentalism and the Post-anthropocentric Paradigm Shift," *Relations: Beyond Anthropocentrism* 4 (2016): 159–73.

70. Manuel Castells, *The Rise of the Network Society*, Cambridge, MA: Blackwell, 1996.

71. Kendall Dent, Nadia Zainuddin, and Leona Tam, "The Good Life: Exploring Values Creation and Destruction in Consumer Well-Being," in *Marketing at the Confluence between Entertainment and Analytics*, edited by Patricia Rossi, pp. 269–72, Cham, Switzerland: Springer, 2017.

72. Luke, *Ecocritique*; Marx, *The German Ideology*; Aasha Sharma, "Green Consumerism: Overview and Further Research Directions," *International Journal of Process Management and Benchmarking* 7, no. 2 (2017): 206–23.

73. Lipschutz with Rowe, *Globalization, Governmentality and Global Politics*.

74. Brian Milani, *Designing the Green Economy: The Postindustrial Alternative to Corporate Globalization*, chapter 11, Lanham, MD: Rowman & Littlefield, 2000.

75. Joel Makower, "Consumer Power," *Our Future, Our Environment*, edited by Noreen Clancy and David Rejeski, Santa Monica, CA: RAND, March 2001, accessed August 18, 2017, http://smapp.rand.org/ise/ourfuture/Welcome/toc.html.

76. Julie H. Guthman, *Agrarian Dreams? The Paradox of Organic Farming in California*, Oakland: University of California Press, 2014, second edition; Catherine R. Greene, "U.S. Organic Farming Emerges in the 1990s: Adoption of Certified Systems," Agriculture information bulletin no. 770, U.S. Dept. of Agriculture, Economic Research Service, Washington, DC, 2001, accessed October 3, 2017, https://www.ers.usda.gov/webdocs/publications/42396/31544_aib770_002.pdf?v=42487.

77. Nicola S. Schutte and Navjot Bhullar, "Approaching Environmental Sustainability: Perceptions of Self-Efficacy and Changeability," *The Journal of Psychology* 151, no. 3 (2017): 321–33; Sonya Sachdeva, Jennifer Jordan, and Nina Mazar, "Green consumerism: moral motivations to a sustainable future," *Current Opinion in Psychology* 6 (2015): 60–65.

78. Thomas Davies, Holly Eva Ryan, and Alejandro Milcíades Peña, "Protest, Social Movements and Global Democracy since 2011: New Perspectives," *Protest, Social Movements and Global Democracy since 2011: New Perspectives*, edited by Thomas Davies, Holly Eva Ryan, and Alejandro Milcíades Peña, 1–29. *Research in Social Movements, Conflicts and Change*, Vol. 39 (2016).

79. Bob Edwards and Melinda Kane, "Resource Mobilization and Social and Political Movements," pp. 205–32, in *Handbook of Political Citizenship and Social Movements*, edited by Hein-Anton van der Heijden, Cheltenham, UK: Edward Elgar, 2014; Bob Edwards and Patrick F. Gillham, "Resource Mobilization Theory," in *Wiley Blackwell Encyclopedia of Social and Political Movements*, Oxford: Blackwell, 2013, accessed August 30, 2017, http://www.academia.edu/download/30829255/Resource_Mobilization_Theory_and_Social_Movements.pdf.

80. Tarrow, *Power in Movement*.

81. Tarrow, *Power in Movement*; Brayden King, "A Social Movement Perspective of Stakeholder Collective Action and Influence," *Business and Society* 47, no. 1 (March 2008): 21–49.

82. Jeffrey M. Ayers, "Framing Collective Action Against Neoliberalism: The Case of the "Anti-Globalization Movement," *Journal of World Systems Research* 1 (2004): 11–34; Cristina Flesher Fominaya, *Social Movements and Globalization: How Protests, Occupations, and Uprisings are Changing the World*, Basingstoke, UK: Palgrave Macmillan, 2014. While the size of global justice mobilizations declined in the post-9/11 security environment, the past decade has also seen other waves of large-scale social and environmental protest ("new" new social movements?), such as blockades of oil pipeline construction in the United States, Black Lives Matter mobilizations against institutionalized racism, Occupy and *indignados* against austerity and economic inequality, and the Ende Gelände and other mine occupations in Germany.

83. George Soros, *Open Society: Reforming Global Capitalism*, New York: Public Affairs, 2000; Joseph E. Stiglitz, *Globalization and Its Discontents*, New York: Norton, 2000; Ayers, "Framing Collective Action."

84. Thaddeus C. Trzyna with Julie Didion, *World Directory of Environmental Organizations*, London: Routledge, 2017, sixth edition, lists more than two thousand, the World Association

of Non-governmental Associations lists almost thirty-eight hundred ("Worldwide NGO Directory," 2017, accessed August 30, 2017, http://www.wango.org/resources.aspx?section=ngodirand sub=regionandregionID=155andcol=ff9900), but these are almost surely an undercount; if we assume one group for every one hundred thousand people, that adds up to seventy thousand worldwide.

85. Trzyna with Didion, *World Directory of Environmental Organizations*; World Association of Non-governmental Associations, "Worldwide NGO Directory"; also Anheier, Glasius, and Kaldor, *Global Civil Society 2001*; Marlies Glasius, Mary Kaldor, and Helmut Anheier, eds., *Global Civil Society 2002*, Oxford: Oxford University Press, 2002. It might be added here that nationalist and xenophobic movements, some of whom invoke themes of "blood and soil," are also regarded in part as responses to globalization and economic change.

86. Friedman, *The Lexus and the Olive Tree*.

87. Lipschutz with Rowe, *Globalization, Governmentality, and Global Politics*.

88. United Nations, "Earth Summit (1992)," 1997, accessed August 30, 2017, http://www.un.org/geninfo/bp/enviro.html; *The Local*, "The Paris COP21 Climate Summit in Numbers," November 30, 2015, accessed August 30, 3017, https://www.thelocal.fr/20151130/cop-21-in-numbers-the-facts-and-figures-to-know.

89. *The Local*, "The Paris COP21 Climate Summit in Numbers."

90. Lloyd Gruber, *Ruling the World: Power Politics and the Rise of Supranational Institutions*, Princeton, NJ: Princeton University Press, 2000.

91. Ronnie D. Lipschutz, "Doing Well by Doing Good? Transnational Regulatory Campaigns, Social Activism, and Impacts on State Sovereignty," in *Challenges to Sovereignty: How Governments Respond*, edited by John Montgomery and Nathan Glazer, pp. 291–320, New Brunswick, NJ: Transaction, 2002.

92. "German Federal Election, 2017," *Wikipedia*, accessed September 26, 2017, https://en.wikipedia.org/wiki/German_federal_election,_2017.

93. Russell J. Dalton, *The Green Rainbow: Environmental Groups in Western Europe*, New Haven, CT: Yale University Press, 1994; Wolfgang Rümdig, ed., *Green Politics Two*, Edinburgh: Edinburgh University Press, 1992.

94. Fritjof Capra and Charlene Spretnak, *Green Politics: The Global Promise*, New York: Dutton, 1984; Thomas Poguntke, "Changing or Getting Changed: The Example of the German Greens since 1979," in *Parties, Governments and Elites*, edited by Philipp Harst, Ina Kubbe and Thomas Poguntke, pp. 87–103, Wiesbaden, FRG: Springer, 2017.

95. Bomberg, *Green Parties and Politics*; Pogunte, "Changing or Getting Changed."

96. Bomberg, *Green Parties and Politics*, pp. 95–96.

97. Bomberg, *Green Parties and Politics*, pp. 95–96.

98. The Greens/European Free Alliance in the European Parliament, "Members," accessed August 19, 2017, https://www.greens-efa.eu/en/our-group/meps/.

99. *Politico*, Europe Edition, "Swedish Greens Get the Coalition Blues," May 9, 2017, accessed August 19, 2017, http://www.politico.eu/article/swedish-greens-get-the-coalition-blues/.

100. Green Party US, "Officeholders," accessed August 19, 2017, http://www.gp.org/officeholders.

101. Lipschutz, with Mayer, *Global Civil Society*.

102. Andreas Reckwitz, "Toward A Theory of Social Practices—A Development in Culturalist Theorizing," *European Journal of Social Theory* 5, no. 2 (2002): 243–63, p. 249.

103. Elizabeth Shove, *Comfort, Cleanliness and Convenience: The Social Organization of Everyday Life*, Oxford: Berg, 2003.

104. Yolande Strengers, "Peak Electricity Demand and Social Practice Theories: Reframing the Role of Change Agents in the Energy Sector," *Energy Policy* 44 (2012): 226–34, p. 229.

105. Catherine Butler, Karen A. Parkhill, and Nicholas F. Pidgeon, "Energy Consumption and Everyday Life: Choice, Values and Agency through a Practice Theoretical Lens," *Journal of Consumer Culture* 16, no. 3 (2016): 887–90, p. 890. For an example of how practice theory applies to laundry, see S. Higginson, E. McKenna, and M. Thomson, "Can Practice Make Perfect (Models)? Incorporating Social Practice Theory into Quantitative Energy Demand Models," paper presented at: Behave 2014—Paradigm Shift: From Energy Efficiency to Energy Reduction through Social Change, 3rd Behave Energy Conference, Oxford, UK, September 3–4, 2014, accessed November 22, 2014, https://dspace.lboro.ac.uk/dspace-jspui/bitstream/2134/15975/1/Higginson,%202014,%20Can%20practice%20make%20perfect%20models.pdf.

106. Rui Neiva and Jonathan L. Gifford, "Declining Car Usage among Younger Age Cohorts: An Exploratory analysis of Private Car and Car-Related Expenditures in the United States," School of Public Policy, George Mason University, GMU School of Public Policy Research Paper No. 2012-10, September 10 2012, accessed July 23, 2015, http://papers.ssrn.com/sol3/papers.cfm?abstract_id=2028230; Elizabeth Rosenthal, "The End of Car Culture," *New York Times*, June 29, 2013, accessed December 28, 2014, http://www.nytimes.com/2013/06/30/sunday-review/the-end-of-car-culture; Jeffrey Ball, "The Proportion of Young Americans Who Drive Has Plummeted—And No One Knows Why," *The New Republic*, March 12, 2014, accessed December 28, 2014, http://www.newrepublic.com/article/116993/millennials-are-abandoning-cars-bikes-carshare-will-it-stick.

107. Ball, "Proportion of Young Americans."

108. See also Lipschutz, "Getting out of the CAR."

109. Colas, *International Civil Society*; Walter L. Adamson, *Hegemony and Revolution: A Study of Antonio Gramsci's Political and Cultural Theory*, Berkeley: University of California Press, 1980.

110. Robert Cox, *Production, Power, and World Order*, New York: Columbia University Press, 1987; Rosenberg, *Empire of Civil Society*.

111. Thomas Piketty, *Capital in the Twenty-First Century*, Cambridge: Harvard University Press, 2014.

112. G. William Domhoff, *Who Rules America? Power and Politics*, Boston: McGraw-Hill, 2002, fourth edition; Thomas R. Dye, *Who's Running America? The Bush Restoration*, Upper Saddle River, NJ: Prentice Hall, 2002, seventh edition; Robert Pollin, *Contours of Descent: U.S. Economic Fractures and the Landscape of Global Austerity*, London: Verso, 2003.

113. Marshall Berman, *All That Is Solid Melts Into Air: The Experience of Modernity*, New York: Simon and Schuster, 1982; Walden Bello, *Dark Victory: The United States and Global Poverty*, Oakland, CA: Food First Books, 1999; Hans-Peter Martin and Harald Schumann, *The Global Trap: Globalization and the Assault on Democracy and Prosperity*, translated by P. Camillar, London: Zed Books, 1997; Eric Sheppard, Philip W. Porter, David R. Faust, and Richa Nagar, *A World of Difference: Encountering and Contesting Development*, New York: Guilford Press, 2009, second edition; Stiglitz, *Globalization*.

114. Manny Fernandez and Richard Fausset, "A Storm Forces Houston, the Limitless City, to Consider Its Limits," *New York Times*, August 30, 2017, accessed August 31, 2017, https://nyti.ms/2wpnVPq. Julie Turkewitz and Audra D. S. Burch, "Storm with 'No Boundaries' Took Aim at Rich and Poor Alike," *New York Times*, August 31, 2017, accessed August 31, 2017, https://nyti.ms/2xBjC3H.

115. Stiglitz, *Globalization*; Sen, *Development as Freedom*; Jagdish Bhagwati, *Free Trade Today*, Princeton, NJ: Princeton University Press, 2002; Dani Rodrik, *The Globalization Paradox*, London: W.W. Norton and Co, 2011.

116. Bob Jessop, "Primacy of the Economy, Primacy of the Political: Critical Theory of Neoliberalism," in *Handbuch Kritische Theorie*, edited by U. Bittlingmayer, Demirović, and T. Freya, Springer Reference, 2016, accessed September 2, 2017, DOI 10.1007/978-3-658-12707-7_46-1.

117. Jonathan A. Fox and L. David Brown, eds., *The Struggle for Accountability: The World Bank, NGOs, and Grassroots Movements*, Cambridge: MIT Press, 1998; Center for International Environmental Law, "NGO Response: Proposed World Bank Standards Represent Dangerous Set-back to Key Environmental and Social Protections," July 22, 2016, accessed September 2, 2017, http://www.ciel.org/news/safeguard-policy-endangers-rights/; Chris Humphrey, "The Problem with Development Banks' Environmental and Social Safeguards," Overseas Development Institute, April 14, 2016, accessed September 2, 2017, https://www.odi.org/comment/10379-problem-development-banks-environmental-social-safeguards-mdbs.

118. Soumodip Sarkar and Oleksiy Osiyevskyy, "Organizational Change and Rigidity during Crisis: A Review of the Paradox," *European Management Journal*, 36 (2–18): 47–58, accessed September 2, 2017, http://www.sciencedirect.com.oca.ucsc.edu/science/article/pii/S026323731730052X/pdfft?md5=ea2293df6c38e294acb0cb7d05d11f67andpid=1-s2.0-S026323731730052X-main.pdf.

119. Frank Biermann and Steffan Bauer, eds., *A World Environmental Organization: Solution or Threat for Effective International Environmental Governance?* London: Routledge, 2005.

120. See, for example, Avraham Ebenstein, Ann Harrison, and Margaret McMillan, "Why are American Workers Getting Poorer? China, Trade and Offshoring," National Bureau of Economic Research, March 2015, Working paper 21027, accessed September 2, 2017, http://www.nber.org/papers/w21027.

121. Margaret Keck and Kathryn Sikkink, *Activists Beyond Borders: Advocacy Networks in International Politics*, Ithaca, NY: Cornell University Press, 1998.

122. Stephen J. Kobrin, "The MAI and the Clash of Globalizations," *Foreign Policy*, fall 1998. For more recent analyses, see: Mario Pianta, "Slowing Trade: Global Activism Against Trade Liberalization," *Global Policy* 5, no. 2 (May 2014): 214–21; Dirk De Bièvre, "A Glass Quite Empty: Issue Groups' Influence in the Global Trade Regime," *Global Policy* 5, no. 2 (May 2014): 222–28.

123. Ronnie D. Lipschutz, "The Clash of Governmentalities: The Fall of the UN Republic and America's Reach for Imperium," *Contemporary Security Policy* 23, no. 3 (Dec. 2002): 214–31; Kara Fox and James Masters, "G20 Protests: Police, Demonstrators Clash in Germany," CNN, July 6, 2017, accessed September 2, 2017, http://www.cnn.com/2017/07/06/europe/hamburg-protests-g20/index.html.

124. Idle No More, "The Story," accessed August 23, 2017, http://www.idlenomore.ca/story.

125. Dina Gilio-Whitaker, "Idle No More and Fourth World Social Movements in the New Millenium," *South Atlantic Quarterly* 114 (2015): 866–77.

126. Felicity Barringer, "Flooded Village Files Suit, Citing Corporate Link to Climate Change," *New York Times*, February 27, 2008, accessed August 23, 2017, http://www.nytimes.com/2008/02/27/us/27alaska.html?_r=1andref=slogin; Felix Gaedtke, "On Thin Ice: Sami Reindeer Herders Fear Climate Change Consequences in the Arctic," *A More Vulnerable World*, December 1, 2015, accessed August 23, 2017, https://climate.earthjournalism.net/2015/12/01/on-thin-ice-sami-reindeer-herders-fear-climate-change-consequences-in-the-arctic/; *Reuters*, "Indigenous Group in Peru Vows to Block Oil Drilling on Amazon Ancestral Land," December 9, 2016, accessed August 23, 2017, https://www.theguardian.com/world/2016/dec/09/peru-amazon-oil-indigenous-achuar-geopark-petroperu.

127. 350.org, "Stop the Keystone XL Pipeline," accessed August 23, 2017, https://350.org/stop-keystone-xl/.

128. DeNeen L. Brown, "Hundreds of Keystone XL Pipeline Opponents Arrested at White House," *Washington Post*, March 2, 2014, accessed August 23, 2017, https://www.washingtonpost.com/blogs/local/wp/2014/03/02/hundreds-of-keystone-xl-pipeline-opponents-arrested-at-white-house/?utm_term=.b4c9cfe6514d.

129. Barack Obama, "Statement by the President on the Keystone XL Pipeline," The White House, November 6, 2015, accessed August 23, 2017, https://obamawhitehouse.archives.gov/the-press-office/2015/11/06/statement-president-keystone-xl-pipeline.

130. EarthJustice, "The Standing Rock Sioux Tribe's Litigation Against the Dakota Access Pipeline," EarthJustice, accessed August 23, 2017, http://earthjustice.org/features/faq-standing-rock-litigation; "Standing with Standing Rock," EarthJustice, accessed August 23, 2017, http://earthjustice.org/features/teleconference-standing-rock.

131. Sophie Lewis, "Veterans Unite for Second 'Deployment' Against the Dakota Access Pipeline," *cnn.com*, February 9, 2017, accessed August 23, 2017, http://www.cnn.com/2017/02/09/us/standing-rock-veterans-trnd/index.html.

132. Phil McKenna, "Keystone XL: Environmental and Native [sic] Groups Sue to Halt Pipeline," *Inside Climate News*, March 30, 2017, accessed September 26, 2017, https://insideclimatenews.org/news/30032017/keystone-xl-pipeline-lawsuit-trump-state-department; Kevin O'Hanlon, "Keystone XL Pipeline Fate in Balance as Nebraska Opens Hearings," *Reuters*, August 7, 2017, accessed September 26, 2017, https://www.reuters.com/article/us-usa-pipeline-keystone/keystone-xl-pipeline-fate-in-balance-as-nebraska-opens-hearings-idUSKBN1AN17P.

133. Christopher M. Matthews and Bradley Olson, "After $3 Billion Spent, Keystone XL Can't Get Oil Companies to Sign On," *FoxBusiness*, June 29, 2017, accessed September 26, 2017, http://www.foxbusiness.com/features/2017/06/29/after-3-billion-spent-keystone-xl-cant-get-oil-companies-to-sign-on.html.

134. C. Lee, "Toxic Waste and Race in the United States," in *Race and the Incidence of Environmental Hazards: A Time for Discourse*, edited by B. I. Bryant and P. Mohai, Boulder, CO: Westview Press, 1992.

135. NAACP, "NAACP Environmental and Climate Justice Program," accessed August 23, 2017, http://www.naacp.org/environmental-climate-justice-about/.

136. David Schlosberg and Lisette B. Collins, "From Environmental to Climate Justice: Climate Change and the Discourse of Environmental Justice," *WIREs Climate Change* (2014), doi: 10.1002/wcc.275.

137. Lipschutz with Rowe, *Globalization, Governmentality and Global Politics*; K. Ravi Raman and Ronnie D. Lipschutz, eds., *Corporate Social Responsibility: Comparative Critiques*, Basingstoke, Houndsmill: Palgrave Macmillan, 2010; Jeremy Moon, *Corporate Social Responsibility: A Very Short Introduction*, Oxford: Oxford University Press, 2014.

138. Lipschutz with Rowe, *Globalization, Governmentality, and Global Politics*; Raman and Lipschutz, *Corporate Social Responsibility*.

139. Jeffrey Leonard, *Pollution and the Struggle for the World Product: Multinational Corporations, Environment, and International Comparative Advantage*, Cambridge: Cambridge University Press, 1988; M. Mani and D. Wheeler, "In Search of Pollution Havens? Dirty Industries in the World Economy, 1960–1995." PRDEI, The World Bank, April 1997. Accessed April 12, 2002, http://www.worldbank.org/research/peg/wps16/index.htm; Poelhekke and van der Ploeg, "Green Havens and Pollution Havens."

140. Edna Bonacich and Richard Appelbaum, *Behind the Label: Inequality in the Los Angeles Apparel Industry*, Berkeley: University of California Press, 2000; Poelhekke and van der Ploeg, "Green Havens and Pollution Havens."

141. Tim Bartley, et al. *Looking Behind the Label: Global Industries and the Conscientious Consumer*, Bloomington: Indiana University Press, 2015.

142. Vinicius Brei and Steffen Böhm, "'1L=10L for Africa': Corporate Social Responsibility and the Transformation of Bottled Water into a 'Consumer Activist' Commodity," *Discourse and Society* (2013): 1–29.

143. Virginia Haufler, *A Public Role for the Private Sector: Industry Self-Regulation in a Global Economy*, Washington, DC: Carnegie Endowment for International Peace, 2001.

144. Joseph Cascio, Gayle Woodside, and Philip Mitchell, *ISO 14000 Guide: The New International Environmental Management Standards*, New York: McGraw-Hill, 1996; Jennifer Clapp, *Toxic Exports: The Transfer of Hazardous Wastes from Rich to Poor Countries*. Ithaca: Cornell University Press, 2001, chapter 6; Jennifer Clapp, "The Privatization of Global Environmental Governance: ISO 14000 and the Developing World," in *The Business of Global Environmental Governance*, edited by David L. Levy and Peter J. Newell, Cambridge: MIT Press, 2005.

145. John G. Ruggie, "The Theory and Practice of Learning Networks: Corporate Social Responsibility and the Global Compact," *Journal of Corporate Citizenship* 5 (Spring 2002): 27–36; United Nations Global Compact, accessed September 3, 2017, https://www.unglobal compact.org/. See also the special section on "UN Global Compact Symposium" in *Journal of Business Ethics* 122, no. 2 (June 2014) for evaluation of the Compact.

146. Ruth Pearson and Gill Seyfang, "New Hope or False Dawn? Voluntary Codes of Conduct, Labour Regulation, and Social Policy in a Globalising World," *Global Social Policy* 1 (Apr. 2001): 49–78.

147. See, for example, Ronald B. Davies and K. C. Vadlamannati, "A Race to the Bottom in Labor Standards? An Empirical Investigation," *Journal of Development Economics* 103 (2013): 1–14; J. Samuel Barkin, "Racing All Over the Place: A Dispersion Model of International Regulatory Competition," *European Journal of International Relations* 21, no. 1 (2015): 171–93.

148. Cory Searcy, "Corporate Sustainability Performance Measurement Systems: A Review and Research Agenda," *Journal of Business Ethics* 107 (2012): 239–53; Karun Kumar, "Sustainability Performance Measurement: An Investigation into Corporate Performance through Environmental Indicators," *International Journal of Management Research and Review* 4, no. 2 (Feb. 2014): 192–206.

149. Aseem Prakash and Matthew Potoski, "Global Private Regimes, Domestic Public Law—ISO 14001 and Pollution Reduction," *Comparative Political Studies* 47, no. 3 (2014): 369–94, p. 369.

150. Tim Bartley, et al., *Looking Behind the Label*. See also the discussion of certification of sustainable forestry management in chapter 5.

151. Anil Markandya, "Eco-Labelling: An Introduction and Review," in *Eco-Labelling and International Trade*, edited by Simonetta Zarrilli, Veena Jha and René Vossenaar, pp. 1–20, Basingstoke, England: Macmillan, 1997; Ronnie D. Lipschutz, "Why Is There No International Forestry Law? An Examination of International Forestry Regulation, Both Public and Private," *UCLA Journal of Environmental Law and Policy* 19 (2001): 155–82.

152. Cascio, Woodside, and Mitchell, *ISO 14000 Guide*. A draft of the ISO 14001 2015 standards explanation can be found at http://ga.hfu.edu.tw/download.php?filename=800_0df4158b.pdf anddir=archiveandtitle=ISO+14001+2015%E5%B9%B4%E8%8B%B1%E6%96%87%E7 %89%88. The actual standards for each industry must be purchased separately.

153. Lars H. Gulbrandsen, *Transnational Environmental Governance: The Emergence and Effects of the Certification of Forests and Fisheries*, Cheltenham, UK: Edward Elgar, 2010.

154. Arno Kourula and Guillaume Delalieux, "The Micro-level Foundations and Dynamics of Political Corporate Social Responsibility: Hegemony and Passive Revolution Through Civil Society," *Journal of Business Ethics* 135 (2016): 769–85; Huan Li, Neha Khanna, and Martina Vidovic, "Third Party Certification and Self-Regulation: Evidence from Responsible Care and Accidents in the US Chemical Industry," paper prepared for presentation at the Agricultural and Applied Economics Association's 2014 AAEA Annual Meeting, Minneapolis, MN, July 27–29, 2014, accessed September 3, 2017, http://ageconsearch.umn.edu/bitstream/170492/2/Accident RC_AgEconSearch_0528.pdf.

155. Greene, "U.S. Organic Farming"; Helga Willer and Julia Lernoud, eds., *The World of Organic Agriculture*, Bonn and Frick: Research Institute of Organic Agriculture and IFOAM—Organics International, 2016, accessed September 4, 2017, http://orgprints.org/31151/1/willer-lernoud-2016-world-of-organic.pdf.

156. Stephanie Strom, "Recalls of Organic Food on the Rise, Report Says," *New York Times*, August 20, 2015, accessed September 4, 2017, https://www.nytimes.com/2015/08/21/business/recalls-of-organic-food-on-the-rise-report-says.html; R. R. Harvey, C. M. Zakhour, and L. H. Gould, "Food Disease Outbreaks Associated with Organic Foods in the United States," *Journal of Food Protection* 79 no. 11 (Nov. 2016): 1953–58.

157. Stephen Kearse, "Why are Salad Greens Always Labeled 'Triple Washed'?" *Slate*, April 27, 2016, accessed September 4, 2017, http://www.slate.com/articles/life/food/2016/04/why_are_salad_greens_always_labeled_triple_washed.html.

158. Marla Cone, "Stalking a Killer in Our Greens," *Los Angeles Times*, August 14, 2007, accessed September 4, 2017, http://www.latimes.com/local/obituaries/la-me-spinach14aug14-story.html; see also Dan Flynn, "E. coli-contaminated Lettuce Came from a California LGMA Grower," *Food Safety News*, January 15, 2013, accessed September 4, 2017, http://www.foodsafetynews.com/2013/01/contaminated-lettuce-came-from-lgma-grower/#.Wa2JKsaQzIU.

159. Archon Fung, Dana O'Rourke, and Charles Sabel, "Realizing Labor Standards: How Transparency, Competition, and Sanctions Could Improve Working Conditions Worldwide," *Boston Review* 26 (Feb./Mar. 2001), accessed August 20, 2017, http://bostonreview.net/archives/BR26.1/fung.html.

160. Bonacich and Appelbaum, *Behind the Label*; Lipschutz with Rowe, *Globalization, Governmentality, and Global Politics*.

161. Keck and Sikkink, *Activists Beyond Borders*; Paul Wapner, *Environmental Activism and World Civic Politics*, Albany: State University of New York Press, 1996; Jackie Smith and Hank Johnston, eds., *Globalization and Resistance: Transnational Dimensions of Social Movements*, Lanham, MD: Rowman & Littlefield, 2002.

162. Lipschutz, with Mayer, *Global Civil Society*, chapter 7; Patrick Novotny, *Where We Live*.

CHAPTER 5: DOMESTIC POLITICS AND GLOBAL ENVIRONMENTAL POLITICS

1. Pamela S. Chasek, David L. Downie, and Janet Welsh Brown, *Global Environmental Politics*, Boulder, CO: Westview Press, 2016, seventh edition.

2. See, for example,, Paul Steinberg, *Who Rules the Earth?*

3. Lipschutz, *When Nations Clash*.

4. Lipschutz, "Environmental History"; Nancy Langston, "Thinking Like a Microbe: Borders and Environmental History," *The Canadian Historical Review* 95, #4 (Dec. 2014): 592–603.

5. See, for example, Oran Young, *International Governance: Protecting the Environment in a Stateless Society*, Ithaca, NY: Cornell University Press, 1994; Robert O. Keohane and Mark A. Levy, eds., *Institutions for Environmental Aid: Pitfalls and Promise*, Cambridge: MIT Press, 1996.

6. Matthew Himley, "Mining History: Mobilizing the Past in Struggles over Mineral Extraction in Peru," *Geographical Review* 104, #2 (April 2014): 174–91.

7. Robert O. Keohane, *International Institutions and State Power*, pp. 1–20.

8. Krasner, *International Regimes*, p. 1.

9. Steinberg, *Who Rules the World*; Nicholas Greenwood Onuf, *World of Our Making: Rule and Rule in Social Theory and International Relations*, London: Routledge, 2013, second edition;

Nicholas Greenwood Onuf, *Making Sense, Making Worlds: Constructivism in Social Theory and International Relations*, London: Routledge, 2013.

10. Hedley Bull, *The Anarchical Society*, New York: Columbia University Press, 1977; Siniša Malešević, *The Sociology of War and Violence*, Cambridge: Cambridge University Press, 2010.

11. Robert L. Friedheim, *Toward a Sustainable Whaling Regime*, Seattle: University of Washington Press, 2001; M. J. Peterson, "Whales, Cetologists, Environmentalists, and the International Management of Whaling," *International Organization* 46 (winter 1992): 187–224.

12. See, for example, Peter Larkin, "An Epitaph for the Concept of Maximum Sustained Yield," *Transactions of the American Fisheries Society* 106 (1977): 1–11; Marc Mangel, Baldo Marinovic, Caroline Pomeroy, and Donald Croll, "Requiem for Ricker: Unpacking MSY," *Bulletin of Marine Science* 70 (2002): 763–81; Carmel Finley, *All the Fish in the Sea: Maximum Sustainable Yield and the Failure of Fisheries Management*, Chicago: University of Chicago Press, 2011.

13. Robert E. Goodin and John S. Dryzek, "Justice Deferred: Wartime Rationing and Postwar Welfare Policy," *Politics and Society* 23, #1 (March 1995): 49–73; Mark Roodhouse, "Rationing Returns: As Solution to Global Warming?" *History and Policy*, March 1, 2007, accessed August 23, 2017, http://www.historyandpolicy.org/policy-papers/papers/rationing-returns-a-solution-to-global-warming.

14. Charlotte Epstein, "Moby Dick or Moby Doll? Discourse, or How to Study the 'Social Construction of' All the Way Down," in *Constructing the International Economy*, edited by Rawi Abdelal, Mark Blyth, and Craig Parsons, pp. 175–93, Ithaca: Cornell University Press, 2010; Monder Khoury, "Whaling in Circles: The Makahs, the International Whaling Commission, and Aboriginal Subsistence Whaling," *Hastings Law Journal* 67 (Dec. 2015): 293–321.

15. See, for example, Aaron Nelsen, "Chile's Fish Supply Decline 'Catastrophic' after Years of Overfishing," *PRI.org*, May 22, 2013, accessed September 28, 2017, https://www.pri.org/stories/2013-05-22/chiles-fish-supply-decline-catastrophic-after-years-overfishing. But see also Sylvia Rowley, "How Dwindling Fish Stocks Got a Reprieve," *New York Times*, April 19, 2016, accessed August 23, 2017, https://opinionator.blogs.nytimes.com/2016/04/19/how-dwindling-fish-stocks-got-a-reprieve/?_r=0.

16. William T. Burke, *The New International Law of Fisheries: UNCLOS 1982 and Beyond*, Oxford: Clarendon Press, 1994; Elizabeth R. DeSombre and J. Samuel Barkin, *Fish*, Cambridge: Polity Press, 2011.

17. DeSombre and Barkin, *Fish*.

18. DeSombre and Barkin, *Fish*.

19. Terry L. Anderson and Donald R. Leal, "Free Market versus Political Environmentalism," *Harvard Journal of Law and Public Policy* 15 (Spring 1992): 297–310; Anderson and Leal, *Free Market Environmentalism for the Next Generation*.

20. Mark Rupert, *Producing Hegemony: The Politics of Mass Production and American Global Power*, Cambridge: Cambridge University Press, 1995; Friedman, *The Lexus and the Olive Tree*; John Whiteley, Helen Ingraham, and Richard Perry, eds., *Water, Place, and Equity*, Cambridge: MIT Press, 2008.

21. Lipschutz, "Environmental History."

22. Rosenberg, *Empire of Civil Society*; Liah Greenfeld, *The Spirit of Capitalism: Nationalism and Economic Growth*, Cambridge: Harvard University Press, 2001.

23. Peter A. Hall and David Soskice, eds., *Varieties of Capitalism: The Institutional Foundations of Comparative Advantage*, Oxford: Oxford University Press, 2001.

24. Benedict Anderson, *Imagined Communities: Reflections on the Origins and Spread of Nationalism*, London: Verso, 1991, revised edition; Ernest Gellner, *Nations and Nationalism*, Ithaca, NY: Cornell University Press, 1983; Eric J. Hobsbawm, *Nations and Nationalism Since 1780: Programme, Myth, Reality*, Cambridge: Cambridge University Press, 1990.

25. David A. Lake and Donald S. Rothchild, eds., *The International Spread of Ethnic Conflict: Fear, Diffusion, and Escalation*, Princeton, NJ: Princeton University Press, 1987; Beverly Crawford and Ronnie D. Lipschutz, eds., *The Myth of "Ethnic Conflict": Politics, Economics, and "Cultural" Violence*, Berkeley: University of California, Berkeley: International and Area Studies Press, 1987.

26. Lipschutz, *When Nations Clash*.

27. O'Connor, "Three Ways to Look at the Ecological History"; Lipschutz, with Mayer, *Global Civil Society*, chapter 5; Lipschutz, "Environmental History."

28. Jim Endersby, "How Botanical Gardens Helped to Establish the British Empire," *Financial Times*, July 25, 2014, accessed September 28, 2017, https://www.ft.com/content/dcd33da0-0e69-11e4-a1ae-00144feabdc0.

29. E. L. Jones, *The European Miracle: Environments, Economies, and Geopolitics in the History of Europe and Asia*, Cambridge: Cambridge University Press, 1981.

30. Greenfeld, *The Spirit of Capitalism*.

31. Mann, *Sources of Social Power*, volume 2, chapter 2.

32. Ernst B. Haas, *Nationalism, Liberalism, and Progress*, volume 1, Ithaca, NY: Cornell University Press, 1997.

33. Anderson, *Imagined Communities*.

34. Lipschutz, "Why Is There No International Forestry Law?"

35. Thomas Dunlap, *Nature and the English Diaspora*, Cambridge: Cambridge University Press, 1999.

36. Ronnie D. Lipschutz, *The Constitution of Imperium*, Boulder, CO: Paradigm Publishers, 2009.

37. Ramesh Mishra, *Globalization and the Welfare State*. Cheltenham, England: Elgar, 1999; Susan Strange, *The Retreat of the State*, Cambridge: Cambridge University Press, 1996; Steven K. Vogel, *Freer Markets, More Rules: Regulatory Reform in Advanced Industrial Countries*, Ithaca, NY: Cornell University Press, 1996.

38. Steven Bernstein, *The Compromise of Liberal Environmentalism*, New York: Columbia University Press, 2001; Steven Bernstein and Hamish van der Ven, "Continuity and Change in Global Environmental Politics," in *International Politics and Institutions in Time*, edited by Orfeo Fioretos, pp. 293–327, Oxford: Oxford University Press, 2017.

39. Sen, *Development as Freedom*.

40. Anatole France, *Le Lys Rouge* (*The Red Lily*). Paris: Calmann-Lévy, 1894.

41. John Gaventa, *Power and Powerlessness*.

42. Adam Taylor, "Why Nicaragua and Syria Didn't Join the Paris Climate Accord," *Washington Post*, May 31, 2017, accessed September 25, 2017, https://www.washingtonpost.com/news/worldviews/wp/2017/05/31/why-nicaragua-and-syria-didnt-join-the-paris-climate-accord/?utm_term=.dd28e07a7a44; personal communication with members of the government of Nicaragua delegation to the UN Framework Convention on Climate Change.

43. Harold D. Lasswell, *Politics: Who Gets What, When, How*, New York: P. Smith, 1936.

44. Louis Hartz, *The Liberal Tradition in America*, part 6, New York: Harcourt Brace, 1955.

45. Hirsch, *Social Limits to Growth*.

46. Hirsch, *Social Limits to Growth*; Daly, *Steady-State Economics*.

47. David E. Camacho, ed., *Environmental Injustices, Political Struggles: Race, Class, and the Environment*, Durham, NC: Duke University Press, 1998.

48. Humberto Llavador, John E. Roemer, and Joaquim Silvestre, *Sustainability for a Warming Planet*, Cambridge, MA: Harvard University Press, 2015. As one may imagine, there has been a vigorous critique of the green economy. See, for example, Thomas Fatheuer, Lili Fuhr, and Barbara Unmüßig, *Inside the Green Economy*, Munich: oekom, 2016.

49. Black, *Development in Theory and Practice*; Tim Di Muzio, *The Tragedy of Human Development: A Genealogy of Capital as Power*, Lanham, MD: Rowman & Littlefield, 2017.

50. World Bank Group, *International Debt Statistics 2017*, Washington, 2017.

51. United Nations, "World Summit on Sustainable Development," accessed August 8, 2017, https://sustainabledevelopment.un.org/milestones/wssd; United Nations, "United Nations Conference on Sustainable Development, Rio+20," accessed August 8, 2017, https://sustainabledevelopment.un.org/rio20.html; Fatheuer, Fuhr, and Unmüßig, *Green Economy*.

52. Lipschutz, with Rowe, *Globalization, Governmentality, and Global Politics*.

53. David Lewis Feldman, *Water*, Cambridge: Polity Press, 2012.

54. World Health Organization and UN Children's Fund, *Progress on Sanitation and Drinking Water—2015 Update and MDG Assessment*, accessed August 8, 2017, http://files.unicef.org/publications/files/Progress_on_Sanitation_and_Drinking_Water_2015_Update_.pdf; Lipschutz and Romano. "The Cupboard is Full."

55. Knowledge@Wharton, "America's Neglected Water Systems Face a Reckoning," accessed August 8, 2017, http://knowledge.wharton.upenn.edu/article/americas-neglected-water-systems-face-a-reckoning/.

56. S-Net Global Water Indexes, "Global Water Industry," accessed August 8, 2017, http://www.snetglobalwaterindexes.com/water-industry-info. Number of people and countries: Brad

Knickerbocker, "Privatizing Water: A Glass Half-Empty?" *Christian Science Monitor*, Oct. 24, 2002, accessed August 8, 2017, http://www.csmonitor.com/2002/1024/p01s02-usec.html.

57. World Health Organization, "What is the minimum quantity of water needed?" 2017, accessed August 25, 2017, http://www.who.int/water_sanitation_health/emergencies/qa/emergencies_qa5 /en/; Brian Reed and Bob Reed, "How much water is needed in emergencies," Prepared for the World Health Organization by the Water, Engineering and Development Centre, Loughborough University, UK, 2013, accessed September 27, 2017, http://www.who.int/water_sanitation _health/publications/2011/WHO_TN_09_How_much_water_is_needed.pdf.

58. Water for Life, "The Human Right to Water and Sanitation," n.d., accessed August 25, 2017, http://www.un.org/waterforlifedecade/human_right_to_water.shtml.

59. See, for example, The USGS Water Science School, "Water Questions and Answers—How Much Water Does the Average Person Use at Home per Day?" U.S. Geological Survey, 2016, accessed August 25, 2017, http://water.usgs.gov/edu/qa-home-percapita.html; Water Information Program, "Water Facts," Southwestern Water Conservation District, n.d., accessed August 25, 2017, https://www.waterinfo.org/resources/water-facts.

60. Abu Shiraz Rahaman, Jeff Everett, and Dean Neu, "Trust, Morality, and the Privatization of Water Services in Developing Countries," *Business and Society Review* 118, #4 (2013): 539–75; Tim Hayward, "A Global Right of Water," *Midwest Studies in Philosophy* 40 (2016): 217–33.

61. Peter Gleick, et al., *The New Economy of Water: The Risks and Benefits of Globalization and Privatization of Fresh Water*, Oakland, CA: Pacific Institute, 2002, http://www.pacinst.org /reports/new_economy_of_water_low_res.pdf.

62. Karl Wittfogel, *Oriental Despotism: A Comparative Study of Total Power*, New Haven, CT: Yale University Press, 1957; Donald Worster, *Rivers of Empire: Water, Aridity, and the Growth of the American West*, New York: Pantheon, 1985; Mark Reisner, *Cadillac Desert: The American West and Its Disappearing Water*, New York: Viking, 1986.

63. Michael Wood, *Legacy: The Search for Ancient Civilization*, New York: Sterling, 1995.

64. Worster, *Rivers of Empire*; James C. Scott, *Seeing Like a State: How Certain Schemes to Improve the Human Condition Have Failed*, New Haven, CT: Yale University Press, 1998.

65. Thayer Scudder, *The Future of Large Dams*, London: Earthscan, 2005; Ken Conca, *Governing Water*, Cambridge: MIT Press, 2006.

66. Green Climate Fund, accessed September 28, 2017, http://www.greenclimate.fund/home.

67. Sandra Postel, *Last Oasis: Facing Water Scarcity*, New York: Norton, 1997; International Rivers, "The State of the World's Rivers," n.d., accessed August 8, 2017, https://www.international ivers.org/sites/default/files/worldsrivers/.

68. CNN.com, "Federal Flood Insurance to Have More Strings," November 10, 1998, accessed August 8, 2017, http://www.cnn.com/US/9811/10/disaster.reform/index.html; Michelle Nijhuis, "Movement to Take Down Thousands of Dams Goes Mainstream," *National Geographic*, January 29, 2015, accessed August 8, 2017, http://news.nationalgeographic.com /news/2015/01/150127-white-clay-creek-dam-removal-river-water-environment/.

69. Postel, *Last Oasis*.

70. Marianne Kjellén and Gordon McGranahan, "Informal Water Vendors and the Urban Poor," International Institute for Environment and Development, London, Human Settlements Discussion Paper Series—Water-3, 2006, accessed August 24, 2017, https://www.sei-international. org/mediamanager/documents/Publications/Water-sanitation/informal_water_vendors.pdf.

71. Paul Rogers, "California Drought: More than 255,000 Homes and Businesses Still Don't Have Water Meters Statewide," *San Jose Mercury News*, August 12, 2016 (updated), accessed August 24, 2017, http://www.mercurynews.com/2014/03/08/california-drought-more-than-255000 -homes-and-businesses-still-dont-have-water-meters-statewide/.

72. Circle of Blue, "The Price of Water," n.d., accessed August 8, 2017, ; U.S. Environmental Protection Agency, "How We Use Water," March 24, 2017, accessed August 24, 2017, https:// www.epa.gov/watersense/how-we-use-water; California Legislative Analysts' Office, "Residential Water Use Trends and Implications for Conservation Policy," California Legislature, Sacramento, March 8, 2017, accessed August 24, 2017, http://www.lao.ca.gov/Publications/Report/3611.

73. Karen Bakker, "Neoliberal Versus Postneoliberal Water: Geographies of Privatization and Resistance," *Annals of the Association of American Geographers* 103, #2 (2013): 253–60; C. Chikozho and K. Kujinga, "Managing Water Supply Systems Using Free-market Economy Approaches: A Detailed Review of the Implications for Developing Countries," *Physics and Chemistry of the Earth* (2016), accessed August 24, 2017, http://www.sciencedirect.com/science/article

/pii/S1474706516300420; Emanuele Lobina, Satoko Kishimoto, and Olivier Petitjean, "Here to Stay: Water Remunicipalisation as a Global Trend," Public Services International Research Unit, Transnational Institute and Multinational Observatory, 2015, accessed August 25, 2017, https://www.tni.org/files/download/heretostay-en.pdf.

74. Anna Lappe, "World Bank Wants Water Privatized, Despite Risks," *Al Jazeera America*, April 17, 2014, accessed August 8, 2017, http://america.aljazeera.com/opinions/2014/4/water-management privatizationworldbankgroupifc.html.

75. Colin Green, "Who Pays the Piper? Who Calls the Tune?" *UNESCO Courier* (Feb. 1999): pp. 22–24.

76. Zach Willey and Adam Diamant, "Water Marketing in the Northwest: Learning by Doing," *Water Strategist* 10 (summer 1996); Paul Nelson, "Citizens, Consumers, Workers, and Activists: Civil Society during and after Water Privatization Struggles," *Journal of Civil Society* 13, #2 (2017): 202–21.

77. Bakker, "Neoliberal vs. Postneoliberal Water."

78. FRONTLINE/World, "Bolivia—Leasing the Rain," June 2002, accessed August 8, 2017, http://www.pbs.org/frontlineworld/stories/bolivia/thestory.html; William Finnegan, "Leasing the Rain," *The New Yorker*, April 8, 2002.

79. FRONTLINE/World, "Bolivia—Leasing the Rain"; Finnegan, "Leasing the Rain."

80. Democracy Center, "The Water Revolt," accessed August 8, 2017, https://democracyctr.org/archive/the-water-revolt/.

81. Finnegan, "Leasing the Rain."

82. The International Centre for Settlement of Investment Disputes, "Aguas del Tunari, S.A. v. Republic of Bolivia: Decision on Respondent's Objections to Jurisdiction," Washington, DC, October 21, 2005, ICSID Case No. ARB/02/3, accessed August 25, 2017, http://www.iisd.org /pdf/2005/AdT_Decision-en.pdf. Investor-state dispute settlement (ISDS) is a relatively recent legal innovation, first codified in the North American Free Trade Agreement and subsequently extended in other international trade agreements. ISDS allows corporations to sue host governments not only for investments lost as a result of regulation but also for loss of future profits due to non-operation. See Shawn Nichols, "Transnational Capital and the Transformation of the State: Investor-State Dispute Settlement (ISDS) in the Transatlantic Trade and Investment Partnership (TTIP)," *Critical Sociology*, 2016, accessed August 25, 2017, http://journals.sagepub.com/doi /full/10.1177/0896920516675205.

83. Democracy Center, "Bechtel vs. Bolivia: Details of the Case and the Campaign," accessed August 8, 2017, https://democracyctr.org/archive/the-water-revolt/bechtel-vs-bolivia-details-of-the -case-and-the-campaign/. Whether Aguas still exists and, if so, who owns and operates it, is unclear.

84. Reisner, *Cadillac Desert.*

85. Agriculture at a Crossroads, "Water," accessed August 8, 2017, http://www.globalagriculture .org/report-topics/water.html.

86. Satoru Miura, et al., "Protective Functions and Ecosystems Services of Global Forests in the Past Quarter-Century," *Forest Ecology and Management* 352 (2015): 35–46.

87. Cathleen A. Fogel, "Greening the Earth with Trees: Science, Storylines, and the Construction of International Climate Change Institutions," PhD dissertation, Environmental Studies, University of California, Santa Cruz, 2003; Susanna Hecht and Alexander Cockburn, *The Fate of the Forest: Developers, Destroyers, and Defenders of the Amazon*, New York: HarperCollins, 1990; Zahra Hirji, "Removing CO_2 From the Air Only Hope for Fixing Climate Change, New Study Says," *Inside Climate News*, October 6, 2016, accessed August 10, 2017, https://insideclimatenews.org /news/04102016/climate-change-removing-carbon-dioxide-air-james-hansen-2-degrees-paris -climate-agreement-global-warming.

88. Victor Menotti, "Forest Destruction and Globalisation," *Ecologist* 29 (May-June 1999): 180–81.

89. Peter Dauvergne and Jane Lister, *Timber*, Cambridge: Polity Press, 2011; Tim Payn, et al., "Changes in Planted Forests and Future Global Implications," *Forest Ecology and Management* 352 (2015): 57–67.

90. Lipschutz, "Why Is There No International Forestry Law?"; Catherine P. MacKenzie, "Lessons from Forestry for International Environmental Law," *Review of European Community and International Environmental Law* 21, #2 (2012): 114–26.

91. Kenneth G. MacDicken, et al., "Global Progress Toward Sustainable Forest Management," *Forest Ecology and Management* 352 (2015): 47–56.

92. Scott, *Seeing Like a State*, p. 14.

93. See, for example, Richard H. Grove, *Green Imperialism: Colonial Expansion, Tropical Island Edens, and the Origins of Environmentalism, 1600–1860*, Cambridge: Cambridge University Press, 1995; Guha, *The Unquiet Woods*.

94. Grove, *Green Imperialism*, chapter 3.

95. Scott, *Seeing Like a State*, pp. 19–20.

96. Scott, *Seeing Like a State*, pp. 11–12.

97. Magnusson and Shaw, *A Political Space*.

98. Guha, *The Unquiet Woods*.

99. Glenn Martin, "California's Rural Economy: Boom Times Long Gone, a Small Town Struggles for Survival," *San Francisco Chronicle*, March 24, 2003, accessed August 10, 2017, http://www.sfgate.com/cgi-bin/article.cgi?file=/chronicle/archive/2003/03/24/MN198336.DTL.

100. See, for example, Scott, *Seeing Like a State*, p. 20.

101. N. Patrick Peritore, *Third World Environmentalism: Case Studies from the Global South*, chapter 4, Gainesville: University Press of Florida, 1999.

102. Madhav Gadgil and Ramachandra Guha, *This Fissured Land: An Ecological History of India*, Berkeley: University of California Press, 1993.

103. Guha, *The Unquiet Woods*.

104. Peritore, *Third World Environmentalism*, p. 71.

105. O. P. Dwivedi, *India's Environmental Policies, Programmes, and Stewardship*, p. 91, Basingstoke, England: Macmillan, 1997.

106. Dwivedi, *India's Environmental Policies, Programmes, and Stewardship*, p. 92; Peritore, *Third World Environmentalism*, p. 69.

107. T. Dash and A. Khotari, "Forest Rights and Conservation in India," in *Right to Responsibility: Resisting and Engaging Development, Conservation and the Law in Asia*, edited by H. Jonas, H. J. Suneetha, and M. Subramanian, pp. 151–74, Tokyo: UNU-IAS and Capetown: Natural Justice, 2015. The process of exclusion is today widespread in Indonesia, Brazil, and other countries with extensive tropical forest cover.

108. Thomas Weber, *Hugging the Trees: The Story of the Chipko Movement*, New York: Viking, 1988; Guha, *The Unquiet Woods*.

109. Guha, *The Unquiet Woods*, "Epilogue."

110. Guha, *The Unquiet Woods*, pp. 201–02.

111. Dash and Khotari, "Forest Rights and Conservation in India," p. 152. See also Indrani Sigamany, "Land Rights and Neoliberalism: An Irreconcilable Conflict for Indigenous Peoples in India?" *International Journal of Law in Context* 13, #3 (2017): 369–87.

112. Derek Hall, Philip Hirsch, and Tania Murray Li, *Powers of Exclusion*, Honolulu: University of Hawai'i Press, 2011.

113. Ronnie D. Lipschutz, "Global Civil Society and Global Environmental Protection: Private Initiatives and Public Goods," in *Environmental Policy Making: Assessing the Use of Alternative Policy Instruments*, edited by Michael Hatch, pp. 45–70. Albany: State University of New York Press, 2005.

114. David Goodman and Anthony Hall, eds., *The Future of Amazonia*, New York: St. Martin's Press, 1990.

115. International Institute for Sustainable Development, "The I.W.G.F. Page," accessed August 10, 2017, http://enb.iisd.org/forestry/iwgf.html.

116. International Institute for Sustainable Development, "A Brief to Global Forest Policy," accessed August 10, 2017, http://enb.iisd.org/process/forest_desertification_land-forestintro.htm.

117. International Institute for Sustainable Development, "A Brief to Global Forest Policy."

118. UN Forum on Forests, "Background Document No. 2: Recent Developments in Existing Forest-Related Instruments, Agreements, and Processes Working Draft," prepared for the Ad hoc expert group on Consideration with a View to Recommending the Parameters of a Mandate for Developing a Legal Framework on All Types of Forests, September 7–10, 2004, accessed August 10, 2017, http://www.un.org/esa/forests/wp-content/uploads/2014/12/background-2.pdf.

119. MacKenzie, "Lessons from Forestry," p. 115.

120. Fogel, "Greening the Earth with Trees," p. 129.

121. Ian Fry, "Twists and Turns in the Jungle: Exploring the Evolution of Land Use, Land-Use Change and Forestry Decisions within the Kyoto Protocol," *RECIEL* 11(2002): 159–68; Esther Turnhout, et al., "Envisioning REDD+"; Asiyanbi, "A Political Ecology of REDD+."

122. Fry, "Twists and Turns in the Jungle."

123. Constance McDermott, B. Cashore, and P. Kanowski, *Global Environmental Forest Policies: An International Comparison*, UK: Earthscan, 2010.

124. Dauvergne and Lister, *Timber*, p. 140; naturally:wood, "British Columbia Forest Facts," August 2010, accessed September 28, 2017, http://www.sfiprogram.org/files/pdf/third-party -certification-aug-2010pdf/.

125. Forest Stewardship Council, "Membership Chambers," accessed August 10, 2017, https:// us.fsc.org/en-us/who-we-are/membership/membership-chambers.

126. Forest Stewardship Council, "FSC A.C. Membership List," accessed August 10, 2017, http://memberportal.fsc.org/.

127. Forest Stewardship Council, "FSC A.C. Statutes," accessed August 10, 2017, https://us.fsc .org/en-us/who-we-are/governance.

128. Forest Stewardship Council, "Facts and Figures August 2017," accessed August 10, 2017, https://ic.fsc.org/en/facts-and-figures.

129. For details of how the certification process works, see Lipschutz with Rowe, *Globalization, Governmentality, and Global Politics*.

130. See, for example, Daniela A. Miteva, Colby J. Loucks, and Subhrendu K. Pattanayaik, "Social and Environmental Impacts of Forest Certification in Indonesia," *PLOS*, July 1, 2015, http://journals.plos.org/plosone/article?id=10.1371/journal.pone.0129675; and World Wildlife Fund, "The Impact of Forest Stewardship Council Certification," October 20, 2014, accessed August 10, 2017, http://d2ouvy59p0dg6k.cloudfront.net/downloads/fsc_research_review.pdf.

131. "Global Forest Atlas," Yale School of Forestry and Environmental Studies, New Haven, CT, 2017, accessed August 26, 2017, http://globalforestatlas.yale.edu/conservation/forest-certification.

132. Nicole Freris and Klemens Laschefski, "Seeing the Wood from the Trees," *The Ecologist* 31, no. 6 (July/August 2001), accessed June 21, 2002, http://www.wald.org/fscamaz/ecol_eng.htm.

133. Lipschutz, "Why Is There No International Forestry Law?"; Lipschutz with Mayer, *Global Civil Society*.

134. Programme for Endorsement of Forest Certification International, "Endorsement and Mutual Recognition of National Systems and their Revision," accessed August 10, 2017, https:// pefc.org/standards-revision/endorsement.

135. Gulbrandsen, *Transnational Environmental Governance*, p. 88.

136. Lipschutz with Rowe, *Globalization, Governmentality, and Global Politics*.

137. For empirical data on consumers' willingness to pay a premium for certified goods, see: Stephanie M. Tully and Russell S. Winer, "The Role of the Beneficiary in Willingness to Pay for Socially Responsible Products. A Meta-analysis," *Journal of Retailing* 90, #2 (2014): 255–74; Jens Hainmueller and Michael J. Hiscox, "Buying Green? Field Experimental Tests of Consumer Support for Environmentalism," December 2015, accessed August 26, 2017, https://papers.ssrn .com/sol3/Delivery.cfm/SSRN_ID2699634_code739013.pdf?abstractid=2062429andmirid=1; Jens Hainmueller, Michael J. Hiscox, and Sandra Sequeira, "Consumer Demand for Fair Trade: Evidence from a Multistore Rield Experiment," *Review of Economics and Statistics* 97, #2 (May 2015): 242–56.

138. Bartley, *Looking Behind the Label*.

139. Bartley, *Looking Behind the Label*, p. 87; Dauvergne and Lister, *Timber*.

140. Stephen Bass, et al., *Certification's Impacts on Forests, Stakeholders, and Supply Chains*, Stevenage, England: Earthprint, 2001, accessed August 10, 2017, http://pubs.iied.org/9013IIED/.

141. Dauvergne and Lister, *Timber*; Peter Dauvergne and Jane Lister, "The Prospects and Limits of Eco-Consumerism: Shopping Our Way to Less Deforestation?" *Organization and Environment* 23 (2010): 132–54; Peter Dauvergne and Jane Lister, "Big Brand Sustainability: Governance Prospects and Environmental Limits," *Global Environmental Change* 22 (2012): 36–45. By value, half of timber goes to paper and paperboard: diapers, milk cartons, toilet paper. Half of paper—in other words, one-quarter of the world's timber—goes to disposable packaging, including cardboard and wood pallets used for shipping.

142. Bartley, *Looking Behind the Label*.

143. Bartley, *Looking Behind the Label*, p. 18.

144. Bartley, *Looking Behind the Label*, p. 4.

145. Lipschutz, "Getting Out of the CAR."

146. Nitzan and Bichler, *Capital as Power*.

147. Dauvergne and Lister, *Timber*; Bartley, et al., *Looking Behind the Label*.

148. Fry, "Twists and Turns in the Jungle."

149. Glen M. McDonald, "Water, Climate Change, and Sustainability in the Southwest," *Proceedings of the National Academy of Sciences* 107 (2010): 21256–62.

150. Karin Bäckstrand and Eva Lövbrand, "Planting Trees to Mitigate Climate Change: Contested Discourses of Ecological Modernization, Green Governmentality and Civic Environmentalism," *Global Environmental Politics* 6 (2006): 50–75.

151. Conference of the Parties, UN Framework Convention on Climate Change, *Report of the Conference of the Parties on Its Twenty-First Session, Held in Paris from 30 November to 13 December 2015—Addendum – Part Two: Action Taken by the Conference of the Parties at Its Twenty-First Session*, Dec 1/CP.21, UN Doc FCCC/CP/2015/10/Add.1 (29 January 2015) (*Adoption of the Paris Agreement*). The agreement is included as an annex to this document (*Paris Agreement*).

152. Benjamin J. Henley and Andrew D. King, "Trajectories toward the 1.5° C Paris Target: Modulation by the Interdecadal Pacific Oscillation," *Geophysical Research Letters* 44, #9 (May 16, 2017): 4256–62.

153. Intergovernmental Panel on Climate Change, *Climate Change 2014: Synthesis Report. Contribution of Working Groups I, II and III to the Fifth Assessment Report of the Intergovernmental Panel on Climate Change*, p. 64, Geneva: Intergovernmental Panel on Climate Change, 2014.

154. Chris Mooney, "We Only Have a Five Percent Chance of Avoiding 'Dangerous' Global Warming, a Study Finds," *Washington Post*, July 31, 2017, accessed September 29, 2017, https://www.washingtonpost.com/news/energy-environment/wp/2017/07/31/we-only-have-a-5-percent-chance-of-avoiding-dangerous-global-warming-a-study-finds/?utm_term=.5931f89e22b3. In the study described in this article, "dangerous" is defined as passing the two-degrees-Celsius threshold. The study gives a 1 percent chance of staying below 1.5°C of warming above pre-industrial levels.

155. Glen P. Peters and Oliver Geden, "Catalysing a Political Shift from Low to Negative Carbon," *Nature Climate Change* (2017), accessed August 27, 2017, http://www.nature.com.oca.ucsc.edu/nclimate/journal/vaop/ncurrent/pdf/nclimate3369.pdf.

156. Sivan Kartha and Kate Dooley, "The Risks of Relying on Tomorrow's 'Negative Emissions' to Guide Today's Mitigation Action," Stockholm Environment Institute, Working paper 2016-08, Stockholm: Stockholm Environment Institute, accessed September 29, 2017, https://www.sei-international.org/mediamanager/documents/Publications/Climate/SEI-WP-2016-08-Negative-emissions.pdf; Kate Dooley and Sivan Kartha. Land-based negative emissions: risks for climate mitigation and impacts on sustainable development. *International Environmental Agreements: Politics, Law and Economics* 81, No. 1 (2017): 79–98.

157. A. G. Bumpus and D. M. Liverman, "Carbon Colonialism? Offsets, Greenhouse Gas Reductions, and Sustainable Development," in *Global Political Ecology*, edited by Richard Peet, Paul Robbins, and Michael J. Watts, London: Routledge, 2011; Asiyanbi, "Political Ecology of REDD+."

158. Intergovernmental Panel on Climate Change, *Synthesis Report*, p. 89.

159. Intergovernmental Panel on Climate Change, *Synthesis Report*, p. 89.

160. Pete Smith, et al., "Biophysical and Economic Limits to Negative CO_2 Emissions," *Nature Climate Change* 6 (2016): 42–46.

161. Intergovernmental Panel on Climate Change, *Synthesis Report*, p. 89.

162. William C. G. Burns "Human Rights Dimensions of Bioenergy with Carbon Capture and Storage: A Framework for Climate Justice in the Realm of Climate Geoengineering," in *Climate Justice: Case Studies in Global and Regional Governance Challenges*, edited by Randall Abate, Washington, DC: Environmental Law Institute, 2016, accessed August 15, 2017, https://ssrn.com/abstract=2871527.

163. Kartha and Dooley, "Negative Emissions."

164. Burns, "Human Rights Dimensions," p. 158, citing Smith, et al., "Biophysical and Economic Limits" and Williamson, "Scrutinize CO_2 Removal Methods," *Nature* 530 (2016): 153–54. For similar calculations, see also Christopher B. Field and Katherine J. Mach, "Rightsizing Carbon Dioxide Removal," *Science* (2017): 706–07.

165. Kevin Anderson and Glen Peters, "The Trouble with Negative Emissions," *Science* 354 (2016): 182–83; Thomas Sikor, et al., "Global Land Governance"; Field and Mach, "Rightsizing."

166. Oliver Geden and Andreas Löschel, "Define Limits for Temperature Overshoot Targets," *Nature Geoscience*, Vol. 10 (Dec. 2017): 881–82.

167. Heather Lovell, Harriet Bulkeley, and Diana Liverman, "Carbon Offsets: Sustaining Consumption?" *Environment and Planning A* 41 (2009): 2357–79.

168. Shelley L. Hurt, "Military's Hidden Hand: Examining the Dual-Use Origins of Biotechnology, 1969-1972," in *State of Innovation: The U.S. Government's Role in Technology Development*, edited by Fred Block and Matt Keller, chapter 3, Boulder: Paradigm Publishers, 2011.

169. Cary Fowler and Pat Mooney, *Shattering: Food, Politics and the Loss of Genetic Diversity*, Tucson: University of Arizona Press, 1990.

170. Jack Ralph Kloppenburg, Jr., *First the Seed: The Political Economy of Plant Biotechnology*, Cambridge: Cambridge University Press, 1988; Cary Fowler, *Unnatural Selection: Technology, Politics, and Plant Evolution*, Yverdon, Switzerland: Gordon and Breach Science Publishers, 1994.

171. Fowler and Mooney, *Shattering*; Kloppenburg, *First the Seed*.

172. David Koepsell, *Who Owns You?: Science, Innovation, and the Gene Patent Wars*, New York: John Wiley and Sons, 2015.

173. United Nations, Declaration of the United Nations Conference on the Human Environment, U.N. Doc. A/Conf.48/14/Rev. 1(1973); 11 ILM 1416 (1972).

174. Scott Shackelford, "The Tragedy of the Common Heritage of Mankind," *Stanford Environmental Law Journal* 27 (2008): 101–57.

175. Convention on Biological Diversity, June 5, 1992, 1760 U.N.T.S. 79, 143; 31 I.L.M. 818 (1992).

176. Kathy McAfee, "Selling Nature to Save It? Biodiversity and the Rise of Green Developmentalism," *Environment and Planning D: Society and Space* 17 (1999): 133–54.

177. Kloppenburg, *First the Seed*.

178. Benjamin Neimark, "Green Grabbing at the 'Pharm' Gate: Rosy Periwinkle Production in Southern Madagascar," *Journal of Peasant Studies* 39 (2012): 423–45.

179. McAfee, "Selling Nature to Save It?"

180. Kelly Servick, "Rise of Digital DNA Raises Biopiracy Fears," *Science*, November 17, 2016, accessed August 16, 2017, http://www.sciencemag.org/news/2016/11/rise-digital-dna-raises-biopiracy-fears; Margo A. Bagley, "Digital DNA: The Nagoya Protocol, Intellectual Property Treaties, and Synthetic Biology," Wilson Center, December 2015, accessed August 16, 2017, https://www.wilsoncenter.org/sites/default/files/digital_dna_final_0.pdf; Edward Hammond, "'Digital DNA' and Biopiracy: Protecting Benefit-Sharing as Synthetic Biology Changes Access to Genetic Resources," TWN Briefings for UN Biodiversity Conference #1, November 24, 2016, accessed August 16, 2017, https://www.biosafety-info.net/article.php?aid=1308.

181. Food and Agriculture Organization, International Treaty on Plant Genetic Resources for Food and Agriculture, "The Global Information System and Genomic Information: Transparency of Rights and Obligations," Background study paper n. 10, IT/GB7/SAC-1/16/ BSP 10, October 2016.

182. Gourevitch, *Politics in Hard Times*, chapter 1; see also Kees Van der Pijl, *Transnational Classes and International Relations*, London: Routledge, 1998.

183. Mirian Hood, *Gunboat Diplomacy, 1895–1905: Great Power Pressure in Venezuela*, London: Allen and Unwin, 1975; Rosenberg, *Empire of Civil Society*.

184. Gilpin, *War and Change*; Stephen Gill, *American Hegemony and the Trilateral Commission*, Cambridge: Cambridge University Press, 1990.

185. John Vidal, "Copenhagen Climate Failure Blamed on 'Danish Text,'" *The Guardian*, May 31, 2010, accessed September 27, 2017, https://www.theguardian.com/environment/2010/may/31/climate-change-copenhagen-danish-text; Per Meilstrup, "The Runaway Summit: The Background Story of the Danish Presidency of COP15, the UN Climate Change Conference," accessed September 27, 2017, http://www.fao.org/fileadmin/user_upload/rome2007/docs/What%20really%20happen%20in%20COP15.pdf; and personal observation by one of the authors in attendance.

186. Daniel Bodansky, "A Tale of Two Architectures: The Once and Future UN Climate Change Regime," *Arizona State Law Journal* 43 (2011): 697–712.

187. See analysis in Matthew Stilwell, "Climate Debt—A Primer," *Development Dialogue* 61 (September 2012): 41–46.

188. World Bank Group, "Climate Finance," accessed August 16, 2017, http://www.worldbank.org/en/topic/climatefinance.

189. BBC News, "Copenhagen Deal Reaction in Quotes," December 19, 2009, accessed September 28, 2017, http://news.bbc.co.uk/2/hi/science/nature/8421910.stm.

190. Sonja Klinsky, et al., "Why Equity is Fundamental in Climate Change Policy Research," *Global Environmental Change* 44 (2017): 170–73.

191. Center for Climate and Energy Solutions, "Alternative Models for the 2015 Climate Change Agreement," website introduction, accessed August 16, 2017, https://www.c2es.org/press-center/article/alternative-models-2015-climate-change-agreement.

192. UN Framework Convention on Climate Change, "The Paris Agreement," 2017, accessed August 28, 2017, http://unfccc.int/paris_agreement/items/9485.php.

193. See Daniel Bodansky, "Legal Options for U.S. Acceptance of a New Climate Change Agreement," Center for Climate and Energy Solutions, May 2015, accessed August 16, 2017, https://www.c2es.org/publications/legal-options-us-acceptance-new-climate-change-agreement.

194. Robert Brulle, "Institutional Delay: Foundation Funding and the Creation of U.S. Climate Change Counter-movement Organizations," *Climatic Change* 122 (2014): 681–94; Coral Davenport and Eric Lipton, "How G.O.P. Leaders Came to View Climate Change as Fake Science," *New York Times*, June 3, 2017, accessed September 29, 2017, https://www.nytimes.com/2017/06/03/us/politics/republican-leaders-climate-change.html.

195. Bodansky, "Legal Options."

196. Daniel Bodansky, "The Legal Character of the Paris Agreement," *Review of European, Comparative & International Environmental Law* 25 (2016): 142–50.

197. UN Framework Convention on Climate Change, *Report of the Conference of the Parties on Its Twenty-First Session*. The Agreement is included as an annex to this document (*Paris Agreement*). The text on voluntary cooperative approaches, which could include market mechanisms, is found in Article 6 of the agreement.

198. Sylvia I. Karlsson-Vinkhuyzen, et al., "Entry into Force and Then?: The Paris Agreement and State Accountability, *Climate Policy*, July 31, 2017.

199. Matthew J. Hoffman, *Climate Governance at the Crossroads: Experimenting with a Global Response after Kyoto*, Oxford: Oxford University Press, 2011. See also the reaction of U.S. cities, states, and citizens to the announcement by the Trump administration that it would withdraw from the Paris Agreement, with the campaign called #wearestillin. "We Are Still In," accessed October 1, 2017, https://www.wearestillin.com/us-action-climate-change-irreversible.

200. Natalia S. Mitrovitskaya, Margaret Clark, and Ronald G. Purver, "North Pacific Fur Seals: Regime Formation as a Means of Resolving Conflict," in *Polar Politics: Creating International Environmental Regimes*, edited by Oran R. Young and Gail Osherenko, pp. 22–55, Ithaca, NY: Cornell University Press, 1993; Edith Brown Weiss, "The Evolution of International Environmental Law," *Japanese Yearbook of International Law* 54 (2011): 1–27.

201. Peterson, "Whales, Cetologists, Environmentalists."

202. Edward L. Miles, *Global Ocean Politics: The Decision Process at the Third United Nations Conference on the Law of the Sea, 1973–1982*, The Hague: Martinus Nijhoff, 1998.

203. James K. Sebenius, *Negotiating the Law of the Sea*, Cambridge: Harvard University Press, 1984.

204. *The Economist*, "Fruits de mer: Plucking Minerals from the Seabed is Back on the Agenda," February 23, 2017, accessed August 28, 2017, https://www.economist.com/news/science-and-technology/21717351-fruits-de-mer-plucking-minerals-seabed-back-agenda.

205. Krasner, *International Regimes*, pp. 1–22, 355–68. Susan Strange, "Cave! Hic dragones: A Critique of Regime Analysis," in Krasner, *International Regimes*, pp. 337–54.

206. Graham Allison, *Essence of Decision: Explaining the Cuban Missile Crisis*, Boston: Little, Brown, 1971; Morton Halperin, *Bureaucratic Politics and Foreign Policy*, Washington, DC: Brookings Institution, 1974; Stephen D. Krasner, "Are Bureaucracies Important? (Or, Allison Wonderland)," *Foreign Policy* 7 (summer 1972): 159–79; Sikina Jinnah, *Post-Treaty Politics*, Cambridge: MIT Press, 2014.

207. World Trade Organization, "Understanding the WTO: The Agreements," accessed August 16, 2017, https://www.wto.org/english/thewto_e/whatis_e/tif_e/agrm1_e.htm.

208. John Braithwaite and Peter Drahos, *Global Business Regulation*, Cambridge: Cambridge University Press, 2000.

209. Nichols, "Transnational Capital and the Transformation of the State."

210. Nichols, "Transnational Capital and the Transformation of the State."

211. Susan K. Sell, *Private Power, Public Law: The Globalization of Intellectual Property Rights*, Cambridge: Cambridge University Press, 2003; Charan Deveraux, Robert Z. Lawrence,

and Michael D. Watson, *Case Studies in U.S. Trade Negotiation, Volume 1: Making the Rules*, Washington, DC: Institute for International Economics, 2006. A group of chief executive officers from industries affected by what might be called intellectual property theft formed the Intellectual Property Committee to lobby the U.S. Government throughout the TRIPS negotiations. The core chief executive officers were from Pfizer, IBM, Merck, General Electric, Warner Communications, Bristol-Myers, Hewlett-Packard, FMC, General Motors, Johnson and Johnson, Monsanto, and Rockwell International.

212. Sell, *Private Power, Public Law*; Deveraux, Lawrence, and Watson, *Case Studies in U.S. Trade Negotiation*.

213. World International Property Organization, "IP Statistics Data Center," accessed August 17, 2017, https://www3.wipo.int/ipstats/index.htm?tab=patent.

214. See, for example, Consumer Project on Technology, "Court Case between 39 Pharmaceutical Firms and the South African Government," accessed October 4, 2017, http://www.cptech.org/ip/health/sa/pharma-v-sa.html.

215. John S. Odell and Susan K. Sell. "Reframing the Issue: The WTO Coalition on Intellectual Property and Public Health, 2001," in *Negotiating Trade: Developing Countries in the WTO and NAFTA*, edited by John S. Odell, pp. 85–114, Cambridge: Cambridge University Press, 2001; World Trade Organization, "Obligations and Exceptions."

216. Coral Davenport, "Administration is Seen as Retreating on Environment in Talks on Pacific Trade," *New York Times*, January 15, 2014, accessed August 28, 2017, https://www.nytimes.com/2014/01/15/us/politics/administration-is-seen-as-retreating-on-environment-in-talks-on-pacific-trade.html?hpwandrref=science; Howard F. Chang, "The Environmental Chapter of the Trans-Pacific Partnership: An Environmental Agreement Within a Trade Agreement," *Trends* 47, #5 (May, June 2015), accessed August 28, 2017, https://www.americanbar.org/publications/trends/2015-2016/may-june-2016/the_environment_chapter_of_the_trans-pacific_partnership.html. As we write this book, the Trump administration looks ready to abandon altogether the long-standing U.S. commitment to free trade and some of the key trade treaties that operationalize that commitment, such as Trans-Pacific Partnership and Transatlantic Trade and Investment Partnership, and possibly the North American Free Trade Agreement, too. How long that position lasts, with what consequences, is anyone's bet at this point.

217. Clapp, *Toxic Exports*.

218. Bomberg, *Green Parties and Politics*.

219. See, for example, Henning Arp, "Technical Regulation and Politics: The Interplay between Economic Interests and Environmental Policy Goals in EC Car Emission Legislation," in *Environmental Policy in the European Union: Actors, Institutions and Processes*, edited by Andrew Jordan, pp. 256–74, London: Earthscan, 2002.

220. Sonia Mazey and Jeremy Richardson, "Environmental Groups and the EC: Challenges and Opportunities," in *Environmental Policy in the European Union—Actors, Institutions and Processes*, edited by Andrew Jordon, pp. 141–56, London: Earthscan, 2002; Bomberg, *Green Parties and Politics*, chapter 6.

221. David Vogel, *Trading Up: Consumer and Environmental Regulation in a Global Economy*, Cambridge: Harvard University Press, 1995.

222. Vogel, *Trading Up*.

223. Duales System Deutschland AG, "Der Grüne Punkt," accessed August 28, 2017, https://www.gruener-punkt.de/en.html; Öko-Institut e.V., "Recycling is the Future—Ecological Achievements and Potential of the Dual Aystem," Freiberg, Germany, June 2016, accessed August 28, 2017, https://www.gruener-punkt.de/fileadmin/layout/redaktion/Mediathek/STUDIE_OEKO_INSTITUT_ENG.pdf.

224. Albert Weale, et al., *Environmental Governance in Europe: An Ever Closer Ecological Union?* p. 420, Oxford: Oxford University Press, 2000.

225. See Adam Baddeley, "In the Rush to Get Waste out of Landfill, the EU Risks Losing Sight of its Recycling Targets," June 20, 2016, accessed August 12, 2017, https://www.euractiv.com/section/sustainable-dev/opinion/in-the-rush-to-get-waste-out-of-landfill-eu-risks-losing-sight-of-its-recycling-targets/.

226. Vogel, *Trading Up*, pp. 82–93.

227. G. Gordon Davis and Jessica Ann Hall, "Circular Economy Legislation – the International Experience," May 1, 2006, accessed September 28, 2017, http://siteresources.worldbank.org/INTEAPREGTOPENVIRONMENT/Resources/CircularEconomy_Legal_IntExperience_Final_EN.doc.

228. Christine Whitehouse, "Driven to Distraction," *Time International* 155 (Feb. 14, 2000): 65–66; Öko-Institut e.V., "Assessment of the Implementation of Directive 2000/53/EC on End-of-life Vehicles (the ELV Directive) with Emphasis on the End-of-life Vehicles of Unknown Whereabouts," Darmstadt, Germany, June 22, 2016, accessed August 28, 2017, http://elv.whereabouts.oeko.info/fileadmin/images/Project_Docs/Assessment_whereabouts.pdf.

229. European Commission–DG Environment, "Ex-post Evaluation of Certain Waste Stream Directives," April 18, 2014, accessed August 28, 2017, http://ec.europa.eu/environment/waste/pdf/target_review/Final%20Report%20Ex-Post.pdf.

230. John McCormick, *Environmental Policy in the European Union*, p. 178, Basingstoke, England: Palgrave, 2001.

231. Keck and Sikkink, *Activists beyond Borders*.

CHAPTER 6: GLOBAL ENVIRONMENTAL POLITICS, SOCIETY, AND YOU

1. John Vidal, "Real Battle for Seattle," *The Guardian*, December 4, 1999, accessed September 2, 2017, https://www.theguardian.com/world/1999/dec/05/wto.globalisation.

2. C. Goldberg, "1500 March in Boston to Protest Biotech Food," *New York Times*, March 27, 2000, accessed September 3, 2017, http://www.nytimes.com/2000/03/27/us/1500-march-in-boston-to-protest-biotech-food.html?mcubz=0.

3. Cited in Goldberg, "1500 March in Boston to Protest Biotech Food."

4. Daniel J. Hicks and Roberta L. Millstein, "GMOs: Non-Health Issues," in *Encyclopedia of Food and Agricultural Ethics*, edited by Paul B. Thompson and David Kaplan, pp. 1–11, Springer, 2016, second edition; Sven-Erik Jacobsen, et al., "Feeding the World: Genetically Modified Crops versus Agricultural Biodiversity," *Agronomy for Sustainable Development* 33 (2013): 651–62; Geoffrey Barrows, Steven Sexton, and David Zilberman, "Agricultural Biotechnology: The Promise and Prospects of Genetically Modified Crops," *Journal of Economic Perspectives* 28, #1 (Winter 2014): 99–119.

5. Andy Coghlan, "More than Half of EU Officially Bans Genetically Modified Crops," *New Scientist*, October 5, 2015, accessed September 5, 2017, https://www.newscientist.com/article/dn28283-more-than-half-of-european-union-votes-to-ban-growing-gm-crops/.

6. Helen Davidson and Oliver Milman, "Global 'March for Science' Protests Call for Action on Climate Change," *The Guardian*, April 22, 2017, accessed September 5, 2017, https://www.theguardian.com/science/2017/apr/22/global-march-for-science-protests-call-for-action-on-climate-change; Chris Mooney, Joe Heim, and Brady Dennis, "Climate March Draws Massive Crowd to D.C. in Sweltering Heat," *Washington Post*, April 29, 2017, accessed September 5, 2017, https://www.washingtonpost.com/national/health-science/climate-march-expected-to-draw-massive-crowd-to-dc-in-sweltering-heat/2017/04/28/1bdf5e66-2c3a-11e7-b605-33413c691853_story.html?utm_term=.91e5b2907270.

7. Andrew Karvonen and Ralf Brand, "Technical Expertise, Sustainability, and the Politics of Specialized Knowledge," in *Global Environmental Governance: Power and Knowledge in a Local-Global World*, edited by Gabriella Kütting and Ronnie D. Lipschutz, pp. 38–58, London: Routledge, 2009.

8. Lipschutz with Rowe, *Globalization, Governmentality, and Global Politics*; Antonio Gramsci, in *Selections from the Prison Notebooks*, edited and translated by Quentin Hoare and Geoffrey N. Smith, p. 414 n83, London: Electric Book Co., 1999, accessed September 5, 2017, http://abahlali.org/files/gramsci.pdf.

9. Smith and Johnston, *Globalization and Resistance*; Robin Broad, ed., *Global Backlash: Citizen Initiatives for a Just World Economy*, Lanham, MD: Rowman & Littlefield, 2002.

10. This is a riff on Lenin's famous work, Что делать? or "What is to be done?" published in 1901, accessed September 5, 2017, https://www.marxists.org/archive/lenin/works/download/what-itd.pdf.

11. Gramsci, *Prison Notebooks*, pp. 670–71.

12. Chantal Mouffe, *The Democratic Paradox*, London: Verso, 2000; Osborne, "Public Political Ecology."

13. See, for example, Nancy Fraser, "Reframing Justice in a Globalizing World," *New Left Review* 36 (November–December 2005): 69–88.

14. Osborne, "Public Political Ecology."
15. Deborah Stone, *Policy Paradox.*
16. Stone, *Policy Paradox*, p. 26.
17. Lasswell, *Politics.*
18. Elizabeth Lunstrum, Pablo Bose, and Anna Zalik, "Environmental Displacement: The Common Ground of Climate Change, Extraction and Conservation," *Area* 48, #2 (2015): 130–33; and other papers in the special issue on environmental displacement that this article introduces.
19. Hardin, "Tragedy of the Commons."
20. Joseph Mazor, "Environmental Ethics," in *Global Environmental Change*, edited by Bill Freedman, pp. 925–31, Dordrecht: Springer, 2014.
21. Thomas Frank and Matt Weiland, eds., *Commodify Your Dissent: The Business of Culture in the New Gilded Age*, New York: Norton, 1997.
22. See, for example, Glen S. Aikenhead and Masakata Ogawa, "Indigenous Knowledge and Science Revisited," *Cultural Studies of Science Education* 2 (2007): 539–91; Patricia Vickers, "Ayaawx: In the Path of Our Ancestors," *Cultural Studies of Science Education* 2 (2007): 592–98.
23. Milton Friedman, *Capitalism and Freedom*, Chicago: University of Chicago Press, 1961.
24. Andrew Feenberg, "The Philosophy of Praxis: Marx, Lukács and the Frankfurt School," Talk at the Vancouver School for Social Research, October 27, 2014, accessed September 7, 2017, https://www.sfu.ca/~andrewf/Philosophy%20of%20Praxis%20preview.pdf.
25. Smith, *Theory of Moral Sentiments.*
26. Massimo Pizzol, et al., "Monetary Valuation in Life Cycle Assessment: A Review," *Journal of Cleaner Production* 86 (2015): 170–79; Katherine Hood, "The Science of Value: Economic Expertise and the Valuation of Human Life in US Federal Regulatory Agencies," *Social Studies of Science* 47, #4 (2017): 441–65.
27. Hans-Jörg Rheinberger, "Culture and Nature in the Prism of Knowledge," *History of Humanities* 1, #1 (2016): 155–81; Andrew Biro, ed., *Critical Ecologies: The Frankfurt School and Contemporary Environmental Crises*, Toronto: University of Toronto Press, 2011.
28. John Meyer, *Political Nature: Environmentalism and the Interpretation of Western Thought*, Cambridge, MA: MIT Press, 2001; Keulartz, *The Struggle for Nature.*
29. The need for a "Green Leviathan" is addressed in Ophuls and Boyan, *Ecology and the Politics of Scarcity Revisited*; Heilbroner, *An Inquiry into the Human Prospect.*
30. William Cronon, ed., *Uncommon Ground: Toward Reinventing Nature*, New York: Norton, 1995.
31. Michael E. Soulé and Gary Lease, eds., *Reinventing Nature? Responses to Postmodern Deconstruction*, Washington, DC: Island Press, 1995.
32. Stone, *Should Trees Have Standing?* Along these lines, see also a recent U.S. lawsuit filed to recognize the Colorado River as a person, to protect the river's "right to exist, flourish, regenerate, be restored, and naturally evolve." Turkewicz, "Corporations Have Rights." https://www.nytimes.com/2017/09/26/us/does-the-colorado-river-have-rights-a-lawsuit-seeks-to-declare-it-a-person.html?mcubz=0.
33. Ulrich Beck, *Risk Society: Towards a New Modernity*, London: Sage, 1992; Ulrich Beck, *Ecological Politics in an Age of Risk*, London: Polity, 1995.
34. Haraway, "Manifesto for Cyborgs."
35. Naomi Klein, *The Shock Doctrine: The Rise of Disaster Capitalism*, Toronto: Random House of Canada, 2007; Naomi Klein, *This Changes Everything: Capitalism vs. The Climate*, New York: Simon and Schuster, 2014.
36. Sen, *Development as Freedom*; Dimitris Stevis and Valerie J. Asseto, eds., *The International Political Economy of the Environment*, Boulder, CO: Lynne Rienner, 2001.
37. Polanyi, *The Great Transformation*, p. 3.
38. Ellen Meiksins Wood, *The Origins of Capitalism: A Longer View*. London: Verso Books, 2002, revised edition.
39. Tim McMahon, "Inflation Adjusted Gasoline Prices," InflationData.com, June 2016, accessed September 8, 2017, https://inflationdata.com/articles/inflation-adjusted-prices/inflation-adjusted-gasoline-prices/.
40. Osborne, "Public Political Ecology," p. 845, italics in original.
41. Paulo Freire, *Pedagogy of the Oppressed*, p. 106, New York: Continuum, 2000.
42. Rhys Jones, Jessica Pykett, and Mark Whitehead, "Governing Temptation: Changing Behaviour in an Age of Libertarian Paternalism," *Progress in Human Geography* 35, #4 (2011):

483–501; Jessica Pykett, "The New Material State: The Gendered Politics of Governing through Behaviour Change," *Antipode* 44, #1 (Jan. 2012): 217–38.

43. Foucault, "Truth and Power," p. 119.

44. Barbara Epstein, "Why Post-Structuralism Is a Dead End for Progressive Thought," *Socialist Review* 25, #2 (1995): 83–119.

45. Arendt, *The Human Condition*, p. 198.

46. Arendt, *The Human Condition*, p. 200. Our emphasis.

47. Arendt, *The Human Condition*, p. 199.

48. Arendt, *The Human Condition*, p. 198; second emphasis added.

49. Peter Dauvergne, "The Sustainability Story: Exposing Truths, Half-Truths, and Illusions," in *New Earth Politics: Essays from the Anthropocene*, edited by Simon Nicholson and Sikina Jinnah, Cambridge: MIT Press, 2016.

50. Dauvergne, "The Sustainability Story."

51. Iris Marion Young, *Justice and the Politics of Difference*, p. 227, Princeton, NJ: Princeton University Press, 1990, quoting Herbert Marcuse, *One-dimension Man: Studies in the Ideology of Advanced Industrial Society*. Boston: Beacon Press, 1964.

52. Young, *Justice and the Politics of Difference*, p. 234.

53. Edella Schlager and William Blomquist, *Embracing Watershed Politics*, Boulder, CO: University Press of Colorado, 2008.

54. Public Citizen, U.S. Non-Profit Water Directory (March 2005), accessed August 20, 2017, https://www.citizen.org/sites/default/files/waterdirectory.pdf.

55. Magnusson and Shaw, *A Political Space*.

56. Lipschutz, with Mayer, *Global Civil Society*, p. 220.

57. Lipschutz with Mayer, *Global Civil Society*, chapters 4 and 5; see also the websites of River Network, www.rivernetwork.org; International Rivers, www.internationalrivers.org; and Global Rivers Environmental Education Network, https://earthforce.org/GMGREEN/.

58. Lipschutz, with Mayer, *Global Civil Society*.

59. Olson, *Logic of Collective Action*; Tarrow, *Power in Movement*; Doug McAdam, John D. McCarthy, and Meyer N. Zald, *Comparative Perspectives on Social Movements—Political Opportunities, Mobilizing Structures, and Cultural Framings*, Cambridge: Cambridge University Press, 1996.

60. Mouffe, *The Democratic Paradox*.

61. Paul Wapner, "Living at the Margins," in *New Earth Politics: Essays from the Anthropocene*, 369–85, edited by Simon Nicholson and Sikina Jinnah, Cambridge: MIT Press, 2016.

Bibliography

Acaroglu, Leyla. "Where Do Old Cellphones Go to Die?" *New York Times*, May 4, 2013. http://www.nytimes.com/2013/05/05/opinion/sunday/where-do-old-cellphones-go-to-die.html.

Achenbach, Joel. "Why Do Many Reasonable People Doubt Science?" *National Geographic*, March 2015. Accessed September 25, 2017, http://ngm.nationalgeographic.com/2015/03/science-doubters/achenbach-text.

Achterbosch, T. J., S. van Berkum, and G. W. Meijerink. *Cash Crops and Food Security: Contributions to Income, Livelihood Risk and Agricultural Innovation.* Wageningen, Netherlands: LEI Wageningen University and Research Centre, 2014. Accessed August 18, 2017, http://edepot.wur.nl/305638.

Adamson, Walter L. 1980. *Hegemony and Revolution: A Study of Antonio Gramsci's Political and Cultural Theory.* Berkeley, CA: University of California Press.

Adams, Suzi, et al. "Social Imaginaries in Debate." *Social Imaginaries* 1, no. 1 (2015): 15–52.

Adbusters uncommercial. "The GDP Goes Up." Accessed September 19, 2017, https://www.youtube.com/watch?v=0hTtAfjBhyoandlist=PLLWWwP1Ombs3_EB0kIm6KdmVrsbAS4anAandindex=2.

Agarwal, Anil, and Sunita Narain. *Global Warming in an Unequal World: A Case of Environmental Colonialism.* New Delhi: Centre for Science and Environment, 1991.

Agnew, John. "Representing Space: Space, Scale and Culture in Social Science." In *Place/Culture/Representation.* Edited by James Duncan and David Ley, 251–71. London: Routledge, 1993.

Agriculture at a Crossroads. "Water." Accessed August 8, 2017, http://www.globalagriculture.org/report-topics/water.html.

Aikenhead, Glen S., and Masakata Ogawa. "Indigenous Knowledge and Science Revisited." *Cultural Studies of Science Education* 2 (2007): 539–91.

"Air Rights New York." Accessed August 8, 2017, http://www.airrightsny.com/.

Akerlof, Karen, et al. "Do People 'Personally Experience' Global Warming, and If So How, and Does It Matter?" *Global Environmental Change* 23 (2013): 81–91.

Alaimo, Stacy, and Susan Heckman, eds. *Material Feminisms*. Bloomington: Indiana University Press, 2007.

Alkon, Alison Hope, and Julian Agyeman, eds. *Cultivating Food Justice: Race, Class, and Sustainability*. Cambridge, MA: MIT Press, 2011.

Allison, Graham. *Essence of Decision: Explaining the Cuban Missile Crisis*. Boston: Little, Brown, 1971.

Altman Daniel. "Just How Far Can Trading of Emissions Be Extended?" *New York Times*, May 31, 2001. Accessed August 15, 2017, http://www.nytimes.com/2002/05/31/business/just-how-far-can-trading-of-emissions-be-extended.html.

Andaloussi, Mehdi Benatiya, and Elisabeth Thuestad Isaksen. "The Environmental and Distributional Consequences of Emissions Markets: Evidence from the Clean Air Interstate Rule." June 25, 2017. Accessed August 16, 2017, https://ssrn.com/abstract=2992180.

Anderson, Kevin. "Talks in the City of Light Generate More Heat." *Nature* 528 (2015). Accessed October 4, 2017, http://www.nature.com/news/talks-in-the-city-of-light-generate-more-heat-1.19074.

Anderson, Kevin, and Glen Peters. "The Trouble with Negative Emissions." *Science* 354 (2016): 182–83.

Anderson, Monica. "Technology Device Ownership: 2015." Pew Research Center, October 29, 2015, p. 3. Accessed January 8, 2017, http://assets.pewresearch.org/wp-content/uploads/sites/14/2015/10/PI_2015-10-29_device-ownership_FINAL.pdf.

Anderson, Terry L., and Donald R. Leal. *Free Market Environmentalism for the Next Generation*. New York: Palgrave Macmillan, 2015.

Anderson, Terry L., and Donald R. Leal. "Free Market versus Political Environmentalism." *Harvard Journal of Law and Public Policy* 15, no. 2 (Spring 1992): 297–310.

Anheier, Helmut, Marlies Glasius, and Mary Kaldor, eds. *Global Civil Society 2001*, Oxford: Oxford University Press, 2001.

Arendt, Hannah. *The Human Condition*. Chicago: University of Chicago Press, 1958, second ed.

Arias, Pedro, et al. "Banana Exporting Countries." In *The World Banana Economy, 1985–2002*. FAO Economic and Social Development Department, Rome, 2003, chapter 2. Accessed June 22, 2017, http://www.fao.org/docrep/007/y5102e/y5102e05.htmno.bm05.

Arp, Henning. "Technical Regulation and Politics: The Interplay between Economic Interests and Environmental Policy Goals in EC Car Emission Legislation." In *Environmental Policy in the European Union: Actors, Institutions and Processes*, edited by Andrew Jordan, 256–74. London: Earthscan, 2002.

Asiyanbi, Adeniyi P. "A Political Ecology of REDD+: Property Rights, Militarised Protectionism, and Carbonised Exclusion in Cross River." *Geoforum* 77 (2016): 146–56.

Associated Press. "Pipeline Exec Compares Dakota Protestors to Terrorists." *New York Times*, February 15, 2017. Accessed August 15, 2017, https://www.nytimes.com/aponline/2017/02/15/us/politics/ap-us-congress-oil-pipeline.html.

Austen, Ian. "Train Blast Kills at Least One and Forces Evacuations in Canada." *New York Times*, July 6, 2013. Accessed August 15, 2017, http://www.nytimes.com/2013/07/07/world/americas/people-missing-and-buildings-destroyed-in-canada-rail-blast.html.

Ayers, Jeffrey M. "Framing Collective Action Against Neoliberalism: The Case of the 'Anti-Globalization Movement.'" *Journal of World Systems Research* 1 (2004): 11–34.

Baard, Patrick, and Karin Edvardsson Björnberg. "Cautious Utopias: Environmental Goal-setting with Long Time Frames." *Ethics, Policy and Environment* 18, no. 2 (2015): 187–201.

Baccini, A., et al. "Tropical Forests are a Net Carbon Source Based on Aboveground Measurements of Gain and Loss." *Science* (2017). Accessed October 4, 2017, http://science.sciencemag.org/content/early/2017/09/27/science.aam5962.

Bacon, Sir Francis. *Novum Organum*, edited by Joseph Devey. New York: Collier and Son, 1902.

Bäckstrand, Karin, and Eva Lövbrand. "Planting Trees to Mitigate Climate Change: Contested Discourses of Ecological Modernization, Green Governmentality and Civic Environmentalism." *Global Environmental Politics* 6 (2006): 50–75.

Baddeley, Adam. "In the Rush to Get Waste Out of Landfill, the EU Risks Losing Sight of its Recycling Targets." *Euractiv*, June 20, 2016. Accessed August 12, 2017, https://www.euractiv.com/section/sustainable-dev/opinion/in-the-rush-to-get-waste-out-of-landfill-eu-risks-losing-sight-of-its-recycling-targets/

Bagehot. "There is No Alternative." *The Economist*, May 7, 2009. Accessed June 29, 2017, http://www.economist.com/node/13610705.

Bagley, Margo A. "Digital DNA: The Nagoya Protocol, Intellectual Property Treaties, and Synthetic Biology." Washington, DC: Wilson Center, December 2015. Accessed August 16, 2017, https://www.wilsoncenter.org/sites/default/files/digital_dna_final_0.pdf;

Bakker, Karen. "Neoliberal Versus Postneoliberal Water: Geographies of Privatization and Resistance." *Annals of the Association of American Geographers* 103, no. 2 (2013): 253–60.

Ball, Jeffrey. "The Proportion of Young Americans Who Drive Has Plummeted—And No One Knows Why." *The New Republic*, March 12, 2014. Accessed December 28, 2014, http://www.newrepublic.com/article/116993/millennials-are-abandoning-cars-bikes-carshare-will-it-stick.

BananaLink.org. "Where Bananas are Grown." n.d. Accessed January 10, 2017, http://www.bananalink.org.uk/content/where-bananas-are-grown.

BananaLink.org. "Why Bananas Matter." n.d. Accessed January 10, 2017, http://www.bananalink.org.uk/why-bananas-matter.

"Banana Market." In *The State of Sustainability Initiatives Review 2014*, edited by Jason Potts, et al., 97–114. Winnipeg: International Institute for Sustainable Development, 2014. Accessed January 10, 2017, https://www.iisd.org/pdf/2014/ssi_2014_chapter_5.pdf.

Bankoff, Greg. "No Such Thing as Natural Disasters." *Harvard International Review*, August 23, 2010. Accessed March 27, 2018, http://hir.harvard.edu/article/?a=2694.

Barboza, Tony. "Will Trump Erect a Roadblock to Southern California's Decades-long Fight against Smog?" *Los Angeles Times*, March 17, 2017. Accessed September 6, 2017, http://www.latimes.com/local/lanow/la-me-ln-vehicles-smog-20170316-story.html.

Barboza, Tony, and Jon Schleuss. "L.A. Keeps Building Near Freeways, Even Though Living There Makes People Sick." *Los Angeles Times*, March 2, 2017. Accessed September 6, 2017, http://www.latimes.com/projects/la-me-freeway-pollution/.

Barkin, J. Samuel. "Racing All Over the Place: A Dispersion Model of International Regulatory Competition." *European Journal of International Relations* 21, no. 1 (2015): 171–93.

Barnes, Kathryn. "How Tropical Bananas are Grown on Santa Barbara's Mesa." *KCRW.com*, February 23, 2017. Accessed September 12, 2017, https://curious.kcrw.com/2017/02/how-tropical-bananas-are-grown-in-santa-barbaras-mesa.

Barringer, Felicity. "Flooded Village Files Suit, Citing Corporate Link to Climate Change." *New York Times*, February 27, 2008. Accessed August 23, 2017, http://www.nytimes.com/2008/02/27/us/27alaska.html?_r=1andoref=slogin.

Barrows, Geoffrey, Steven Sexton, and David Zilberman. "Agricultural Biotechnology: The Promise and Prospects of Genetically Modified Crops." *Journal of Economic Perspectives* 28, no. 1 (Winter 2014): 99–119.

Bartley, Tim, et al. *Looking Behind the Label*. Bloomington: Indiana University Press, 2015.

Bartolini, Stefano, and Francesco Sarracino. "Happy for How Long? How Social Capital and Economic Growth Relate to Happiness Over Time." *Ecological Economics* 108 (2014): 242–56.

Bass, Stephen, et al. *Certification's Impacts on Forests, Stakeholders, and Supply Chains*. Stevenage, England: Earthprint, 2001. Accessed August 10, 2017, http://pubs.iied.org/9013IIED/.

BBC News. "Copenhagen Deal Reaction in Quotes." December 19, 2009. Accessed September 28, 2017, http://news.bbc.co.uk/2/hi/science/nature/8421910.stm.

Beck, Ulrich. *Ecological Politics in an Age of Risk*. London: Polity, 1995.

Beck, Ulrich. *Risk Society: Towards a New Modernity*. London: Sage, 1992.

Beckert, Sven. *Empire of Cotton: A Global History*. New York: Alfred A. Knopf, 2014.

Bellamy, Alex J. "The Responsibility to Protect Turns Ten." *Ethics and International Affairs* 29, no. 2 (2015): 161–85.

Bello, Walden. *Dark Victory: The United States and Global Poverty*. Oakland, CA: Food First Books, 1999.

Belluck, Pam. "In Breakthrough, Scientists Edit a Dangerous Mutation from Genes in Human Embryos." *New York Times*, August 2, 2017. Accessed August 14, 2017, https://www.nytimes.com/2017/08/02/science/gene-editing-human-embryos.html.

Benn, Stanley I., and Gerald F. Gaus. "The Liberal Conception of the Public and the Private." In *Public and Private in Social Life*, edited by Stanley I. Benn and Gerald F. Gaus, 31–66. London: Croom Helm, 1983.

Bennett, Drake. "Do Natural Disasters Stimulate Economic Growth?" *New York Times*, July 8, 2008. Accessed September 14, 2017, http://www.nytimes.com/2008/07/08/business/worldbusiness/08iht-disasters.4.14335899.html.

Bennis, Will M., Douglas L. Medin, and Daniel M. Bartels. "The Costs and Benefits of Calculation and Moral Rules." *Perspectives on Psychological Science* 5, no. 2 (2010): 187–202.

Berman, Marshall. *All That is Solid Melts Into Air: The Experience of Modernity*. New York: Simon and Schuster, 1982.

Bernstein, Steven. *The Compromise of Liberal Environmentalism*. New York: Columbia University Press, 2001.

Bernstein, Steven, and Hamish van der Ven. "Continuity and Change in Global Environmental Politics." In *International Politics and Institutions in Time*, edited by Orfeo Fioretos, 293–327. Oxford: Oxford University Press, 2017.

Berta, Nathalie, Emmanuelle Gautherat, and Ozgur Gun. *Transactions in the European Carbon Market: A Bubble of Compliance in a Whirlpool of Speculation.* Paris: Documents de travail du Centre d'Economie de la Sorbonne, 2015.

Berta, Nathalie, Emmanuelle Gautherat, and Ozgur Gun. "Transactions in the European Carbon Market: A Bubble of Compliances in a Whirlpool of Speculation." *Cambridge Journal of Economics* 41 (2017): 575–93.

Bethge, Philip. "Modern Day Flintstones: A Stone Age Subculture Takes Shape in the US." *Speigel Online*, February 11, 2010. Accessed February 14, 2017, http://www.spiegel.de/international/zeitgeist/a-677121.html.

Bhagwati, Jagdish. *Free Trade Today.* Princeton, NJ: Princeton University Press, 2002.

Biermann, Frank, and Steffan Bauer, eds. *A World Environmental Organization: Solution or Threat for Effective International Environmental Governance?* London: Routledge, 2005.

Bièvre, Dirk De. "A Glass Quite Empty: Issue Groups' Influence in the Global Trade Regime." *Global Policy* 5, no. 2 (May 2014): 222–28.

Birdsall, Nancy. "Oops: Economists in Confused Search for the Middle Class in the Developing World." Center for Global Development, May 31, 2012. Accessed June 22, 2016, http://www.cgdev.org/blog/oops-economists-confused-search-middle-class-developing-world.

Biro, Andrew, ed. *Critical Ecologies: The Frankfurt School and Contemporary Environmental Crises.* Toronto: University of Toronto Press, 2011.

Black, Jan Knippers. *Development in Theory and Practice: Bridging the Gap.* Boulder: Westview, 1991.

Blaikie, Piers. *The Political Economy of Soil Erosion in Developing Countries.* New York: Wiley, 1985.

Blaikie, Piers, and Harold Brookfield. *Land Degradation and Society.* London: Methuen, 1987.

Bodansky, Daniel. "The Legal Character of the Paris Agreement." *Review of European, Comparative & International Environmental Law* 25 (2016): 142–50.

Bodansky, Daniel. "Legal Options for U.S. Acceptance of a New Climate Change Agreement." Center for Climate and Energy Solutions, May 2015. Accessed August 16, 2017, https://www.c2es.org/publications/legal-options-us-acceptance-new-climate-change-agreement.

Bodansky, Daniel. "A Tale of Two Architectures: The Once and Future UN Climate Change Regime." *Arizona State Law Journal* 43 (2011): 697–712.

Bomberg, Elizabeth. *Green Parties and Politics in the European Union.* London: Routledge, 1998.

Bonacich, Edna, and Richard Appelbaum. *Behind the Label: Inequality in the Los Angeles Apparel Industry.* Berkeley: University of California Press, 2000.

Bonderud, Doug. "Microwave Repair and What It Costs." *Angies' List.* Accessed August 4, 2017, https://www.angieslist.com/articles/microwave-repair-and-what-it-costs.htm.

Bongiovanni, Marissa, and Erin Clover Kelly. "Ecosystem Service Commodification: Lessons from California." *Global Environmental Politics* 16, no. 4 (Nov. 2016): 90–110.

Bonneville Environmental Foundation. "Carbon Offsets—International REDD+." Accessed August 25, 2017, https://store.b-e-f.org/products/carbon-offsets-international-redd/.

Bontron, Cécile. "Rare-Earth Mining in China Comes at a Heavy Cost for Local Villages." *The Guardian*, August 7, 2012. Accessed February 10, 2017, https://www.theguardian.com/environment/2012/aug/07/china-rare-earth-village-pollution.

Bookchin, Murray. *The Ecology of Freedom*. Montreal: Black Rose Books, 1992, revised ed.

Bennis, Will M., Douglas L. Medin, and Daniel M. Bartels. "The Costs and Benefits of Calculation and Moral Rules." *Perspectives on Psychological Science 5*, no. 2 (2010): 187–202.

Borghesi, S., M. Montini, and A. Barreca. "Comparing the EU ETS with Its Followers." In *The European Emission Trading System and Its Followers*, edited by S. Borghesi, M. Montini, and A. Barreca, 71–90. Cham, Switzerland: Springer, 2016.

Bourdieu, Pierre. *The Logic of Practice*, translated by Richard Nice. Stanford, CA: Stanford University Press, 1990.

Bourdieu, Pierre. *Outline of a Theory of Practice*. Cambridge: Cambridge University Press, 1977.

Boyle, Kevin J. "Contingent Valuation in Practice." In *A Primer on Nonmarket Valuation*, edited by Patricia A. Champ, Thomas C. Brown, and Kevin J. Boyle, 83–132. Dordrecht, Netherlands: Springer, 2017, second edition.

Braithwaite, John, and Peter Drahos. *Global Business Regulation*. Cambridge: Cambridge University Press, 2000.

Bramwell, Anna. *Ecology in the 20th Century: A History*. New Haven: Yale University Press, 1989.

Brei, Vinicius, and Steffen Böhm. "'1L=10L for Africa': Corporate Social Responsibility and the Transformation of Bottled Water into a 'Consumer Activist' Commodity." *Discourse and Society* (2013): 1–29.

Bretton Woods Project, et al. "Bottom Lines, Better Lives?" April 22, 2010. Accessed August 12, 2017, http://www.brettonwoodsproject.org/2010/04/art-566197/.

Briceno, Tania, and Sigrid Stagl. "The Role of Social Processes for Sustainable Consumption." *Journal of Cleaner Production* 14 (2006): 1541–51.

Bricker, Jesse, Rodney Ramcharan, and Jacob Krimmel. "Signaling Status: The Impact of Relative Income on Household Consumption and Financial Decisions." Federal Reserve Board, Washington, DC, 2014-76. Accessed September 26, 2014, https://www.federalreserve.gov/econresdata/feds/2014/files/201476pap.pdf.

Brisack, Jan. "Edward I: Environmentalist by Accident." *PlanetGreen.org*, March 19, 2012. Accessed August 1, 2017, http://www.planetgreen.org/2012/03/edward-i-environmentalist-by-a.html.

Broad, Robin, ed. *Global Backlash: Citizen Initiatives for a Just World Economy*. Lanham, MD: Rowman & Littlefield, 2002.

Bromley, Daniel W., ed. *Making the Commons Work*. San Francisco: ICS Press, 1992.

Brown, DeNeen L. "Hundreds of Keystone XL Pipeline Opponents Arrested at White House." *Washington Post*, March 2, 2014. Accessed August 23, 2017, https://www.washingtonpost.com/blogs/local/wp/2014/03/02/hundreds-of-keystone-xl-pipeline-opponents-arrested-at-white-house/?utm_term=.b4c9cfe6514d.

Brown, Harrison. *The Challenge of Man's Future*. New York: Viking, 1954.

Brulle, Robert. "Institutional Delay: Foundation Funding and the Creation of U.S. Climate Change Counter-movement Organizations." *Climatic Change* 122 (2014): 681–94.

Bryson, Jay H., and Erik Nelson. "How Exposed is the U.S. Economy to China?" Wells Fargo Economics Group, August 13, 2015. Accessed August 4, 2017, http://www

.realclearmarkets.com/docs/2015/08/US%20Exposure%20to%20China%20
_%20Aug%202015.pdf.

Buck, Susan J. B. "No Tragedy on the Commons." *Environmental Ethics* 7 (Spring 1985): 49–61.

Buckley, Chris. "Xi Jinping is Set for a Big Gamble with China's Carbon Trading Market." *New York Times*, June 23, 2017. Accessed September 3, 2017, https://www.nytimes.com/2017/06/23/world/asia/china-cap-trade-carbon-greenhouse.html?_r=0.

Bull, Hedley. *The Anarchical Society.* New York: Columbia University Press, 1977.

Bumpus, A. G., and D. M. Liverman. "Carbon Colonialism? Offsets, Greenhouse Gas Reductions, and Sustainable Development." In *Global Political Ecology*, edited by Richard Peet, Paul Robbins, and Michael J. Watts, 203–24. London: Routledge, 2011.

Burgess, Rod, Marisa Carmona, and Theo Kolstee, eds. *The Challenge of Sustainable Cities: Neoliberalism and Urban Strategies in Developing Countries.* London: Zed Books, 1997.

Burke, William T. *The New International Law of Fisheries: UNCLOS 1982 and Beyond.* Oxford: Clarendon Press, 1994.

Burns, William C. G. "Human Rights Dimensions of Bioenergy with Carbon Capture and Storage: A Framework for Climate Justice in the Realm of Climate Geoengineering." In *Climate Justice: Case Studies in Global and Regional Governance Challenges*, edited by Randall Abate, 149–76. Washington, DC: Environmental Law Institute, 2016. Accessed August 15, 2017, https://ssrn.com/abstract=2871527

Burrows, Nichole. "How Consumers Drive the Economy." *Nassau Guardian*, June 11, 2014. Accessed August 4, 2017, http://www.thenassauguardian.com/opinion/op-ed/48000-how-consumers-drive-the-economy-.

Büscher, Bram, and Robert Fletcher, "Nature is Priceless, Which is Why Turning it into 'Natural Capital' is Wrong." *The Conversation*, September 21, 2016. Accessed December 3, 2017, https://theconversation.com/nature-is-priceless-which-is-why-turning-it-into-natural-capital-is-wrong-65189.

Butler, Catherine, Karen A. Parkhill, and Nicholas F. Pidgeon. "Energy Consumption and Everyday Life: Choice, Values and Agency through a Practice Theoretical Lens." *Journal of Consumer Culture* 16, no. 3 (2016): 887–90.

Butters, Julie. "This is Dysprosium—If We Run Out of It, Say Goodbye to Smartphones, MRI Scans and Hybrid Cars." *Phys.Org*, June 6, 2016. Accessed August 1, 2017, https://phys.org/news/2016-06-dyprosiumif-goodbye-smartphones-mri-scans.html.

California Legislative Analysts' Office. "Residential Water Use Trends and Implications for Conservation Policy." California Legislature, Sacramento, March 8, 2017. Accessed August 24, 2017, http://www.lao.ca.gov/Publications/Report/3611.

Camacho, David E., ed. *Environmental Injustices, Political Struggles: Race, Class, and the Environment.* Durham, NC: Duke University Press, 1998.

Cames, Martin, et al. "How Additional is the Clean Development Mechanism?" Öko-Institut e.V., Berlin, March 2016. Accessed August 16, 2017, https://ec.europa.eu/clima/sites/clima/files/ets/docs/clean_dev_mechanism_en.pdf

Capra, Fritjof, and Charlene Spretnak. *Green Politics: The Global Promise.* New York: Dutton, 1984.

Carbon Trade Watch. "Carbon Trading, REDD+ and the Push for Pricing Forests." June 19, 2012. Accessed September 29, 2017, http://www.carbontradewatch.org /articles/carbon-trading-redd-and-the-push-for-pricing-forests.html.

Carr, Matthew. "The Cost of Carbon: Putting a Price on Pollution." *Bloomberg*, June 1, 2017. Accessed September 3, 2017, https://www.bloomberg.com/quicktake /carbon-markets-2-0.

Carr, Matthew, and Anna Hirtenstein. "How Countries Cut Carbon by Putting a Price on It: Quick Take Q&A." *Bloomberg*, April 11, 2017. Accessed September 3, 2017, https://www.bloomberg.com/news/articles/2017-04-11/how-countries-cut -carbon-by-putting-a-price-on-it-quicktake-q-a.

Carroll, N, et al., eds. *Conservation and Biodiversity Banking: A Guide to Setting Up and Running Biodiversity Credit Trading Systems*. London, UK: Earthscan, 2008.

Cascio, Joseph, Gayle Woodside, and Philip Mitchell. *ISO 14000 Guide: The New International Environmental Management Standards*. New York: McGraw Hill, 1996.

Cashore, Benjamin. "What Should Canada Do When the Softwood Lumber Agreement Expires?" Policy.ca, April 20, 2001. Accessed April 16, 2002, http://www.policy .ca/PDF/20010205.pdf.

Castells, Mauel. *The Rise of the Network Society*. Cambridge, MA: Blackwell, 1996.

Center for Climate and Energy Solutions. "Alternative Models for the 2015 Climate Change Agreement." Accessed August 16, 2017, https://www.c2es.org /press-center/article/alternative-models-2015-climate-change-agreement.

Center for International Environmental Law. "NGO Response: Proposed World Bank Standards Represent Dangerous Set-back to Key Environmental and Social Protections." July 22, 2016. Accessed September 2, 2017, http://www.ciel.org /news/safeguard-policy-endangers-rights/.

Cerny, Philip G. "Structuring the Political Arena: Public Goods, States, and Governance in a Globalizing World." In *Global Political Economy: Contemporary Theories*, edited by Ronen Palan, 21–35. London: Routledge.

CGIAR. "Banana and Plantain." n.d. Accessed Jan. 15, 2017, http://www.cgiar.org /our-strategy/crop-factsheets/bananas/.

Chan, Kai M. A., et al. "Why Protect Nature? Rethinking Values and the Environment." *PNAS* 113, no. 6 (Feb. 9, 2016): 1462–65.

Chang, Howard F. "The Environmental Chapter of the Trans-Pacific Partnership: An Environmental Agreement within a Trade Agreement." *Trends* 47, no. 5 (May/ June 2015). Accessed August 28, 2017, https://www.americanbar.org/publications /trends/2015-2016/may-june-2016/the_environment_chapter_of_the_trans -pacific_partnership.html.

Chang, Ha-Joon. *Economics: The User's Guide*. New York: Bloomsbury Press, 2014.

Chang, Ha-Joon. *23 Things They Don't Tell You about Capitalism*. New York: Bloomsbury Press, 2012.

Chasek, Pamela S., David L. Downie, and Janet Welsh Brown. *Global Environmental Politics*. Boulder, CO: Westview Press, 2016, seventh edition.

Cheat Neutral YouTube advertisement. Accessed August 25, 2017, https://www.you tube.com/watch?v=f3_CYdYDDpk.

Chen, Brian X., Farhad Manjoo, and Vindu Goel. "Apples New iPhone Release: What to Expect." *New York Times*, September 12, 2017. Accessed September 12, 2017, https://www.nytimes.com/2017/09/12/technology/apple-iphone-event.html?_r=0.

Chiang, John. "Winter is Coming, but 1 Idea May Help Save Planet Earth." *San Francisco Chronicle*, September 13, 2017. Accessed September 13, 2017, http://www

.sfchronicle.com/opinion/openforum/article/Winter-is-coming-but-1-idea-may -help-save-Planet-12195996.php.

Chikozho, C., and K. Kujinga. "Managing Water Supply Systems Using Free-market Economy Approaches: A Detailed Review of the Implications for Developing Countries." *Physics and Chemistry of the Earth* (2016). Accessed August 24, 2017, http://www.sciencedirect.com/science/article/pii/S1474706516300420.

Choice Behavior Insights. "The Myth of 5,000 Ads." Hill Holliday, n.d., Accessed August 3, 2017, http://cbi.hhcc.com/writing/the-myth-of-5000-ads/.

Chomba, Susan, Juliet Kariuki, Jens Friis Lund, and Fergus Sinclair. "Roots of Inequity: How the Implementation of REDD+ Reinforces Past Injustices." *Land Use Policy* 50 (2016): 202–13.

Circle of Blue. "The Price of Water." n.d. Accessed August 8, 2017, http://www.circleof blue.org/waterpricing/.

Clapp, Jennifer. "The Privatization of Global Environmental Governance: ISO 14000 and the Developing World." In *The Business of Global Environmental Governance*, edited by David L. Levy and Peter J. Newell, 223–48. Cambridge: MIT Press, 2005.

Clapp, Jennifer. *Toxic Exports: The Transfer of Hazardous Wastes from Rich to Poor Countries*. Ithaca: Cornell University Press, 2001.

Clapp, Jennifer, and Peter Dauvergne. *Paths to a Green World: The Political Economy of the Environment*. Cambridge: MIT Press, 2005.

Clark, John. "A Social Ecology." In *Environmental Philosophy: From Animal Rights to Radical Ecology*, edited by Michael E. Zimmerman, et al., 416–40. Upper Saddle River, NJ: Prentice Hall, 1998, second edition.

Clifford, Stephanie. "That 'Made in U.S.A.' Premium." *New York Times*, November 30, 2013. Accessed August 8, 2017, http://www.nytimes.com/2013/12/01/business /that-made-in-usa-premium.html.

Climate Equity Reference Calculator. Accessed October 5, 2017, https://calculator .climateequityreference.org/.

ClimateFairshares.org. Accessed October 6, 2017, http://climatefairshares.org.

Climate and Security Advisory Group. "Briefing Book for a New Administration." Center for Climate and Security, Washington, DC, September 14, 2016. Accessed February 24, 2017, https://climateandsecurity.files.wordpress.com/2016/09/climate-and -security-advisory-group_briefing-book-for-a-new-administration_2016_11.pdf.

Clinginsmith, David, and Roman Sheremta. "Status and the Demand for Visible Goods: Experimental Evidence on Conspicuous Consumption." Munich Personal RePEc Archive, December 4, 2015. Accessed August 3, 2017, https://mpra.ub .uni-muenchen.de/68202/.

CNN.com. "Massive Iceberg Breaks Away from Antarctica." July 12, 2017. Accessed August 17, 2017, http://www.cnn.com/2017/07/12/world/larsen-c-antarctica/index .html.

CNN.com. "Federal Flood Insurance to Have More Strings." November 10, 1998. Accessed August 8, 2017, http://www.cnn.com/US/9811/10/disaster.reform/index.html.

CNNMoney. "Median Income Jumps 5.2% in 2015." September 13, 2016. Accessed July 12, 2017, money.cnn.com/2016/09/13/news/economy/median-income-census/.

CNNMoney. "America's Upper Middle Class is Thriving." June 21, 2016. Accessed July 12, 2017, money.cnn.com/2016/06/21/news/economy/upper-middle-class/.

Coase, Ronald. "The Problem of Social Cost." *Journal of Law and Economics* 3 (1960): 1–44.

Coghlan, Andy. "More than Half of EU Officially Bans Genetically Modified Crops." *New Scientist*, October 5, 2015. Accessed September 5, 2017, https://www.new scientist.com/article/dn28283-more-than-half-of-european-union-votes-to-ban -growing-gm-crops/.

Cohn, Victor. *1999: Our Hopeful Future*. Indianapolis: Bobbs-Merrill, 1956.

Colas, Alejandro. *International Civil Society: Social Movements in World Politics*. Cambridge, MA: Blackwell, Polity Press, 2002.

Coleman, James S. *Foundations of Social Theory*. Cambridge, MA: Belknap Press of Harvard University Press, 1990.

Commoner, Barry. *The Closing Circle: Nature, Man, and Technology*. New York: Knopf, 1971.

"The Commons and the Common." *Review of Radical Political Economics* 48, no. 1 (March 2016), special issue.

Conca, Ken. *Governing Water*. Cambridge, MA: MIT Press, 2006.

Cone, Marla. "Stalking a Killer in Our Greens." *Los Angeles Times*, August 14, 2007. Accessed September 4, 2017, http://www.latimes.com/local/obituaries/la-me-spin-ach14aug14-story.html.

Consumer Project on Technology. "Court Case between 39 Pharmaceutical Firms and the South African Government." Accessed October 4, 2017, http://www.cptech. org/ip/health/sa/pharma-v-sa.html.

Convention on Biological Diversity. "About the Nagoya Protocol." n.d. Accessed August 15, 2017, https://www.cbd.int/abs/about/.

Convention on Biological Diversity, June 5, 1992, 1760 U.N.T.S. 79, 143; 31 I.L.M. 818 (1992).

Corama, Vlad C., Åsa Moberg, and Lorenz M. Hilty. "Dematerialization Through Electronic Media." In *ICT Innovations for Sustainability*, edited by L. M. Hilty and B. Aebischer, 405–21. Cham, Switzerland: Springer, 2015.

Corson, Catherine, and Kenneth Iain MacDonald. "Enclosing the Global Commons: The Convention on Biological Diversity and Green Grabbing." *The Journal of Peasant Studies* 39, no. 2 (2012): 263–83.

Cosgrove, Denis. "Geography Is Everywhere: Culture and Symbolism in Human Land-scapes." In *Horizons in Human Geography*, edited by Derek R. Gregory and Rex Walford, 118–35. London: Macmillan, 1985.

Costanza, Robert, et al. "Changes in the Global Value of Ecosystem Services." *Global Environmental Change* 26 (2014): 152–58.

Costanza, Robert, et al. "The Value of the World's Ecosystem Services and Natural Capital." *Nature* 387, no. 6630 (May 15, 1997): 253–60.

Cox, Robert. *Production, Power and World Order*. New York: Columbia University Press, 1987.

Cramer, C. "*Homo Economicus* Goes to War: Methodological Individualism, Rational Choice and the Political Economy of War." *World Development* 30, no. 11 (Nov. 2002): 1845–64.

Crawford, Beverly, and Ronnie D. Lipschutz, eds. *The Myth of "Ethnic Conflict": Politics, Economics, and "Cultural" Violence*. Berkeley: UC Berkeley International and Area Studies Press, 1998.

Credit Suisse Research Institute. "Global Wealth Report 2016." Zurich, 2016. Accessed June 13, 2017, http://publications.credit-suisse.com/tasks/render/file /index.cfm?fileid=AD783798-ED07-E8C2-4405996B5B02A32E.

"Critical Theory." Stanford Encyclopedia of Philosophy. Accessed August 24, 2017, https://plato.stanford.edu/entries/critical-theory/.

Crocker, Lawrence. "Marx's Use of Contradiction." *Philosophy and Phenomenological Research* 40, no. 4 (June 1980): 558–63.

Cronon, William. *Nature's Metropolis: Chicago and the Great West*. New York: Norton, 1991.

Cronon, William, ed. *Uncommon Ground: Toward Reinventing Nature*. New York: Norton, 1995.

Crozier, Michel, Samuel P. Huntington, and Joji Watanuki. *The Crisis of Democracy: Report on the Governability of Democracies to the Trilateral Commission*. New York: New York University Press, 1975.

Curran, Dean. "The Treadmill of Production and the Positional Economy of Consumption." *Canadian Review of Sociology* 54, no. 1 (Feb. 2017): 28–47.

Dahl, Robert. "The Concept of Power." *Systems Research and Behavioral Science* 2, no. 3 (1957): 210–15.

Dalsgaard, Steffan. "The Commensurability of Carbon: Making Value and Money of Climate Change." *HAU: Journal of Ethnographic Theory* 3, no. 1 (2013): 80–98.

Dalton, Russell J. *The Green Rainbow: Environmental Groups in Western Europe*. New Haven, CT: Yale University Press, 1994.

Daly, Herman. *From Uneconomic Growth to a Steady-State Economy*. Cheltenham: Edward Elgar, 2014.

Daly, Herman. *Steady-State Economics*. Washington, DC: Island Press, 1991, second edition.

Daly, Herman. *Toward a Steady-State Economy*. San Francisco: W.H. Freeman, 1973.

Daly, Herman, and John Cobb, Jr. *For the Common Good—Redirecting the Economy Toward Community, the Environment, and a Sustainable Future*. Boston: Beacon Press, 1989.

Dash, T., and A. Khotari. "Forest Rights and Conservation in India." In *Right to Responsibility: Resisting and Engaging Development, Conservation and the Law in Asia*, edited by H. Jonas, H. J. Suneetha, and M. Subramanian, 151–74. Tokyo: UNU-IAS and Capetown: Natural Justice, 2015.

Dauvergne, Peter. "The Sustainability Story: Exposing Truths, Half-Truths, and Illusions." In *New Earth Politics: Essays from the Anthropocene*, edited by Simon Nicholson and Sikina Jinnah, 387–404. Cambridge: MIT Press, 2016.

Dauvergne, Peter, *Environmentalism of the Rich*. Cambridge: MIT Press, 2016.

Dauvergne, Peter. "The Problem of Consumption." *Global Environmental Politics* 10, no. 2 (May 2010): 1–10.

Dauvergne, Peter. *The Shadows of Consumption: Consequences for the Global Environment*. Cambridge: MIT Press, 2008.

Dauvergne, Peter, and Jane Lister. "Big Brand Sustainability: Governance Prospects and Environmental Limits." *Global Environmental Change* 22 (2012): 36–45.

Dauvergne, Peter, and Jane Lister. *Timber*. Cambridge: Polity Press, 2011.

Dauvergne, Peter, and Jane Lister. "The Prospects and Limits of Eco-Consumerism: Shopping Our Way to Less Deforestation?" *Organization and Environment* 23 (2010): 132–54.

Davenport, Coral, and Eric Lipton. "How G.O.P. Leaders Came to View Climate Change as Fake Science." *New York Times*, June 3, 2017. Accessed September 29,

2017, https://www.nytimes.com/2017/06/03/us/politics/republican-leaders-climate
-change.html.

Davenport, Coral. "Scott Pruitt is Carrying Out His E.P.A. Agenda in Secret, Critics
Say." *New York Times*, August 11, 2017. Accessed August 14, 2017, https://www
.nytimes.com/2017/08/11/us/politics/scott-pruitt-epa.html.

Davenport, Coral. "Administration is Seen as Retreating on Environment in Talks
on Pacific Trade." *New York Times*, January 15, 2014. Accessed August 28,
2017, https://www.nytimes.com/2014/01/15/us/politics/administration-is-seen-as
-retreating-on-environment-in-talks-on-pacific-trade.html?hpwandrref=science.

David, Avril. "Can Carbon Sequestration Be Leased?" *Ecosystem Marketplace*, July 24,
2009. Accessed August 15, 2017, http://www.ecosystemmarketplace.com/articles
/can-carbon-sequestration-be-leased/.

Davidson, Helen, and Oliver Milman. "Global 'March for Science' Protests Call for
Action on Climate Change." *The Guardian*, April 22, 2017, Accessed September
ber 5, 2017, https://www.theguardian.com/science/2017/apr/22/global-march-for
-science-protests-call-for-action-on-climate-change.

Davidson, Jacob. "Here's How Many Internet Users There Are." *Time.com Money*,
May 26, 2015. Accessed August 17, 2017, http://time.com/money/3896219
/internet-users-worldwide/.

Davidson, Mark D. "On the Relation between Ecosystem Services, Intrinsic Value, Existence Value and Economic Valuation." *Ecological Economics* 95 (2013): 171–77.

Davies, Ronald B., and K. C. Vadlamannati. "A Race to the Bottom in Labor Standards? An Empirical Investigation." *Journal of Development Economics* 103
(2013): 1–14.

Davies, Thomas, Holly Eva Ryan, and Alejandro Milcíades Peña. "Protest, Social
Movements and Global Democracy since 2011: New Perspectives." In *Protest,
Social Movements and Global Democracy since 2011: New Perspectives*, edited by
Thomas Davies, Holly Eva Ryan, and Alejandro Milcíades Peña, 1–29. *Research
in Social Movements, Conflicts and Change*, Vol. 39 (2016).

Davis, G. Gordon, and Jessica Ann Hall. "Circular Economy Legislation—the International Experience." May 1, 2006. Accessed September 28, 2017, http://site
resources.worldbank.org/INTEAPREGTOPENVIRONMENT/Resources
/CircularEconomy_Legal_IntExperience_Final_EN.doc.

Dawson, Jane I. *Eco-Nationalism: Anti-Nuclear Activism and National Identity in Russia, Lithuania, and Ukraine*. Durham, NC: Duke University Press, 1996.

de Freytas-Tamura, Kimiko. "Plastics Pile Up as China Refuses to Take the West's Recycling," *New York Times*, January 18, 2018. Accessed March 21, 2018, https://
www.nytimes.com/2018/01/11/world/china-recyclables-ban.html.

Dean, Mitchell. *Governmentality: Power and Rule in Modern Society*. London: Sage,
1999.

DeGroot, Gerard J., ed. *Student Protest: The Sixties and After*. London: Longman,
1998.

Democracy Center. "The Water Revolt." Accessed August 8, 2017, https://democracy
ctr.org/archive/the-water-revolt/.

Democracy Center. "Bechtel vs. Bolivia: Details of the Case and the Campaign."
Accessed August 8, 2017, https://democracyctr.org/archive/the-water-revolt/bechtel
-vs-bolivia-details-of-the-case-and-the-campaign/.

Deane-Drummond, Celia. "Natural Law Revisited: Wild Justice and Human Obligations for Other Animals." *Journal of the Society of Christian Ethics* 35, no. 2 (2015): 159–73.

Deleuze, Gilles, and Felix Guattari. *A Thousand Plateaus: Capitalism and Schizophrenia.* Minneapolis, MN: University of Minnesota Press, 1987.

Demaria, Federico, et al. "What is Degrowth? From an Activist Slogan to a Social Movement." *Environmental Values* 22 (2013): 195–215.

Dempsey, Jessica. "The Financialization of Nature Conservation?" In *Money and Finance After the Crisis; Critical Thinking for Uncertain Times*, edited by Brett Christophers, Andrew Leyshon and Geoff Mann, 191–216. New York: Wiley Online Library, 2017.

Denham, Tim, Edmond De Langh, and Luc Vrydaghs. "Special Issue: History of Banana Domestication." *Ethnobotany Research and Applications* 7 (2009): 163–380.

Dent, Kendall, Nadia Zainuddin, and Leona Tam. "The Good Life: Exploring Values Creation and Destruction in Consumer Well-Being." In *Marketing at the Confluence between Entertainment and Analytics*, edited by Patricia Rossi, 269–72. Cham, Switzerland: Springer, 2017.

de-Shalit, Avner. *Why Posterity Matters—Environmental Policies and Future Generations.* London: Routledge, 1995.

DeSombre, Elizabeth R., and J. Samuel Barkin. *Fish.* Cambridge: Polity Press, 2011.

Deveraux, Charan, Robert Z. Lawrence, and Michael D. Watson. *Case Studies in U.S. Trade Negotiation, Volume 1: Making the Rules.* Washington, DC: Institute for International Economics, 2006.

Dietz, Simon, Ben Groom, and William A. Pizer. "Weighing the Costs and Benefits of Climate Change to Our Children." *The Future of Children* 26, no. 1 (Spring 2016): 133–55.

Di Muzio, Tim. *The Tragedy of Human Development: A Genealogy of Capital as Power.* Lanham, MD: Rowman & Littlefield, 2017.

DivSeek. "Working Groups." Accessed August 15, 2017, http://www.divseek.org/.

Domhoff, G. William. *Who Rules America? Power and Politics.* Boston: McGraw-Hill, 2002, fourth edition.

Domo. "Data Never Sleeps 5.0." *Domo.com*, July 7, 2017. Accessed August 17, 2017, https://web-assets.domo.com/blog/wp-content/uploads/2017/07/17_domo_data -never-sleeps-5-01.png.

Dooley, Kate, and Sivan Kartha. Land-based negative emissions: risks for climate mitigation and impacts on sustainable development. *International Environmental Agreements: Politics, Law and Economics* 81, No. 1 (2017): 79–98.

Dotson, Kristie. "Tracking Epistemic Violence, Tracking Practices of Silencing." *Hypatia* 26, no. 2 (Spring 2011): 236–57.

Duales System Deutschland AG. "Der Grüne Punkt." Accessed August 28, 2017, https://www.gruener-punkt.de/en.html.

Dunlap, Thomas R. *Nature and the English Diaspora.* Cambridge: Cambridge University Press, 1999.

Dwivedi, O. P. *India's Environmental Policies, Programmes and Stewardship.* Houndsmill, Basingstoke, UK: Macmillan, 1997.

Dye, Thomas R. *Who's Running America? The Bush Restoration.* Upper Saddle River, NJ: Prentice Hall, 2002, seventh edition.

EarthJustice. "The Standing Rock Sioux Tribe's Litigation Against the Dakota Access Pipeline." Accessed August 23, 2017, http://earthjustice.org/features/faq -standing-rock-litigation.

EarthJustice. "Standing with Standing Rock." Accessed August 23, 2017, http://earth justice.org/features/teleconference-standing-rock.

Ebenstein, Avraham, Ann Harrison, and Margaret McMillan. "Why are American Workers Getting Poorer? China, Trade and Offshoring." National Bureau of Economic Research, March 2015, Working paper 21027. Accessed September 2, 2017, http://www.nber.org/papers/w21027.

The Economist. "Fruits de mer: Plucking Minerals from the Seabed is Back on the Agenda." February 23, 2017. Accessed August 28, 2017, https://www.economist. com/news/science-and-technology/21717351-fruits-de-mer-plucking-minerals -seabed-back-agenda.

The Economist. "Planet of the Smartphones." February 26, 2015. Accessed February 10, 2017, http://www.economist.com/news/leaders/21645180-smartphone -ubiquitous-addictive-and-transformative-planet-phones.

The Economist. "Sins of Emission." August 3, 2006. Accessed August 16, 2017, http:// www.economist.com/node/7252897.

Economist Free Exchange. "Do Economists All Favour a Carbon Tax?" September 19, 2011. Accessed August 2, 2017, http://www.economist.com/blogs/free exchange/2011/09/climate-policy.

Edwards, Bob, and Melinda Kane. "Resource Mobilization and Social and Political Movements." In *Handbook of Political Citizenship and Social Movements*, edited by Hein-Anton van der Heijden, 205–32. Cheltenham, UK: Edward Elgar, 2014.

Edwards, Bob, and Patrick F. Gillham. "Resource Mobilization Theory." In *Wiley Blackwell Encyclopedia of Social and Political Movements*. Oxford: Blackwell, 2013. Accessed August 30, 2017, http://www.academia.edu/download/30829255 /Resource_Mobilization_Theory_and_Social_Movements.pdf.

Edwards, Michael. *Civil Society* Cambridge: Polity Press, 2014.

Edwards, Michael, ed. *The Oxford Handbook of Civil Society*. Oxford: Oxford University Press, 2011.

Ehrenberg, Michael. "The History of Civil Society Ideas." In *The Oxford Handbook of Civil Society*, edited by Michael Edwards. Oxford: Oxford University Press, 2011. Accessed September 25, 2017, http://www.oxfordhandbooks.com/view/10.1093 /oxfordhb/9780195398571.001.0001/oxfordhb-9780195398571-e-2?print=pdf.

Ehrenreich, Barbara, and John Ehrenreich. *Long March, Short Spring: The Student Uprising at Home and Abroad*. New York: Monthly Review Press, 1969.

Ehrlich, Paul. *The Population Bomb*. New York: Ballantine, 1968.

Elliot, Robert. *Faking Nature: The Ethics of Environmental Restoration*. London: Routledge, 1997.

Eilperin, Juliet, and Brady Dennis. "Obama Names Five New National Monuments, Including Southern Civil Rights Sites." *The Washington Post*, January 12, 2017. Accessed August 8, 2017, https://www.washingtonpost.com/national/health -science/obama-names-five-new-national-monuments-including-southern-civil -rights-sites/2017/01/12/7f5ce78c-d907-11e6-9a36-1d296534b31e_story.html.

eMarketer. "Worldwide Ad Spending Growth Revised Downward." April 21, 2016. Accessed September 14, 2017, https://www.emarketer.com/Article/World wide-Ad-Spending-Growth-Revised-Downward/1013858.

Endersby, Jim. "How Botanical Gardens Helped to Establish the British Empire." *Financial Times*, July 25, 2014. Accessed September 28, 2017, https://www.ft.com/content/dcd33da0-0e69-11e4-a1ae-00144feabdc0.

Environmental Defense Fund. "World's Carbon Markets." 2017. Accessed September 30, 2017, https://www.edf.org/worlds-carbon-markets.

Epstein, Barbara. "Why Post-Structuralism is a Dead End for Progressive Thought." *Socialist Review* 25, no. 2 (1995): 83–119.

Epstein, Charlotte. "Moby Dick or Moby Doll? Discourse, or How to Study the 'Social Construction of' All the Way Down." In *Constructing the International Economy*, edited by Rawi Abdelal, Mark Blyth, and Craig Parsons, 175–93. Ithaca: Cornell University Press, 2010.

Ercan, Mina, et al. "Life Cycle Assessment of a Smartphone." Fourth International Conference on ICT for Sustainability, August 29–September 1, 2016, Amsterdam. Accessed February 10, 2017, http://www.atlantis-press.com/php/download_paper.php?id=25860375.

Escobar, Arturo. "Degrowth, Postdevelopment, and Transitions: A Preliminary Conversation." *Sustainability Science* 10 (2015): 451–62.

Esty, Daniel C., et al. "State Failure Task Force Report: Phase II Findings." *Environmental Change and Security Project Report* 5 (Summer 1999). Accessed June 29, 2017, https://www.wilsoncenter.org/sites/default/files/Phase2.pdf.

European Commission. *Communication from the Commission to the European Parliament, the Council, the European Economic and Social Committee and the Committee of the Regions, Closing the Loop—An EU Action Plan for the Circular Economy*. COM(2015) 614 final (Brussels, 2015). Accessed September 28, 2017, http://eur-lex.europa.eu/legal-content/EN/TXT/PDF/?uri=CELEX:52015DC0614andfrom=en.

European Commission–DG Environment. "Ex-post Evaluation of Certain Waste Stream Directives." April 18, 2014. Accessed August 28, 2017, http://ec.europa.eu/environment/waste/pdf/target_review/Final%20Report%20Ex-Post.pdf.

Experts Exchange. "Processing Power Compared." n.d. Accessed August 4, 2017, http://pages.experts-exchange.com/processing-power-compared/.

"Factors of Production." In *Principles of Economics*, chapter 2.1. Minneapolis: University of Minnesota Libraries Publishing, 2011. Accessed August 2, 2017, https://open.lib.umn.edu/principleseconomics/chapter/2-1-factors-of-production/.

Fairhead, James, Melissa Leach, and Ian Scoones. "Green Grabbing: A New Appropriation of Nature?" *The Journal of Peasant Studies* 39, no. 2 (2012): 237–61.

Fairphone. "A Better Phone is a Phone Better Made." Accessed August 7, 2017, https://www.fairphone.com/en/our-goals/.

Fang, Kai, Reinout Heijungs, and Geer R. de Snoo. "Theoretical Exploration for the Combination of the Ecological, Energy, Carbon, and Water Footprints: Overview of a Footprint Family." *Ecological Indicators* 36 (2014): 508–18.

Farber, Daniel A. "The Conservative as Environmentalist: From Goldwater and the Early Regan to the 21st Century." University of California Berkeley Public Law Research Paper, July 14, 2017. Accessed August 30, 2017, https://papers.ssrn.com/sol3/Delivery.cfm/SSRN_ID3000045_code102259.pdf?abstractid=2919633andmirid=1.

Fatheuer, Thomas, Lili Fuhr, and Barbara Unmüßig. *Inside the Green Economy*. Munich: oekom, 2016.

Federici, Silvia. "The Reproduction of Labour-power in the Global Economy, Marxist Theory and the Unfinished Feminist Revolution." Reading for January 27, 2009

University of California Santa Cruz seminar "The Crisis of Social Reproduction and Feminist Struggle." Accessed August 14, 2017, https://caringlabor.wordpress .com/2010/10/25/silvia-federici-the-reproduction-of-labour-power-in-the-global -economy-marxist-theory-and-the-unfinished-feminist-revolution.

Feenberg, Andrew. "The Philosophy of Praxis: Marx, Lukács and the Frankfurt School." Talk at the Vancouver School for Social Research, October 27, 2014. Accessed September 7, 2017, https://www.sfu.ca/~andrewf/Philosophy%20of%20 Praxis%20preview.pdf.

Feeny, David, Fikret Berkes, Bonnie J. McCay, and James M. Acheson. "The Tragedy of the Commons: Twenty-Two Years Later." *Human Ecology* 18, no. 1 (1990): 1–19.

Fehling, Maya, Brett D. Nelson, and Sridhar Venkatapuram. "Limitations of the Millennium Development Goals: A Literature Review." *Global Public Health* 8, no. 1 (2013): 1109–22.

Feldman, David Lewis. *Water*. Cambridge: Polity Press, 2012.

Feldstein, Martin S., Ted Halstead, and N. Gregory Mankiw (Climate Leadership Council). "A Conservative Case for Climate Action." Accessed September 30, 2017, https://www.nytimes.com/2017/02/08/opinion/a-conservative-case-for-climate -action.html?action=clickandcontentCollection=Scienceandmodule=Related Coverageandregion=Marginaliaandpgtype=article.

Fenzi, Marianna, and Christophe Bonneuil. "From 'Genetic Resources' to 'Ecosystems Services': A Century of Science and Global Policies for Crop Diversity Conservation." *Culture, Agriculture, Food and Environment* 38, no. 2 (Dec. 2016): 72–83.

Fernandez, Manny, and Richard Fausset. "A Storm Forces Houston, the Limitless City, to Consider Its Limits." *New York Times*, August 30, 2017. Accessed August 31, 2017, https://nyti.ms/2wpnVPq.

Ferrando. Francesca. "The Party of the Anthropocene: Post-humanism, Environmentalism and the Post-anthropocentric Paradigm Shift." *Relations: Beyond Anthropocentrism* 4 (2016): 159–73.

Feyerabend, Paul. *Against Method: Outline of an Anarchistic Theory of Knowledge*. London: Verso, 1993.

Field, Christopher B., and Katherine J. Mach. "Rightsizing Carbon Dioxide Removal." *Science* (2017): 706–07.

Fields, Desiree. "Unwilling Subjects of Financialization." *International Journal of Urban & Regional Research* 41, #4 (2017): 588–603.

Fikes, Bradley J., and Morgan Cook. "Saving Water Adds Up to Rate Hikes." *San Diego Union-Tribune*, July 27, 2015. Accessed September 28, 2017, http://www.sandiego uniontribune.com/news/drought/sdut-drought-water-prices-rise-2015jul27-story .html.

"Final Declaration of the Peoples' Summit at Rio+20." June 22, 2012. Accessed October 5, 2017, http://www.mstbrazil.org/news/final-declaration-peoples%E2%80%99 -summit-rio20.

Finley, Carmel. *All the Fish in the Sea: Maximum Sustainable Yield and the Failure of Fisheries Management*. Chicago: University of Chicago Press, 2011.

Finnegan, William. "Leasing the Rain." *The New Yorker*, April 8, 2002. Accessed October 5, 2017, https://www.newyorker.com/magazine/2002/04/08/leasing-the-rain.

Fisher, Edward F., and Peter Benson. *Broccoli and Desire*. Stanford: Stanford University Press, 2006.

Flomenhoft, Gary. "Earth Inc., Shareholders' Report." Gund Institute for Ecological Economics, University of Vermont, 2003. Accessed September 12, 2017, http://www.earthinc.org/earthinc.php?page=shareholder_report.

Flynn, Dan. "E. coli-contaminated Lettuce Came from a California LGMA Grower." *Food Safety News*, January 15, 2013. Accessed September 4, 2017, http://www.foodsafetynews.com/2013/01/contaminated-lettuce-came-from-lgma-grower/#.Wa2JKsaQzIU.

Flynn, Leisa Reinecke, Ronald E. Goldsmith, and Wesley Pollitte. "Materialism, Status Consumption and Market Involved Consumers." *Psychology and Marketing* 33, no. 9 (Sept. 2016): 761–76.

Fogel, Cathleen A. "The Local, the Global, and the Kyoto Protocol." In *Earthly Politics: Local and Global in Environmental Governance*, edited by Sheila Jasanoff and Marybeth Long Martello, 103–25. Cambridge, MA: MIT Press, 2003.

Fogel, Cathleen A. "Greening the Earth with Trees: Science, Storylines, and the Construction of International Climate Change Institutions." PhD dissertation, Environmental Studies, University of California, Santa Cruz, 2003.

Fominaya, Cristina Flesher. *Social Movements and Globalization: How Protests, Occupations, and Uprisings are Changing the World*. Basingstoke, UK: Palgrave Macmillan, 2014.

Fontaine, Philippe. "Making Use of the Past: Theorists and Historians on the Economics of Altruism." *European Journal of the History of Economic Thought* 7 (2000): 407–22.

Food and Agriculture Organization. *The Future of Food and Agriculture—Trends and Challenges*. Rome, 2017. Accessed July 3, 2017, http://reliefweb.int/sites/reliefweb.int/files/resources/a-i6583e.pdf.

Food and Agriculture Organization. "Global Forests Resources Assessment, Maps and Figures, 2015." Accessed August 25, 2017, http://www.fao.org/forest-resources-assessment/current-assessment/maps-and-figures/en/.

Food and Agriculture Organization International Treaty on Plant Genetic Resources for Food and Agriculture. "The Global Information System and Genomic Information: Transparency of Rights and Obligations." Background study paper n. 10, IT/GB7/SAC-1/16/ BSP 10, October 2016.

Food and Water Watch. "Our Right to Water." May 2012. Accessed August 12, 2017, https://www.foodandwaterwatch.org/sites/default/files/our_right_to_water_report_may_2012.pdf

Forest Stewardship Council. "Membership Chambers." Accessed August 10, 2017, https://us.fsc.org/en-us/who-we-are/membership/membership-chambers.

Forest Stewardship Council. "FSC A.C. Membership List." Accessed August 10, 2017, http://memberportal.fsc.org/.

Forest Stewardship Council. "FSC A.C. Statutes." Accessed August 10, 2017, https://us.fsc.org/en-us/who-we-are/governance.

Forest Stewardship Council. "Facts and Figures August 2017." Accessed August 10, 2017, https://ic.fsc.org/en/facts-and-figures.

Forster, E. M. "The Machine Stops." *The Oxford and Cambridge Review*, November 1909. Accessed September 26, 2017, http://www.ele.uri.edu/faculty/vetter/Other-stuff/The-Machine-Stops.pdf.

Foster, Laura A. "Privatizing Hoodia—Patent Ownership, Benefit-sharing, and Indigenous Knowledge in South Africa." In *Privatization, Vulnerability and Social*

Responsibility: A Comparative Perspective, edited by Martha Albertson Fineman, Titti Mattsson, and Ulrika Andersson, 108–23. London: Routledge, 2017.

Foucault, Michel. "Nietzsche, Genealogy, History." In *The Essential Foucault*, edited by Paul Rabinow and Nikolas Rose, 351–69. New York: The New Press, 2003.

Foucault, Michel. "Governmentality." In *The Foucault Effect: Studies in Governmentality*, edited by Graham Burchell, Colin Gordon, and Peter Miller, 87–104. Chicago: University of Chicago Press, 1991.

Foucault, Michel. "Truth and Power." In *Power/Knowledge*, translated by Colin Gordon, 109–33. New York: Pantheon, 1980.

Fowler, Cary. *Unnatural Selection: Technology, Politics, and Plant Evolution*. Yverdon, Switzerland: Gordon and Breach Science Publishers, 1994.

Fowler, Cary, and Pat Mooney. *Shattering: Food, Politics and the Loss of Genetic Diversity*. Tucson: University of Arizona Press, 1990.

Fox, Jonathan A., and L. David Brown, eds. *The Struggle for Accountability: The World Bank, NGOs, and Grassroots Movements*. Cambridge, MA: MIT Press, 1998.

Fox, Kara, and James Masters. "G20 Protests: Police, Demonstrators Clash in Germany." CNN, July 6, 2017. Accessed September 2, 2017, http://www.cnn.com /2017/07/06/europe/hamburg-protests-g20/index.html.

France, Anatole. *Le Lys Rouge (The Red Lily)*. Paris: Calmann-Lévy, 1894.

Frank, Charles. "Pricing Carbon: A Carbon Tax or Cap-and-Trade." Brookings Institution, August 12, 2014. Accessed September 3, 2017, https://www.brookings.edu /blog/planetpolicy/2014/08/12/pricing-carbon-a-carbon-tax-or-cap-and-trade/.

Frank, Thomas, and Matt Weiland, eds. *Commodify Your Dissent: The Business of Culture in the New Gilded Age*. New York: Norton, 1997.

Franklin, Rachel S., and Matthias Ruth. "Growing Up and Cleaning Up: The Environmental Kuznets Curve Redux." *Applied Geography* 32 (2012): 29–39.

Fraser, Nancy. "Reframing Justice in a Globalizing World." *New Left Review* 36 (November-December 2005): 69–88.

Freire, Paulo. *Pedagogy of the Oppressed*. New York: Continuum, 2000.

Fremstad, Anders. "Does Craigslist Reduce Waste? Evidence from California and Florida." *Ecological Economics* 132 (2017): 135–42.

Freris, Nicole, and Klemens Laschefski. "Seeing the Wood from the Trees." *The Ecologist* 31, no. 6 (July/August 2001). Accessed June 21, 2002, www.wald.org/fscamaz /ecol_eng.htm.

Friedman, Milton. *Capitalism and Freedom*. Chicago: University of Chicago Press, 1962.

Friedheim, Robert L. *Toward a Sustainable Whaling Regime*. Seattle: University of Washington Press, 2001.

Friedman, Thomas L. *The Lexus and the Olive Tree: Understanding Globalization*. New York: Farrar Straus Giroux, 1999.

FRONTLINE/World. "Bolivia—Leasing the Rain." June 2002. Accessed August 8, 2017, http://www.pbs.org/frontlineworld/stories/bolivia/thestory.html.

Fry, Ian. "Twists and Turns in the Jungle: Exploring the Evolution of Land Use, Land-Use Change and Forestry Decisions within the Kyoto Protocol." *RECIEL* 11(2002): 159–68.

Fung, Archon, Dana O'Rourke, and Charles Sabel. "Realizing Labor Standards: How Transparency, Competition, and Sanctions Could Improve Working Conditions

Worldwide." *Boston Review* 26 (Feb./Mar. 2001). Accessed August 20, 2017, http://bostonreview.net/archives/BR26.1/fung.html.

Gaard, Greta. "Ecofeminism Revisited: Rejecting Essentialism and Re-Placing Species in a Material Feminist Environmentalism." *Feminist Foundations* 23, no. 2 (Summer 2011): 26–53.

Gadgil, Madhav, and Ramachandra Guha. *This Fissured Land: An Ecological History of India*. Berkeley: University of California Press, 1993.

Gaedtke, Felix. "On Thin Ice: Sami Reindeer Herders Fear Climate Change Consequences in the Arctic." *A More Vulnerable World*, December 1, 2015. Accessed August 23, 2017, https://climate.earthjournalism.net/2015/12/01/on-thin-ice-sami -reindeer-herders-fear-climate-change-consequences-in-the-arctic.

Galli, Allesandro, et al. "Questioning the Ecological Footprint." *Ecological Indicators* 69 (2016): 224–32.

Galtung, Johan. *Human Rights in Another Key*. Cambridge, MA: Blackwell, Polity Press, 1995.

Gaventa, John. *Power and Powerlessness: Quiescence and Rebellion in an Appalachian Valley*. Urbana-Champaign: University of Illinois Press, 1980.

Geden, Oliver, and Andreas Löschel. "Define Limits for Temperature Overshoot Targets." *Nature Geoscience*, November 27, 2017, https:doi.org/10.1038/s41561 -017-0026-z.

Gellner, Ernest. *Nations and Nationalism*. Ithaca, NY: Cornell University Press, 1983.

Gereffi, Gary, and Olga Memedovic. "The Global Apparel Value Chain: What Prospects for Upgrading by Developing Countries?" Vienna: UNIDO, 2003. Accessed August 17, 2017, https://www.unido.org/uploads/tx_templavoila/Global_apparel _value_chain.pdf.

"German Federal Election, 2017." *Wikipedia*. Accessed September 26, 2017, https:// en.wikipedia.org/wiki/German_federal_election,_2017.

Gerth, H. H., and C. Wright Mills, eds. *From Max Weber: Essays in Sociology*. New York: Oxford University Press, 1946.

Gezon, Lisa L., and Susan Paulson, eds. "Degrowth, Culture and Power." Special Section of the *Journal of Political Ecology* 24 (2017): 425–666.

Gilio-Whitaker, Dina. "Idle No More and Fourth World Social Movements in the New Millenium." *South Atlantic Quarterly* 114 (2015): 866–77.

Giljum, Stefan, and Friedrich Hinterberger. "The Limits of Resource Use and Their Economic and Policy Implications." In *Factor X: Policy, Strategies and Instruments for a Sustainable Resource Use*, edited by Michael Angrick, Andreas Burger, and Harry Lehmann, 3–16. Dordrecht, Netherlands: Springer Science, 2014.

Gill, Stephen. *American Hegemony and the Trilateral Commission*. Cambridge: Cambridge University Press, 1990.

Gillis, Justin. "The Real Unknown of Climate Change: Our Behavior." *New York Times*, September 18, 2017. Accessed September 23, 2017, https://nyti.ms/2y99xuj.

Gilpin, Robert. *The Political Economy of International Relations*. Princeton, NJ: Princeton University Press, 1987.

Gilpin, Robert. *War and Change in World Politics*. Cambridge: Cambridge University Press, 1981.

Gladwell, Malcolm. "The Science of Shopping." *New Yorker*, November 1996. Accessed October 5, 2017, http://gladwell.com/the-science-of-shopping/.

Glasius, Marlies, Mary Kaldor, and Helmut Anheier, eds. *Global Civil Society 2002*. Oxford: Oxford University Press, 2002.

Glazer, Penina M., and Myron P. Glazer. *The Environmental Crusaders: Confronting Disaster and Mobilizing Community*. University Park, PN: The Pennsylvania State University Press, 1998.

Gleick, Peter. "Water and U.S. National Security." *War Room*, June 15, 2017. Accessed June 29, 2017, https://warroom.armywarcollege.edu/articles/water-u-s-national-security.

Gleick, Peter, et al. "The New Economy of Water: The Risks and Benefits of Globalization and Privatization of Fresh Water." Oakland, CA: Pacific Institute, 2002. Accessed February 19, 2003, http://www.pacinst.org/wp-content/uploads/2013/02/new_economy_of_water3.pdf.

GlobalEconomy.com. "Global exports." 2017. Accessed August 17, 2017, http://www.theglobaleconomy.com/guide/article/65/.

Global Ecovillage Network. "What is an Ecovillage?" Accessed July 17, 2017, https://ecovillage.org/about/gen/.

Global Ecovillage Network. "Four Dimensions of Sustainability." Accessed July 17, 2017, https://ecovillage.org/projects/dimensions-of-sustainability/.

Global Footprint Network. "2017 Edition National Footprint Accounts: Ecological Footprint and Biocapacity." Accessed August 4, 2017, data.footprintnetwork.

"Global Forest Atlas." Yale School of Forestry and Environmental Studies, New Haven, CT, 2017. Accessed August 26, 2017, http://globalforestatlas.yale.edu/conservation/forest-certification.

Global Greens. "Member Parties." Accessed August 30, 2017, https://www.globalgreens.org/member-parties.

Global Rivers Environmental Education Network. https://earthforce.org/GMGREEN/.

Global Witness. *Defenders of the Earth*, July 13, 2017. Accessed August 10, 2017, https://www.globalwitness.org/en/campaigns/environmental-activists/defenders-earth/.

Goeller, H. E., and Alvin M. Weinberg. "The Age of Substitutability." *Science* 191, no. 4228 (Feb. 20, 1976): 683–89.

Goldberg, C. "1500 March in Boston to Protest Biotech Food." *New York Times*, March 27, 2000. Accessed September 3, 2017, http://www.nytimes.com/2000/03/27/us/1500-march-in-boston-to-protest-biotech-food.html?mcubz=0.

The Goldman Environmental Prize. "Berta Cáceres." Accessed August 8, 2017, http://www.goldmanprize.org/recipient/berta-caceres/.

Goldstein, Joseph. "The New Age Caveman and the City." *New York Times*, January 8, 2010. Accessed February 14, 2017, http://www.nytimes.com/2010/01/10/fashion/10caveman.html.

Golub, Alla A., et al. "Global Climate Policy Impacts on Livestock, Land Use, Livelihoods, and Food Security." *PNAS* 110, no. 52 (Dec. 24, 2013): 20894–99.

Gonzalez, Carmen G. "Food Justice: An Environmental Justice Critique of the Global Food System." In *International Environmental Law and the Global South*, }edited by Shawkat Alam, et al., 401–34. Cambridge: Cambridge University Press, 2015.

Goodin, Robert E., and John S. Dryzek. "Justice Deferred: Wartime Rationing and Postwar Welfare Policy." *Politics and Society* 23, no. 1 (March 1995): 49–73.

Goodman, David, and Anthony Hall. *The Future of Amazonia: Destruction or Sustainable Development?* New York: St. Martin's, 1990.

Goudie, Andrew. *The Human Impact on the Natural Environment*. Cambridge, MA: MIT Press, 2000.

Gould, Stephen J. *Wonderful Life: The Burgess Shale and the Nature of History*. New York: Norton, 1990.

Goulder, Lawrence H., and Andrew Scheim. "Carbon Taxes vs. Cap and Trade: A Critical Review." NBER Working Paper no. 19338, August 2013. Accessed August 15, 2017, http://www.nber.org/papers/w19338.pdf.

Gourevitch, Peter. *Politics in Hard Times*. Ithaca, NY: Cornell University Press, 1986.

Graham, Alistair. "The Hypocrisy of Land-use Accounting." *Ecosystem Marketplace*, April 5, 2011. Accessed August 15, 2017, http://www.ecosystemmarketplace.com/articles/the-hypocrisy-of-land-use-accounting/.

Gramsci, Antonio. *Selections from the Prison Notebooks*, edited and translated by Quentin Hoare and Geoffrey N. Smith. London: Electric Book Co., 1999. Accessed September 5, 2017, http://abahlali.org/files/gramsci.pdf.

Green Climate Fund. Accessed September 28, 2017, http://www.greenclimate.fund/home.

Green, Colin. "Who Pays the Piper? Who Calls the Tune?" *UNESCO Courier* (Feb. 1999): 22–24.

Greene, Catherine R. "U.S. Organic Farming Emerges in the 1990s: Adoption of Certified Systems." Washington, DC: U.S. Dept. of Agriculture, Economic Research Service, 2001. Accessed October 3, 2017, https://www.ers.usda.gov/webdocs/publications/42396/31544_aib770_002.pdf?v=42487.

Greene, Jay. "The Environmental Pitfalls at the End of an iPhone's Life." *CNET.com*, September 26, 2012. Accessed October 5, 2017, https://www.cnet.com/news/the-environmental-pitfalls-at-the-end-of-an-iphones-life/.

Greene, Shane. "Indigenous People Incorporated? Culture as Politics, Culture as Property in Pharmaceutical Bioprospecting." *Current Anthropology* 45, no. 2 (April 2004): 211–37.

Greenfeld, Liah. *The Spirit of Capitalism: Nationalism and Economic Growth*. Cambridge: Harvard University Press, 2001.

Green Party US. "Officeholders." Accessed August 19, 2017, http://www.gp.org/officeholders.

Greens/European Free Alliance in the European Parliament. "Members." Accessed August 19, 2017, https://www.greens-efa.eu/en/our-group/meps/.

Greer, Allen. "Commons and Enclosure in the Colonization of North America." *American History Review* (2012): 365–86.

Gresser, Charis, and Sophia Tickell. *Mugged: Poverty in Your Coffee Cup*. London: Oxfam International, 2002.

Grinin, Leonid E., and Andrey V. Korotavey. "Origins of Globalization in the Framework of the AfroEurasian World-system History." *Journal of Globalization Studies* 5, no. 1 (May 2014): 32–64.

Grote, Ulrike. "Can We Improve Global Food Security? A Socio-economic and Political Perspective." *Food Security* 6 (2014): 187–200.

Grove, Richard H. *Green Imperialism: Colonial Expansion, Tropical Island Edens, and the Origins of Environmentalism, 1600–1860*. Cambridge: Cambridge University Press, 1995.

Grove, Richard H. "Origins of Western Environmentalism." *Scientific American* 267, no. 1 (July 1992): 42–47.

Gruber, Lloyd. *Ruling the World: Power Politics and the Rise of Supranational Institutions*. Princeton, NJ: Princeton University Press, 2000.

Guha, Ramachandra. *The Unquiet Woods: Ecological Change and Peasant Resistance in the Himalaya*. Berkeley, CA: University of California Press, 2000, expanded edition.

Gulbrandsen, Lars H. *Transnational Environmental Governance: The Emergence and Effects of the Certification of Forests and Fisheries*. Cheltenham, UK: Edward Elgar, 2010.

Guthman, Julie H. *Agrarian Dreams? The Paradox of Organic Farming in California*. Oakland: University of California Press, 2014, second edition.

Haas, Ernst B. *Nationalism, Liberalism, and Progress*. Ithaca, NY: Cornell University Press, 1997, volume 1.

Haberman, Clyde. "The Unrealized Horrors of Population Explosion." *New York Times*, May 31, 2015. Accessed June 29, 2017, https://www.nytimes.com/2015/06/01/us /the-unrealized-horrors-of-population-explosion.html.

Haggerty, K. D., and R. V. Ericson. "The Surveillant Assemblage." *British Journal of Sociology* 51, no. 4 (2000): 605–22.

Hainmueller, Jens, and Michael J. Hiscox. "Buying Green? Field Experimental Tests of Consumer Support for Environmentalism." December 2015. Accessed August 26, 2017, https://papers.ssrn.com/sol3/Delivery.cfm/SSRN_ID2699634_code739013 .pdf?abstractid=2062429andmirid=1.

Hainmueller, Jens, Michael J. Hiscox, and Sandra Sequeira. "Consumer Demand for Fair Trade: Evidence from a Multistore Field Experiment." *Review of Economics and Statistics* 97, no. 2 (May 2015): 242–56.

Hák, Tomáš, Svatava Janoušková, and Bedřich Moldan. "Sustainable Development Goals: A Need for Relevant Indicators." *Ecological Indicators* 60 (2016): 565–73.

Hale, Galina, and Bart Hobijn. "The U.S. Content of 'Made in China.'" *FRBSF Economic Letter*, August 8, 2011. Accessed August 4, 2017, http://www.frbsf.org /economic-research/publications/economic-letter/2011/august/us-made-in-china/.

Hall, Derek, Philip Hirsch, and Tania Murray Li. *Powers of Exclusion*. Honolulu: University of Hawai'i Press, 2011.

Hall, Peter A., and David Soskice, eds. *Varieties of Capitalism: The Institutional Foundations of Comparative Advantage*. Oxford: Oxford University Press, 2001.

Halperin, Morton. *Bureaucratic Politics and Foreign Policy*. Washington, DC: Brookings Institution, 1974.

Hammer, Joshua. "Is a Lack of Water to Blame for the Conflict in Syria?" *Smithsonian Magazine*, June 2013. Accessed June 29, 2017, http://www.smithsonianmag.com /innovation/is-a-lack-of-water-to-blame-for-the-conflict-in-syria-72513729/.

Hammond, Edward. "'Digital DNA' and Biopiracy: Protecting Benefit-Sharing as Synthetic Biology Changes Access to Genetic Resources." TWN Briefings for UN Biodiversity Conference no. 1, November 24, 2016. Accessed August 16, 2017, https://www.biosafety-info.net/article.php?aid=1308.

Haraway, Donna. "Manifesto for Cyborgs: Science, Technology, and Socialist Feminism in the 1980s." *Socialist Review* 80 (1985): 65–108.

Harbers, Hans, and Gerard de Vries. "Empirical Consequences of the 'Double Hermeneutic.'" *Social Epistemology* 7, no. 2 (1993): 183–92.

Hardin, Garrett. 1974. "Lifeboat Ethics: The Case Against the Poor." *Psychology Today* 8 (September): 38–43, 124–26.

Hardin, Garrett. 1967. "The Tragedy of the Commons." *Science* 162 (13 December): 1243–48.

Hardin, Russell. *Collective Action*. Baltimore: Johns Hopkins University Press, 1982.

Harman, Chris. *The Fire Last Time: 1968 and After*. London: Bookmarks, 1998, second edition.

Harris, Jonathan, and Brian Roach. *Environmental and Natural Resource Economics: A Contemporary Approach*. London: Routledge, 2018, fourth edition.

Hart, Craig A. *Climate Change and the Private Sector*. London: Routledge, 2013.

Hartz, Louis. *The Liberal Tradition in America*. New York: Harcourt Brace, 1955.

Harvey, Alan L., RuAngelie Edrada-Ebel, and Ronald J. Quinn. "The Re-emergence of Natural Products for Drug Discovery in the Genomics Era." *Nature Reviews* 14 (Feb. 2015): 110–20.

Harvey, David. *Spaces of Hope*. Berkeley, CA: University of California Press, 2000.

Harvey, R. R., C. M. Zakhour, and L. H. Gould. "Food Disease Outbreaks Associated with Organic Foods in the United States." *Journal of Food Protection* 79 no. 11 (Nov. 2016): 1953–58.

Haufler, Virginia. *A Public Role for the Private Sector—Industry Self-Regulation in a Global Economy*. Washington, DC: Carnegie Endowment for International Peace, 2001.

Hays, Samuel P. *Beauty, Health, and Permanence: Environmental Politics in the United States, 1955–1985*. Cambridge: Cambridge University Press, 1987.

Hayward, Tim. "A Global Right of Water." *Midwest Studies in Philosophy* 40 (2016): 217–33.

Healy, Hali, Joan Martinez-Alier, and Giogos Kallis. "From Ecological Modernization to Socially Sustainable Economic Degrowth: Lessons from Ecological Economics." In *The International Handbook of Political* Ecology, edited by Raymond L. Bryant, 577–90. Cheltenham, UK: Edward Elgar, 2015.

Hecht, Susanna, and Alexander Cockburn. *The Fate of the Forest: Developers, Destroyers and Defenders of the Amazon*. New York: HarperCollins, 1990.

Hegel, G. W. F. *The Philosophy of History*, translated by J. Sirbee. New York: Dover, 1837/1956.

Heilbroner, Robert L. *An Inquiry into the Human Prospect: Looked at Again for the 1990s*. New York: Norton, 1991.

Hein, Jonas, and Jean Carlo Rodriguez. "Carbon and Biodiversity Offsetting: A Way Towards Sustainable Development?" *The Current Column*. Bonn: German Development Institute, March 21, 2016.

Hellweg, Stefanie, and Lorenç Milà I. Canals. "Emerging Approaches, Challenges and Opportunities in Life Cycle Assessment." *Science* 344, no. 6188 (June 6, 2014): 1109–13.

Henley, Benjamin J., and Andrew D. King. "Trajectories Toward the 1.5° C Paris Target: Modulation by the Interdecadal Pacific Oscillation." *Geophysical Research Letters* 44, no. 9 (May 16, 2017): 4256–62.

Henriksson, Greger, and Miriam Börjesson Rivera. "Why Do We Buy and Throw Away Eelectronics?" Paper presented to ISDRC 2014: Resilience—The New Research Frontier, Trondheim, 2014. Accessed August 6, 2015, https://www.diva-portal.org/smash/get/diva2:737102/FULLTEXT01.pdf.

Herper, Matthew. "The Cost of Creating a New Drug Now $5 Billion, Pushing Big Pharma to Change." *Forbes*, August 11, 2013. Accessed August 14, 2017, https://

www.forbes.com/sites/matthewherper/2013/08/11/how-the-staggering-cost-of-inventing-new-drugs-is-shaping-the-future-of-medicine/no. 12c3d92213c3.

Hickman, Leo. "Wind Myths: Turbines Kill Birds and Bats." *The Guardian*, February 27, 2012. Accessed September 14, 2017, https://www.theguardian.com/environment/2012/feb/27/wind-energy-myths-turbines-bats.

Hicks, Daniel J., and Roberta L. Millstein. "GMOs: Non-Health Issues." In *Encyclopedia of Food and Agricultural Ethics*, edited by Paul B. Thompson and David Kaplan, 1–11. Springer International, 2016, second edition.

Higginson, S., E. McKenna, and M. Thomson. "Can Practice Make Perfect (Models)? Incorporating Social Practice Theory into Quantitative Energy Demand Models." Paper presented at: Behave 2014—Paradigm Shift: From Energy Efficiency to Energy Reduction through Social Change, 3rd Behave Energy Conference, Oxford, UK, September 3–4, 2014. Accessed November 22, 2014, https://dspace.lboro.ac.uk/dspace-jspui/bitstream/2134/15975/1/Higginson,%202014,%20Can%20practice%20make%20perfect%20models.pdf.

Highfield, Roger. "Men are Mice: Thereby Hangs a Tail." *The Telegraph*, December 5, 2002. Accessed August 14, 2017, http://www.telegraph.co.uk/news/uknews/1415272/Men-are-mice-thereby-hangs-a-tail.html.

Hilty, L. M., and B. Aebischer, eds. "Part II—The Energy Cost of Information Processing." In *ICT Innovations for Sustainability*, 171–205. Cham, Switzerland: Springer, 2015.

Himley, Matthew. "Mining History: Mobilizing the Past in Struggles over Mineral Extraction in Peru." *Geographical Review* 104, no. 2 (April 2014): 174–91.

Hirji, Zahra. "Removing CO_2 From the Air Only Hope for Fixing Climate Change, New Study Says." *Inside Climate News*, October 6, 2016. Accessed August 10, 2017, https://insideclimatenews.org/news/04102016/climate-change-removing-carbon-dioxide-air-james-hansen-2-degrees-paris-climate-agreement-global-warming.

Hirsch, Fred. *Social Limits to Growth*. Cambridge, MA: Harvard University Press, 1976.

Hobbes, Thomas. *Leviathan*. New York: Macmillan, 1962; Oakeshott edition.

Hobsbawm, Eric J. *Nations and Nationalism Since 1780—Programme, Myth, Reality*. Cambridge: Cambridge University Press, 1990.

Hobson, Kersty, and Nicholas Lynch. "Diversifying and De-growing the Circular Economy: Radical Social Transformation in a Resource-scarce World." *Futures* 82 (2016): 15–25.

Hobson, Kersty, and Nicholas Lynch. "Ecological Modernization, Techno-politics and Social Life Cycle Assessment: A View from Human Geography." *International Journal of Life Cycle Assessment* (2015). Accessed September 12, 2017, https://link.springer.com/article/10.1007/s11367-015-1005-5.

Hoffman, Matthew J. *Climate Governance at the Crossroads: Experimenting with a Global Response after Kyoto*. Oxford: Oxford University Press, 2011.

Hollander, Jack M. *The Real Environmental Crisis: Why Poverty, Not Affluence, Is the Environment's Number One Enemy*. Berkeley: University of California Press, 2003.

Holt-Giménez, Eric, and Miguel A. Altieri. "Agroecology, Food Sovereignty, and the New Green Revolution." *Agroecology and Sustainable Food Systems* 37 (2013): 90–102.

Holz, Christian, Sivan Kartha, and Tom Athanasiou. "Fairly Sharing 1.5: National Fair Shares of a 1.5 °C-compliant Global Mitigation Effort." *International Environmental Agreements* (2017). Accessed October 1, 2017, https://link.springer.com/article/10.1007/s10784-017-9371-z.

Hood, Katherine. "The Science of Value: Economic Expertise and the Valuation of Human Life in US Federal Regulatory Agencies." *Social Studies of Science* 47, no. 4 (2017): 441–65.

Hood, Mirian. *Gunboat Diplomacy, 1895–1905: Great Power Pressure in Venezuela.* London: Allen and Unwin, 1975.

Hoornweg, Daniel, and Perinaz Bhada-Tata. "What a Waste—A Global Review of Solid Waste Management." Urban Development and Local Government Unit, World Bank, March 2012, no. 15, Annex J. Accessed August 3, 2017, https://siteresources.worldbank.org/INTURBANDEVELOPMENT/Resources/336387-1334852610766/What_a_Waste2012_Final.pdf.

Howard, Brandon. "Timeline: History of the Dakota Access Pipeline." *Chicago Tribune*, January 24, 2017. Accessed February 16, 2017, http://www.chicagotribune.com/news/nationworld/ct-dakota-access-pipeline-timeline-dapl-20161219-html story.html.

Howitt, Richard. *Rethinking Resource Management—Justice, Sustainability and Indigenous Peoples.* London: Routledge, 2001.

Hu, Winnie. "Your Uber Car Creates Congestion. Should You Pay a Fee to Ride?" *New York Times*, December 26, 2017. Accessed March 21, 2018, https://www.nytimes.com/2017/12/26/nyregion/uber-car-congestion-pricing-nyc.html.

Huggan, Graham, and Helen Tiffin. "Green Postcolonialism." *Interventions* 9, no. 1 (2007): 1–11.

Humphrey, Chris. "The Problem with Development Banks' Environmental and Social Safeguards." Overseas Development Institute, April 14, 2016. Accessed September 2, 2017, https://www.odi.org/comment/10379-problem-development-banks-environmental-social-safeguards-mdbs.

Huntington, Samuel P. *American Politics: The Promise of Disharmony.* Cambridge, MA: Belknap Press of Harvard University Press, 1981.

Hurt, Shelley L. "Military's Hidden Hand: Examining the Dual-Use Origins of Biotechnology, 1969–1972." In *State of Innovation: The U.S. Government's Role in Technology Development*, edited by Fred Block and Matt Keller, 31–56. Boulder: Paradigm Publishers, 2011.

Hurt, Shelley L., and Ronnie D. Lipschutz, eds. *Hybrid Rule and State Formation: Public-Private Power in the 21st Century.* London: Routledge, 2015.

Idle No More. "The Story." Accessed August 23, 2017, http://www.idlenomore.ca/story.

Inglehart, Ronald. *Culture Shift in Advanced Industrial Society.* Princeton, NJ: Princeton University Press, 1990.

Intelligence Community Assessment. "Global Water Security." National Intelligence Council, Washington, DC, February 2, 2012, ICA 2012-08. Accessed June 29, 2017, https://fas.org/irp/nic/water.pdf.

Intergovernmental Group on Bananas and Tropical Fruits. "Banana Market Review and Banana Statistics 2012–2013." FAO, Rome, 2014, p. 3. Accessed January 10, 2017, http://www.fao.org/docrep/019/i3627e/i3627e.pdf.

Intergovernmental Group on Bananas and Tropical Fruits. "The Changing Role of Multinational Companies in the Global Banana Trade." Food and Agriculture Organization, Rome, 2014. Accessed January 10, 2017, http://www.fao.org/docrep/019/i3746e/i3746e.pdf.

Intergovernmental Panel on Climate Change. *Climate Change 2014: Synthesis Report.* Geneva, Switzerland, 2015.

Intergovernmental Panel on Climate Change. *Climate Change 2014: Synthesis Report. Contribution of Working Groups I, II and III to the Fifth Assessment Report of the Intergovernmental Panel on Climate Change.* Geneva, Switzerland, 2014.

Intergovernmental Panel on Climate Change Working Group III: Mitigation. "Historic Trends and Driving Forces." Section 3.8.3, 2001. Accessed August 16, 2017, http://www.ipcc.ch/ipccreports/tar/wg3/index.php?idp=125.

International Centre for Settlement of Investment Disputes. "Aguas del Tunari, S.A. v. Republic of Bolivia: Decision on Respondent's Objections to Jurisdiction." Washington, DC, October 21, 2005, ICSID Case no. ARB/02/3. Accessed August 25, 2017, http://www.iisd.org/pdf/2005/AdT_Decision-en.pdf.

International Civil Aeronautics Organization Environment. "The Carbon Offsetting and Reduction Scheme for International Aviation." Accessed September 29, 2017, https://www.icao.int/environmental-protection/Pages/market-based-measures.aspx.

International Emissions Trading Association. "The World's Carbon Markets—A Case Study Guide to Emissions Trading." 2016. Accessed August 15, 2017, http://www.ieta.org/The-Worlds-Carbon-Markets.

International Institute for Sustainable Development. "The I.W.G.F. Page." Accessed August 10, 2017, http://enb.iisd.org/forestry/iwgf.html.

International Institute for Sustainable Development. "A Brief to Global Forest Policy." Accessed August 10, 2017, http://enb.iisd.org/process/forest_desertification_land-forestintro.htm.

International Rivers. "The State of the World's Rivers." n.d. Accessed August 8, 2017, https://www.internationalrivers.org/sites/default/files/worldsrivers/.

International Union for the Conservation of Nature and Natural Resources, UN Environment Programme, and World Wildlife Fund. *World Conservation Strategy: Living Resource Conservation for Sustainable Development.* Moreges, Switzerland: International Union for the Conservation of Nature and Natural Resources/World Wildlife Fund; Nairobi, Kenya: UN Environment Programme, 1980.

James, Rhydian Fôn, and Molly Scott Cato. "A Bioregional Economy: A Green and Post-capitalist Alternative to an Economy of Accumulation." *Local Economy* 29, no. 3 (2014): 173–80.

Jansen, Kees. "The Debate on Food Sovereignty Theory: Agrarian Capitalism, Dispossession and Agroecology." *The Journal of Peasant Studies* 42, no. 1 (2015): 213–32.

Jardim, Elizabeth. "From Smart to Senseless: The Global Impact of 10 Years of Smartphones." Greenpeace, Washington, DC, 2017. Accessed August 4, 2017, http://www.greenpeace.org/usa/wp-content/uploads/2017/03/FINAL-10YearsSmartphones-Report-Design-230217-Digital.pdf.

Jasanoff, Sheila. "Future Imperfect: Science, Technology, and the Imaginations of Modernity." In *Dreamscapes of Modernity: Sociotechnical Imaginaries and the Fabrication of Power*, edited by S. Jasanoff and S.-H. Kim, 1–33. Chicago: University of Chicago Press, 2015.

Jessop, Bob. "Primacy of the Economy, Primacy of the Political: Critical Theory of Neoliberalism." In *Handbuch Kritische Theorie*, edited by Ullrich Bauer, Uwe H. Bittlingmayer, Alex Demirovic, Tatjana Freytag. Springer Reference, 2016. Accessed September 2, 2017, https://link.springer.com/referenceworkentry/10.1007/978-3-658-12707-7_46-1.

Jinnah, Sikina. *Post-Treaty Politics.* Cambridge: MIT Press, 2014.

Johnson, Sheree. "New Research Sheds Light on Daily Ad Exposures." *SJ Insights*, September 29, 2014. Accessed August 3, 2017, https://sjinsights.net/2014/09/29/new-research-sheds-light-on-daily-ad-exposures/.

Jones, Andrew P., Stuart A. Thompson, and Jessia Ma. "You Fix It: Can You Stay Within the World's Carbon Budget." *New York Times*, August 29, 2017. Accessed September 14, 2017, https://www.nytimes.com/interactive/2017/08/29/opinion/climate-change-carbon-budget.html.

Jones, E. L. *The European Miracle: Environments, Economies and Geopolitics in the History of Europe and Asia.* Cambridge: Cambridge University Press, 1981.

Jones, Rhys, Jessica Pykett, and Mark Whitehead. "Governing Temptation: Changing Behaviour in an Age of Libertarian Paternalism." *Progress in Human Geography* 35, no. 4 (2011): 483–501.

Justia. "Mother Natures' Perfect Food: Premium Bananas from Ecuador—Trademark Details." Accessed June 21, 2017, https://trademarks.justia.com/863/62/mother-nature-s-perfect-food-premium-bananas-from-86362788.html.

Kan, Haidong. "Globalisation and Environmental Health in China." *The Lancet* 384 (Aug. 30, 2014): 721–23.

Karlsson-Vinkhuyzen, Sylvia I., et al., "Entry into Force and Then?: The Paris Agreement and State Accountability, *Climate Policy*, July 31, 2017, DOI: 10.1080/14693062.2017.1331904.

Kartha, Sivan, and Kate Dooley. "The Risks of Relying on Tomorrow's 'Negative Emissions' to Guide Today's Mitigation Action." Stockholm Environment Institute, Working paper 2016-08. Stockholm: SEI. Accessed September 29, 2017, https://www.sei-international.org/mediamanager/documents/Publications/Climate/SEI-WP-2016-08-Negative-emissions.pdf.

Karvonen, Andrew, and Ralf Brand. "Technical Expertise, Sustainability, and the Politics of Specialized Knowledge." In *Global Environmental Governance: Power and Knowledge in a Local-Global World*, edited by Gabriella Kütting and Ronnie D. Lipschutz, 38–59. London: Routledge, 2009.

Kay, Kelly. "Breaking the Bundle of Rights: Conservation Easements and the Legal Geographies of Individuating Nature." *Environment and Planning A* 48, no. 3 (2015): 504–22.

Kearse, Stephen. "Why are Salad Greens Always Labeled 'Triple Washed'?" *Slate*, April 27, 2016. Accessed September 4, 2017, http://www.slate.com/articles/life/food/2016/04/why_are_salad_greens_always_labeled_triple_washed.html.

Keck, Margaret, and Kathryn Sikkink. *Activists Beyond Borders: Advocacy Networks in International Politics.* Ithaca, NY: Cornell University Press, 1998.

Kelly, William J. "U.S. Trade Deals from the 90's Set up China as a Pollution Haven." *Inside Climate News*, March 6, 2014. Accessed August 17, 2017, https://insideclimatenews.org/news/20140306/us-trade-deals-90s-set-china-pollution-haven.

Kenny, C. "Why Ending Extreme Poverty Isn't Good Enough." *Bloomberg Businessweek*, April 29, 2013. Accessed 22 June 2016, http://www.bloomberg.com/news/articles/2013-04-28/why-endingextreme-poverty-isnt-good-enough.

Keohane, Robert O. *International Institutions and State Power*. Boulder, CO: Westview, 1989.

Keohane, Robert O., and M. A. Levy, eds. *Institutions for Environmental Aid: Pitfalls and Promise*. Cambridge, MA: MIT Press, 1996.

Keulartz, Jozef. *The Struggle for Nature: A Critique of Radical Ecology*, translated by Rob Keuitenbrouwer. London: Routledge, 1998.

Khoury, Monder. "Whaling in Circles: The Makahs, the International Whaling Commission, and Aboriginal Subsistence Whaling." *Hastings Law Journal* 67 (Dec. 2015): 293–321.

King, Brayden. "A Social Movement Perspective of Stakeholder Collective Action and Influence." *Business and Society* 47, no. 1 (March 2008): 21–49.

Kjellén, Marianne, and Gordon McGranahan. "Informal Water Vendors and the Urban Poor." International Institute for Environment and Development, London, Human Settlements Discussion Paper Series—Water-3, 2006. Accessed August 24, 2017, https://www.sei-international.org/mediamanager/documents/Publications /Water-sanitation/informal_water_vendors.pdf.

Klare, Michael T. *The Race for What's Left—The Global Scramble for the World's Last Resources*. New York: Metropolitan Books, 2012.

Klare, Michael T. *Resource Wars: The New Landscape of Global Conflict*. New York: Metropolitan Books, 2001.

Klein, Joanna. "Long Before Making Enigmatic Earthworks, People Reshaped Brazil's Rain Forest." *New York Times*, February 10, 2017. Accessed February 11, 2017, https://www.nytimes.com/2017/02/10/science/amazon-earthworks-geoglyphs -brazil.html.

Klein, Naomi. *This Changes Everything: Capitalism vs. The Climate*. New York: Simon and Schuster, 2014.

Klein, Naomi. *The Shock Doctrine: The Rise of Disaster Capitalism*. Toronto: Random House Canada, 2007.

Klinsky, Sonja, et al. "Why Equity is Fundamental in Climate Change Policy Research." *Global Environmental Change* 44 (2017): 170–73.

Kloppenburg, Jack Ralph, Jr. *First the Seed: The Political Economy of Plant Biotechnology*. Cambridge: Cambridge University Press, 1988.

Knickerbocker, Brad. "Privatizing Water: A Glass Half-Empty?" *Christian Science Monitor*, October 24, 2002. Accessed August 8, 2017, http://www.csmonitor.com /2002/1024/p01s02-usec.html.

Knowledge@Wharton. "America's Neglected Water Systems Face a Reckoning." Accessed August 8, 2017, http://knowledge.wharton.upenn.edu/article/americas -neglected-water-systems-face-a-reckoning/.

Kobrin, Stephen J. "The MAI and the Clash of Globalizations." *Foreign Policy*, Fall 1998.

Koepsell, David. *Who Owns You?: Science, Innovation, and the Gene Patent Wars*. New York: John Wiley and Sons, 2015.

Konisky, David M., Llewelyn Hughes, and Charles H. Kaylor. "Extreme Weather Events and Climate Change Concern." *Climatic Change* 134 (2016): 533–47.

Kosoy, Nicolás, and Esteve Corbera. "Payments for Ecosystem Services as Commodity Fetishism." *Ecological Economics* 69 (2010): 1228–36.

Kourula, Arno, and Guillaume Delalieux. "The Micro-level Foundations and Dynamics of Political Corporate Social Responsibility: Hegemony and Passive Revolution Through Civil Society." *Journal of Business Ethics* 135 (2016): 769–85.

Krasner, Stephen D., ed. *International Regimes*. Ithaca, NY: Cornell University Press, 1983.

Krasner, Stephen D. "Are Bureaucracies Important? (Or, Allison Wonderland)." *Foreign Policy* 7 (Summer 1972): 159–79.

Kriesberg, Jennifer Cobb. "A Globe, Clothing Itself with a Brain." *Wired*, June 1, 1995. Accessed August 24, 2017, https://www.wired.com/1995/06/teilhard/.

Kropotkin, Peter. *Mutual Aid: A Factor of Evolution*, edited by Paul Avrich. London: Allen Lane, 1972.

Kuehls, Thom. "The Environment of Sovereignty." In *A Political Space: Reading the Global through Clayoquot Sound*, edited by Warren Magnusson and Karena Shaw, 179–98. Minneapolis, MN: University of Minnesota Press, 2003.

Kuehls, Thom. *Beyond Sovereign Territory: The Space of Ecopolitics*. Minneapolis, MN: University of Minnesota Press, 1996.

Kumar, Karun. "Sustainability Performance Measurement: An Investigation into Corporate Performance through Environmental Indicators." *International Journal of Management Research and Review* 4, no. 2 (Feb. 2014): 192–206.

Kurz, Tim, et al. "Habitual Behaviors or Patterns of Practice? Explaining and Changing Repetitive Climate-relevant Actions." *WIRE Climate Change* 6 (2015): 113–28.

Kütting, Gabriella, and Ronnie D. Lipschutz, eds. *Global Environmental Governance: Power and Knowledge in a Local-Global World*. London: Routledge, 2009.

Lake, David A., and Donald S. Rothchild, eds. *The International Spread of Ethnic Conflict: Fear, Diffusion, and Escalation*. Princeton, NJ: Princeton University Press, 1998.

Land Trust Alliance. "2016 Annual Report." Washington, DC Accessed August 9, 2017, http://s3.amazonaws.com/landtrustalliance.org/2016AnnualReport.pdf.

Lang, Chris. "A Plan to Burn the Planet: The Aviation Industry is in Talks with the World Bank about Using REDD to Offset its Emissions." REDD-Monitor.org, May 17, 2017. Accessed August 16, 2017, http://www.redd-monitor.org/2017/05/17/a-plan-to-burn-the-planet-the-aviation-industry-is-in-talks-with-the-world-bank-about-using-redd-to-offset-its-emissions/.

Langston, Nancy. "Thinking Like a Microbe: Borders and Environmental History." *The Canadian Historical Review* 95, no. 4 (Dec. 2014): 592–603.

Lapore, Jill. "The Disruption Machine—What the Gospel of Innovation Gets Wrong." *The New Yorker*, June 23, 2014. Accessed September 6, 2017, http://www.newyorker.com/magazine/2014/06/23/the-disruption-machine.

Lappe, Anna. "World Bank Wants Water Privatized, Despite Risks." *Al Jazeera America*, April 17, 2014. Accessed August 8, 2017, http://america.aljazeera.com/opinions/2014/4/water-managementprivatizationworldbankgroupifc.html.

Larkin, Peter. "An Epitaph for the Concept of Maximum Sustained Yield." *Transactions of the American Fisheries Society* 106 (1977): 1–11.

Lasswell, Harold D. *Politics: Who Gets What, When, How*. New York: P. Smith, 1936.

Latour, Bruno. *Pandora's Hope: Essays on the Reality of Science Studies*. Cambridge, MA: Harvard University Press, 2009.

Latour, Bruno. *Reassembling the Social: An Introduction to Actor-Network-Theory*. Oxford: Oxford University Press, 2007.

Latour, Bruno. *Politics of Nature: How to Bring the Sciences into Democracy*. Cambridge, MA: Harvard University Press, 2004.

Latour, Bruno. *Science in Action: How to Follow Scientists and Engineers Through Society*. Cambridge, MA: Harvard University Press, 1987.

Latour, Bruno, and Steve Woolgar. *Laboratory Life: The Construction of Scientific Facts*. Princeton: Princeton University Press, 1979.

Lazarus, Richard J. "A Different Kind of 'Republican Moment' in Environmental Law." *Minnesota Law Review* 87 (2003): 999–1035.

Lear, Linda. *Rachel Carson: Witness for Nature*. New York: Henry Holt, 1997.

Lee, C. "Toxic Waste and Race in the United States." In *Race and the Incidence of Environmental Hazards: A Time for Discourse*, edited by B. I. Bryant and P. Mohai, 10–16, 22–27. Boulder, CO: Westview Press, 1992.

Lenin, V. I. Что делать? ("What is to be done?") 1901. Accessed September 5, 2017, https://www.marxists.org/archive/lenin/works/download/what-itd.pdf.

Leonard, Jeffrey. *Pollution and the Struggle for the World Product: Multinational Corporations, Environment, and International Comparative Advantage*. Cambridge: Cambridge University Press, 1988.

Lepawsky, Josh. "The Changing Geography of Global Trade in Electronic Discards: Time to Rethink the E-waste Problem." *The Geographical Journal* 181, no. 2 (June 2015): 147–59.

Le Quéré, Corinne, et al. "Global Carbon Budget 2016." *Earth Systems Science* Data 8 (2016): 605–49.

Letter from the Council of Europe and the European Commission to Yvo de Boer, executive secretary of the UN Framework Convention on Climate Change. "Expression of Willingness to be Associated with the Copenhagen Accord and Submission of the Quantified Economy-wide Emissions Reduction Targets for 2020." Accessed August 16, 2017, https://unfccc.int/files/meetings/cop_15/copenhagen_accord /application/pdf/europeanunioncphaccord_app1.pdf.

Lewis, David. *The Brain Sell: When Science Meets Shopping*. London: Brealey, 2013.

Lewis, Sophie. "Veterans Unite for Second 'Deployment' Against the Dakota Access Pipeline." CNN.com, February 9, 2017. Accessed August 23, 2017, http://www .cnn.com/2017/02/09/us/standing-rock-veterans-trnd/index.html.

Li, Huan, Neha Khanna, and Martina Vidovic. "Third Party Certification and Self-Regulation: Evidence from Responsible Care and Accidents in the US Chemical Industry." Paper prepared for presentation at the Agricultural and Applied Economics Association's 2014 AAEA Annual Meeting, Minneapolis, MN, July 27–29, 2014. Accessed September 3, 2017, http://ageconsearch.umn.edu/bitstream/170492/2 /AccidentRC_AgEconSearch_0528.pdf.

Lin, Jintai, et al. "China's International Trade and Air Pollution in the United States." *PNAS* 111, no. 5 (Feb. 4, 2014): 1736–41.

Lipschutz, Ronnie D. "Can Climate Change Us?" *Development and Change* 48, no. 3(2017): 623–35, esp., 627–29.

Lipschutz, Ronnie D. "Property and Privatization." In *Research Handbook on Climate Governance*, edited by Karin Bäckstrand and Eva Lövbrand, 537–46. Cheltenham, UK: Edward Elgar, 2016.

Lipschutz, Ronnie D. "The Assemblage of American Imperium: Hybrid Power, World War and World Government(ality) in the Twenty-first Century." In *Legacies of Empire: Imperial Roots of the Contemporary Global Order*, edited by Sandra Halperin and Ronen Palan, 221–42. Cambridge: Cambridge University Press, 2015.

Lipschutz, Ronnie D. "From Stakeholders to Shareholders: Fostering Global Economic Democracy through Usufruct Rights." Paper presented to the Lund Conference on Earth System Governance: Towards Just and Legitimate Earth System Governance—Addressing Inequalities, Lund University, April 18–20, 2012.

Lipschutz, Ronnie D. "Getting Out of the CAR: DeCARbonization, Climate Change and Sustainable Society." *International Journal of Sustainable Society* 4, no. 4 (2012): 336–56.

Lipschutz, Ronnie D. *Political Economy, Capitalism and Popular Culture.* Lanham, MD: Rowman & Littlefield, 2010.

Lipschutz, Ronnie D. "The Sustainability Debate: Déjà Vu All Over Again?" *Global Environmental Politics* 9, no. 4 (Nov. 2009): 136–41.

Lipschutz, Ronnie D. *The Constitution of Imperium.* Boulder, CO: Paradigm Publishers, 2009.

Lipschutz, Ronnie D. "Global Civil Society and Global Environmental Protection: Private Initiatives and Public Goods." In *Environmental Policy Making: Assessing the Use of Alternative Policy Instruments*, edited by Michael Hatch, 45–70. Albany: State University of New York Press, 2005.

Lipschutz, Ronnie D. "Power, Politics, and Global Civil Society." *Millennium* 33, no. 3 (2005): 747–69.

Lipschutz, Ronnie D. "The Environment and Global Governance." In *Global Governance in the Twenty-First Century*, edited by John N. Clarke and Geoffrey R. Edwards, 204–36. Basingstoke, England: Palgrave Macmillan, 2004.

Lipschutz, Ronnie D. "The Clash of Governmentalities: The Fall of the UN Republic and America's Reach for Imperium." *Contemporary Security Policy* 23, no. 3 Dec. 2002: 214–31.

Lipschutz, Ronnie D. "Doing Well by Doing Good? Transnational Regulatory Campaigns, Social Activism, and Impacts on State Sovereignty." In *Challenges to Sovereignty: How Governments Respond*, edited by John Montgomery and Nathan Glazer, 291–320. New Brunswick, NJ: Transaction, 2002.

Lipschutz, Ronnie D. "Environmental History, Political Economy and Policy: Rediscovering Lost Frontiers in Environmental Research." *Global Environmental Politics* 1, no. 3 (August 2001): 72–91.

Lipschutz, Ronnie D. "Why Is There No International Forestry Law? An Examination of International Forestry Regulation, both Public and Private." *UCLA Journal of Environmental Law and Policy* 19, no. 1 (2000/2001): 155–82.

Lipschutz, Ronnie D. *After Authority: War, Peace and Global Politics in the 21st Century.* Albany, NY: State University of New York Press, 2000.

Lipschutz, Ronnie D. "Bioregionalism, Civil Society, and Global Environmental Governance." In *Bioregionalism*, edited by Michael Vincent McGinnis, 102–20. London: Routledge, 1999.

Lipschutz, Ronnie D. "Reconstructing World Politics: The Emergence of Global Civil Society." *Millennium* 21, no. 3 (Winter 1992/1993): 389–420.

Lipschutz, Ronnie D. "Wasn't the Future Wonderful? Resources, Environment, and the Emerging Myth of Global Sustainable Development." *Colorado Journal of International Environmental Law and Policy* 2 (1991): 35–54.

Lipschutz, Ronnie D. *When Nations Clash: Raw Materials, Ideology and Foreign Policy.* New York: Ballinger/Harper and Row, 1989.

Lipschutz, Ronnie D., Dominique de Wit, and Kevin Bell. "Practicing Energy, or Energy Consumption as a Social Practice." Paper presented to the 2015 Behavior, Energy and Climate Change Conference, Sacramento, CA, October 18–21. Accessed August 20, 2015, http://escholarship.org/uc/item/1vs503px.

Lipschutz, Ronnie D., and Sarah T. Romano. "The Cupboard is Full: Public Finance for Public Services in the Global South." Municipal Services Project, Queens University, Kingston, Ontario, Occasional Paper 16, March 2012. Accessed October 5, 2017, http://www.municipalservicesproject.org/sites/municipalservicesproject.org/files/publications/Lipschutz-Romano_The_Cupboard_is_Full_May2012_FINAL.pdf.

Lipschutz, Ronnie D., with James K. Rowe. *Globalization, Governmentality, and Global Politics: Regulation for the Rest of Us?* London: Routledge, 2005.

Lipschutz, Ronnie D., with Judith Mayer. *Global Civil Society and Global Environmental Governance: The Politics of Nature from Place to Planet.* Albany, NY: State University of New York Press, 1996.

Lipschutz, Ronnie D., and Judith Mayer. 1993. "Not Seeing the Forest for the Trees: Rights, Rules, and the Renegotiation of Resource Management Regimes." In *The State and Social Power in Global Environmental Politics*, edited by Ronnie D. Lipschutz and Ken Conca, 246–73. New York: Columbia University Press.

Lipschutz, Ronnie D., and Ken Conca, eds. *The State and Social Power in Global Environmental Politics.* New York: Columbia University Press, 1993.

Litfin, Karen. *Ecovillages: Lessons for Sustainable Community.* Cambridge: Polity, 2013.

Litfin, Karen. *Ozone Discourses: Science and Politics in Global Environmental Cooperation.* New York: Columbia University Press, 1994.

Llavador, Humberto, John E. Roemer, and Joaquim Silvestre. *Sustainability for a Warming Planet.* Cambridge, MA: Harvard University Press, 2015.

Lo, Alex Y., and C. Y. Jim. "Protest Response and Willingness to Pay for Culturally Significant Urban Trees: Implications for the Contingent Valuation Method." *Ecological Economics* 114 (2015): 58–66.

Lobina, Emanuele, Satoko Kishimoto, and Olivier Petitjean. "Here to Stay: Water Remunicipalisation as a Global Trend." Public Services International Research Unit, Transnational Institute and Multinational Observatory, 2015. Accessed August 25, 2017, https://www.tni.org/files/download/heretostay-en.pdf.

The Local. "The Paris COP21 Climate Summit in Numbers." November 30, 2015. Accessed August 30, 3017, https://www.thelocal.fr/20151130/cop-21-in-numbers-the-facts-and-figures-to-know.

Locke, John. *Two Treatises of Government.* Cambridge: Cambridge University Press, 1960.

Lohmann, Larry. "Uncertainty Markets and Carbon Markets: Variations on Polanyian Themes." *New Political Economy* 15, no. 2 (2010): 225–54.

Los Angeles Times. "Here are the National Monuments Being Reviewed under Trump's Order." May 5, 2017. Accessed August 8, 2017, http://www.latimes.com/nation/la-na-national-monuments-pictures-20170426-htmlstory.html.

Los Angeles Times. "Photos: Los Angeles Under Cover—Smog Through the Years." April 28, 2017. Accessed September 6, 2017, http://www.latimes.com/local/la-me-air-pollution-0428-pictures-photogallery.html.

Lovell, Heather, Harriet Bulkeley, and Diana Liverman. "Carbon Offsets: Sustaining Consumption?" *Environment and Planning A* 41 (2009): 2357–79.

Lovelock, James E., and Lynn Margulis. "Atmospheric Homeostasis by and for the Biosphere: The Gaia Hypothesis." *Tellus.* Series A. 26 (1974): 2–10, Stockholm: International Meteorological Institute.

Luke, Timothy W. "On the Political Economy of Clayoquot Sound: The Uneasy Transition from Extractive to Attractive Models of Development." In *A Political Space: Reading the Global through Clayoquot Sound*, edited by Warren Magnusson and Karena Shaw, 91–112. Minneapolis: University of Minnesota Press, 2003.

Luke, Timothy W. *Ecocritique: Contesting the Politics of Nature, Economy, and Culture*. Minneapolis, MN: University of Minnesota Press, 1997.

Lunstrum, Elizabeth, Pablo Bose, and Anna Zalik. "Environmental Displacement: The Common Ground of Climate Change, Extraction and Conservation." *Area* 48, no. 2 (2015): 130–33.

MacDicken, Kenneth G., et al. "Global Progress Toward Sustainable Forest Management." *Forest Ecology and Management* 352 (2015): 47–56.

MacKenzie, Catherine P. "Lessons from Forestry for International Environmental Law." *Review of European Community and International Environmental Law* 21, no. 2 (2012): 114–26.

Mackey, Brendan. "Counting Trees, Carbon and Climate Change." *Significance* 11, no. 1 (Feb. 2014): 19–23.

Mackinder, Halford. "The Geographical Pivot of History." *The Geographical Journal* 23, no. 4 (April 1904): 421–37.

Macpherson, C. B. *The Political Theory of Possessive Individualism: Hobbes to Locke*. Oxford: Oxford University Press, 1962.

Magnusson, Warren, and Karena Shaw, eds. *A Political Space: Reading the Global through Clayoquot Sound*. Minneapolis: University of Minnesota Press, 2003.

MahiFX. "MahiFX Infographic: The $5.3 trillion Forest Market Explained." *Cision PR Newswire*, October 24, 2013. Accessed August 17, 2017, http://www.prnewswire.com/news-releases/mahifx-infographic-the-53-trillion-forex-market-explained-229134291.html.

Makower, Joel. "Consumer Power." In *Our Future, Our Environment*, edited by Noreen Clancy and David Rejeski, Santa Monica, CA: RAND, March 2001. Accessed August 18, 2017, http://smapp.rand.org/ise/ourfuture/Welcome/toc.html.

Malešević, Siniša. *The Sociology of War and Violence*. Cambridge: Cambridge University Press, 2010.

Malthus, Thomas. *An Essay on the Principle of Population*. Amherst, NY: Prometheus Books, 1998.

Mandel, James T., C. Josh Dolan, and Jonathan Armstrong. "A Derivative Approach to Endangered Species Conservation." *Frontiers in Ecology and the Environment* 8, no. 1 (Feb. 2010): 44–49.

Mangell, Marc, Baldo Marinovic, Caroline Pomeroy, and Donald Croll. "Requiem for Ricker: Unpacking MSY." *Bulletin of Marine Science* 70 (2002): 763–81.

Mani, M., and D. Wheeler. "In Search of Pollution Havens? Dirty Industries in the World Economy, 1960–1995." PRDEI, The World Bank, April 1997. Accessed April 12, 2002, http://www.worldbank.org/research/peg/wps16/index.htm.

Manjoo, Farhad. "A Wild Idea: Making our Smartphones Last Longer." *New York Times*, March 12, 2014. Accessed February 10, 2017, https://www.nytimes.com/2014/03/13/technology/personaltech/the-radical-concept-of-longevity-in-a-smartphone.html?_r=0.

Mann, Michael. *The Sources of Social Power: The Rise of Classes and Nation-states, 1760–1914*. Cambridge: Cambridge University Press, 1993, volume 2.

Marcuse, Herbert. *One-dimensional Man: Studies in the Ideology of Advanced Industrial Society*. Boston: Beacon Press, 1964.

Markandya, Anil. 1997. "Eco-Labelling: An Introduction and Review." In *Eco-Labelling and International Trade*, edited by Simonetta Zarrilli, Veena Jha and René Vossenaar, 1–20. Houndsmill, Basingstoke, UK: Macmillan, 1997.

Martin, Glenn. "California's Rural Economy: Boom Times Long Gone, a Small Town Struggles for Survival." *San Francisco Chronicle*, March 24, 2003. Accessed August 10, 2017, http://www.sfgate.com/cgi-bin/article.cgi?file=/chronicle/archive/2003/03/24/MN198336.DTL.

Martin, Hans-Peter, and Harald Schumann. *The Global Trap: Globalization and the Assault on Democracy and Prosperity*, translated by P. Camillar. London: Zed Books, 1997.

Martorana, Megan. "Pareto Efficiency." *Region Focus* (Federal Reserve Bank of Richmond), Winter 2007, p. 8. Accessed August 2, 2017, https://www.richmondfed.org/~/media/richmondfedorg/publications/research/region_focus/2007/winter/pdf/jargon_alert.pdf.

Marx, Karl. *Capital: The Process of Production of Capital*, volume 1, chapter 7. Accessed August 18, 2017, https://www.marxists.org/archive/marx/works/1867-c1/ch07.htm.

Marx, Karl. *The German Ideology*. London: Lawrence and Wishart, 1938.

Marx, Karl. "On the Jewish Question." In: Karl Marx and Frederich Engels, *Marx and Engels 1843–1844—Collected Works, Vol. 3*. London: Lawrence and Wishart, 1975.

Marx, Karl. "The Eighteenth Brumaire of Louis Napoleon." In *The Karl Marx Library, Vol. 1*, edited by Saul K. Padover. McGraw Hill: New York, 1972.

Marx, Karl. "The Eighteenth Brumaire of Louis Bonaparte." In *Collected Works*, 99–197. New York: International Publishers, 1979.

Marx, Karl. *Critique of the Gotha Programme*. Peking: Foreign Languages Press, 1972. Accessed July 18, 2017, https://archive.org/stream/CritiqueOfTheGotha Programme/cgp_djvu.txt.

Massey, Doreen. "Places and Their Pasts." *History Workshop Journal* 39 (1995): 182–92.

Maya Global 2012. "A Founder's Story: Chapter 6." November 17, 2012. Accessed January 10, 2017, http://www.mayaglobal2012.com/a-founders-story-chapter-6/.

Mazey, Sonia, and Jeremy Richardson. "Environmental Groups and the EC: Challenges and Opportunities." In *Environmental Policy in the European Union: Actors, Institutions and Processes*, edited by Andrew Jordan, 141–56. London: Earthscan, 2002.

Mazor, Joseph. "Environmental Ethics." In *Global Environmental Change*, edited by Bill Freedman, 925–31. Dordrecht: Springer, 2014.

McAdam, Doug, John D. McCarthy, and Meyer N. Zald. *Comparative Perspectives on Social Movements: Political Opportunities, Mobilizing Structures, and Cultural Framings*. Cambridge: Cambridge University Press, 1996.

McAfee, Kathy. "Selling Nature to Save It? Biodiversity and the Rise of Green Developmentalism." *Environment and Planning D: Society and Space* 17 (1999): 133–54.

McCormick, John. *Understanding the European Union: A Concise Introduction*. London: Palgrave, 2017, seventh edition.

McCormick, John. *Environmental Policy in the European Union*. Houndsmill, Basingstoke: Palgrave, 2001.

McDermott, Constance L. "REDDuced: From Sustainability to Legality to Units of Carbon: The Search for Common Interests in International Forest Governance." *Environmental Science and Policy* 35 (2014): 12–19.

McDermott, C., B. Cashore, and P. Kanowski. *Global Environmental Forest Policies: An International Comparison.* UK: Earthscan, 2010.

McDonald, Glen M. "Water, Climate Change, and Sustainability in the Southwest." *Proceedings of the National Academy of Sciences* 107 (2010): 21256–62.

McDougall, Phillips. "The Cost and Time Involved in the Discovery, Development and Authorisation of a New Plant Biotechnology Derived Trait." Consultancy Study for Crop Life International, September 2011. Accessed August 14, 2017, https://croplife.org/wp-content/uploads/2014/04/Getting-a-Biotech-Crop-to-Market-Phillips-McDougall-Study.pdf.

McGinnis, Michael Vincent, ed. *Bioregionalism.* London: Routledge, 1999.

McGonigle, Ian Vincent. "Patenting Nature or Protecting Culture? Ethnopharmacology and Indigenous Intellectual Property Rights." *Journal of Law and the Biosciences* (Feb. 2016): 217–26.

McKenna, Phil. "Keystone XL: Environmental and Native [sic] Groups Sue to Halt Pipeline." *Inside Climate News*, March 30, 2017. Accessed September 26, 2017, https://insideclimatenews.org/news/30032017/keystone-xl-pipeline-lawsuit-trump-state-department.

McMahon, Tim. "Inflation Adjusted Gasoline Prices." InflationData.com, June 2016. Accessed September 8, 2017, https://inflationdata.com/articles/inflation-adjusted-prices/inflation-adjusted-gasoline-prices/.

McMichael, Philip. "Global Development and The Corporate Food Regime." In *New Directions in the Sociology of Global Development*, edited by Frederick H. Buttel and Philip McMichael, 265–99. Emerald Group Publishing Limited, 2005.

McMurtry, John. *Value Wars: The Global Market versus the Life Economy.* London: Zed Books, 2002.

Matthews, Christopher M., and Bradley Olson. "After $3 billion spent, Keystone XL can't get oil companies to sign on." *FoxBusiness*, June 29, 2017. Accessed September 26, 2017, http://www.foxbusiness.com/features/2017/06/29/after-3-billion-spent-keystone-xl-cant-get-oil-companies-to-sign-on.html.

Meilstrup, Per. "The Runaway Summit: The Background Story of the Danish Presidency of COP15, the UN Climate Change Conference." Food and Agriculture Organization, Rome. Accessed September 27, 2017, http://www.fao.org/fileadmin/user_upload/rome2007/docs/What%20really%20happen%20in%20COP15.pdf

Meinert, Lawrence D., Gilpin R. Robinson, Jr., and Nedal T. Nassar. "Mineral Resources: Reserves, Peak Production and the Future." *Resources* 5, no. 1 (2016). Accessed August 1, 2017, http://www.mdpi.com/2079-9276/5/1/14/htm.

Menotti, Victor. "Forest Destruction and Globalisation." *The Ecologist* 29, no. 3 (May-June 1999): 180–81.

Meyer, Lukas H., and Dominic Roser. "Climate Justice and Historical Emissions." *Critical Review of International Social and Political Philosophy* 13, no. 1 (March 2010): 229–53.

Meyer, John. *Political Nature: Environmentalism and the Interpretation of Western Thought.* Cambridge, MA: MIT Press, 2001.

Milani, Brian. *Designing the Green Economy: The Postindustrial Alternative to Corporate Globalization.* Lanham, MD: Rowman & Littlefield, 2000.

Miles, Edward L. *Global Ocean Politics: The Decision Process at the Third United Nations Conference on the Law of the Sea, 1973–1982.* The Hague: Martinus Nijhoff, 1998.

Milkoreit, Manjana. "Hot Deontology and Cold Consequentialism: An Empirical Exploration of Ethical Reasoning among Climate Change Negotiators." *Climatic Change* 130 (2015): 397–409.

Miller, Daniel Miller, ed. *Acknowledging Consumption: A Review of New Studies.* London: Routledge, 1995.

Minx, Jan C., et al. "Fast Growing Research on Negative Emissions." *Environmental Research Letters* (2017). Accessed October 6, 2017, http://iopscience.iop.org /article/10.1088/1748-9326/aa5ee5/pdf.

Mitchell, Ronald B., et al. "International Environmental Agreements Database Project, 1850–2016." Eugene, OR: University of Oregon, 2017. Accessed July 3, 2017, https://iea.uoregon.edu/.

Mishra, Ramesh. *Globalization and the Welfare State.* Cheltenham, England: Elgar, 1999.

Miteva, Daniela A., Colby J. Loucks, and Subhrendu K. Pattanayaik. "Social and Environmental Impacts of Forest Certification in Indonesia." *PLOS,* July 1, 2015. Accessed October 6, 2017, http://journals.plos.org/plosone/article?id=10.1371 /journal.pone.0129675.

Mitrovitskaya, Natalia S., Margaret Clark, and Ronald G. Purver. "North Pacific Fur Seals: Regime Formation as a Means of Resolving Conflict." In *Polar Politics— Creating International Environmental Regimes,* edited by Oran R. Young and Gail Osherenko, 22–55. Ithaca, NY: Cornell University Press, 1993.

Moon, Jeremy. *Corporate Social Responsibility: A Very Short Introduction.* Oxford: Oxford University Press, 2014.

Mooney, Chris. "We Only Have a Five Percent Chance of Avoiding 'Dangerous' Global Warming, a Study Finds." *Washington Post,* July 31, 2017. Accessed September 29, 2017, https://www.washingtonpost.com/news/energy-environment /wp/2017/07/31/we-only-have-a-5-percent-chance-of-avoiding-dangerous-global -warming-a-study-finds/?utm_term=.5931f89e22b3.

Mooney, Chris, Joe Heim, and Brady Dennis. "Climate March Draws Massive Crowd to D.C. in Sweltering Heat." *Washington Post,* April 29, 2017. Accessed September 5, 2017, https://www.washingtonpost.com/national/health-science/climate-march -expected-to-draw-massive-crowd-to-dc-in-sweltering-heat/2017/04/28/1bdf5e66 -2c3a-11e7-b605-33413c691853_story.html?utm_term=.91e5b2907270.

Moore, Jason W. "Ecology, Capital, and the Nature of Our Times: Accumulation and Crisis in the Capitalist World-Ecology." *American Sociological Association* 17, no. 1 (2011): 107–46.

Moore, Niamh. "Eco/feminism and Rewriting the Ending of Feminism: From the Chipko Movement to Clayoquot Sound." *Feminist Theory* 12, no. 1 (2011): 3–21.

Morrison, Kathleen D. "The Human Face of the Land: Why the Past Matters for India's Future." Nehru Memorial Museum and Library, History and Society Occasional Paper, New Series 27, 2013. Accessed February 13, 2017, http://125.22.40 .134:8082/jspui/bitstream/123456789/819/1/27_Kathleen_D_Morrison.pdf.

Mortimer-Sandilands, Catriona. "Queer Ecology." In *Keywords for Environmental Studies,* edited by Joni Adamson, William A. Gleason, and David N. Pellow. New

York: NYU Press, 2016. Accessed July 24, 2017, http://keywords.nyupress.org/environmental-studies/essay/queer-ecology/.

Mortimer-Sandilands, Catriona. "Unnatural Passions? Notes Toward a Queer Ecology." *Invisible Culture* 9 (2005). Accessed July 24, 2017, https://www.rochester.edu/in_visible_culture/Issue_9/issue9_sandilands.pdf.

Moscati, Ivan. "Were Jevons, Menger, and Walras Really Cardinalists? On the Notion of Measurement in Utility Theory, Psychology, Mathematics, and Other Disciplines, 1870–1910." *History of Political Economy* 45, no. 3 (2013): 373–414.

Mouffe, Chantal. *The Democratic Paradox*. London: Verso, 2000.

Mukherjee, Pablo. "Surfing the Second Waves: Amitav Ghosh's Tide Country." *New Formations* 59 (Autumn 2006): 144–57.

Murphy, Hannah, and Aynsley Kellow. "Forum Shopping in Global Governance: Understanding States, Business and NGOs in Multiple Arenas." *Global Policy* 4, no. 2 (May 2013): 139–49.

Murphy, Kate. "America is Running Out of Bomb-Sniffing Dogs." *New York Times*, August 4, 2017. Accessed August 4, 2017, https://nyti.ms/2ur2ZEL.

Murray, Cameron K. "What if Consumers Decided to All 'Go Green'? Environmental Rebound Effects from Consumption Decisions." *Energy Policy* 54 (2013): 240–56.

NAACP (National Association for the Advancement of Colored People). "NAACP Environmental and Climate Justice Program." Accessed August 23, 2017, http://www.naacp.org/environmental-climate-justice-about/.

Nag, Anirban, and Jamie McGeever. "Foreign Exchange, the World's Biggest Market, is Shrinking." *Reuters*, February 11, 2016, Accessed August 17, 2017, http://www.reuters.com/article/us-global-fx-peaktrading-idUSKCN0VK1UD.

Nanda, Ved P. *International Environmental Law and Policy*. Irving-on-Hudson, NY: Transnational, 1995.

naturally:wood. "British Columbia Forest Facts." August 2010. Accessed September 28, 2017, http://www.sfiprogram.org/files/pdf/third-party-certification-aug-2010pdf/.

Neimanis, Astrida, Cecilia Åsberg, and Suzi Hayes. "Posthumanist Imaginaries." In *Research Handbook on Climate Governance*, edited by Karin Bäckstrand and Eva Lövbrand, 480–90. Cheltenham: Edward Elgar, 2015.

Neimark, Benjamin. "Green Grabbing at the 'Pharm' Gate: Rosy Periwinkle Production in Southern Madagascar." *Journal of Peasant Studies* 39 (2012): 423–45.

Neimark, Benjamin D. "Industrializing Nature, Knowledge, and Labour: The Political Economy of Bioprospecting in Madagascar." *Geoforum* 75 (2012): 274–85.

Neiva, Rui, and Jonathan L. Gifford. "Declining Car Usage among Younger Age Cohorts: An Exploratory analysis of Private Car and Car-Related Expenditures in the United States." School of Public Policy, George Mason University, GMU School of Public Policy Research Paper No. 2012-10, September 10, 2012. Accessed July 23, 2015, http://papers.ssrn.com/sol3/papers.cfm?abstract_id=2028230.

Nelsen, Aaron. "Chile's Fish Supply Decline 'Catastrophic' after Years of Overfishing." *PRI.org*, May 22, 2013. Accessed September 28, 2017, https://www.pri.org/stories/2013-05-22/chiles-fish-supply-decline-catastrophic-after-years-overfishing.

Nelson, Paul. "Citizens, Consumers, Workers, and Activists: Civil Society during and after Water Privatization Struggles." *Journal of Civil Society* 13, no. 2 (2017): 202–21.

New York Times. "Furor on Memo at World Bank." February 7, 1992. Accessed July 12, 2017, http://www.nytimes.com/1992/02/07/business/furor-on-memo-at-world-bank.html.

New York Times. "Miles of Ice Collapsing into the Sea." May 18, 2017. Accessed August 17, 2017, https://www.nytimes.com/interactive/2017/05/18/climate/ant arctica-ice-melt-climate-change.html?_r=0.

Nichols, Claude. "Chapter 8—Strengthening Business-to-Business Relationships via Supply Chain and Customer Relationship Management Supply chain management (SCM) systems." SlidePlayer, 2014. Accessed August 17, 2017, http://slideplayer .com/slide/7073422/.

Nichols, Shawn. "Transnational Capital and the Transformation of the State: Investor-State Dispute Settlement (ISDS) in the Transatlantic Trade and Investment Partnership (TTIP)." *Critical Sociology* (2016). Accessed August 25, 2017, http:// journals.sagepub.com/doi/full/10.1177/0896920516675205.

Nightingale, Andrea J. "Environment and Gender." In *The International Encyclopedia of Geography: People, the Earth, Environment, and Technology*, edited by Douglas Richardson, et al. New York: John Wiley & Sons, Ltd, 2017.

Nijar, Gurdial Singh. *The Nagoya Protocol on Access and Benefit Sharing of Genetic Resources: An Analysis*. Kuala Lumpur: CEBLAW, 2011.

Nijhuis, Michelle. "Movement to Take Down Thousands of Dams Goes Mainstream." *National Geographic*, January 29, 2015. Accessed August 8, 2017, http://news.nationalgeographic.com/news/2015/01/150127-white-clay-creek-dam -removal-river-water-environment/.

Nitzan, Jonathan, and Shimshon Bichler, *Capital as Power: A Study of Order and Creorder*. London: Routledge, 2009.

Nixon, Rob. "Slow Violence, Gender, and the Environmentalism of the Poor." *Journal of Commonwealth and Postcolonial Studies* 13, no. 2/14, no. 1 (2006–2007): 14–37.

Novotny, Patrick. *Where We Live, Work and Play: The Environmental Justice Movement and the Struggle for a New Environmentalism*. Westport, CT: Praeger, 2000.

Nuccitelli, Dana. "95% Consensus of Expert Economists: Cut Carbon Pollution." *The Guardian*, January 4, 2016. Accessed August 2, 2017, https://www .theguardian.com/environment/climate-consensus-97-per-cent/2016/jan/04 /consensus-of-economists-cut-carbon-pollution.

Nye, David E., ed. *Technologies of Landscape: From Reaping to Recycling*. Amherst, MA: University of Massachusetts Press, 1999.

Obama, Barack. "Statement by the President on the Keystone XL Pipeline." The White House, November 6, 2015. Accessed August 23, 2017, https://obamawhitehouse .archives.gov/the-press-office/2015/11/06/statement-president-keystone-xl -pipeline.

O'Connor, James. "On the Two Contradictions of Capitalism." *Capitalism Nature Socialism* 2, no. 3 (1991): 107–9.

O'Connor, James. 1998. "Three Ways to Look at the Ecological History and Cultural Landscapes of Monterey Bay." In *Natural Causes: Essays in Ecological Marxism*, edited by James O'Connor, 71–93. New York: Guilford Press, 1998.

Odell, John S., and Susan K. Sell. "Reframing the Issue: The WTO Coalition on Intellectual Property and Public Health, 2001." In *Negotiating Trade: Developing Countries in the WTO and NAFTA*, edited by John S. Odell, 85–114. Cambridge: Cambridge University Press, 2001.

Ogden, Lesley Evans. "The Complex Business of Sustainable Exploitation of Wildlife." *BioScience* 67, no. 8 (August 2017): 691–97.

O'Hanlon, Kevin. "Keystone XL Pipeline Fate in Balance as Nebraska Opens Hearings." *Reuters*, August 7, 2017. Accessed September 26, 2017, https://www.reuters .com/article/us-usa-pipeline-keystone/keystone-xl-pipeline-fate-in-balance-as -nebraska-opens-hearings-idUSKBN1AN17P.

Okereke, Chukwumerije. "Climate Justice and the International Regime." *WIREs Climate Change* 1 (May/June 2010): 462–74.

Öko-Institut e.V. "Recycling is the Future—Ecological Achievements and Potential of the Dual System." Freiberg, Germany, June 2016. Accessed August 28, 2017, https://www.gruener-punkt.de/fileadmin/layout/redaktion/Mediathek/STUDIE _OEKO_INSTITUT_ENG.pdf.

Öko-Institut e.V. "Assessment of the Implementation of Directive 2000/53/EC on End-of-life Vehicles (the ELV Directive) with Emphasis on the End-of-life Vehicles of Unknown Qhereabouts." Darmstadt, Germany, June 22, 2016. Accessed August 28, 2017, http://elv.whereabouts.oeko.info/fileadmin/images/Project_Docs/Assessment _whereabouts.pdf.

Olson, Mancur. *The Logic of Collective Action*. Cambridge, MA: Harvard University Press, 1965.

Onuf, Nicholas Greenwood. *World of Our Making: Rule and Rule in Social Theory and International Relations*. London: Routledge, 2013, second edition.

Onuf, Nicholas Greenwood. *Making Sense, Making Worlds: Constructivism in Social Theory and International Relations*. London: Routledge, 2013.

Ophuls, William, and A. Stephen Boyan. *Ecology and the Politics of Scarcity Revisited: The Unraveling of the American Dream*. San Francisco: Freeman, 1992.

Organisation for Economic Co-operation and Development. "Development Aid Rises Again in 2015, Spending on Refugees Doubles." n.d. Accessed August 27, 2017, http://www.oecd.org/dac/development-aid-rises-again-in-2015-spending-on -refugees-doubles.htm.

O'Riordan, Tim, James Cameron, and Andrew Jordan, eds. *Reinterpreting the Precautionary Principle*. London: Cameron May, 2001.

O'Rourke Group Partners LLC, April 2011, as reported by Rosemary Westwood. "What Does That $14 Shirt Really Cost?" *Maclean's*, May 1, 2013. Accessed August 8, 2017, http://www.macleans.ca/economy/business/what-does-that-14-shirt-really -cost/.

Osborne, Tracy. "Public Political Ecology: A Community of Praxis for Earth Stewardship." *Journal of Political Ecology* 24 (2017): 843–60.

Ostrom, Elinor. "Collective Action and the Evolution of Social Norms." *Journal of Natural Resources Policy Research* 6, no. 4 (2014): 235–52.

Ostrom, Elinor. *Governing the Commons: The Evolution of Institutions for Collective Action*. Cambridge: Cambridge University Press, 1990.

Outka, Uma. "Fairness in the Low-Carbon Shift: Learning from Environmental Justice." *Brooklyn Law Review* 82, no. 2 (forthcoming 2017). Accessed August 16, 2017, https://ssrn.com/abstract=2989256.

Owley, Jessica. "Cultural Heritage Conservation Easements: Heritage Protection with Property Law Tools." *Land Use Policy* 49 (2015): 177–82.

Pandey, Posh Raj, Ratnakar Adhikari, and Swarnim Waglé. "Nepal's Accession to the World Trade Organization: Case Study of Issues Relevant to Least Developed Countries." UN Department of Economic and Social Affairs, CDP Background Paper no. 23 ST/ESA/2014/CDP/23 (November 2014).

Partnership for Responsible Growth. "About the Partnership." Accessed September 30, 2017, https://www.partnershipforresponsiblegrowth.org/about/.

Pateman, Carole. 1983. "Feminist Critiques of the Public/Private Dichotomy." In *Public and Private in Social Life*, edited by Stanley I. Benn and Gerald F. Gaus, 281–303. London: Croom Helm, 1983.

Paterson, Matthew. *Understanding Global Environmental Politics: Domination, Accumulation, Resistance*. Houndsmill, Basingstoke, UK: Macmillan, 2000.

Patton, P. "MetamorphoLogic: Bodies and Powers in *A Thousand Plateaus*." *Journal of the British Society for Phenomenology* 25, no. 2 (1994): 157–69.

Paulson, Henry M., Jr. "The Coming Climate Crash." *New York Times*, June 22, 2014. Accessed September 30, 2017, https://www.nytimes.com/2014/06/22/opinion/sunday/lessons-for-climate-change-in-the-2008-recession.html.

Pawliczek, J., and S. Sullivan. "Conservation and Concealment in SpeciesBanking.com, USA: An Analysis of Neoliberal Performance in the Species Offsetting Industry." *Environmental Conservation* 38, no. 4 (2011): 435–44.

Payn, Tim, et al. "Changes in Planted Forests and Future Global Implications." *Forest Ecology and Management* 352 (2015): 57–67.

Pearce, David W., and R. Kerry Turner. *Economics of Natural Resources and the Environment*. Baltimore: Johns Hopkins University Press, 1990.

Pearce, Fred. *Going Negative: How Carbon Sinks Could Cost the Earth*. Brussels: FERN, 2016.

Pearson, Ruth, and Gill Seyfang. "New Hope or False Dawn? Voluntary Codes of Conduct, Labour Regulation and Social Policy in a Globalising World." *Global Social Policy* 1, no. 1 (April 2001): 49–78.

Peck, Felicia A. "Carbon Chains: An Elemental Ethnography." PhD dissertation, Politics, University of California, Santa Cruz, 2016.

Peet, Richard, Paul Robbins, and Michael J. Watts, eds. *Global Political Ecology*. London: Routledge, 2011.

PEFC International. "Endorsement and Mutual Recognition of National Systems and their Revision." Accessed August 10, 2017, https://pefc.org/standards-revision/endorsement.

PEFC International. "Membership." Accessed March 26, 2018, https://www.pefc.org/about-pefc/membership.

Pelizzon, Alessandro, and Monica Gagliano. "The Sentience of Plants: Animal Rights and Rights of Nature Intersecting?" *Australian Animal Protection Law Journal* 11, no. 5 (2015): 5–13.

Pellow, David N. *Garbage Wars: The Struggle for Environmental Justice in Chicago*. Cambridge: MIT Press, 2002.

Pepper, David. *Eco-socialism: From Deep Ecology to Social Justice*. London: Routledge, 1993.

Peritore, N. Patrick. *Third World Environmentalism: Case Studies from the Global South*. Gainesville, FL: University Press of Florida, 1999.

Persson, Erik. "What are the Core Ideas behind the Precautionary Principle?" *Science of the Total Environment* 557-558 (2016): 134–41.

Peters, Glen P., and Oliver Geden. "Catalysing a Political Shift from Low to Negative Carbon." *Nature Climate Change* (2017). Accessed August 27, 2017, http://www.nature.com.oca.ucsc.edu/nclimate/journal/vaop/ncurrent/pdf/nclimate3369.pdf.

Peters, Glen P., and Edgar G. Hertwich. "Pollution Embodied in Trade: The Norwegian Case." *Global Environmental Change* 16 (2006): 379–87.

Peterson, M. J. "Whales, Cetologists, Environmentalists, and the International Management of Whaling." *International Organization* 46, no. 1 (Winter, 1992): 187–224.

Petras, James, and Henry Veltmeyer, eds. *The New Extractivism: A Post-Neoliberal Development Model or Imperialism of the Twenty-First Century?* London: Zed, 2014.

Petroff, Alanna. "There's a Global Banana Crisis." CNN.com, April 20, 2016. Accessed January 10, 2017, http://money.cnn.com/2016/04/19/news/bananas-banana-crisis -disease/.

Pianta, Mario. "Slowing Trade: Global Activism Against Trade Liberalization." *Global Policy* 5, no. 2 (May 2014): 214–21.

Piatti-Farnell, Lorna. *Banana: A Global History*. London: Reaktion Books, 2016.

Pidcock, Roz. "Analysis: What Global Emissions in 2016 Mean for Climate Change Goals." CarbonBrief, November 15, 2016. Accessed August 16, 2017, https:// www.carbonbrief.org/what-global-co2-emissions-2016-mean-climate-change.

Piketty, Thomas. *Capital in the Twenty-First Century*. Cambridge: Harvard University Press, 2014.

Pinder, John, and Simon Usherwood. *The European Union: A Very Short Introduction*. Oxford: Oxford University Press, 2013, third edition.

Pizzol, Massimo, et al. "Monetary Valuation in Life Cycle Assessment: A Review." *Journal of Cleaner Production* 86 (2015): 170–79.

Plenke, Max. "14 Shocking Photos Show What Your Smartphone is Really Doing to the World." *Mic.com*, May 19, 2015. Accessed February 10, 2017, https://mic.com /articles/118620/14-powerful-photos-show-what-your-smartphone-is-really -doing-to-the-worldno.AyCVpJzjW.

Poelhekke, Steven, and Frederick van der Ploeg. "Green Havens and Pollution Havens." *The World Economy* 38, no. 7 (July 2015): 1159–78.

Poguntke, Thomas. "Changing or Getting Changed: The Example of the German Greens since 1979." In *Parties, Governments and Elites*, edited by Philipp Harst, Ina Kubbe, and Thomas Poguntke, 87–103. Wiesbaden, FRG: Springer, 2017.

Polanyi, Karl. *The Great Transformation*. Boston: Beacon, 2001, second edition.

Politico, Europe Edition. "Swedish Greens Get the Coalition Blues." May 9, 2017. Accessed August 19, 2017, http://www.politico.eu/article/swedish-greens-get-the -coalition-blues/.

Pollin, Robert. *Contours of Descent: U.S. Economic Fractures and the Landscape of Global Austerity*. London: Verso, 2003.

Porter, Eduardo. "Does a Carbon Tax Work? Ask British Columbia." *New York Times*, March 1, 2016. Accessed September 3, 2017, https://www.nytimes.com/2016/03/02 /business/does-a-carbon-tax-work-ask-british-columbia.html?mcubz=0and_r=0.

Postel, Sandra. *Last Oasis: Facing Water Scarcity*. New York: Norton, 1997.

Powell, Lewis J., Jr. "Attack on American Free Enterprise System." U.S. Chamber of Commerce, August 23, 1971. Accessed August 29, 2017, http://law2.wlu.edu /powellarchives/page.asp?pageid=1251.

Prakash, Aseem, and Matthew Potoski. "Global Private Regimes, Domestic Public Law—ISO 14001 and Pollution Reduction." *Comparative Political Studies* 47, no. 3 (2014): 369–94.

Prell, Christina, and Kuishuang Feng. "Unequal Carbon Exchanges—The Environmental and Economic Impacts of Iconic U.S. Consumption Items." *Journal of Industrial Ecology* 20, no. 3 (2015): 537–46.

Princen, Thomas. "Consumption and Its Externalities: Where Economy Meets Ecology." *Global Environmental Politics* 1, no. 3 (August 2001): 11–30.

Princen, Thomas, Michael Maniates, and Ken Conca. *Confronting Consumption.* Cambridge, MA: MIT Press, 2002.

Property and Environment Research Center. "Free Market Environmentalism." 2017. Accessed August 17, 2017, https://www.perc.org/about-perc/free-market-environmentalism.

Pryde, Philip R. *Environmental Management in the Soviet Union.* Cambridge: Cambridge University Press, 1991.

Public Citizen. "U.S. Non-Profit Water Directory." March 2005. Accessed August 20, 2017, https://www.citizen.org/sites/default/files/waterdirectory.pdf.

Putnam, Robert. *Bowling Alone: The Collapse and Revival of American Community.* New York: Simon and Schuster, 2000.

Pykett, Jessica. "The New Material State: The Gendered Politics of Governing through Behaviour Change." *Antipode* 44, no. 1 (Jan. 2012): 217–38.

Radicati Group. "Email Statistics Report, 2015–2019." March 2015. Accessed August 17, 2017, http://www.radicati.com/wp/wp-content/uploads/2015/02/Email-Statistics-Report-2015-2019-Executive-Summary.pdf.

Raffensperger, Carolyn, and Joel A. Tickner. *Protecting Public Health and the Environment: Implementing the Precautionary Principle.* Washington, DC: Island Press, 1999.

Ragnarsdóttir, K. V., H. U. Sverdrup, and D. Koca. "Assessing Long Term Sustainability of Global Supply of Natural Resources and Materials." In *Sustainable Development—Energy, Engineering and Technologies—Manufacturing and Environment,* edited by Chaouki Ghenai, 83–116. Rijeka, Croatia: InTech. Accessed August 1, 2017, https://www.intechopen.com/books/sustainable-development-energy-engineering-and-technologies-manufacturing-and-environment.

Rahaman, Abu Shiraz, Jeff Everett, and Dean Neu. "Trust, Morality, and the Privatization of Water Services in Developing Countries." *Business and Society Review* 118, no. 4 (2013): 539–75.

Rainforest Action Network, et al. *Banking on Climate Change: Fossil Fuel Finance Report Card 2017.* Accessed August 22, 2017, https://d3n8a8pro7vhmx.cloudfront.net/rainforestactionnetwork/pages/17879/attachments/original/1499895600/RAN_Banking_On_Climate_Change_2017.pdf?1499895600.

Raman, K. Ravi, and Ronnie D. Lipschutz, eds. *Corporate Social Responsibility: Comparative Critiques.* Basingstoke, Houndsmill: Palgrave Macmillan, 2010.

Rawls, John. A *Theory of Justice.* Cambridge, MA: Harvard University Press, 1971.

Reckwitz, Andreas. "Toward A Theory of Social Practices—a Development in Culturalist Theorizing." *European Journal of Social Theory* 5, no. 2 (2002): 243–63.

Rea, Christopher M. "Commodifying Conservation." *Contexts* 14, no. 1 (Feb. 2013): 72–73.

REDD Offset Working Group. "California, Acre and Chiapas." Accessed August 25, 2017, https://www.arb.ca.gov/cc/capandtrade/sectorbasedoffsets/row-final-recommendations.pdf.

Redford, Kent H., William Adams, and Georgina M. Mace. "Synthetic Biology and Conservation of Nature: Wicked Problems and Wicked Solutions." *PLOS Biology* 11, no. 4 (April 2013). Accessed August 15, 2017, http://journals.plos.org/plos biology/article?id=10.1371/journal.pbio.1001530.

Reed, Brian, and Bob Reed. "How Much Water is Needed in Emergencies." Prepared for the World Health Organization by the Water, Engineering and Development Centre, Loughborough University, UK, 2013. Accessed September 27, 2017, http://www.who.int/water_sanitation_health/publications/2011/WHO_TN_09_How_much_water_is_needed.pdf.

Rehfuess, E. A., et al. "Enablers and Barriers to Large-scale Uptake of Improved Solid Fuel Stoves: A Systematic Review." *Environmental Health Perspectives* 122 (2014): 120–30.

Reisner, Mark. *Cadillac Desert: The American West and its Disappearing Water*. New York: Viking, 1986.

Requia, Weeberb J., et al. "Carbon Dioxide Emissions of Plug-in Hybrid Electric Vehicles: A Life-cycle Analysis in Eight Canadian Cities." *Renewable and Sustainable Energy Reviews* 78 (2017): 1390–96.

Reuters. "Indigenous Group in Peru Vows to Block Oil Drilling on Amazon Ancestral Land." December 9, 2016. Accessed August 23, 2017, https://www.theguardian.com/world/2016/dec/09/peru-amazon-oil-indigenous-achuar-geopark-petroperu.

Rheinberger, Hans-Jörg. "Culture and Nature in the Prism of Knowledge." *History of Humanities* 1, no. 1 (2016): 155–81.

Rice, Zak Cheney. "7 Stunning Statistics Show the True Cost of Owning a Smartphone." *Mic*, April 14, 2014. Accessed August 17, 2017, https://mic.com/articles/87617/7-stunning-statistics-show-the-true-cost-of-owning-a-smartphoneno.vLP2W0cdT.

Ritchie, Hannah. "How Much Will It Cost to Mitigate Climate Change?" *Our World in Data*, March 27, 2017. Accessed August 27, 2017, https://ourworldindata.org/how-much-will-it-cost-to-mitigate-climate-change/.

Rivera, Julio L., and Amrine Lallmahomed. "Environmental Implications of Planned Obsolescence and Product Lifetime: A Literature Review." *International Journal of Sustainable Engineering* 9, no. 2 (2016): 119–29.

Rivoli, Pietra. *The Travels of a T-shirt in the Global Economy*. Hoboken, NJ: Wiley, 2015, second edition.

Robinson, Kim Stanley. *Pacific Edge*. New York: Tor, 1990.

Rocha, Marcia, et al. "Implications of the Paris Agreement for Coal Use in the Power Sector." *Climate Analytics*, 2016. Accessed August 16, 2017, http://climateanalytics.org/files/climateanalytics-coalreport_nov2016_1.pdf.

Rocheleau, Dianne E. "Political Ecology in the Key of Policy: From Chains of Explanation to Webs of Relation." *Geoforum* 39 (2008): 716–27.

Rocheleau, Dianne E., Barbara Thomas-Slayter, and Esther Wangari, eds. *Feminist Political Ecology: Global Issues and Local Experiences*. London: Routledge, 1996.

Rockstrom, Johan, et al. "Planetary Boundaries: Exploring the Safe Operating Space for Humanity." *Ecology and Society* 14 (2009). Accessed August 30, 2017, http://www.ecologyandsociety.org/vol14/iss2/art32/.

Rodrik, Dani. *The Globalization Paradox*. London: Norton, 2011.

Rogelj, Joeri, et al. "Paris Agreement Climate Proposals Need a Boost to Keep Warming Well below 2°C." *Nature* 534 (June 10, 2016): 631–39.

Rogers, Paul. "California Drought: More than 255,000 Homes and Businesses Still Don't Have Water Meters Statewide." *San Jose Mercury News*, August 12, 2016 (updated). Accessed August 24, 2017, http://www.mercurynews.com/2014/03/08/california-drought-more-than-255000-homes-and-businesses-still-dont-have-water-meters-statewide.

Rohrig, Brian. "Smartphones Smart Chemistry." *ChemMatters* (April/May 2015): 10. Accessed February 10, 2017, https://www.acs.org/content/dam/acsorg/education/resources/highschool/chemmatters/archive/chemmatters-april2015-smartphones.pdf.

Roman, Sabin, Seth Bullock, and Markus Brede. "Coupled Societies are More Robust Against Collapse: A Hypothetical Look at Easter Island." *Ecological Economics* 132 (2017): 264–78.

Roodhouse, Mark. "Rationing Returns: A Solution to Global Warming?" *History and Policy*, March 1, 2007. Accessed August 23, 2017, http://www.historyandpolicy.org/policy-papers/papers/rationing-returns-a-solution-to-global-warming.

Rootes, Christopher, ed. *Environmental Movements: Local, National and Global.* London: Frank Cass, 1999.

Rose, Fred. *Coalitions Across the Class Divide: Lessons from the Labor, Peace, and Environmental Movements.* Ithaca, NY: Cornell University Press, 2000.

Rose, Janna. "Biopiracy: When Indigenous Knowledge is Patented for Profit." *The Conversation*, March 7, 2016. Accessed August 15, 2017, http://theconversation.com/biopiracy-when-indigenous-knowledge-is-patented-for-profit-55589.

Rosenberg, Justin. *The Empire of Civil Society.* London: Verso, 1994.

Rosenthal, Elizabeth. "The End of Car Culture." *New York Times*, June 29, 2013. Accessed December 28, 2014, http://www.nytimes.com/2013/06/30/sunday-review/the-end-of-car-culture.

Ross, Eric B. *The Malthus Factor.* London: Zed, 1998.

Rothfeder, Jeffrey. "The Great Unraveling of Globalization." *Washington Post*, April 24, 2015. Accessed August 17, 2017, https://www.washingtonpost.com/business/reconsidering-the-value-of-globalization/2015/04/24/7b5425c2-e82e-11e4-aae1-d642717d8afa_story.html?utm_term=.4d4271023479.

Rothschild, Emma. *Economic Sentiments: Adam Smith, Condorcet, and the Enlightenment.* Cambridge: Harvard University Press, 2001.

Rousseau, Jean-Jacques. *The Social Contract and Discourses*, translated by G. D. H. Cole. New York: Dutton, 1950.

Rowe, Briony MacPhee. "Anthropocentrism vs. Biocentrism." In *Green Issues and Debates: An A-to-Z Guide*, edited by Howard S. Schiffman, 30–35. Los Angeles: Sage Publishing, 2011.

Rowley, Sylvia. "How Dwindling Fish Stocks Got a Reprieve." *New York Times*, April 19, 2016. Accessed August 23, 2017, https://opinionator.blogs.nytimes.com/2016/04/19/how-dwindling-fish-stocks-got-a-reprieve/?_r=0.

Rubenstein, Steve. "Sierra Sequoia's Toppling: One Less Tree to Walk Through." *San Francisco Chronicle*, January 9, 2017. Accessed January 9, 2017, http://www.sfchronicle.com/science/article/Sierra-sequoia-s-toppling-One-less-tree-to-10846026.php.

Ruggie, John. G. "The Theory and Practice of Learning Networks: Corporate Social Responsibility and the Global Compact." *Journal of Corporate Citizenship* 5 (Spring 2002): 27–36.

Rümdig, Wolfgang, ed. *Green Politics Two*. Edinburgh: Edinburgh University Press, 1992.

Rupert, Mark. *Producing Hegemony: The Politics of Mass Production and American Global Power*. Cambridge: Cambridge University Press, 1995.

Sachdeva, Sonya, Jennifer Jordan, and Nina Mazar. "Green Consumerism: Moral Motivations to a Sustainable Future." *Current Opinion in Psychology* 6 (2015): 60–65.

Sachs, J., et al. "SDG Index and Dashboards Report 2017." New York: Bertelsmann Stiftung and Sustainable Development Solutions Network, 2017. Accessed July 11, 2017, http://www.sdgindex.org/assets/files/2017/2017-SDG-Index-and-Dash boards-Report--full.pdf.

Said, Edward. *Culture and Imperialism*. New York: Vintage, 1994.

Said, Edward. *Orientalism*. New York: Pantheon Books, 1978.

Salleh, Ariel. *Ecofeminism as Politics: Nature, Marx and the Postmodern*. London: Zed Books, 1997.

Samerski, Silja. "Tools for Degrowth? Ivan Illich's Critique of Technology Revisited." *Journal of Cleaner Production* (in press). Accessed August 18, 2017, http://www .sciencedirect.com/science/article/pii/S0959652616316377?via%3Dihub.

Sandbu, Martin. "Critics Question Success of UN's Millennium Development Goals." *Financial Times*, September 14, 2015. Accessed August 18, 2017, https://www .ft.com/content/51d1c0aa-5085-11e5-8642-453585f2cfcd.

Sandler, Todd. "Collective Action: Fifty Years Later." *Public Choice* 164 (2015): 195–216.

Sarkar, Soumodip, and Oleksiy Osiyevskyy. "Organizational Change and Rigidity during Crisis: A Review of the Paradox." *European Management Journal*, 36 (2018): 47–58. Accessed September 2, 2017, http://www.sciencedirect.com.oca.ucsc.edu /science/article/pii/S026323731730052X/pdfft?md5=ea2293df6c38e294acb0cb7 d05d11f67andpid=1-s2.0-S026323731730052X-main.pdf.

Satoru Miura, et al. "Protective Functions and Ecosystems Services of Global Forests in the Past Quarter-century." *Forest Ecology and Management* 352 (2015): 35–46.

Shackelford, Scott. "The Tragedy of the Common Heritage of Mankind." *Stanford Environmental Law Journal* 27 (2008): 101–57.

Schandl, Heinz, et al. "Global Material Flows and Resource Productivity." UNEP International Resource Panel, 2016, 99–141. Accessed August 3, 2017, http://www .resourcepanel.org/file/423/download?token=Av9xJsGS.

Schell, Jonathan. *The Fate of the Earth*. New York: Knopf, 1982.

Schlager, Edella, and William Blomquist. *Embracing Watershed Politics*. Boulder, CO: University Press of Colorado, 2008.

Schlosberg, David, and Lisette B. Collins. "From Environmental to Climate Justice: Climate Change and the Discourse of Environmental Justice." *WIREs Climate Change* (2014). Accessed October 6, 2017, http://onlinelibrary.wiley.com /doi/10.1002/wcc.275/epdf.

Schmalensee, Richard, and Robert N. Stavins. "The SO_2 Allowance Trading System: The Ironic History of a Grand Policy Experiment." *Journal of Economic Perspectives* 27, no. 1 (Winter 2013): 103–21.

Schrijver, Nico. "Managing the Global Commons: Common Good or Common Sink?" *Third World Quarterly* 37, no. 7 (2016): 1252–67.

Schutte, Nicola S., and Navjot Bhullar. "Approaching Environmental Sustainability: Perceptions of Self-Efficacy and Changeability." *The Journal of Psychology* 151, no. 3 (2017): 321–33.

Scott, James C. *Seeing Like a State: How Certain Schemes to Improve the Human Condition Have Failed*. New Haven, CT: Yale University Press, 1998.

Scott, Vivian, et al. "Fossil Fuels in a Trillion Tonne World." *Nature Climate Change* 5 (May 2015): 419–23.

Scudder, Thayer. *The Future of Large Dams*. London: Earthscan, 2005.

Searcy, Cory. "Corporate Sustainability Performance Measurement Systems: A Review and Research Agenda." *Journal of Business Ethics* 107 (2012): 239–53.

Sebenius, James K. *Negotiating the Law of the Sea*. Cambridge, MA: Harvard University Press, 1984.

Sekulova, Filka, et al. "Special Issue: Degrowth: From Theory to Practice." *Journal of Cleaner Production* 38, (Jan. 2013): 1–98.

Selby, Jan, et al. "Climate Change and the Syrian Civil War Revisited." *Political Geography* 60 (2017): 232–44.

Sell, Susan K. *Private Power, Public Law: The Globalization of Intellectual Property Rights*. Cambridge: Cambridge University Press, 2003.

Sen, Amartya. *Development as Freedom*. New York: Random House, 1999.

Sen, Amartya. *Poverty and Famines: An Essay on Entitlement and Deprivation*. Oxford: Oxford University Press, 1981.

Sen, Amartya, and Jean Dreze, eds. *The Amartya Sen and Jean Dreze Omnibus*. New Delhi: Oxford University Press, 1999.

Seppelt, Ralf, et al. "Synchronized Peak-rate Years of Global Resources Use." *Ecology and Society* 19, no. 4 (2014): 50.

Servick, Kelly. "Rise of Digital DNA Raises Biopiracy Fears." *Science*, November 17, 2016. Accessed August 16, 2017, http://www.sciencemag.org/news/2016/11/rise-digital-dna-raises-biopiracy-fears.

Sewell, William H., Jr. "A Theory of Structure: Duality, Agency and Transformation." *American Journal of Sociology* 98, no. 1 (July 1992): 1–29.

Shani, Amir. "From Threat to Opportunity: The Free Market Approach for Managing National Parks and Protected Areas." *Tourism Recreation Research* 38, no. 3 (2013): 371–74.

Shapiro, Beth. *How to Clone a Mammoth*. Princeton, NJ: Princeton University Press, 2015.

Sharma, Aasha. "Green Consumerism: Overview and Further Research Directions." *International Journal of Process Management and Benchmarking* 7, no. 2 (2017): 206–23.

Sheppard, Eric, Philip W. Porter, David R. Faust, and Richa Nagar. *A World of Difference: Encountering and Contesting Development*. New York: Guilford Press, 2009, second edition.

Shermer, Michael. "Scientific Naturalism: A Manifesto for Enlightenment Humanism." *Theology and Science* 15, no. 3 (2017): 220–30.

Shove, Elizabeth. *Comfort, Cleanliness and Convenience: The Social Organization of Everyday Life*. Oxford: Berg, 2003.

Shwom, Rachael L. "A Middle Range Theorization of Energy Politics: The Struggle for Energy Efficient Appliances." *Environmental Politics* 20:5 (2011): 705–26.

Sigamany, Indrani. "Land Rights and Neoliberalism: An Irreconcilable Conflict for Indigenous Peoples in India?" *International Journal of Law in Context* 13, no. 3 (2017): 369–87.

Sikor, Thomas, et al. "Global Land Governance: From Territory to Flow?" *Current Opinion in Environmental Sustainability* 5 (2013): 522–27.

Sinclair, Upton. *Oil! A Novel*, New York: Grossett and Dunlap, 1927.

Singer, Peter. "Famine, Affluence and Morality." *Philosophy and Public Affairs* 1, no. 3 (Spring, 1972): 229–43.

Sivak, Michael, and Brandon Schoettle. "Recent Changes in the Age Composition of Drivers in 15 Countries." Transportation Research Institute, University of Michigan, Report no. UMTRI-2011-43, October 2011. Accessed July 22, 2015, http:// deepblue.lib.umich.edu/bitstream/handle/2027.42/86680/102764.pdf.

Smil, Vaclav. *Making the Modern World: Materials and Dematerialization.* Chichester, West Sussex: Wiley and Sons, 2014.

Smith, Adam. *The Theory of Moral Sentiments.* New York: A.M. Kelley, 1777/1966.

Smith, Adam. *The Wealth of Nations.* New York: Knopf, 1776/1991.

Smith, Jackie, and Hank Johnston, eds. *Globalization and Resistance: Transnational Dimensions of Social Movements.* Lanham, MD: Rowman & Littlefield, 2002.

Smith, Pete, et al. "Biophysical and Economic Limits to Negative CO_2 Emissions." *Nature Climate Change* 6 (2016): 42–46.

S-Net Global Water Indexes. "Global Water Industry." Accessed August 8, 2017, http:// www.snetglobalwaterindexes.com/water-industry-info.

Soulé, Michael E., and Gary Lease, eds. *Reinventing Nature? Responses to Postmodern Deconstruction.* Washington, DC: Island Press, 1995.

Soros, George. *Open Society: Reforming Global Capitalism.* New York: Public Affairs, 2000.

Souder, William. *On a Farther Shore: The Life and Legacy of Rachel Carson.* New York: Broadway Books, 2013.

Spash, Clive L., and Iulie Aklaken. "Re-establishing an Ecological Discourse in the Policy Debate Over How to Value Ecosystems and Biodiversity." *Journal of Environmental Management* 159 (2015): 245–53.

Spengler, Joseph J. *French Predecessors of Malthus: A Study of Eighteenth-Century Wage and Population Theory.* Durham, NC: Duke University Press, 1942.

Special Correspondent. "When the Big Dams Came Up." *The Hindu*, March 20, 2015. Accessed September 28, 2017, http://www.thehindu.com/todays-paper/tp -national/when-the-big-dams-came-up/article7013509.ece.

Speciesbanking.com. "Saving Biodiversity by Integrating it into All Spheres of Society." Accessed August 16, 2017, http://us.speciesbanking.com/.

Sprout, Harold, and Margaret Sprout. 1965. *The Ecological Perspective on Human Affairs.* Princeton, NJ: Princeton University Press.

Spykman, Nicholas J. 1942. *America's Strategy in World Politics.* New York: Harcourt, Brace.

Sroz, PennElys. "Biocultural Engineering Design: An Anishinaabe Analysis for Building Sustainable Nations." *American Indian Culture and Research Journal* 38, no. 4 (2014): 105–26.

Stabinsky, Doreen. "Rich and Poor Countries Face Off Over 'Loss and Damage' Caused by Climate Change." *The Conversation*, December 7, 2015. Accessed August 10, 2017, https://theconversation.com/rich-and-poor-countries-face-off-over-loss-and -damage-caused-by-climate-change-51841.

Statista.com. "Number of Smartphones Sold to End Users Worldwide from 2007 to 2015." Statista 2017. Accessed February 10, 2017, https://www.statista.com /statistics/263437/global-smartphone-sales-to-end-users-since-2007/.

Statista.com. "Leading Airlines Worldwide in 2016, by Passenger Kilometers Flown (in Billions." Accessed August 17, 2017, https://www.statista.com/statistics/270986 /airlines-by-passenger-kilometers-flown/.

Stein, Rachel. "Introduction." In *New Perspectives on Environmental Justice: Gender, Sexuality, and Activism*, edited by Rachel Stein, et al., 1–17. New Brunswick, NJ: Rutgers University Press, 2004.

Steinberg, Paul F. *Who Rules the Earth? How Social Rules Shape Our Planet and Our Lives*. Oxford: Oxford University Press, 2015.

Stern, David I. "The Environmental Kuznets Curve after 25 Years." *Journal of Bioeconomics* 19 (2017): 7–28.

Stevis, Dimitris, and Valerie J. Asseto, eds. *The International Political Economy of the Environment*. Boulder, CO: Lynne Rienner.

Stiglitz, Joseph E. *Globalization and Its Discontents*. New York: Norton, 2002.

Stilwell, Matthew. "Climate Debt—A Primer." *Development Dialogue* 61 (September 2012): 41–46.

Stone, Christopher D. 1974. *Should Trees Have Standing? Toward Legal Rights for Natural Objects*. Los Altos, CA: W. Kaufmann.

Stone, Deborah. *Policy Paradox: The Art of Political Decision Making*. New York: Norton, 2011, third edition.

Strange, Susan. *The Retreat of the State*. Cambridge: Cambridge University Press, 1996.

Strange, Susan. "*Cave! Hic dragones:* A Critique of Regime Analysis." In *International Regimes*, edited by Stephen D. Krasner, 337–54. Ithaca, NY: Cornell University Press, 1983.

Strategic-metal.com. "Scandium Price." Strategic Metals Investments Ltd., 2017. Accessed September 11, 2017, http://strategic-metal.com/products/scandium /scandium-price/.

Strengers, Yolande. "Peak Electricity Demand and Social Practice Theories: Reframing the Role of Change Agents in the Energy Sector." *Energy Policy* 44 (2012): 226–34.

Strom, Stephanie. "Recalls of Organic Food on the Rise, Report Says." *New York Times*, August 20, 2015. Accessed September 4, 2017, https://www.nytimes.com /2015/08/21/business/recalls-of-organic-food-on-the-rise-report-says.html.

Sukhdev, Pavan, et al. "The Economics of Ecosystems and Biodiversity—An Interim Report." European Communities, 2008. Accessed August 17, 2017, http://www .teebweb.org/media/2008/05/TEEB-Interim-Report_English.pdf.

Sullivan, Sian. "Banking Nature? The Spectacular Financialisation of Environmental Conservation. *Antipode* 45, no. 1 (2013): 198–217.

Sustainable Travel International. "Carbon Offsets." Accessed August 16, 2017, http:// sustainabletravel.org/our-work/our-approach/carbon-offsets/.

Swift, Jonathan. "A Modest Proposal: For Preventing the Children of Poor People in Ireland from Being a Burden to their Parents or Country, and for Making Them Beneficial to the Public." (1729). Accessed September 2, 2017, http://art-bin.com /art/omodest.html.

Szczerba, Robert J. "Ingredients of 'Nature's Perfect Foods' Will Terrify You." *Forbes*, July 24, 2014. Accessed June 21, 2017, https://www.forbes.com/sites/robertszczerba /2014/07/24/ingredients-of-natures-perfect-foods-will-terrify-you/no.6fab7e 59551d.

Tabuchi, Hiroko. "Behind the Quiet State-by-State Fight Over Electric Vehicles." *New York Times*, March 11, 2017. Accessed March 16, 2017, https://www.nytimes .com/2017/03/11/business/energy-environment/electric-cars-hybrid-tax-credits .html.

't Hoen, Ellen M. F. "TRIPS, Pharmaceutical Patens and Access to Essential Medicines: Seattle, Doha and Beyond." *Chicago Journal of International Law* 3, no. 1 (2002): Article 6. Accessed August 28, 2017, http://chicagounbound.uchicago.edu /cgi/viewcontent.cgi?article=1171andcontext=cjil.

Tang, John P. "Pollution Havens and the Trade in Toxic Chemicals; Evidence from U.S. Trade Flows." *Ecological Economics* 112 (2015): 150–60.

Tao, Lin, et al. "Nature's Contribution to Today's Pharmacopeia." *Nature Biotechnology* 32, no. 10 (Oct. 2014): 979–80.

Tarrow, Sidney. *Power in Movement: Social Movements and Contentious Politics.* Cambridge: Cambridge University Press, 2011, third edition, revised and updated.

Taylor, Adam. "Why Nicaragua and Syria Didn't Join the Paris Climate Accord." *Washington Post*, May 31, 2017. Accessed September 25, 2017, https://www .washingtonpost.com/news/worldviews/wp/2017/05/31/why-nicaragua-and-syria -didnt-join-the-paris-climate-accord/?utm_term=.dd28e07a7a44.

Than, Ker. "7 Takeaways From the Supreme Court's Gene Patent Decision." *National Geographic*, June 15, 2013. Accessed August 14, 2017, http://news.nationalgeo-graphic.com/news/2013/06/130614-supreme-court-gene-patent-ruling-human -genome-science/.

Thompson, Derek. "The Dubious Future of the American Car Business—in 14 Charts." *The Atlantic*, September 6, 2013. Accessed December 28, 2014, http://www .theatlantic.com/business/archive/2013/09/the-dubious-future-of-the-american -car-business-in-14-charts/279422/.

Thoreau, Henry David. "Walking." *The Atlantic Monthly*, June 1862. Accessed July 14, 2017, https://www.theatlantic.com/magazine/archive/1862/06/walking/304674/.

350.org. "Stop the Keystone XL Pipeline." Accessed August 23, 2017, https://350.org /stop-keystone-xl/.

Trump, Donald J. "Presidential Memorandum Regarding Construction of the Dakota Access Pipeline." Office of the Press Secretary, The White House, January 24, 2017. Accessed February 16, 2017, https://www.whitehouse.gov/the-press-office /2017/01/24/presidential-memorandum-regarding-construction-dakota-access -pipeline.

Trzyna, Thaddeus C., with Julie Didion. *World Directory of Environmental Organizations.* London: Routledge, 2017, sixth edition.

Tully, Stephanie M., and Russell S. Winer. "The Role of the Beneficiary in Willingness to Pay for Socially Responsible Products. A Meta-analysis." *Journal of Retailing* 90, no. 2 (2014): 255–74.

Turkewitz, Julie. "Corporations Have Rights. Why Shouldn't Rivers? *New York Times*, September 26, 2017. Accessed September 28, 2017, https://nyti.ms/2jZ2Jx1.

Turkewitz, Julie, and Audra D. S. Burch. "Storm with 'No Boundaries' Took Aim at Rich and Poor Alike." *New York Times*, August 31, 2017. Accessed August 31, 2017, https://nyti.ms/2xBjC3H.

Turnhout, Esther, et al. "Envisioning REDD+ in a Post-Paris Era: Between Evolving Expectations and Current Practice." *WIRE's Climate Change* 8, no. 1 (2017). Accessed October 1, 2017, http://onlinelibrary.wiley.com/doi/10.1002/wcc.425/full.

Turnhout, Esther, Margaret M. Skutsch, and Jessica de Koning. "Carbon Accounting." In *Research Handbook in Climate Governance*, edited by Karen Bäckstrand and Eva Lövbrand, 366–76. Cheltenham, UK: Edward Elgar, 2015.

Umeek of Ahousaht (E. Richard Atleo). "Commentary: Discourses in and about Clayoquot Sound: A First Nations Perspective." In *A Political Space: Reading the Global through Clayoquot Sound*, edited by Warren Magnusson and Karena Shaw, 199–208. Minneapolis: University of Minnesota Press, 2003.

Underwriters Laboratory. "The Life Cycle of Materials in Mobile Phones." 2011. Accessed August 4, 2017, http://services.ul.com/wp-content/uploads/sites/4/2014/05/ULE_CellPhone_White_Paper_V2.pdf.

United Air Lines. "CarbonChoice Carbon Offset Program." Accessed August 16, 2017, https://www.united.com/web/en-US/content/company/globalcitizenship/environment/carbon-offset-program.aspx.

United Nations. "Sustainable Development Goals Kick Off With Start of New Year." December 30, 2015. Accessed August 17, 2017, http://www.un.org/sustainabledevelopment/blog/2015/12/sustainable-development-goals-kick-off-with-start-of-new-year/.

United Nations. "The Millennium Development Goals Report 2015." New York. Accessed August 18, 2017, http://www.un.org/millenniumgoals/2015_MDG_Report/pdf/MDG%202015%20rev%20(July%201.pdf.

United Nations. "The Future We Want." Outcome document of the United Nations Conference on Sustainable Development, Rio de Janeiro, Brazil, June 20–22, 2012. Accessed September 3, 2017, https://sustainabledevelopment.un.org/content/documents/733FutureWeWant.pdf.

United Nations. "World Summit on Sustainable Development." Accessed August 8, 2017, https://sustainabledevelopment.un.org/milesstones/wssd.

United Nations, Declaration of the United Nations Conference on the Human Environment, UN Doc. A/Conf.48/14/Rev. 1(1973); 11 ILM 1416 (1972). Accessed October 6, 2017, http://www.un-documents.net/unchedec.htm.

United Nations. "Earth Summit (1992)." 1997. Accessed August 30, 2017, http://www.un.org/geninfo/bp/enviro.html.

"United Nations Conference on Sustainable Development, Rio+20." Accessed August 8, 2017, https://sustainabledevelopment.un.org/rio20.html.

UN Conference on Trade and Development. "World Investment Report 2017." Geneva, 2017. Accessed August 27, 2017, http://unctad.org/en/PublicationsLibrary/wir2017_en.pdf.

UN Department of Economic and Social Affairs. "Sustainable Development Goals." Accessed September 3, 2017, https://sustainabledevelopment.un.org/?menu=1300.

UN Development Programme. *Human Development Report 2006*. New York: Palgrave Macmillan, 2006. Accessed August 25, 2017, http://hdr.undp.org/sites/default/files/reports/267/hdr06-complete.pdf.

UN Forum on Forests. "Background Document no. 2: Recent Developments in Existing Forest-Related Instruments, Agreements, and Processes Working Draft." Prepared for the Ad hoc expert group on Consideration with a View to Recommending the Parameters of a Mandate for Developing a Legal Framework on All Types of Forests, September 7–10, 2004. Accessed August 10, 2017, http://www.un.org/esa/forests/wp-content/uploads/2014/12/background-2.pdf.

UN Framework Convention on Climate Change. "The Paris Agreement." 2017. Accessed August 28, 2017, http://unfccc.int/paris_agreement/items/9485.php.

UN Framework Convention on Climate Change. *Report of the Conference of the Parties on Its Twenty-First Session, Held in Paris from 30 November to 13 December 2015—Addendum—Part Two: Action Taken by the Conference of the Parties at Its Twenty-First Session*, Dec 1/CP.21, UN Doc FCCC/CP/2015/10/Add.1 (29 January 2015) (*Adoption of the Paris Agreement*). Accessed October 6, 2017, https://digitallibrary.un.org/record/831052?ln=en.

UN Framework Convention on Climate Change. "Fact Sheet—The Need for Mitigation." Accessed August 10, 2017, https://unfccc.int/files/press/backgrounders/application/pdf/press_factsh_mitigation.pdf.

UN Framework Convention on Climate Change. "Adoption of the Paris Agreement." Article 2.1.a, p. 22, December 12, 2017, Draft decision -/CP.21, FCCC/CP/2015/L9/Rev.1. Accessed August 17, 2017, https://unfccc.int/resource/docs/2015/cop21/eng/l09r01.pdf.

UN General Assembly Resolution 664/292. "The Human Right to Water and Sanitation." July 28, 2010. Accessed October 6, 2017, https://documents-dds-ny.un.org/doc/UNDOC/GEN/N09/479/35/PDF/N0947935.pdf?OpenElement.

UN Global Compact. "United Nations Global Compact." Accessed September 3, 2017, https://www.unglobalcompact.org/.

"UN Global Compact Symposium." *Journal of Business Ethics* 122, no. 2 (June 2014).

UPOV (International Union for the Protection of New Varieties of Plants). "UPOV." Accessed August 16, 2017, http://www.upov.int/portal/index.html.en.

U.S. Bureau of Economic Analysis. "Table 1.1.5. Gross Domestic Product." U.S. Dept. of Commerce, July 28, 2017. Accessed August 2, 2017, https://www.bea.gov/iTable/iTable.cfm?ReqID=9andstep=1no. reqid=9andstep=3andisuri=1and903=5.

U.S. Energy Information Administration. "International Energy Outlook 2016: Chapter 8. Transportation sector energy consumption." September 2017. Accessed August 17, 2017, https://www.eia.gov/outlooks/ieo/transportation.php.

U.S. Environmental Protection Agency. "How We Use Water." March 24, 2017. Accessed August 24, 2017, https://www.epa.gov/watersense/how-we-use-water.

U.S. Environmental Protection Agency. "The Social Cost of Carbon." n.d. Accessed August 16, 2017, https://19january2017snapshot.epa.gov/climatechange/social-cost-carbon_.html.

U.S. Environmental Protection Agency. "Waters of the United States (WOTYS) Rulemaking." n.d. Accessed August 14, 2017, https://www.epa.gov/wotus-rule.

U.S. Fish and Wildlife Service. "For Landowners—Conservation Banking." Accessed August 16, 2017, https://www.fws.gov/endangered/landowners/conservation-banking.html.

U.S. Geological Survey Water Science School. "Water Questions and Answers—How Much Water Does the Average Person Use at Home per Day? U.S. Geological Survey, 2016." Accessed August 25, 2017, http://water.usgs.gov/edu/qa-home-per capita.html.

U.S. Supreme Court. "Citizens United v. Federal Election Commission, (2010)." *FindLaw*, January 10, 2010. Accessed September 28, 2017, http://caselaw.findlaw.com/us-supreme-court/08-205.html.

Van Dender, Kurt, and Martin Clever. "Recent Trends in Car Usage in Advanced Economies—Slower Growth Ahead?" International Transport Forum, Organisation for Economic Co-operation and Development, No 2013-9, April 2013. Accessed July 22, 2015, http://www.internationaltransportforum.org/jtrc/DiscussionPapers /DP201309.pdf.

Van der Pijl, Kees. *Transnational Classes and International Relations.* London: Routledge, 1998.

Vedantam, Shankar. "Does Studying Economics Make You Selfish?" *NPR*, February 21, 2017. Accessed August 29, 2017, http://www.npr.org/2017/02/21/516375434 /does-studying-economics-make-you-selfish.

Veenhoven, Ruut, and Floris Vergunst. "The Easterlin Illusion: Economic Growth Does Go with Greater Happiness." *International Journal of Happiness and Development* 1, no. 4 (2014): 311–43.

Vlachou, Andriana, and Georgios Pantelias. "The EU's Emissions Trading System, Part 1: Taking Stock." *Capitalism Nature Socialism* 28, no. 2 (2017): 84–102.

Vlachou, Andriana, and Georgios Pantelias. "The EU's Emissions Trading System, Part 2: A Political Economy Critique." *Capitalism Nature Socialism* 28, no. 3 (2017): 108–27.

van Wendel de Joode, Berna, et al. "Indigenous Children Living Nearby Plantations with Chlorpyrifos-treated Bags have Elevated 3,5,6-trichloro-2-pyridinol (TCPy) Urinary Concentrations." *Environmental Research* 117 (2012): 17–26.

Velmurugan, Manivannan Senthil. "Environmental and Health Aspects of Mobile Phone Production and Use: Suggestions for Innovation and Policy." *Environmental Innovation and Societal Transitions* 21 (2016): 69–79.

Vickers, Patricia. "Ayaawx: In the Path of Our Ancestors." *Cultural Studies of Science Educaion* 2 (2007): 592–98.

Vidal, John. "Copenhagen Climate Failure Blamed on 'Danish Text.'" *The Guardian*, May 31, 2010. Accessed September 27, 2017, https://www.theguardian.com /environment/2010/may/31/climate-change-copenhagen-danish-text.

Vidal, John. "Real Battle for Seattle." *The Guardian*, December 4, 1999. Accessed September 2, 2017, https://www.theguardian.com/world/1999/dec/05/wto.globalisation.

Vogel, David. *Trading Up: Consumer and Environmental Regulation in a Global Economy.* Cambridge, MA: Harvard University Press, 1995.

Vogel, Steven K. *Freer Markets, More Rules: Regulatory Reform in Advanced Industrial Countries.* Ithaca, NY: Cornell University Press, 1996.

Wackernagel, Mathis, and William Rees. "Ecological Footprints for Beginners." In *The Ecological Design and Planning Reader*, edited by Foster O. Ndubisi, 501–05. Washington, DC: Island Press, 2014.

Wall, Derek. *The Commons in History.* Cambridge, MA: MIT Press, 2014.

Walmart. "Sustainability." Accessed August 14, 2017, http://walmartstores.com /sustainability.

Waltz, Kenneth. *Theory of International Politics.* Reading, MA: Addison-Wesley, 1979.

Waltz, Kenneth N. *Man, the State and War.* New York: Columbia University Press, 1958.

Wapner, Paul. "Living at the Margins." In *New Earth Politics: Essays from the Anthropocene*, edited by Simon Nicholson and Sikina Jinnah, 369–85. Cambridge: MIT Press, 2016.

Wapner, Paul. *Environmental Activism and World Civic Politics.* Albany, NY: State University of New York Press, 1996.

Water for Life. "The Human Right to Water and Sanitation." n.d. Accessed August 25, 2017, http://www.un.org/waterforlifedecade/human_right_to_water.shtml.

Water Information Program. "Water Facts." Southwestern Water Conservation District, n.d. Accessed August 25, 2017, https://www.waterinfo.org/resources/water-facts.

Watts, Michael. "Political Ecology." In *A Companion to Economic Geography*, edited by Eric Sheppard and Trevor Barnes, 257–74. Oxford: Blackwell, 2000.

Weale, Albert, et al. *Environmental Governance in Europe: An Ever Closer Ecological Union?* Oxford: Oxford University Press, 2000.

wearestillin. "We Are Still In." Accessed October 1, 2017, https://www.wearestillin.com/us-action-climate-change-irreversible.

Weber, Max. *The Protestant Ethic and the Spirit of Capitalism*, translated by Talcott Parsons. New York: Scribner, 1958.

Weber, Thomas. *Hugging the Trees: The Story of the Chipko Movement.* New York: Viking, 1988.

Weiss, Edith Brown. "The Evolution of International Environmental Law." *Japanese Yearbook of International Law* 54 (2011): 1–27.

Weiss, Edith Brown. *In Fairness to Future Generations: International Law, Common Patrimony, and Intergenerational Equity.* Dobbs Ferry, NY: Transnational, 1989.

Wesseling, Catharina, et al. "Hazardous Pesticides in Central America." *International Journal of Occupational and Environmental Health* 7, no. 4 (2001): 287–94.

White, Lynn, Jr. "The Historical Roots of Our Ecologic Crisis." *Science* 155 (1967): 1203–07.

Whitehouse, Christine. "Driven to Distraction." *Time International* 155 no. 6 (Feb. 14, 2000): 65–66.

Whiteley, John, Helen Ingraham, and Richard Perry, eds. *Water, Place, and Equity.* Cambridge: MIT Press, 2008.

The Wilderness Society. "America's Public Lands." Washington, DC, n.d. Accessed August 9, 2017, http://wilderness.org/sites/default/files/Fact%20Sheet%20America%27s%20Public%20Lands%20.pdf.

Willer, Helga, and Julia Lernoud, eds. *The World of Organic Agriculture.* Bonn and Frick: Research Institute of Organic Agriculture and IFOAM–Organics International, 2016. Accessed September 4, 2017, http://orgprints.org/31151/1/willer-lernoud-2016-world-of-organic.pdf.

Willey, Zach, and Adam Diamant. "Water Marketing in the Northwest: Learning by Doing." *Water Strategist* 10 (Summer 1996).

Williamson, Phil. "Scrutinize CO_2 Removal Methods." *Nature* 530 (2016): 153–54.

Wilson, L. Don. "If All You Have is a Hammer!" *Dental Economics* 102, no. 4 (2013). Accessed June 29, 2017, http://www.dentaleconomics.com/articles/print/volume-102/issue-4/feature/if-all-you-have-is-a-hammer.html.

Wittfogel, Karl. *Oriental Despotism: A Comparative Study of Total Power.* New Haven, CT: Yale University Press, 1957.

Wolin, Sheldon. "Fugitive Democracy." In *Democracy and Difference*, edited by Seyla Benhabib, 31–45. Princeton, NJ: Princeton University Press, 1996.

Wood, Ellen Meiksins. *The Origins of Capitalism: A Longer View.* London: Verso Books, 2002, revised edition.

Wood, Michael. *Legacy: The Search for Ancient Civilization.* New York: Sterling, 1995.

Working Group on the "Anthropocene." "What is the 'Anthropocene'—Current Definition and Status?" Subcommission on Quaternary Stratigraphy. Accessed February 13, 2017, https://quaternary.stratigraphy.org/workinggroups/anthropocene/.

World Association of Non-governmental Associations. "Worldwide NGO Directory." 2017. Accessed August 30, 2017, http://www.wango.org/resources.aspx?section=ngodirandsub=regionandregionID=155andcol=ff9900.

World Bank. "Air Transport, Passengers Carried." Accessed August 17, 2017, http://data.worldbank.org/indicator/IS.AIR.PSGR.

World Bank. "Household Final Consumption Expenditure, etc. (% of GDP)." Data for 2016. Accessed August 2, 2017, http://data.worldbank.org/indicator/NE.CON.PETC.ZS.

World Bank. "Pricing Carbon." Accessed September 3, 2017, http://www.worldbank.org/en/programs/pricing-carbon.

World Bank DataBank. "Gross Domestic Product 2016." Accessed August 14, 2017, http://databank.worldbank.org/data/download/GDP.pdf.

World Bank Group. *International Debt Statistics 2017*. Washington, DC, 2017.

World Bank Group. *State and Trends of Carbon Pricing 2016*. Accessed September 3, 2017, https://openknowledge.worldbank.org/bitstream/handle/10986/25160/9781464810015.pdf?sequence=7andisAllowed=y.

World Bank Group. "State and Trends of Carbon Pricing." Washington, DC, October 2016. Accessed August 15, 2017, https://openknowledge.worldbank.org/bitstream/handle/10986/25160/9781464810015.pdf?sequence=7andisAllowed=y.

World Bank Group. "Climate Finance." Accessed August 16, 2017, http://www.worldbank.org/en/topic/climatefinance

World Commission on Environment and Development. *Our Common Future*. Oxford: Oxford University Press, 1987.

World Economic Forum. "The Future Availability of Natural Resources—A New Paradigm for Global Resource Availability." Geneva, Switzerland, November 2014. Accessed August 1, 2017, http://www3.weforum.org/docs/WEF_FutureAvailabilityNaturalResources_Report_2014.pdf.

World Health Organization. "What is the Minimum Quantity of Water Needed?" 2017. Accessed August 25, 2017, http://www.who.int/water_sanitation_health/emergencies/qa/emergencies_qa5/en.

World Health Organization and UN Children's Fund. *Progress on Sanitation and Drinking Water—2015 Update and MDG Assessment*. Accessed August 8, 2017, http://files.unicef.org/publications/files/Progress_on_Sanitation_and_Drinking_Water_2015_Update_.pdf.

World Intellectual Property Organization. "In Focus." Accessed August 27, 2017, http://www.wipo.int/portal/en/index.html.

World Intellectual Property Organization. Background Brief. "The WIPO Intergovernmental Committee on Intellectual Property and Genetic Resources, Traditional Knowledge and Folklore." Geneva, 2016. Accessed August 15, 2017, http://www.wipo.int/edocs/pubdocs/en/wipo_pub_tk_2.pdf.

World Intellectual Property Organization. "IP Statistics Data Center." Accessed August 17, 2017, https://www3.wipo.int/ipstats/index.htm?tab=patent.

World Trade Organization. "Understanding the WTO: The Agreements." Accessed August 16, 2017, https://www.wto.org/english/thewto_e/whatis_e/tif_e/agrm1_e.htm.

World Trade Organization. "Obligations and Exceptions." September 2006. Accessed September 29, 2017, https://www.wto.org/english/tratop_e/trips_e/factsheet_pharm 02_e.htm.

World Wildlife Fund. "The Impact of Forest Stewardship Council Certification." October 20, 2014. Accessed August 10, 2017, http://d2ouvy59p0dg6k.cloudfront.net /downloads/fsc_research_review.pdf.

Worster, Donald. *Rivers of Empire: Water, Aridity and the Growth of the American West*. New York: Pantheon, 1985.

Yanmei, Li, et al. "Sources and Flows of Embodied CO-2 Emissions in Import and Export Trade of China." *Chinese Geographical Science* 24, no. 2 (2014): 220–30.

Young, Iris Marion. *Justice and the Politics of Difference*. Princeton, NJ: Princeton University Press, 1990.

Young, Oran R. *International Governance: Protecting the Environment in a Stateless Society*. Ithaca, NY: Cornell University Press, 1994.

Zakreski, Sheldon. "Avoiding a Turbulent Ride with ICAO." Climate Trust, May 15, 2017. Accessed September 29, 2017, https://climatetrust.org/avoiding-a -turbulent-ride-with-icao-scorcher/.

Zhou, Wen. "The Patent Landscape of Genetically Modified Organisms." *Signal to Noise Special Edition: GMOs and Our Food*, Harvard Graduate School of Arts and Sciences." August 10, 2015. Accessed August 14, 2017, http://sitn.hms.harvard .edu/flash/2015/the-patent-landscape-of-genetically-modified-organisms/.

Index

About the Authors

Ronnie D. Lipschutz is professor of politics and, from 2012 to 2018, was provost of Rachel Carson College at the University of California, Santa Cruz. He is the inaugural holder (2017–2019) of the Robert Headley Presidential Chair in Integral Ecology and Environmental Justice in Rachel Carson College. He received his PhD in Energy and Resources from University of California, Berkeley, in 1987 and an SM in Physics from Massachusetts Institute of Technology in 1978. He has taught and conducted research at University of California, Santa Cruz, since 1990. He has written numerous books and articles about global politics, the environment, energy and resources, popular culture, and surveillance. As Provost of Rachel Carson College, he created and directed a Minor in Sustainability Studies, designed to train and prepare students for the green economy and society of the future, with a focus on STEM skills and complex problem-solving.

Doreen Stabinsky is professor of global environmental politics at College of the Atlantic in Bar Harbor, Maine. Her research focuses on the impacts of climate change on agriculture and food security, uses of land in climate mitigation and carbon markets, and on the emerging issue of loss and damage from climate change impacts. Stabinsky is also an independent consultant and serves as advisor to governments and international nongovernmental organizations on issues at the intersection of climate change, food security, and loss and damage. In 2015 to 2016, she held the first Zennström Visiting Professorship in Climate Change Leadership at Uppsala University, Sweden. She is co-editor (with Stephen B. Brush) of *Valuing Local Knowledge: Indigenous People and Intellectual Property Rights*.